PLANETARY ATMOSPHERES

INTERNATIONAL ASTRONOMICAL UNION

UNION ASTRONOMIQUE INTERNATIONALE

SYMPOSIUM No. 40

HELD IN MARFA, TEXAS, U.S.A., OCTOBER 26–31, 1969

PLANETARY ATMOSPHERES

EDITED BY

CARL SAGAN

*Laboratory for Planetary Studies, Center for Radiophysics and Space Research,
Cornell University, Ithaca, N.Y., U.S.A.*

TOBIAS C. OWEN

*State University of New York at Stony Brook,
Department of Earth and Space Sciences, Stony Brook, N.Y., U.S.A.*

AND

HARLAN J. SMITH

Dept. of Astronomy, University of Texas at Austin, Austin, Tex., U.S.A.

D. REIDEL PUBLISHING COMPANY

DORDRECHT-HOLLAND

1971

Published on behalf of
the International Astronomical Union
by
D. Reidel Publishing Company, Dordrecht, Holland

Library of Congress Catalog Card Number 77–140566
ISBN-13: 978-94-010-3065-6   e-ISBN-13: 978-94-010-3063-2
DOI: 10.1007/978-94-010-3063-2

PREFACE

IAU Symposium 40, on Planetary Atmospheres, was held at Marfa, Texas, in the Paisano Hotel, October 26–31, 1969, hosted by The University of Texas at Austin and the McDonald Observatory of the University of Texas.

The Organizing Committee consisted of: Dr. Tobias Owen (Illinois Institute of Technology), Dr. Carl Sagan (Cornell University), Dr. John Hall (Lowell Observatory), Dr. Arvydas Kliore (Jet Propulsion Laboratory), Academician Georgi I. Petrov (Institute of Cosmic Research, Soviet Academy of Sciences), Dr. V. I. Moroz (Sternberg Institute), Dr. H. C. van de Hulst (Leiden University).

Local arrangements were under the care of Dr. Harlan Smith (McDonald Observatory), assisted especially by Dr. Ronald Schorn (Jet Propulsion Laboratory), Mr. Curtis Laughlin (McDonald Observatory), and Miss Virginia Church (University of Texas); also by Dr. Joseph Chamberlain (Kitt Peak National Observatory), Dr. Maurice Marin (McDonald Observatory), Mr. and Mrs. J. Bergstrahl and Mrs. K. MacFarlane (McDonald Observatory), and Dr. Donald Rea (NASA Headquarters).

The scientific program was divided into three parts; Venus, Mars, and the outer planets. Detailed programs for these sections, and their subsequent editing, were respectively in the hands of the undersigned.

It was possible to hold the Symposium because of generous sponsorship by the Executive Committee of the International Astronomical Union, including the provision of a travel grant for assisting younger astronomers to attend, and also including the active attention and assistance of the Assistant General Secretary, Dr. C. de Jager.

Costs of the meeting were covered primarily by grants from the National Science Foundation and the Martin-Marietta Company, and by support from the University of Texas. The help of these groups is gratefully acknowledged.

Because of difficulties in tape recording comments after papers at such a meeting, only a small fraction of the very interesting discussions which marked the Marfa meeting is reproduced in this volume. We are grateful to Mrs. M. Syzmanski for helping to recover some of the discussions, to Dr. Dale Cruikshank for translation from Russian to English, and to the Editor of *Science* for permission to reproduce two papers on the Mariner 6 and 7 results.

C. SAGAN
T. OWEN
H. SMITH

TABLE OF CONTENTS

PART I / VENUS

LIST OF PARTICIPANTS

Anderson, J., Jet Propulsion Laboratory, 4800 Oak Grove Drive, Pasadena, Calif. 91103, U.S.A.

Barger, Allen, Martin Marietta Corporation, Box 179, Denver, Colo. 80201, U.S.A.

Barker, Dr. Edwin S., McDonald Observatory, University of Texas, Fort Davis, Texas, U.S.A.

Barth, Dr. Charles A., Laboratory for Atmospheric and Space Physics, University of Colorado, Boulder, Colo. 80302, U.S.A.

Baum, Dr. William A., Planetary Research Center, Lowell Observatory, Flagstaff, Ariz. 86001, U.S.A.

Belton, Dr. Michael J. S., Kitt Peak National Observatory, Box 4130, Tucson, Ariz. 85717, U.S.A.

Bender, Welcome W., Martin Marietta Corporation, Box 179, Denver, Colo. 80201, U.S.A.

Benedict, Dr. William, Department of Physics, University of Maryland, College Park, Md. 20742, U.S.A.

Binder, Dr. Alan B., IIT Research Institute, 10 West 35th St., Chicago, Ill. 60616, U.S.A.

Boyce, Dr. Peter B., Lowell Observatory, Box 1269, Flagstaff, Ariz. 86001, U.S.A.

Brandt, Dr. John C., Goddard Space Flight Center, Greenbelt, Md. 20771, U.S.A.

Broadfoot, Dr. Lyle, Kitt Peak National Observatory, Box 4130, Tucson, Ariz. 85717, U.S.A.

Brunk, Dr. William E., Office of Space Science and Applications, NASA Headquarters, Code SL, Washington, D.C. 20546, U.S.A.

Cameron, Dr. A. G. W., Yeshiva University, Amsterdam Ave. and 186th St., New York, N.Y. 10033, U.S.A.

Capen, Charles F., Jet Propulsion Laboratory, 4800 Oak Grove Drive, Pasadena, Calif. 91103, U.S.A.

Carter, Dr. Virginia, Space Physics Laboratory, Aerospace Corp., Box 95085, Los Angeles, Calif. 90045, U.S.A.

Chamberlain, Dr. Joseph W., Kitt Peak National Observatory, Box 4130, Tucson, Ariz. 85717, U.S.A.

Clements, Dr. Arthur E., Lunar and Planetary Laboratory, University of Arizona, Tucson, Ariz. 85721, U.S.A.

Cloud, Dr. Preston E., Department of Geology, University of California, Santa Barbara, Calif. 93016, U.S.A.

Cloutier, Dr. Paul A., Department of Space Physics, Rice Univ., 6100 South Main, Houston, Tex. 77001, U.S.A.

Coffeen, Dr. David L., Lunar and Planetary Laboratory, University of Arizona, Tucson, Ariz. 85721, U.S.A.

Coughlin, K., Martin Marietta Corporation, Box 179, Denver, Colo. 80201, U.S.A.

Cruikshank, Dr. Dale P., Lunar and Planetary Laboratory, University of Arizona, Tucson, Ariz. 85721, U.S.A.

Dalgarno, Dr. A., Harvard College Observatory, Harvard Square, Cambridge, Mass. 02138, U.S.A.

Dermott, Dr. S. F., Royal Military College of Science, Physics Branch, Shrivenham, Swindon, Wilts., England

De Vaucouleurs, Dr. Gerard, Department of Astronomy, University of Texas at Austin, Austin, Tex. 78712, U.S.A.

di Benedetto, Dr. Felice, Operational Director, Laboratory of Astrophysics, Casella Postale, 67, 00044 Frascati (Roma), Rome, Italy

Dollfus, Dr. Audouin, Observatoire de Paris, Section d'Astrophysique, Meudon (S. et O.), France

Doose, Dr. Lyn, Lunar and Planetary Laboratory, University of Arizona, Tucson, Ariz. 85721, U.S.A.

Ducsai, Steve, Martin Marietta Corporation, Box 179, Denver, Colo. 80201, U.S.A.

Egan, Dr. Walter G., Research Department, Grumman Aerospace Corporation, Bethpage, N.Y. 11714, U.S.A.

Farmer, Dr. Charles B., Jet Propulsion Laboratory, 4800 Oak Grove Drive, Pasadena, Calif. 91103, U.S.A.

Fastie, William G., Department of Physics, Johns Hopkins University, Baltimore, Md. 21218, U.S.A.

Galkin, Dr. L., Crimean Astrophysical Observatory, P/O Nauchny, Crimea, U.S.S.R.

Gehrels, Dr. Tom, Lunar and Planetary Laboratory, University of Arizona, Tucson, Ariz. 85721, U.S.A.

Golitsyn, G., Institute of Atmospheric Physics, Soviet Academy of Sciences, Leninsky Prospekt, 14, Moscow, B-71, U.S.S.R.

Greyber, Dr. Howard, Martin Marietta Corporation, Box 179, Denver, Colo. 80201, U.S.A.

Hall, Dr. John S., Director, Lowell Observatory, Flagstaff, Ariz. 86002, U.S.A.

Hanel, Dr. Rudy A., Goddard Space Flight Center, Greenbelt, Md. 20771, U.S.A.

Herman, Dr. Jay R., Ionospheric and Radiophysics Branch, Goddard Space Flight Center, Greenbelt, Md. 20771, U.S.A.

Hord, Charles W., Laboratory for Atmospheric and Space Physics, University of Colorado, Boulder, Colo. 80302, U.S.A.

Hudson, Dr. R. D., Space Physics Division, NASA, Manned Spacecraft Division, Houston, Tex. 77058, U.S.A.

Hunten, Dr. D. M., Kitt Peak National Observatory, Box 4130, Tucson, Ariz. 85717, U.S.A.

Irvine, Dr. William M., Astronomy Department, Massachusetts University, Amherst, Mass. 01002, U.S.A.

Jensen, J., Martin Marietta Corporation, Box 179, Denver, Colo. 80201, U.S.A.

Johnson, Dr. Francis S., University of Texas at Dallas, Box 30365, Dallas, Tex. 75230, U.S.A.

Kliore, Dr. Arvydas, Jet Propulsion Laboratory, 4800 Oak Grove Drive, Pasadena, Calif. 91103, U.S.A.

Kroupenio, N., Institute for Cosmic Research, Soviet Academy of Sciences, Moscow, U.S.S.R.

Kuiper, Dr. Gerard, University of Arizona, Park Avenue at 3rd Street, Tucson, Ariz. 85721, U.S.A.

Leighton, Dr. Robert B., Bridge Laboratory of Physics, California Institute of Technology, Pasadena, Calif. 91109, U.S.A.

Libby, Dr. Willard F., Department of Space Sciences, Los Angeles, Calif. 90024, U.S.A.

Liddel, Dr. Urner, Lunar and Planetary Programs, Office of Space Science and Applications, NASA Headquarters, Code SL, Washington, D.C. 20546, U.S.A.

Little, Dr. Steve J., Department of Astronomy, University of Texas at Austin, Austin, Tex. 78712, U.S.A.

Marin, Maurice, McDonald Observatory, Fort Davis, Tex. 79734, U.S.A.

Mead, Dr. Jaylee, Theoretical Studies Branch, Goddard Space Flight Center, Greenbelt, Md. 20771, U.S.A.

Moos, W., Baltimore, Md.

Morrison, Dr. David, Institute for Astronomy, University of Hawaii, 2525 Correa Rd., Honolulu, Hawaii 98622, U.S.A.

Münch, Dr. Guido, California Institute of Technology, 1201 E. California Blvd., Pasadena, Calif. 91109, U.S.A.

McElroy, Dr. Michael B., Kitt Peak National Observatory, Box 4130, Tucson, Ariz. 85717, U.S.A.

McGovern, Dr. Wayne, Institute of Atmospheric Physics, The University of Arizona, Tucson, Ariz. 05727, U.S.A.

Owen, Dr. Tobias C., IIT Research Institute, 10 East 35th Street, Chicago, Ill. 60616, U.S.A.

Pearce, Dr. Jeffrey B., Laboratory for Atmospheric and Space Physics, University of Colorado, Boulder, Colo. 80302, U.S.A.

Petrov, Dr. G. I., Institute for Cosmic Research, Academy of Sciences of the U.S.S.R., Leninsky Prospekt 14, Moscow B-71, U.S.S.R.

Poll, Dr. J. D., Department of Physics, University of Toronto, Toronto, Ont., Canada

Pollack, Dr. James B., Laboratory for Planetary Studies, Cornell University, Ithaca, N.Y. 14850, U.S.A.

Potter, Dr. John, Lockheed Electronic Co., 16811 El Camino Real, Houston, Tex. 77058, U.S.A.

Price, Dr. Michael J., Kitt Peak National Observatory, Box 4130, Tucson, Ariz. 85717, U.S.A.

Rasool, Dr. S. I., Goddard Institute for Space Studies, 2880 Broadway, New York, N.Y. 10025, U.S.A.

Rea, Dr. Donald S., NASA Headquarters, Washington, D.C. 20546, U.S.A.

Roosen, Dr. Robert G., Goddard Space Flight Center, Greenbelt, Md. 20771, U.S.A.

Sadin, Steve, Martin Marietta Corp., Box 179, Denver, Colo. 80201, U.S.A.

Sagan, Dr. Carl, Laboratory for Planetary Studies, Center for Radiophysics and Space Research, Cornell University, Ithaca, N.Y. 14850, U.S.A.

Samuelson, Dr. Robert E., Goddard Space Flight Center, Code 622, Greenbelt, Md. 20771, U.S.A.

Schorn, Dr. Ronald A., Jet Propulsion Laboratory, 4800 Oak Grove Drive, Pasadena, Calif. 91103, U.S.A.

Seidel, Dr. Boris, Jet Propulsion Laboratory, 4800 Oak Grove Drive, Pasadena, Calif. 91103, U.S.A.

Shimizu, Dr. Mikio, Ochanomizu University, Department of Physics, Tokyo, Japan

Sill, Dr. Godfrey T., Lunar and Planetary Laboratory, The University of Arizona, Tucson, Ariz. 85721, U.S.A.

Sinton, Dr. William M., Department of Astronomy, University of Hawaii, Honolulu, Hawaii 96822, U.S.A.

Slanger, Dr. Tom G., Stanford Research Institute, Menlo Park, Calif. 94025, U.S.A.

Smith, Dr. Bradford A., The Observatory, New Mexico State University, Las Cruces, N.M. 88001, U.S.A.

Smith, Dr. Harlan, Department of Astronomy, University of Texas at Austin, Austin, Tex. 78712, U.S.A.

Snyder, Dr. Conway W., Viking Orbiter Scientist, Jet Propulsion Laboratory, Pasadena, Calif. 91103, U.S.A.

Squires, Dr. P., Desert Research Institute, University of Nevada, Reno, Nev. 89507, U.S.A.

Staley, Dr. D. O., Institute of Atmospheric Physics, University of Arizona, Tucson, Ariz. 85721, U.S.A.

Stallkamp, Dr. John, Jet Propulsion Laboratory, 4800 Oak Grove Drive, Pasadena, Calif. 91103, U.S.A.

Stewart, Dr. A. Ian, Laboratory for Atmospheric and Space Physics, University of Colorado, Boulder, Colo. 80302, U.S.A.

Streett, Dr. William B., Department of Physics, U.S. Military Academy, West Point, N.Y. 10996, U.S.A.

Suess, Dr. Hans E., Department of Chemistry, University of California, La Jolla, Calif. 92037, U.S.A.

Thomas, Dr. Gary E., Laboratory for Atmospheric and Space Physics, University of Colorado, Boulder, Colo. 80302, U.S.A.

Thomson, Allen B., Lunar and Planetary Laboratory, Tucson, Ariz. 85721, U.S.A.

Trafton, Dr. Laurence, Department of Astronomy, University of Texas at Austin, Austin, Tex. 78712, U.S.A.

Traught, Steve, Martin Marietta Corp., Box 179, Denver, Colo. 80201, U.S.A.

Tull, Dr. Robert G., Department of Astronomy, University of Texas at Austin, Austin, Tex. 78712, U.S.A.

Uesugi, Dr. Akira, Department of Astronomy, University of Massachusetts, Amherst, Mass, 01002, U.S.A.

Van de Hulst, Dr. H. C., University Observatory, Leiden, The Netherlands

Westphal, James A., Department of Geological Sciences, California Institute of Technology, Pasadena, Calif. 91107, U.S.A.

Wildey, Dr. Robert L., USGS Branch of Astrogeology, 601 E. Cedar Avenue, Flagstaff, Ariz. 86001, U.S.A.

Wildt, Dr. Rupert, Department of Astronomy, Yale University, New Haven, Conn. 06520, U.S.A.

Wood, George P., Yorktown, Va. 23490, U.S.A.

Woodman, Jerry, McDonald Observatory, Fort Davis, Texas, U.S.A.

Woszczyk, Dr. Andzrej, McDonald Observatory, Fort Davis, Texas, U.S.A.

Young, Dr. Andrew T., Jet Propulsion Laboratory, 4800 Oak Grove Drive, Pasadena, Calif. 91103, U.S.A.

Young, Dr. Louis Gray, Jet Propulsion Laboratory, 4800 Oak Grove Drive, Pasadena, Calif., U.S.A.

Young, Dr. Richard S., NASA Headquarters, Washington, D.C., U.S.A.

PART I

VENUS

THE CHEMICAL COMPOSITION OF THE ATMOSPHERE
OF VENUS

A. P. VINOGRADOV, YU. A. SURKOV, B. M. ANDREICHIKOV,
O. M. KALINKINA, and I. M. GRECHISCHEVA

Vernadskii Institute of Geochemistry, Soviet Academy of Sciences, Moscow, U.S.S.R.

Abstract. The chemical composition sensors on the Venera-4, -5, and -6 spacecraft are described. The mixing ratio by volume of carbon dioxide is determined to be 97 (+3, −4)%; nitrogen, less than 2%, oxygen, less than about 0.1%; and water vapor at the 25 °C temperature level, on the order of 10 mg/l, an amount large enough to imply that the clouds of Venus are composed of condensed water.

Introduction

In two earlier papers, [1, 2] we had an opportunity to present briefly the measurements of the composition of the Venus atmosphere on the basis of data acquired with the interplanetary station* Venera-4. In this paper we present data, partly already known [3], obtained with the Venera-5 and Venera-6 spacecraft, and derive the chemical composition of the atmosphere on the basis of all the measurements available.

Until the launch of Venera-4, we had remarkably contradictory information on the chemical composition of the atmosphere of Venus. This information, as is known, was obtained from either spectral observations from the earth, or by means of various computations, comparisons, and the construction of models of the Venus atmosphere [4].

Data on the constituents CO_2, H_2O, O_2, N_2, and other gases had a very wide range of values. Therefore, as we noted before, in the first experiments on Venera-4 we were restricted to making threshold determinations of these components. The results of the measurements were presented at a symposium on the atmospheres of the planets, and are found in a series of reports [5].

In the subsequent investigations with the Venera-5 and Venera-6 spacecraft, thanks to the data from Venera-4, we were able to make determinations of these gases with greater precision and within narrower limits than before for the concentrations of CO_2, H_2O, N_2, and O_2. It was thereby determined from the data of Venera-4 that the optimum regimes for the operation of the instruments could be established in accordance with the presumed conditions on the planet.

2. Organization of the Experiments

The Venera-4 spacecraft was launched in June, 1967. Four months later it made the first smooth descent into the atmosphere of Venus and made investigations of the physical, chemical, and structural parameters in the atmosphere of the planet. Measurements were made during the descent of the spacecraft by parachute. Those quantities

---

* *Translator's note:* The Soviets refer to their spacecraft in this paper as 'automatic interplanetary stations', 'interplanetary stations', 'landing stations', etc. For clarity, these terms are uniformly translated herein as 'space probe' or 'spacecraft'.

Sagan et al. (eds.), Planetary Atmospheres, 3–16.

measured were temperature, pressure, and composition, over the range of 0.7 atmospheres pressure and 25 °C to 18 atm pressure and 280 °C. In that experiment the first direct measurements in the atmosphere of a planet were made. This allowed us to restrict considerably the circle of different suppositions about the atmosphere of Venus, its origin and evolution, and to form definite ideas [1, 2]. We had placed on Venera-4 apparatus capable of a wide range of measurements which relatively coarsely, but reliably, determined the basic parameters of the atmosphere. The wide range was necessary because of the indeterminate nature of our knowledge of physical conditions on Venus, as we have already mentioned.

In January, 1969, two more space probes were sent to Venus – Venera-5 and Venera-6. The purpose of the launching of these probes was to obtain further and more detailed studies of the atmosphere of the planet. As is known, these space probes penetrated deeper into the atmosphere of Venus. Measurements were made during parachute descent, and the temperature and pressure were observed to vary from 25 °C and 0.6 atm to 320 °C and 27 atm. Apart from the investigations of temperature, pressure, and composition, the depth of the atmosphere was determined with these two probes.

As it turned out, the measured change in temperature with height in the entire interval studied is close to the adiabatic gradient. If it is assumed that the temperature varies according to the adiabatic law down to the very surface, then at the surface level as determined by the altimeter on Venera-6, the temperature and pressure must be

Fig. 1.   General view of the gas analysis apparatus.

400 °C and 60 atm, and at the surface level determined by the altimeter on Venera-5 they must be 530 °C and 140 atm [6].

These differences in temperature and pressure in the regions of the descents of Venera-5 and Venera-6 follow from the differences (on the order of 14 km) of the level of the surface as measured by the altimeters on the probes, and reveal, it appears, the existence of large nonuniformities in the relief of the planet's surface.

At the present time we have results of measurements in two positions. Therefore, we may speak of some mean level of the surface, to which must be accorded a pressure of about 100 atm and temperature about 500 °C. Then the levels of the surface measured by Venera-5 and Venera-6 would correspond to differences in altitude of −7 km and +7 km, respectively. This mean level, fortunately, agrees with the data of radioastronomical measurements [7]. In addition, it can be refined, of course, with additional data on different regions of the surface.

On all three spacecraft we installed special gas analysis apparatus for investigations of the composition of the Venus atmosphere, and to determine the concentrations of carbon dioxide, oxygen, water, nitrogen, together with inert gases (Figure 1). On each probe were two instruments which worked in determining the altitude (pressure). The command to the two instruments came from a program-time system. The weight of each instrument was about 1 kg, using electrical power of about 30 W in 10 sec time.

In the course of the operation of all the probes in the Venus atmosphere, the concentration of carbon dioxide gas, oxygen, water, and nitrogen (together with the inert gases) was determined several times. This information was repeatedly transmitted from each instrument.

3. Methods of Measurement, and the Results

In the determination of the concentrations of the basic components of the atmosphere, a manometric method was mainly used. This system is remarkably simple and reliable. The sensing elements were of the pressure membrane type. Depending on the construction, these permit measurement of absolute pressure or differences in pressure between two chambers arising from absorption of the component being studied. For each component of the atmosphere, a suitable absorbent material was selected. With this method the concentration of carbon dioxide gas, nitrogen, oxygen, and water vapor was determined. In addition, in the instruments an electrolytic method was used for determination of small amounts of moisture, and a thermochemical device for small quantities of oxygen. Measurements of a few of the components of the atmosphere were repeated a second time for additional reliability. The choice of chemical absorbents was made according to the following criterion: absorbents with sufficient substance and selectivity must provide complete absorption of the gas in a fixed time interval. At the same time the speed of the reaction of absorption must not be so great as to cause noticeable changes in the composition of the gas as it was being taken into the instrument before the entrance closed.

The reagents should be stable over a sufficient range of temperature, determined by the technical conditions of storage and operation of the instrument, and should satisfy the indicated demands on the speed of reaction in the given range.

The reagents should not react with one another or change their characteristics upon long storage in vacuum conditions.

Gaseous products should not arise upon reaction because these would interfere with the determination of a component in a given volume.

The reagents should be able to withstand a considerable loading (this makes the use of liquids and pastes more difficult).

A. DETERMINATION OF CO$_2$ GAS

In selecting the CO$_2$ adsorber, the adsorptive properties of ascharite, sodium hydroxide, potassium hydroxide, and calcium hydroxide were compared. It was completely clear that acid fumes (HF, HCl) would be adsorbed by this method. However, it is known that the observed HF and HCl in the atmosphere of Venus constitute,

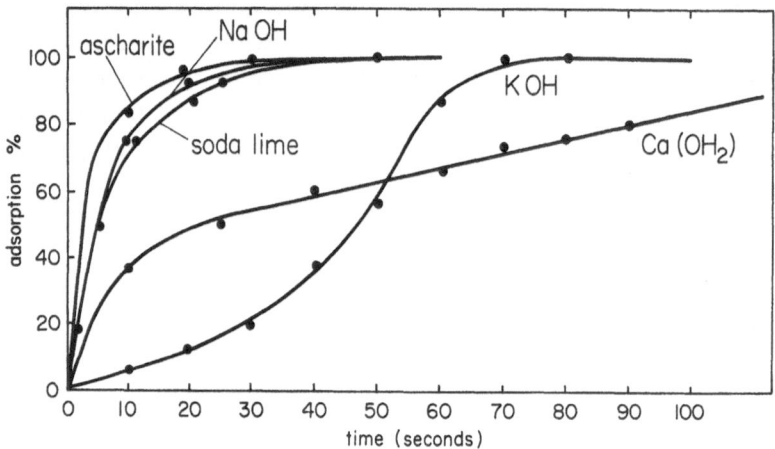

Fig. 2. Adsorption of CO$_2$ by different adsorbent materials.

respectively, not more than 5×10^{-9} and 6×10^{-7} of the quantity of CO$_2$; i.e. six to eight orders of magnitude less. Thus they can be neglected.

Figure 2 shows the dependence of adsorption of CO$_2$ on time for various adsorbers. From the figure it can be seen that calcium hydroxide cannot be used because of the slowness of the reaction. Potassium hydroxide has a clearly expressed delayed reaction, depending on the moistness of the reagent. The adsorptive qualities of ascharite, sodium hydroxide, and soda lime are almost equal to one another. In order to use them, a minimum time of specimen intake would have been necessary. The most suitable adsorber for our use turned out to be potassium hydroxide. In order to eliminate the dependence of the speed of the reaction on the degree of desiccation, a lithium adsorber of CO$_2$ was added to the KOH (about 3%). The degree of desiccation of the reagent did not then influence the adsorption of CO$_2$, and the adsorption was completed in 35–40 sec.

The CO$_2$ content of the atmosphere of Venus was determined by the difference in pressures between compartments of a cell, in one of which was placed the CO_2 adsorber. In order to avoid errors associated with the adsorption of water vapor by KOH, $CaCl_2$ was placed in the other compartment to adsorb moisture. The difference in pressure was registered using a sensor employing a membrane, as indicated above. The results of the determination of CO_2 by these methods are listed in Table I.

TABLE I

Results of the determination of CO_2 in the atmosphere of Venus

Number	Pressure (atm)	Temperature of the atmosphere of Venus, °C	Results of measurements in volume %	Spacecraft
1	0.6	~ 25	97 ± 4	Venera-5
2	0.7	~ 25	90 ± 10	Venera-4
3	2.0	85	> 56	Venera-6
4	5.0	150	> 60	Venera-5
5	10.0	220	> 30	Venera-6

As can be seen from the table, the atmosphere of the planet contains up to 97 (+3, −4)% carbon dioxide gas. The large error in determining the carbon dioxide content by Venera-4 (90 ± 10%) was caused by the fact that the ambient pressure at the moment of sampling was not measured and was determined by extrapolation.

In order to reduce errors in the determinations, Venera-5 and Venera-6 were provided with means to measure the ambient pressure directly at the point where the samples for analyses were taken. This allowed us to reduce the error in determination to about 4%.

Some instruments worked at pressures significantly surpassing the assumed working range of the pressure sensors, and so threshold values of the CO_2 content were obtained which do not contradict the quantitative determinations. The reduction to the threshold determination of CO_2 with height is connected with the increase of pressure at the same sensitivity.

Thus, by direct quantitative determinations we can accept a mean content of carbon dioxide in the atmosphere of Venus of 97 (+3, −4)%.

B. DETERMINATION OF MOLECULAR NITROGEN

Nitrogen was determined by two methods. On Venera-4 it was determined by the difference in pressure between two compartments, in one of which all active gases, together with nitrogen, were adsorbed and in the other of which CO_2, O_2, and water vapor were adsorbed. In Venera-5 and Venera-6 the determination was made by means of a measurement of the pressure remaining after adsorption of the basic constituents of the atmosphere – CO_2, H_2O, and O_2.

Adsorption of N_2 is usually accomplished by its ability to form nitrides. On investigating the metals that easily form stable nitrides, the most acceptable was found to be zirconium. It is sufficiently stable in ordinary conditions and when heated reacts well with various gases.

The optimum temperatures for the adsorption by zirconium of the most frequently encountered gases are: for hydrogen 300 °C; for oxygen 400 °C; for water vapor 250–350 °C; for nitrogen 800 °C; for carbon dioxide 800 °C; and for carbon monoxide 800 °C.

A quartz capillary tube, inside of which a tungsten spiral was placed for heating, was used as a heater.

In order to obtain a sufficient layer of the adsorber, zirconium powder (grain size 1–5 μ) was cemented on the quartz capillary with a 20% solution of sodium silicate. To speed and intensify the reaction an initiating mixture was used. The best results were obtained with aluminum and barium peroxide in a ratio of 2:1. The weight of the initiating mixture comprised about one-fifth of the weight of the zirconium adsorber.

A test showed that the presence of small quantities of CO_2, O_2, and water vapor did not affect the operation of the zirconium adsorber. As O_2, CO_2, and H_2O would be adsorbed simultaneously with nitrogen, adsorbers for these gases were placed in the comparison chamber of the gas analyser. For CO_2 and water, potassium hydroxide was used, and for oxygen, phosphorus.

This method was used to determine the nitrogen content only on the first spacecraft, Venera-4, since we had at first allowed for a large nitrogen content in the atmosphere of the planet. More precise data on nitrogen were completely lacking at that time. The measurements on Venera-4 showed that nitrogen is scarce [1, 2]. Because of this, in subsequent experiments (on Venera-5 and Venera-6) the quantity of nitrogen (together with inert gases possibly present) was determined by the pressure remaining after adsorption of CO_2, O_2, and water vapor. This method was chosen because zirconium adsorbers in mixtures with CO_2 present do not operate with sufficient stability.

The results of all determinations of nitrogen in the atmosphere of Venus are given in Table II.

TABLE II

Results of the determination of N_2 in the atmosphere of Venus (together with possible existing inert gases)

Number	Pressure (atm)	Temperature of the atmosphere of Venus, °C	Results of measurements in volume %	Spacecraft
1	0.6	~ 25	< 3.5	Venera-5
2	0.7	~ 25	< 7[a]	Venera-4
3	2	85	< 2.5[a]	Venera-4
4	2	85	< 9.5	Venera-6
5	5	150	< 4	Venera-5
6	5	150	< 4	Venera-5
7	10	220	< 2.5	Venera-6
8	10	220	< 2	Venera-6

[a] Measurement of nitrogen determined by adsorption by zirconium. The others by the pressure method.

As can be seen from the table, nitrogen was determined eight times and all the measurements indicate its low abundance. As errors of measurement are possible, owing to incomplete adsorption, the given measurements can be taken as reliable upper limits on the nitrogen content.

Thus, we can adopt a nitrogen content in the atmosphere of Venus of not more than 2%. In addition it should be noted that the quantity of inert gases, mainly argon, is evidently very small. In the atmosphere of the earth the content of argon, the most widely distributed of the inert gases, is around 1% of the amount of nitrogen in the atmosphere.

C. DETERMINATION OF OXYGEN

As indicated above, two methods for the determination of oxygen were used – the manometric and the thermochemical.

The simplest and best studied adsorber of oxygen is white phosphorus. The

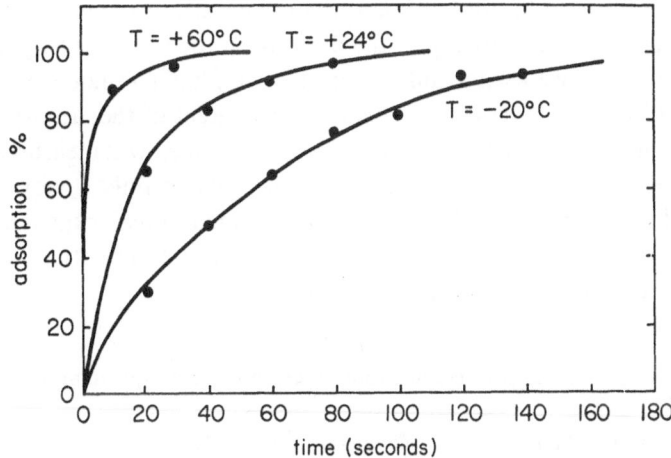

Fig. 3.   Adsorption of O_2 by white phosphorus, applied to pumice at different temperatures.

oxygen adsorption curves for white phosphorus deposited on pumice are shown in Figure 3. However, the phosphorus adsorption reaction has a strong dependence on temperature and at low temperatures proceeds with insufficient speed. This circumstance forced us to investigate the possibility of heating the experimental apparatus.

As an adsorber we used red phosphorus, which upon sublimation changes over to the white modification, energetically adsorbing oxygen. This method insures both the inertness of the reagent during the time of its storage and the independence of the speed of the reaction on the temperature of the surrounding medium. To insure the mechanical strength of the adsorber, a binder was used (a solution of potassium bichromate and sodium silicate).

This method was used only in the first experiment on Venera-4 when the possibility of a large oxygen content in the atmosphere of the planet had not been excluded.

However, Venera-4 showed that the oxygen content in the atmosphere of Venus is small. Because of this the following spacecraft used only the thermochemical method, intended for threshold detection of small quantities of oxygen.

In view of the great chemical activity of oxygen, one can select a whole series of solids that react with oxygen when heated.

The refractory metals are of greatest interest, since there is a great difference among their temperatures of oxidation and fusion. For the determination of oxygen they can be used in the form of a fine wire (a filament or coil). The resistance of the heated wire increases on reaction with oxygen as a result of an increase in temperature caused by a noticeable thermal effect of the reaction, and as a result of a decrease in the wire's cross-section on being oxidized.

The use of a refractory wire in itself makes its heating easy. The most suitable metal was found to be tungsten. Its reaction with oxygen in the interval 700–1000 °C has a sufficiently distinctive character. In pure nitrogen and in carbon dioxide, as our experiments showed, burn-out of tungsten coils does not occur in the course of extended periods of time. The addition of a small amount of oxygen (partial pressure of a few torr) leads to the functioning of the detector in the required length of time. For oxygen determination we used a coil of tungsten wire having diameter 12 μ.

The results of the measurements of the oxygen content of the atmosphere of Venus are given in Table III. As can be seen from the table, oxygen was measured seven times. All measurements (except one) indicate the possibility of only a very small amount of oxygen in the atmosphere of Venus. One measurement (greater than 0.4%) obtained on Venera-4 could be connected with a less precise determination than the other measurements made on Venera-5 and Venera-6.

TABLE III

Results of the determination of O_2 in the atmosphere of Venus

Number	Pressure (atm)	Temperature of the atmosphere of Venus, °C	Results of measurements in Volume %	Spacecraft
1	0.7	~ 25	> 0.4	Venera-4
2	2	85	< 1.5[a]	Venera-4
3	2	85	< 0.3	Venera-6
4	5	150	> 0.1	Venera-5
5	5	150	< 0.1	Venera-5
6	10	220	< 0.1	Venera-6
7	10	220	< 0.2	Venera-6

[a] Data obtained by the manometric method. All others by thermochemical means.

Thus the content of oxygen in the atmosphere of Venus does not exceed 0.1%. This content is the limit of sensitivity of our instruments. Because of this it was not possible to carry out determinations of still smaller quantities of O_2. The limit given here does not contradict data on the content of various forms of oxygen in the ionosphere of Venus.

D. THE DETERMINATION OF MOISTURE

Two methods were used in the determination of moisture; the manometric method for large quantities, and the electrolytic method for small quantities.

In order to choose the adsorbant for use in the manometric method, comparisons were made of the effectiveness of various substances such as P_2O_5 (phosphorus anhydride), KOH, $CaCl_2$, and others. In the present instruments the desiccant $CaCl_2$ was used. Its desiccating properties have been well studied. The advantage of $CaCl_2$ lies in its inertness relative to other gaseous components and their adsorbants. The pure calcium chloride used was first fused, ground up, and then sifted without exposure to moisture. In use, the sifted fraction with grain sizes 1–3 mm achieved practically complete adsorption of moisture in the instrument in 30–60 sec. (Figure 4).

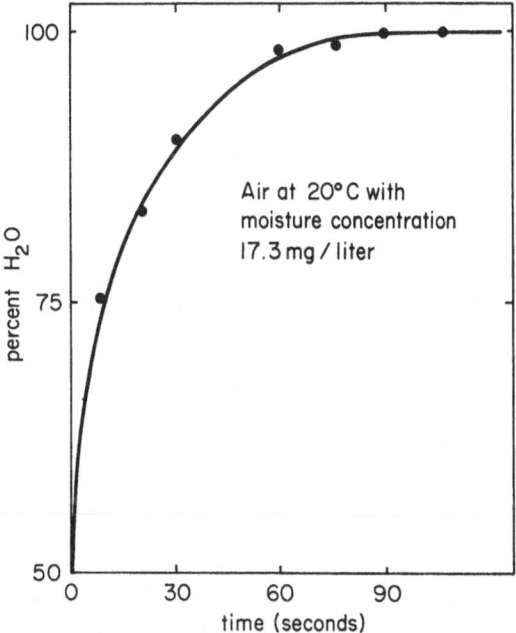

Fig. 4.   Adsorption of H_2O by calcium chloride.

In the electrolytic method a sensor that operated on the principle of a change in electrical resistance of a layer of conducting substance as it adsorbs moisture was used. The moisture sensor was made in the form of a glass coil with a double winding of platinum wires in grooves on its exterior surface. As a moisture adsorbing substance to act as the conducting layer, we used hydrated phosphorus anhydride in a 20% solution of P_2O_5 in water. This was applied to the working surface of the sensor and desiccated by electrolysis. When the operating voltage is turned on, metaphosphoric acid electrolyzes to practically complete dryness, decomposing according to the reaction

$$2\,HPO_3 \rightarrow H_2 + \tfrac{1}{2}\,O_2 + P_2O_5$$

In this reaction the concentration of water decreases and that of P_2O_5 increases, causing an increase in the resistance of the conducting layer and a decrease in the current. At sufficiently high concentration of P_2O_5 the layer is practically an insulator. On absorption of moisture, metaphosphoric acid is formed anew, accompanied by a fall in the resistance, thus determining the moistness of the surrounding medium.

Since P_2O_5 is a most effective adsorber of small amounts of moisture, during its long storage time the sensor was placed over an alkali to maintain its resistivity in the proper condition for effective operation. On attaining an equilibrium condition, the resistance of the coil remained within the assigned limits. The behavior of the moisture sensor under various conditions is shown in Figure 5.

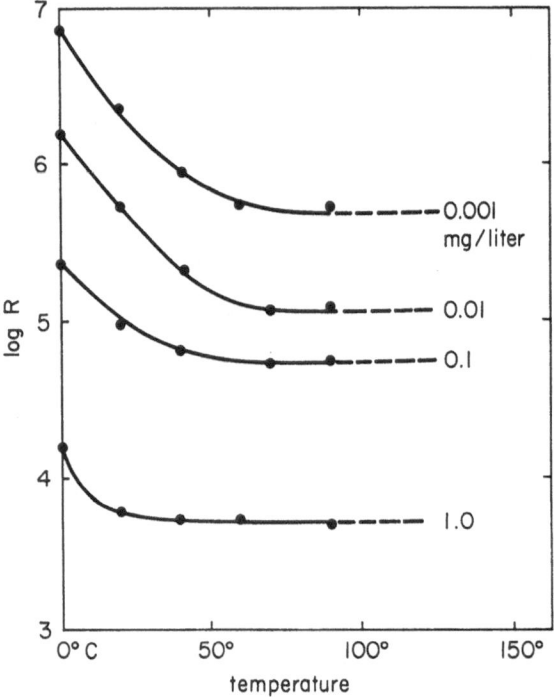

Fig. 5.   Measurements of the resistance of the humidity sensor with temperature for different amounts of moisture.

The moisture sensor was situated under a glass membrane where the atmosphere entered the instrument, and during the descent of the spacecraft (after opening the instrument) the measurements of moisture were made directly in the atmosphere of the planet. The results of the measurements are presented in Table IV.

As is seen in Table IV, all the measured data attest to the existence of water in the atmosphere of Venus. Also seen is the slight tendency towards diminished water concentration with reduction of pressure. This tendency was noted from the measured humidity during the descent of the spacecraft in the atmosphere of Venus.

If we presume that the concentration of water vapor at the level of condensation is approximately 11 mg/l, then the temperature of the lower part of the cloud layer must

TABLE IV

Results of the determination of the concentration of H_2O in the atmosphere of Venus

Number	Pressure (atm)	Temperature of the atmosphere of Venus, °C	Results of measurement, mg/l	Adsorber	Spacecraft
1	0.6	~ 25	> 4	P_2O_5	Venera-5
2	0.6	~ 25	~ 11[a]	$CaCl_2$	Venera-5
3	0.7	~ 25	> 0.7	P_2O_5	Venera-4
4	2	85	< 8[a]	$CaCl_2$	Venera-4
5	2	85	~ 6[a]	$CaCl_2$	Venera-6
6	5	150	> 0.7	P_2O_5	Venera-5
7	10	220	> 0.7	P_2O_5	Venera-6

[a] These measurements were made by the manometric method, while the others were made by the electrolytic method.

be about 286 K (dew point). Then the atmospheric pressure at the level of condensation may be evaluated by extrapolation from the measured quantities at higher layers in the atmosphere. For the above-mentioned chemical composition of the atmosphere and the quasi-linear variation of temperature from the level of the first humidity measurement ($P_1 = 0.6$ atm, $T_1 \simeq 25$ °C) to the level of condensation ($T_2 = 13$ °C), the pressure at the level of condensation is about 0.5 atm.

Using the known barometric relationship, we may also evaluate the height of the lower boundary of the cloud layer above the level of the first measurement and the temperature gradient at that point, and these correspond to ~1.3 km and 7.5 °/km.

Upon analysis of the atmosphere inside the cloud layer, we may make suppositions that the relationship of the pressure of water vapor to the pressure inside the cloud layer remains constant. The release of latent heat, originating upon condensation of water vapor will change the temperature gradient in accordance with the relationship

$$\frac{dT}{dh} = \left(\frac{dT}{dh}\right)_{\text{adiabatic}} \cdot \frac{a + M_{H_2O}(LE/PRT)}{1 + \frac{M^2_{H_2O}}{M} \cdot \frac{L}{PC_p} \cdot \frac{dE}{dT}},$$

where a = the ratio of the temperature gradient under the clouds to the dry adiabatic gradient, M_{H_2O} = molecular weight of water vapor, L = latent heat of the phase change, E = vapor tension of saturated vapor, M = molecular weight of the atmosphere, C_p = specific heat of the atmosphere at constant pressure.

On the basis of the measurements of temperature and pressure in the high layers of the atmosphere, we can determine the quantity of condensible water vapor from the formula

$$Q = \frac{M_{H_2O}}{RT}(e - E),$$

where e is the partial pressure of water vapor in the atmosphere.

The change in humidity of the clouds with height is shown in Figure 6. From the figure it is seen that the thickness of the clouds is about 15 km. Up to the height of

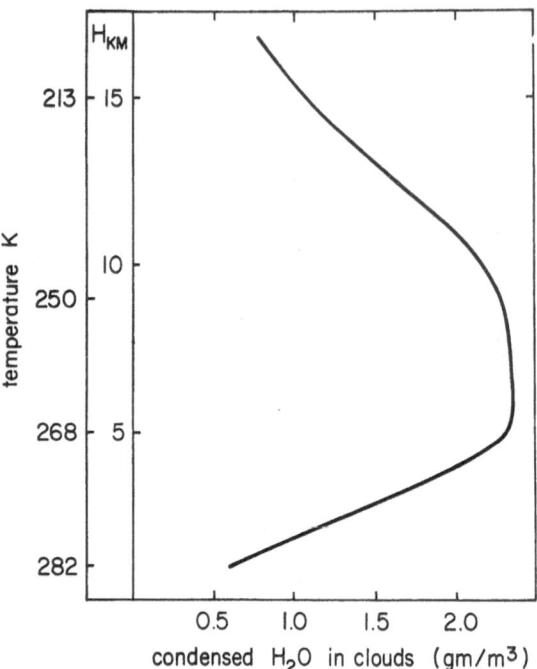

Fig. 6.   Distribution of condensable H_2O in the cloud layer with respect to altitude.

7 km from the level of condensation, the clouds are composed, in all probability, of droplets of water, and at higher levels they consist of ice crystals. The maximum moisture content of the clouds is about 2.4 gm/m^3 at an atmospheric temperature from $-5\,°C$ to $-25\,°C$ and is situated 5–10 km above the level of condensation. The upper boundary of the clouds must be diffuse. Obviously, in the sub-cloud layer there exists a region of intense circulation of moisture. In this region there is falling condensation in the cloud layer. The water drops fall until they reach a region of higher temperature, are vaporized, and again moved upward. In such a case the amount of water in the atmosphere must diminish with depth.

4. Composition and Origin of the Atmosphere of Venus

As is known, from the Venera-4 spacecraft we determined the following composition of the Venus atmosphere:

CO_2 $90 \pm 10\%$

O_2 0.4–1.5%

N_2 No more than 7% (probably no more than 2.5%)

H_2O 1–8 mg/l

From the results of more precise and repeated determinations from the Venera-5 and Venera-6 spacecraft as discussed above, the composition is made more exact and can be written as follows:

CO_2 $97 \pm 4\%$

O_2 No more than 0.1%

N_2 No more than 2%

H_2O 6–11 mg/l (at 0.6 to 2 atm pressure)

Thus, the atmosphere is composed mostly of carbon dioxide gas. The concentration of nitrogen (together with the inert gases which may comprise only a small fraction of the amount of nitrogen) is not in excess of 2%. The concentration of oxygen appears to be appreciably less than we found from Venera-4 and does not exceed 0.1%. This quantity differs from the value determined from radioastronomical measurements of the concentration of molecular and other forms of oxygen in the ionosphere. We discussed earlier the nature of the distribution of water in the atmosphere of Venus, regarding the concentration of humidity. From all the experiments that we were able to put on the space probes, we were unable to detect in the atmosphere any other gases in any noticeable quantities. Naturally, this does not exclude the possible existence of CO, HF, HCl, and Ar in the Venus atmosphere, and possibly other gases, but the quantities must be very small. As is known, their presence is shown in the data from ionospheric investigations.

Returning to the problem of the origin of the atmosphere of Venus [8], it may be said that the new data on the atmospheric pressure and temperature [6] show that the origin is connected with thermal effects. We had already considered the formation of large quantities of CO_2 in the Venus atmosphere as following from the proximity of the planet to the sun, and the high equilibrium temperature. It seemed to us that in these conditions there occurs intensive erosion of the rocks and the composition of the atmosphere is determined in large measure by the wollastonite equilibrium

$$CaSiO_3 + CO_2 \rightleftarrows CaCO_3 + SiO_2.$$

Many other reactions for the formation of wollastonite are known, for which we have experimental data on the equilibria at various temperatures and pressures. This reaction at a pressure of about 50 atm goes in the direction of the formation of wollastonite and CO_2 already at temperature higher than 350 °C.

We have already given an analysis of the possible concentration of CO_2 on the earth and Venus. If we assume that the general reserve of CO_2 on both planets is the same, then the quantity of CO_2 on Venus, from other data, should be about 50 atm. This, in general, is not in disagreement with the experimental determination of the amount of CO_2, as well as the pressure and temperature we observe at the surface of Venus.

Oxygen in the atmosphere of Venus can form as a result of photodissociation of carbon dioxide gas and water. However, the low concentration of oxygen shows that it strongly recombines with the rock materials. Hydrogen, resulting from the photodissociation will disappear. It is known that Venus has a hydrogen corona, but it is not large. That is, there is a mechanism for the loss of water, or more exactly, hydrogen.

The atmosphere, as we saw, has an insignificant amount of water. The maximum H_2O concentration according to our data lies in the cloud layer. One may presume

3—P.A.

that some quantity of water in the cloud layer occurs in the solid phase, but it requires an unambiguous determination.

The source of nitrogen–ammonia, apparently, as on the earth is degassing, a result of volcanic activity, which on Venus is very likely. Future experiments, obviously, can confirm it. We have already remarked that conditions on Venus, in comparison with those on the earth, are favourable for the rapid formation of nitrogen from ammonia.

As computations indicate, the possible share of heat that is brought up from the interior of the planet to the surface of Venus is extremely insignificant by comparison with the solar heat absorbed by the atmosphere owing to the carbon dioxide gas and water. Therefore, it is precisely this heating effect that is the main mechanism that supports and self-heats Venus and forms the heavy carbon dioxide atmosphere.

References

[1] Vinogradov, A. P., Surkov, Yu. A., Florenskii, K. P., and Andreichikov, B. M.: 1968, 'Determination of the Chemical Composition of the Atmosphere of Venus with the Interplanetary Station Venera-4', *Dokl. Akad. Sci. U.S.S.R.* **179**, 37. (In Russian.)

[2] Vinogradov, A. P., Surkov, Yu. A., and Florenskii, K. P.: 1968, 'Chemical Composition of the Atmosphere of Venus Based on the Data of the Interplanetary Station Venera-4', *J. Atmospheric Sci.* **25**, 535–536.

[3] Vinogradov, A. P., Surkov, Yu. A., and Andreichikov, B. M.: 1970, 'Investigations of the Composition of the Atmosphere of Venus with the Automatic Space Stations Venera-5 and Venera-6', *Dokl. Acad. Sci. U.S.S.R.* **190**, 552. (In Russian.)

Moroz, V. I.: 1967, *Physics of the Planets*, Nauka, Moscow (English translation available in NASA Technical Translation TT F-515.

[5] Papers from the Second Arizona Conference on Planetary Atmospheres, *J. Atmospheric Sci.* **25**, 533–671, 1968. Conference held March 11–13, 1968.

[6] 'Soviet Interplanetary Stations Venera-5 and Venera-6', *Pravda* 4 June, 1969. (In Russian.)

[7] Kuzmin, A. D.: 1967, *Radiophysical Investigations of Venus*, Levels of Science, Physics Series, VINITI Publishers, Moscow. (In Russian.)

[8] Vinogradov, A. P.: 1969, 'The Atmospheres of the Planets of the Solar System', *Vestnik* (Moscow State University), No. 4, Geological Series.

EXOSPHERIC TEMPERATURE OF VENUS FROM MARINER 5

CHARLES A. BARTH

Dept. of Astro-Geophysics and Laboratory for Atmospheric and Space Physics, University of Colorado, Boulder, Colo., U.S.A.

Abstract. The Lyman alpha measurements made on the sunlit side of Venus by Mariner 5 showed that the distribution of the radiating hydrogen atoms was governed by two scale heights. A revised model is presented in which the inner or cold distribution is attributed to hydrogen atoms in thermal equilibrium with the thermosphere and the outer or hot distribution is due to hydrogen atoms produced by the photodissociation of molecular hydrogen. The consequence of this interpretation is that the temperature of the dayside thermosphere of Venus is 325 K.

1. Introduction

The ultraviolet photometer data from Mariner 5 shows that the Lyman alpha emitters in the upper atmosphere of Venus must be present either in two forms or at two different temperatures (Barth, 1968). Among the several models suggested, one requires the presence of both molecular and atomic hydrogen (Barth *et al.*, 1968) while the other has hydrogen atoms present in both a thermal and a 'hot' population. At the time of the original suggestion (Barth, 1968), there was no plausible physical model for producing the 'hot' atoms. The present paper offers a model in which the molecular hydrogen model and the 'hot' atom model are combined. In this theory, the 'hot' atoms arise from the photodissociation of molecular hydrogen.

Since there is water vapor in the atmosphere of Venus, there should be molecular hydrogen as well, just as there is molecular hydrogen in the earth's upper atmosphere. Lewis (1968, 1970) has calculated that under chemical equilibrium conditions in the dense, lower atmosphere of Venus, the reaction $CO + H_2O = CO_2 + H_2$ leads to a mixing ratio for molecular hydrogen of the order of 10^{-6}. In the upper atmosphere, diffusive separation should set in and molecular hydrogen may become the major atmospheric constituent over a limited altitude range. In the region where the molecular hydrogen is no longer screened from the solar ultraviolet radiation by carbon dioxide, photodissociation will occur.

2. Photodissociation of Molecular Hydrogen

Molecular hydrogen is photodissociated by ultraviolet radiation shortward of 845 Å into two hydrogen atoms.

$$H_2 + h\nu \ (\lambda < 845 \text{ Å}) \rightarrow H(1s) + H(2s, 2p) \tag{1}$$

This direct photodissociation produces half of the atoms in the $1s$ state and the other half in either the $2s$ or $2p$ depending upon the particular photodissociation continua

Sagan et al. (eds.), Planetary Atmospheres, 17–22.

followed. Dalgarno and Allison (1969) have calculated the cross sections for photo-dissociations proceeding by way of the $B^1\Sigma_u^+$, $C^1\pi_u$, and $B'^1\Sigma_u^+$ states.

Photodissociations that are caused by photons with energy greater than 14.7 eV (the photodissociation threshold) will produce atoms with excess kinetic energy. In this way, the photodissociation of molecular hydrogen in the exosphere of Venus, where collisions do not occur, will produce a population of hot atoms.

The rate of production of hydrogen atoms from the photodissociation of molecular hydrogen is given by the following expression, since in the exosphere, densities are sufficiently low that there is no attenuation of the solar radiation:

$$[\dot{H}] = 2[H_2] \sum_{\lambda < 845\,\text{Å}} \sigma_\lambda(H_2)F_\lambda \tag{2}$$

where the square brackets designate volume densities of the chemical constituent within the bracket, $\sigma_\lambda(H_2)$ is the photodissociation cross-section of molecular hydro-gen, and F_λ is the solar flux. Since the last two quantities are functions of wavelength, their product needs to be summed for wavelengths less than 845 Å, the photodissocia-tion threshold. The total flux of photodissociated hydrogen atoms may be obtained by integrating the rate of photodissociation from the base of the exosphere to infinity. Using the continuity equation, this flux of newly formed atoms may be equated to a volume density of hydrogen atoms times an average velocity.

$$\int_h^\infty [\dot{H}]\,dz = [H]\bar{v} \tag{3}$$

This average velocity represents the excess kinetic energy shared by the atoms in the photodissociation process, and is a function of the wavelength of the photon. The flux of photodissociated atoms may be equated to the photodissociation process by integrating the molecular hydrogen density over altitude. This integral may be approximated by the local molecular hydrogen density times the molecular hydrogen scale height, $\mathscr{H}(H_2)$.

The density of freshly photodissociated hydrogen atoms at the base of the exosphere may be related to the density of molecular hydrogen at the same level through the following equation

$$[H] = 2 \sum \left(\frac{\sigma_\lambda(H_2)F_\lambda}{\bar{v}}\right) \mathscr{H}(H_2)[H_2] \tag{4}$$

The quantities in the large parentheses including the average velocity are all functions of the wavelength of the photodissociating photon and should be calculated separately and then summed. Using the solar fluxes of Hinteregger et al. (1965), adjusted by a factor of 1.5 to make them appropriate to October 1967, and the photodissociation cross sections of Dalgarno and Allison (1969), the atom production rates and the average excess energy of the freshly dissociated atoms are given in Table I. In Equation (4), the density of atomic hydrogen that has been converted from molecular hydrogen is given in terms of the molecular hydrogen density. This 'conversion factor' has been

TABLE I

Wavelength	Average excess energy	Solar flux Venus, 1967	Production rate	Conversion factor
Å	eV	10^9 cm^{-2} sec^{-1}	10^{-9} p sec^{-1}	10^{-7}
840–830	0.08	2.1	19.3	7.93
832–835	0.10	1.5	13.8	5.01
830–820	0.17	1.9	14.4	4.05
820–810	0.27	1.7	9.9	2.19
810–796	0.38	2.0	9.2	1.71
790.1	0.51	1.0	2.8	0.45
787.7	0.53	0.92	2.6	0.40
780.3	0.60	0.44	1.0	0.14
796–780	0.53	2.1	5.9	0.93
770.4	0.71	1.2	2.0	0.27
765.1	0.76	0.60	0.8	0.10
780–760	0.71	2.1	3.6	0.50
		Total	83	24

calculated as a function of average excess energy and is listed in Table I also. The conversion factor summed over all energies which gives the amount of atomic hydrogen at a given altitude that has been converted from molecular hydrogen is 2.4×10^{-6}. Each of the subgroups of hydrogen atoms will be distributed with their own scale height depending on their average kinetic energy. This calculated value for the conversion factor is a minimum since some processes that photodissociate molecular hydrogen have been neglected. Mentall and Gentieu (1969) have found that absorption into the $D^1\pi_u$ state leads to predissociation producing additional hydrogen atoms. A more important one is the photo-destruction mechanism proposed by Stecher and Williams (1967) in which ultraviolet radiation shortward of 1108 Å is absorbed into discrete excited states and then reradiated into the continuum of the ground state producing dissociation. Dalgarno and Allison (1969) have estimated that this process may produce ten times as many atoms as the photodissociation process calculated above.

3. The Model

This model identifies the Lyman alpha emission from the outer atmosphere of Venus as arising from the resonance scattering of solar Lyman alpha radiation by atomic hydrogen. The emissions originating from planetocentric distances less than 9000 km come from thermal hydrogen atoms; i.e., atoms that have undergone collisions below the exobase and are in thermal equilibrium with the constituents of the thermosphere. The Lyman alpha emissions arising from planetocentric distances greater than 9000 km originate from 'hot' hydrogen atoms, i.e., atoms that have been produced by the photodissociation of molecular hydrogen in the exosphere and have not undergone any collisions since their formation. This identification of the observations with 'hot' and thermal hydrogen atoms is illustrated in Figure 1 which has been adopted from a figure used for an alternate model (Barth, 1968).

The identification of the lower altitude regime of the data with thermal hydrogen atoms leads to the requirement that the thermospheric temperature of Venus be 325 K. The bottom of the exosphere, the exobase, will be located at the level above which collisions essentially do not take place. The cross section for thermalizing collisions between hydrogen atoms and hydrogen molecules may be estimated to be 3.5×10^{-16} cm². This means that the volume density of molecular hydrogen at the exobase can be no more than 2×10^8 molecules cm^{-3} since the molecular hydrogen scale height at 325 K is 160 km. The carbon dioxide volume density at this same level needs to be 4×10^9 molecules cm^{-3} or less. The distribution of these additional constituents in the model are illustrated in Figure 2 where the curve marked atomic hydrogen contains both the thermal and 'hot' populations from Figure 1.

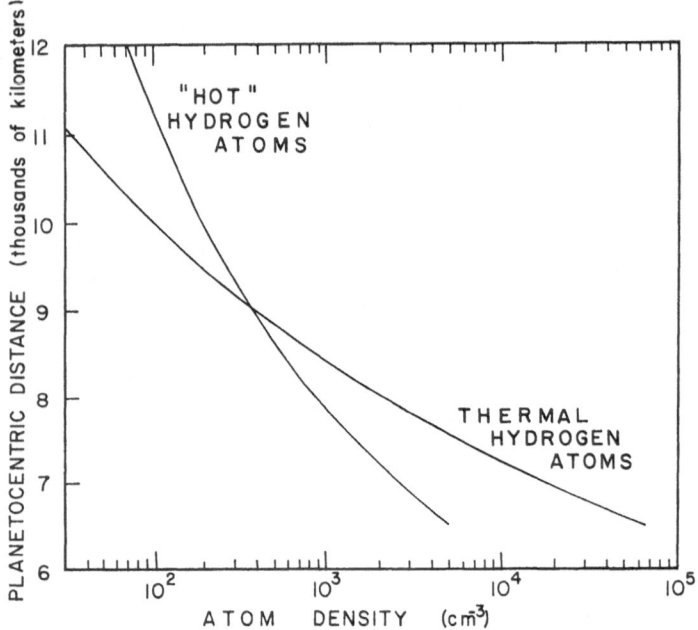

Fig. 1.   Distribution of 'hot' and thermal hydrogen atoms in model.

4. Test of Model

When the conversion factor determined in the Table is applied to molecular hydrogen density at the exobase, the density of 'hot' hydrogen atoms at this level is calculated to be 5×10^2 atoms cm^{-3}. The values of excess kinetic energy listed in the table indicate that the bulk of these atoms will have scale heights that exceed the 640 km scale height of the hot atom distributions in the model shown in Figure 1. This means that the 'hot' atom distribution at planetocentric distances less than 9000 km actually is less than that indicated in Figure 1 and at distances greater than 9000 km the distribution is greater than that shown in Figures 1 and 2.

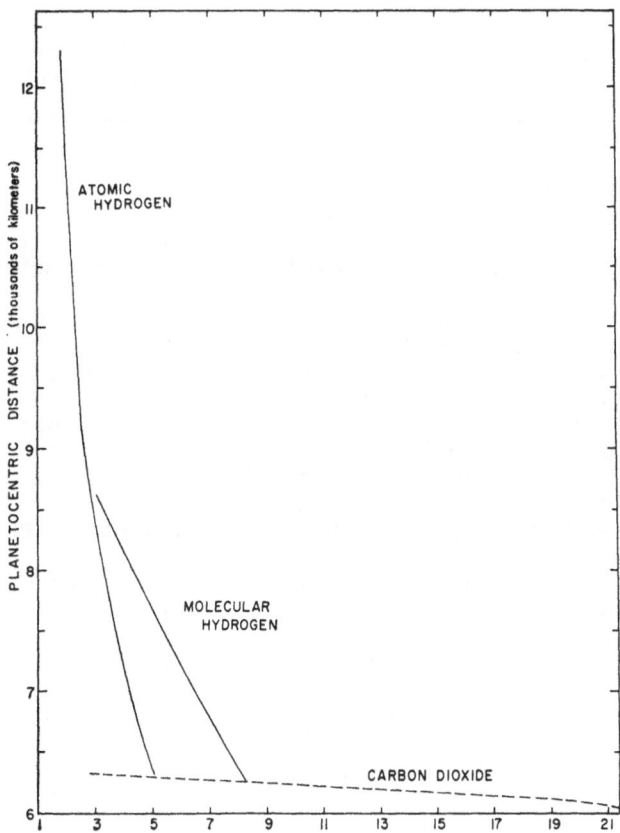

Fig. 2. Distribution of atomic and molecular hydrogen and carbon dioxide in model. Atomic hydrogen distribution includes both the 'hot' and thermal atoms shown in Figure 1.

When the Stecher-Williams process is taken into account, the conversion factor is ten times larger and the density of 'hot' atoms shown in Figure 1 can be matched using a scale height of 640 km. To exactly match the distribution in the figure, the bulk of the 'hot' atoms would need to have an average energy that is twice the kinetic energy of the thermal atoms.

5. Summary

The molecular hydrogen model of Barth (1968) failed to take into account the resonance scattering of Lyman alpha radiation by the freshly created hydrogen atoms. When these 'hot' atoms are included, the Lyman alpha emissions at distances greater than 9000 km are accounted for. The Lyman alpha emissions at distances less than 9000 km are then attributed to thermal hydrogen atoms with the consequences that the Venus thermosphere is very cold, 325 K. As compared with the earlier model, the amount of molecular hydrogen in this model is greatly reduced with the consequence that the Lyman alpha radiation that is radiated during the photodissociation process no longer plays a role in the explanation of the observations.

Acknowledgments

The author appreciates the comments and suggestions made by Professor A. Dalgarno following the presentation of this paper at the symposium. This work has been supported by the National Aeronautics and Space Administration, NASA grant NGL-06-003-052.

References

Barth, C. A.: 1968, 'Interpretation of the Mariner 5 Lyman Alpha Measurements', *J. Atmospheric Sci.* **25**, 564.

Barth, C. A., Wallace, L., and Pearce, J. B.: 1968, 'Mariner 5 Measurement of Lyman Alpha Radiation Near Venus', *J. Geophys. Res.* **73**, 2541.

Dalgarno, A. and Allison, A. C.: 1969, 'Photodissociation of Molecular Hydrogen on Venus', *J. Geophys. Res.* **74**, 4178.

Hinteregger, H. E., Hall, L. A., and Schmidtke, G.: 1965, 'Solar XUV Radiation and Neutral Particle Distribution in July 1963 Thermosphere', *Space Res.* **5**, 1175.

Lewis, John S.: 1968, 'An Estimate of the Surface Conditions of Venus', *Icarus* **8**, 434.

Lewis, John S.: 1970, 'Geochemistry of the Volatile Elements on Venus', *Icarus* **11**, 367.

Mentall, J. E. and Gentieu, E. P.: 1969, 'Lyman-α Fluorescence From the Photodissociation of H_2', GSFC X-616-69-497.

Stecher, Theodore, P. and Williams, David A.: 1969, 'Photo-destruction of Hydrogen Molecules in H_1 Regions', *Astrophys. J.* **149**, L29.

MODELS OF THE VENUS IONOSPHERE

J. R. HERMAN, R. E. HARTLE, and S. J. BAUER

Laboratory for Planetary Atmospheres, Goddard Space Flight Center, NASA,
Greenbelt, Md., U.S.A.

Theoretical modeling of the daytime Venus ionosphere can be used to augment the measurements of Mariner V made during the 1967 fly-by mission of Venus. The models discussed here are obtained by solving the equations of heat conduction for the electron, ion, and neutral gases along with the momentum and chemical equations for the charged particle densities [1, 2, 3]. When the model is brought into conformity with as much of the data as is possible, constraints can be placed on some of the unknown parameters such as the electron and ion temperatures, and the strength of the magnetic field in the topside Venus ionosphere.

Most of the boundary conditions needed to construct the ionosphere models can be obtained directly from the Mariner V measurements [4]. The upper boundary is selected to be at the altitude of the observed abrupt termination of the electron density profile near 500 km. Since the peak electron density of 5.2×10^5 cm^{-3} occurs at an altitude of 135 to 140 km, the lower boundary, or reference altitude, is placed at 100 km. It has been suggested by several authors that the abrupt termination of the electron density profile in the topside ionosphere, the ionopause, arises from the fact that the magnetic fields carried along by the solar wind are forced to pile up on top of the highly conducting ionosphere. This magnetic obstacle then forms a natural upper boundary for the ionosphere and interacts with the super-alfvenic solar wind to form a bow shock that has been observed at about 50 000 km from Venus.

If the momentum and energy equations are applied across the bow shock, the resulting density just within the bow shock is about 12 cm^{-3} with a corresponding proton temperature of approximately 4×10^6 K. Near the ionopause these values can be used to obtain a pressure balance that requires a magnetic field build up to about 50 γ. Since the ionopause is interpreted as the interface between the solar wind and the Venus ionosphere, a balance must be made between the total solar wind pressure P_w and the total charged particle pressure P_c immediately below the ionopause. This pressure balance, $P_c = P_w \cong KNMV^2 \cos^2\psi$, calculated at $\psi = 45°$ from the subsolar wind point, forms one of the boundary conditions for our model. The precise value of the pressure depends upon the value of the accomodation coefficient K [5]. Since the most likely value lies near one, we have adopted $K = 1$ to obtain $P_c = 8.78 \times 10^{-9}$ dyne-cm^{-2} for a solar wind velocity $V = 590$ km/sec and a density of $N = 3$ cm^{-3}. Figure 1 illustrates this schematically.

The presence or absence of a magnetic field has important consequences for the thermal structure of the charged particles. There are two extremes: (1) where the presence of an essentially horizontal magnetic field inhibits thermal conduction across field lines and (2) where due to the complete absence of a magnetic field or to the presence of a tilted magnetic field and possible turbulence, the thermal structure is

Sagan et al. (eds.), Planetary Atmospheres, 23–27.

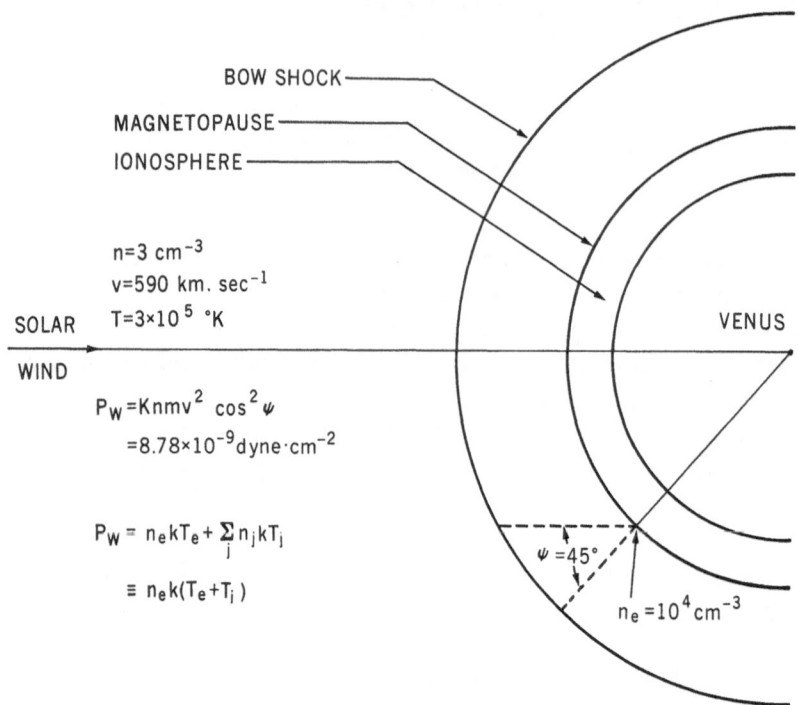

Fig. 1.　Schematic representation of the solar wind interaction with the Venus ionosphere.

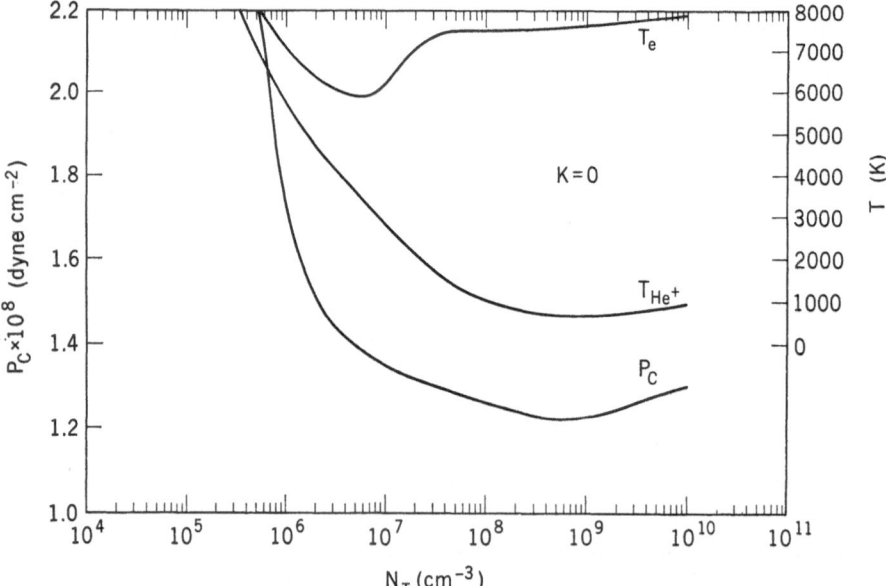

Fig. 2.　The charged particle pressure P_c, electron temperature T_e, and major ion temperature T_{He^+} just below the ionopause as a function of the neutral helium concentration N_T at the reference altitude 100 km. The thermal conductivity $K=0$.

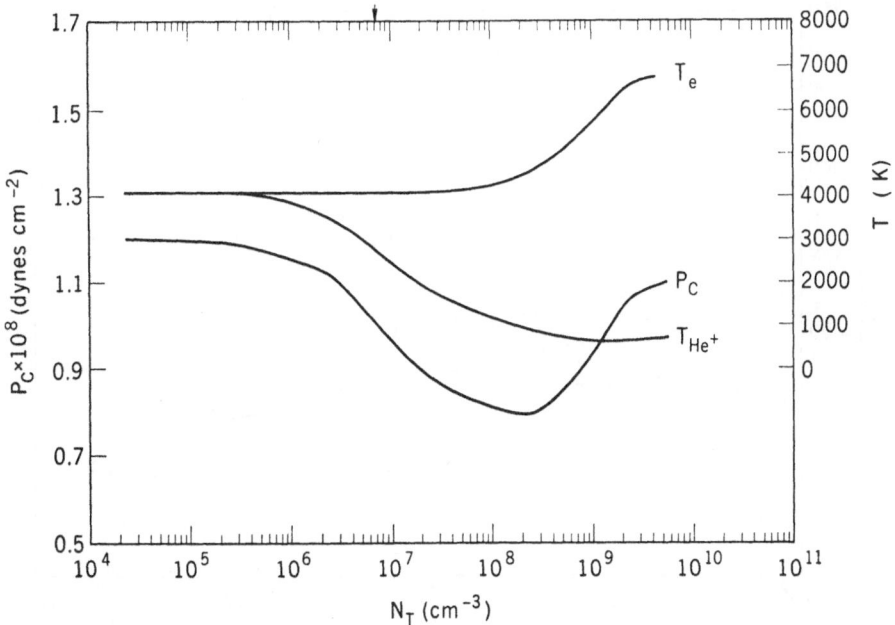

Fig. 3.   The same conditions as in Figure 2, except $K = 0.006\,K_{\parallel}$.

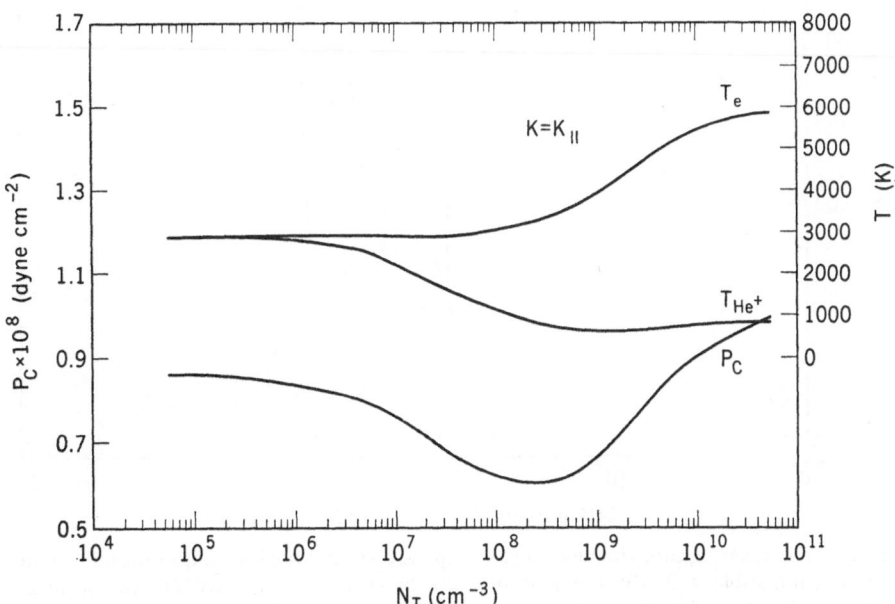

Fig. 4.   The same conditions as in Figure 2, except $K = K_{\parallel}$.

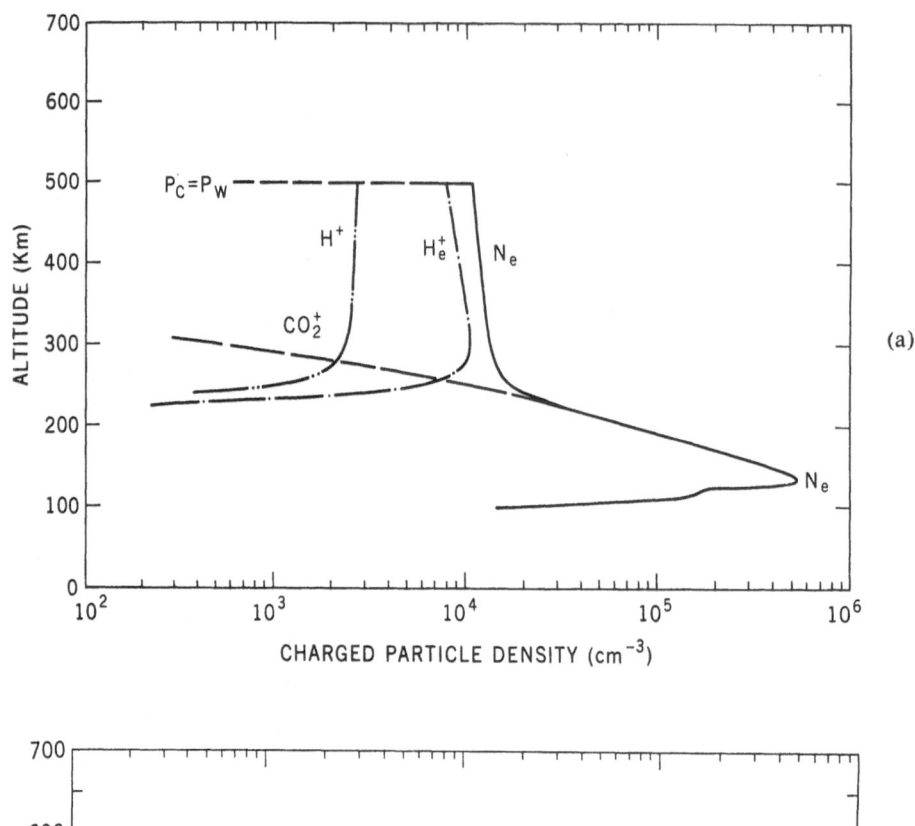

(a)

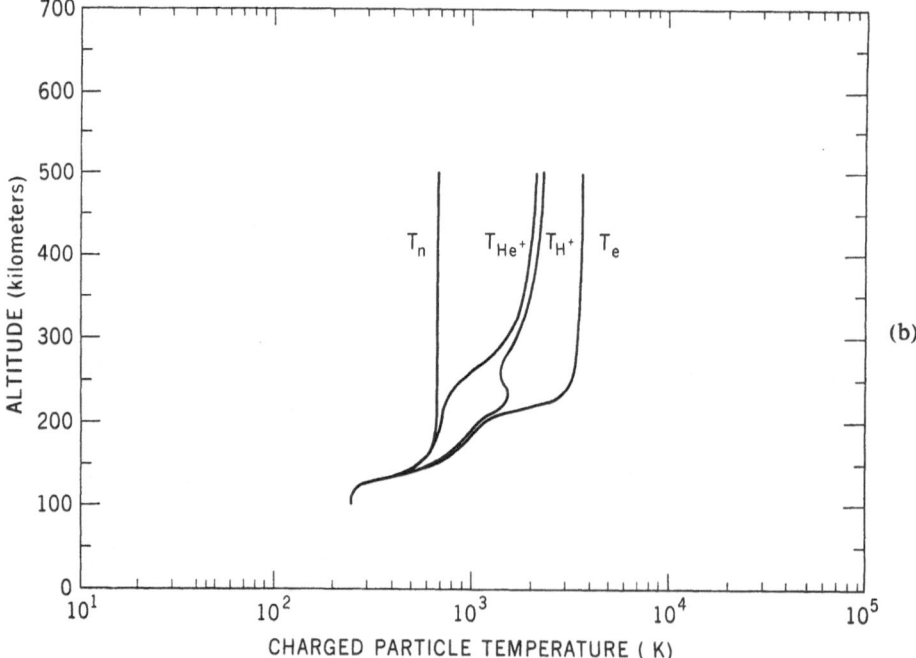

(b)

Fig. 5a, b. The electron, ion, and neutral gas temperatures, and the ion composition for boundary conditions compatible with the required pressure balance, $P_c = P_w \simeq KNMV^2 \cos^2 \psi$, and when $K = 0.006\ K_{\parallel}$. At 100 km $(N_2) = 5 \times 10^8$, $(CO_2) = 2 \times 10^{13}$, $(H_2) = 4 \times 10^6$, $(He) = 5 \times 10^7$, $(H) = 2 \times 10^4$, $(D) = 1 \times 10^6$, with a hydrostatic distribution above 100 km.

strongly controlled by thermal conduction. If a series of solutions are constructed as shown in Figures 2, 3, and 4, the density of neutral helium N_T at 100 km, the strength of the magnetic field within the ionosphere, and the effective tilt* (in terms of magnetic dip angle) can be estimated. Since a magnetic field within the ionosphere exerts a pressure $B^2/8 \pi$, the charged particle pressure need only make up the difference to balance the solar wind pressure. For example, for fields of 10, 20, and 30 γ, P_c (500 km) must be 8.4×10^{-9}, 7.2×10^{-9}, and 5.2×10^{-9} dyne-cm^{-2} respectively. As can be seen from the figures, no solutions exist for dip angles less than about 3.5°, which is equivalent to the thermal conductivity K less than $0.0037\, K_{\parallel}$. Because of the essentially horizontal nature of the induced magnetic field and the small upper limit placed on a possible intrinsic planetary magnetic moment [6], large effective dip angles are not possible. Thus, the magnetic field within the ionosphere has a strength between 10 and 20 γ with an effective magnetic dip angle near 4.5° ($K=0.006\, K_{\parallel}$). For these values, the neutral helium density at 100 km must lie in the range 3×10^7 to 6×10^8 cm^{-3} as can be seen from Figure 3.

Selecting boundary values corresponding to the acceptable range of solutions in Figure 3, the thermal structure and ion composition of the daytime Venus ionosphere can be obtained. The solutions shown in Figure 5 predict that the electron temperature T_e and the major ion temperature T_{He^+} are not in thermal equilibrium with the neutral gas temperature T_n except at altitudes near 100 km. In the region near 500 km, $T_e = 3700$ K, $T_{\text{He}^+} = 2100$ K, and $T_n = 660$ K.

References

[1] Herman, J. R. and Chandra, S.: 1969, *Planetary Space Sci.* **17**, 815.
[2] Bauer, S. J., Hartle, R. E., and Herman, J. R.: 1970, *Nature* **225**, 533.
[3] McElroy, M. B.: 1969, *J. Geophys. Res.* **74**, 29.
[4] Kliore, A., Levy, G., Cain, D., Fjeldbo, G., and Rasool, S.: 1967, *Science* **158**, 1683.
[5] Schield, M. A.: 1969, *J. Geophys. Res.* **74**, 1275.
[6] Dolginov, S. S., Yeroshenko, E. G., and Zhuzgov, L. N.: 1968, *Kosm. Issled. Moscow.* NASA Translation ST-LPS-PMF-10730.

* The effective magnetic dip angle may be composed in part of: (1) an actual geometrical tilt derived from the sum of the intrinsic and induced fields; (2) bulk motion of the high β Venus ionosphere that produces significant distortion in the magnetic field; (3) the possible presence of turbulence that can change the effective thermal conductivity $K = \Omega K_{\parallel}$.

VENUS: DETERMINATION OF ATMOSPHERIC PARAMETERS FROM THE MICROWAVE SPECTRUM

JAMES B. POLLACK* and DAVID MORRISON†

Laboratory for Planetary Studies, Center for Radiophysics and Space Research, Cornell University, Ithaca, N.Y., U.S.A.

Abstract. The microwave spectrum of Venus is compared with the structure and composition of the Venus atmosphere as determined by Veneras 4, 5, and 6 and Mariner V. The results are consistent with a radar radius of 6049.5 ± 3 km, surface pressures of 95 ± 20 atm, and a surface temperature of 770 ± 25 K, as well as a water vapor volume mixing ratio of 0.65 ± 0.35%.

The high atmospheric temperatures measured by the Mariner V and Venera 4, 5, and 6 space probes as well as many theoretical investigations (see e.g. Barrett and Staelin (1964)) have indicated that the microwave spectrum of Venus is produced by thermal emission from a hot surface overlaid by an atmosphere that is optically thick at millimeter wavelengths. In this paper we further investigate this suggestion, comparing theoretical spectra computed for a variety of models with a critically selected set of observations. When combined with recent data obtained by radar and by direct measurement by spacecraft, this analysis yields improved values for some parameters describing the troposphere and surface of Venus.

To define the microwave spectrum at wavelengths from 2 mm to 21 cm, we have selected 21 recent measurements of high quality. In addition, we have used the 43-meter telescope of the National Radio Astronomy Observatory to make two new determinations of the brightness temperatures: 700 ± 45 K at 6.0 cm wavelength and 495 ± 35 K at 1.9 cm wavelength. (These uncertainties are computed standard errors and include a 5% uncertainty in the assumed flux scale.) At wavelengths of 2 cm and larger, we have defined the spectrum using only measurements made with respect to standard celestial sources, and we have normalized all of these data to the flux scale of Kellermann *et al.* (1969). At shorter wavelengths, we have generally relied on the estimates of absolute antenna gain made by the observers. All of the data used in our spectrum have been published since 1963.

This observational spectrum, which is illustrated in Figure 1 below, has significantly smaller scatter than previous compilations, such as those of Dickel (1967). We estimate that these data define the spectrum with an absolute accuracy of ± 50 K and a relative accuracy of ± 30 K. The observations confirm the peak in the brightness temperature near 6 cm suggested by Dickel (1967) and show a rapid decline in temperature between 4 cm and 1.5 cm wavelengths.

For comparison with the observations, we have computed theoretical spectra using techniques similar to those described by Pollack and Wood (1969) and by Wood *et al.* (1969). We have explicitly accounted for variation of the microwave emissivity over the

* Present address: Space Sciences Division, NASA Ames Research Center, Moffett Field, California 94035.
† Present address: Institute for Astronomy, University of Hawaii, Honolulu, Hawaii 96822.

disk, and we have computed the opacity contribution of water vapor using corrections to the absorption coefficients (measured in air and nitrogen) to allow for the greater absorption that takes place in a predominantly CO_2 atmosphere. Our models of the temperature and pressure structure of the atmosphere are based on the measurements of Venera 4, 5, and 6 (Avduevsky *et al.*, 1969) and of Mariner V (Kliore *et al.*, 1969). At levels below those measured directly, we have extrapolated the temperature and pressure to the surface. The nature of this extrapolation and the composition of the atmosphere are constrained to match the observed radius of 6051 ± 5 km (Anderson *et al.*, 1968) and the radar measurements of a vertical optical depth of the atmosphere of 1.0 ± 0.1 at 3.8 cm wavelength (Muhleman, 1969). In our computations, we have considered variations in radius and surface dielectric constant, variations in the atmospheric content of CO_2, H_2O, and suspended dust, and departures from an adiabatic

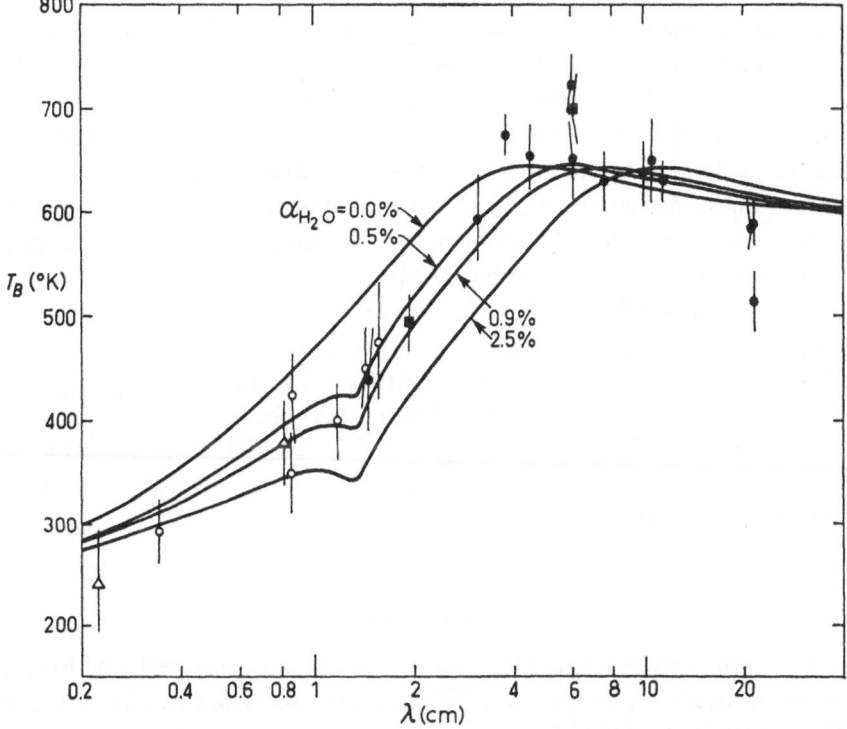

Fig. 1.   A comparison of theoretical models containing various abundances of water vapor with observed microwave spectrum of Venus. The measured values of the brightness temperature are indicated by triangles, squares, or circles. The vertical lines associated with these points give the estimated errors.

structure in the lower few kilometers of the atmosphere. A good fit to the data is provided by the following 'standard' model: 90% CO_2 mixing ratio, 0.7% H_2O mixing ratio, no significant opacity contribution by dust, surface pressure 90 atm, surface temperature 760 K, dielectric constant 4.8, planetary radius 6051 km, and strictly adiabatic lower atmosphere. In view of the recent 11 cm interferometric study of Venus made by Sinclair *et al.* (1969), we do not expect any of these parameters to vary significantly with latitude and longitude on the planet.

In Figure 1 we compare the observed brightness temperatures with theoretical spectra in which the H_2O mixing ratio is varied. The other parameters of the model are the same as for the 'standard' model. It is apparent that CO_2 alone does not supply sufficient opacity to match the observed spectra, and this conclusion is not altered if we consider CO_2 mixing ratios up to 100%. Krupenio, in this symposium, has indicated that the radar data also require an additional opacity source. We note that the argument from the microwave spectrum is less ambiguous than that made from radar data alone, since the microwave emission between 1 and 3 cm arises primarily in parts of the troposphere where direct measurements of atmospheric structure are available. Alterations in the parameters of the model at the base of the troposphere, which can significantly change the computed radar cross section, do not influence the predicted spectra in this wavelength region; only by changing the composition can we produce agreement between the models and the microwave observations. We wish to stress, however, that in the absence of observational evidence for the 1.35 cm resonant absorption of H_2O, we cannot assert that the microwave spectrum demonstrates the presence of H_2O in the troposphere of Venus. However, the amount of H_2O required to fit the data – 0.4 to 1.2% volume mixing ratio – lies close to the lower limits of the *in situ* measurements of H_2O obtained from the Soviet Venera entry probes (Avduevsky *et al.*, 1969). In addition, for this range of mixing ratios, condensation of water will occur close to the top of the troposphere.

Our preliminary calculations indicate that the observed spectrum cannot be matched by models in which there is a substantial (> 4 km) isothermal layer at the base of the troposphere or in which there is a major contribution to the microwave opacity by suspended dust particles. In addition we obtain values for the radar radius, surface pressure, and surface temperature of 6049.5 ± 3 km, 95 ± 20 atm, and 770 ± 25 K, respectively. Finally further analysis reduces the range of allowable water vapor volume mixing ratios to $(0.65 \pm 0.35)\%$.

Acknowledgment

This report is a summary of a paper to be published elsewhere (Pollack and Morrison, 1970). We acknowledge the support of NSF grant GA-10836 and NASA grants NGR 33-010-082 and NGL 12-001-057.

References

Anderson, J. D., Cain, D. L., Efron, L., Goldstein, R. M., Melbourne, W. G., O'Handley, D. A., Pease, G. E., and Tausworth, R. S.: 1968, *J. Atmospheric Sci.* **25**, 1171.

Avduevsky, V. S., Marov, M. Ya., and Rozhdestvensky, M. K.: 1969, 'The Tentative Model of the Atmosphere of the Planet Venus Based on the Results of Measurements of Space Probes Venera 5 and Venera 6', presented at the IAU-URSI symposium on planetary atmospheres and surfaces; to be published in *Radio Sci.* **5**, 333.

Barrett, A. H. and Staelin, D. H.: 1964, *Space Sci. Rev.* **3**, 109.

Dickel, J.: 1967. *Icarus* **6**, 417.

Kellermann, K. I., Pauliny-Toth, I. I. K., and Williams, P. J. S.: 1969, *Astrophys. J.* **157**, 1.

Kliore, A., Cain, D., Fjeldbo, G., and Rasool, S.: 1969, *Space Research*, Vol. IX, North-Holland Publ. Co., Amsterdam.

Muhleman, D. O.: 1969, *Astron. J.* **74**, 57.

Pollack, J. B. and Wood, A. T.: 1969, *Science* **161**, 1125.

Pollack, J. B. and Morrison, D.: 1970, *Icarus* **12**, 376.

Sinclair, A. C. E., Basart, J. P., Buhl, D., Gale, W. A., and Liwshitz, M.: 1969, *Radio Sci.* **5**, 347.

Wood, A. T., Wattson, R. B., and Pollack, J. B.: 1969, *Science* **162**, 114.

PECULIARITIES OF MM AND CM RADIOWAVE PROPAGATION IN THE VENUS ATMOSPHERE

N. N. KROUPENIO

Institute for Space Research, Academy of Sciences of the U.S.S.R., Moscow

Abstract. The total vertical absorption of mm and cm radiowaves in the Venus atmosphere is calculated from direct measurements of temperature, pressure and chemical composition, provided by the automatic interplanetary stations Venera-4, Venera-5 and Venera-6. The results are in good agreement with radar measurements of Venus and with extrapolations of direct measurements of temperature and pressure. The data fit a model with surface pressure 90 atm and a water vapor content of a few milligrams per liter. Vertical radiowave attenuations at various wavelengths are computed.

The flights of the automatic interplanetary stations (AIS), Venera 4, Venera 5, and Venera 6 allowed us to carry out a series of interesting experiments in the atmosphere of Venus. AIS Venera 4, Venera 5, and Venera 6 have measured the atmospheric chemical composition, temperature, and pressure in the range of altitudes greater than 30 km.

The space vehicle Mariner VI accomplished a radio occultation experiment at the wavelength 12.5 cm, and from that the atmosphere altitude profile up to a pressure of about 5 atm has been determined. By an extrapolation of the results of Venera 4 and of Mariner V data into the lower altitude regions, pressure on the Venus surface was estimated, and according to the calculations turned out to be about 100 atm [1]. Values of the atmospheric parameters near the Venus surface were determined more precisely after the flights of AIS Venera 5 and Venera 6.

The results of direct measurements of the Venus atmospheric parameters allowed us to determine the conditions of propagation of mm and cm radiowaves in the planet's atmosphere. Gas analyzers installed on board Venera 5 and Venera 6 operating at 450 mm Hg and 5 atm (Venera 5), and at 2 atm and 10 atm (Venera 6) have determined the following chemical composition of the atmosphere:

CO_2 $97 \pm 4\%$
N_2 with inert gases $\leq 2\%$
O_2 less than 0.1%.

Water vapor at pressures of 0.6–2.0 atm – 0.6–1.1% water vapor at pressures of 5–10 atm – 0.07%. [2]

AIS Venera 5 and Venera 6 have measured temperatures and pressures in the altitude range greater than 30 km where the temperature changed from 300 to 600 K and pressure from 0.5 to 27 atm [3].

Considerable pressures and temperatures existing in the Venus atmosphere with an abundance of carbon dioxide and a small admixture of both water vapor and oxygen cause a series of interesting phenomena connected with mm and cm radiowaves propagation in the Venus atmosphere. Among these are:

– strong attenuation of radiowaves in this band, which increases with shorter wavelengths;

Sagan et al. (eds.), Planetary Atmospheres, 32–35.
All Rights Reserved. Copyright © 1971 by the I.A.U.

– lack of a sharply prominent water vapor absorption line in the lower atmosphere;
– thermal radiation formed in the planetary atmosphere at mm and shortwave parts of the cm radioband.

For calculations of the radiowave absorption coefficient a model of the Venus atmosphere was computed based on AIS Venera 5 and Venera 6 measurement data [2, 3] extrapolated into the lower altitudes region under the assumption of a near-adiabatic altitude-variation of temperature.

Figure 1 gives the altitude dependences of pressure and temperature for this model. Radiowave absorption was calculated as follows. The carbon dioxide absorption coefficient was calculated in accordance with Ho, Kaufman and Thaddeus' semi-empirical formula, the water vapor absorption coefficient was calculated from A. P. Naumov and S. A. Jevakin's quantum-mechanical formula [5], bearing in mind the errors in our

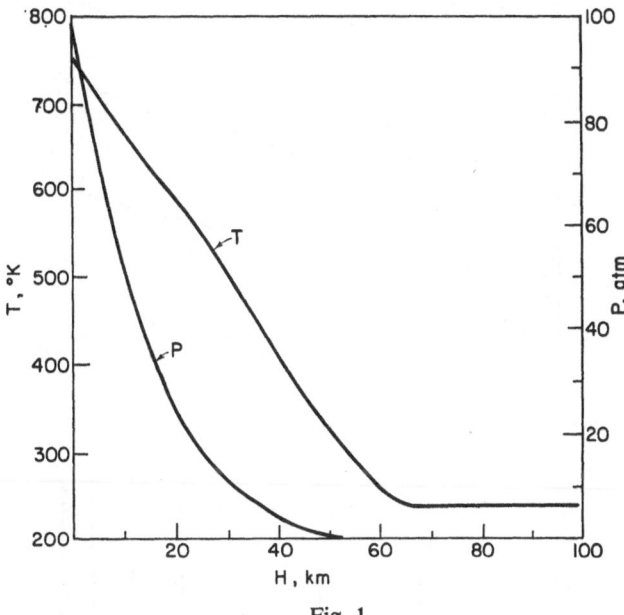

Fig. 1.

knowledge of the chemical composition of the atmosphere. As the molecular oxygen contribution to the optical thickness of Venus atmosphere beyond the O_2 absorption line does not exceed 2% at the mm radioband and 5% at the longwave part of the cm radioband, an absorption due to this component was not taken into account in the calculations [6]. Absorption was calculated for an oxygen abundance of 0.4%.

A complete vertical attenuation of radiowaves in the Venus atmosphere was calculated for a series of pressure values in the range of 0 to 100 atm and corresponding temperatures were determined by the atmosphere model considered. These calculations were performed for a carbon dioxide percentage of 97% and water vapor content of 0.1 to 0.8%.

Figure 2 gives the results of the calculation for 0.1% H_2O. According to these data the dependencies of complete vertical absorption of radiowaves on the trajectory 'space vehicle surface' were calculated as a function of the vehicle altitude above the Venus surface.

These dependencies are presented in Figure 3. Zero altitude in these calculations corresponds to a surface pressure equal to 96.7 atm [3].

From Figures 2 and 3 one can easily conclude that it is inexpedient to use wavelengths shorter than 5 cm both for communication between the ground control station

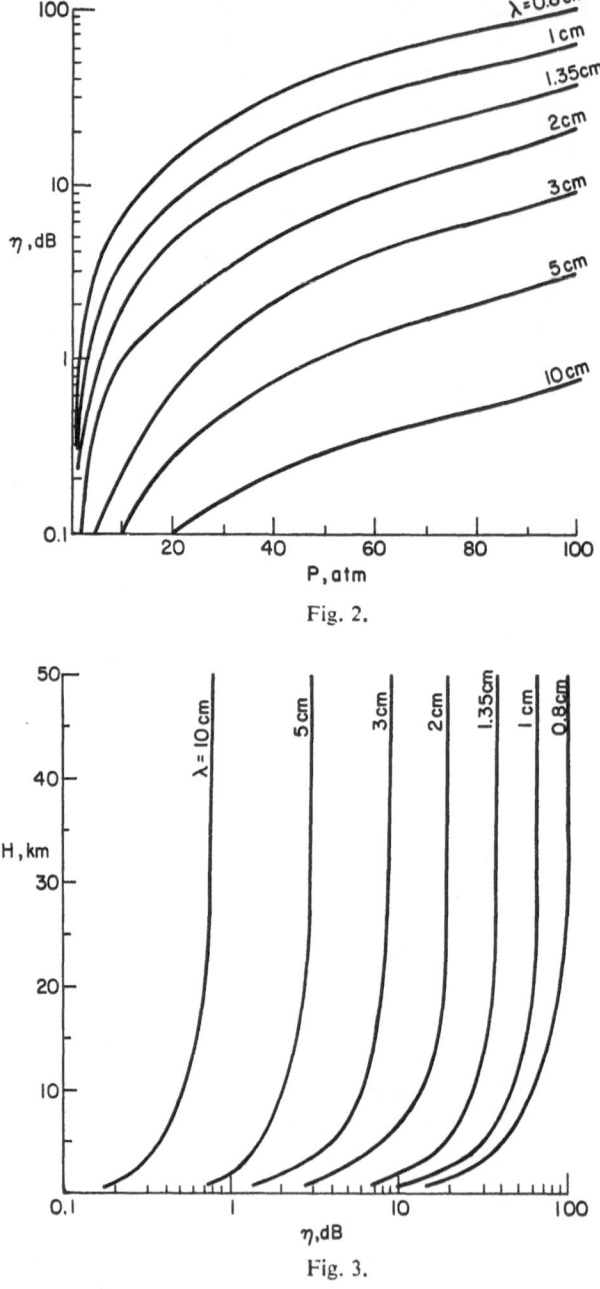

Fig. 2.

Fig. 3.

and the space vehicle during its landing on the Venus surface, and for the operation of radiotechnical equipment of the space vehicle landing system, because with shorter wavelengths it is necessary to increase the power of the transmitters and the sensitivity of the receivers. The examination of the curves in Figure 2 shows that the values of complete vertical absorption of radiowaves with wavelengths at pressures more than 40 atm are described by a law close to λ^{-3}. A similar dependence is obtained in the case of sound wave propagation in an acoustically thick medium. Because of this, regularities existing in hydroacoustics can probably be applied to radiowave propagation in media with pressures more than 40 atm. This conclusion is a preliminary one and should be verified by means of appropriate experimental studies of radiowave propagation in similar media using the experience accumulated not only by radiophysics but also by seismologists and hydroacoustic specialists.

It is of interest to compare the calculated data with the results of Venus radar measurements at wavelength 3, 8, 12.6, and 23 cm [7, 8]. If the planetary radar cross-section is assumed to be invariant with wavelength in cm and dm bands, as can take place, for example, when the density and conductivity of the surface layer material does not vary with depth, then the measured attenuation of the reflected signal towards shorter wavelengths can be determined only by the absorption in the atmosphere.

In this case the results of Venus radar measurements from the ground stations can be correlated under the following surface pressures:

– up to 70 atm for a uniform water vapour distribution in the atmosphere at a concentration of 8 mg/l.

– up to 90 atm for a uniform water vapor distribution in the atmosphere with a concentration of 1 mg/l; or for water vapor with a concentration of 8 mg/l in a thin layer of the atmosphere at altitudes where the temperature does not exceed 400 K.

– up to 110 atm when water exists only in the upper atmosphere in the solid phase.

A comparison of these three models with the results of direct measurements performed by Venera 5 and Venera 6 reveals that the best correlation occurs in the case of the second model.

Thus, a model of the Venus atmosphere obtained as a result of AIS experiments is in good agreement with the results of Venus radar measurements made from the earth.

References

[1] Kliore, A.: 1968, Report at Symposium on the Moon and Planets, Kiev, U.S.S.R.
[2] Vinogradov, A. P., Surkov, Yu. A., Andreitchikov, B. M.: 1970, *Dokl. Akad. Nauk S.S.S.R.* **190**, 552.
[3] Avduevsky, V. S., Marov, M. Ya., Rozhdestvensky, M. K., Borodin, N. F., Koryagin, V. P.: 1970, *Dokl. Akad. Nauk S.S.S.R.* (in press).
[4] Ho, W., Kaufman, I. A., Thaddeus, P.: 1966, *J. Geophys. Res.* **71**, 5091.
[5] Naumov, A. P., Jevakin, S. A.: 1968, *Izv. Vysshikh. Ucheln Zavedenii, Radiofiz.* **6**, 674.
[6] Kroupenio, N. N., Naumov, A. P.: 1968, Report at Symposium on the Moon and Planets, Kiev, U.S.S.R.
[7] Evans, J. V.: 1965, *Astron. J.* **73**, 125.
[8] Evans, J. V.: Report at Symposium on the Moon and Planets, Kiev, U.S.S.R.

SOME CONSEQUENCES OF CRITICAL REFRACTION IN THE VENUS ATMOSPHERE

CONWAY W. SNYDER

Jet Propulsion Laboratory, Pasadena, Calif., U.S.A.

Abstract. Mariner V and the three successful Venera probes have shown the density of the atmosphere of Venus to be so high that critical refraction occurs at an altitude near 35 km. Somewhat bizarre optical effects are to be expected in such an atmosphere. Using a spherically symmetric model based upon the measurements of the refractivity at the top of the atmosphere by the Mariner V S-band occultation experiment, the Venera 4 data in its altitude range, and adiabatic extrapolation to the surface, calculations of a variety of ray paths have been made with a double-precision computer program. Phenomena discussed include the magnification of the planet, the elevation of the horizon, the apparent motion and distortion of the sun, and the possibility of 'seeing' completely around the planet.

Both the Mariner V mission and the Venera 4 mission demonstrate conclusively the existence of the phenomenon of critical refraction in the Venus atmosphere. This means that the atmosphere is so dense that there is a level at which a horizontal ray of electromagnetic radiation would have a curvature equal to the distance from that level to the center of the planet and hence it would be bent into a circle and returned to its original point. Horizontal rays above this level would be less refracted and would not meet the surface of the planet if extended in either direction, whereas horizontal rays below the critical level would encounter the planet's surface at both ends. Since the somewhat bizarre effects that can be produced by critical refraction has been commented upon rather inaccurately in both the Soviet and American press, it seems appropriate to consider the matter quantitatively so as to put the phenomena in the proper perspective.

For these calculations, I have used a model of the Venus atmosphere which is based upon the measurements of the refractivity at the top of the atmosphere by the Mariner V S-band occultation experiment and fits smoothly on to the Venera 4 model down to the level at which the probe ceased communicating. Using Kliore's latest range scale, this occurred at 6078 km. Since the Venera 4 data indicates a near adiabatic lapse rate at this level, I have extrapolated on down along an adiabat to the range of 6050 km corresponding to the radar value for the radius of the planet. At this level, the model gives a temperature close to 770 K, a pressure of 135 atm, and a density 78 times as great as the earth's atmosphere at the surface. The atmosphere is assumed to consist practically entirely of CO_2 and the refractive index corresponding to S-band microwaves has been used.

The critical level in this atmosphere occurs at 6087.8 km. Since this is probably 20 to 25 km below the top of the opaque clouds, certain of the interesting phenomena that are possible will not actually occur with visible light, although they may occur in the microwave region where the transparency of the atmosphere is higher. For this

Sagan et al. (eds.), *Planetary Atmospheres*, 36–38.
All Rights Reserved. Copyright © 1971 by the I.A.U.

discussion, I shall simply imagine that the atmosphere is infinitely transparent and examine what phenomena we might see.

As we approach the planet, the first thing we discover is that the apparent radius of the planet is 6099.5 km. This is a sort of magnification effect of the atmosphere and is independent of the actual size of the planet. It comes about because a ray which approaches the planet in a direction to pass by it a distance slightly larger than 6099.5 km from the center will be refracted inward, become tangential just outside the critical level, and pass back out through the atmosphere again. However, a ray that is very slightly closer to the planet will be refracted inward, will spend a very long time near the critical level, but will never quite become tangential, and will then spiral in and strike the surface. The apparent size of the planet is independent of our distance from the planet so long as we are outside the atmosphere, but as soon as we have a little of the atmosphere behind us, the magnification changes. Thus when we actually reach the range of 6099.5 km where the surface had appeared to be, we will find that the surface is now at 6096 km. When we reach 6096 km we will find it to be at 6094 km, and when we reach 6094 km, the planet will still be slightly smaller than 6092 km. Somewhere along here the planet's apparent size begins to increase, since when we get to the limiting altitude the planet looks infinitely large. This is because we can see the surface by looking horizontally along the critical ray in any direction or more exactly by looking an infinitesimal angle below the horizontal. The planet in fact appears to be a very flat bowl with its edges extremely far away and just at the level where we are sitting.

Let us consider for a moment the outermost grazing rays that can actually penetrate the atmosphere from outside and reach the surface. It reaches the 19 bar level at 6078 km with a slope of 1.97°. If the surface is actually 28 km lower down, as I suspect, the ray then continues downward going another 3.4° around the planet before reaching 6050 km which it approaches with a slope of 9.4°. It has been stated that it would appear to an observer on Venus that he was always at the bottom of a hole. For the 19 bar atmosphere, a more apt description would be at the center of a very flat saucer. If he looks outward along a ray that is more than about 10° upward, the amount of bending in that ray is almost negligible. Even at a 4° elevation the effect is small and looking upward 2° he can still see out into space.

If the planet radius is 6050 km then the effect is very much larger. Looking upward at an elevation angle of 45° the ray is deflected downward by more than 1°. At 30° it is deflected by more than 2° and at 10° it is deflected downward by more than 12°. Thus our flat saucer becomes more like a bowl. The theoretical horizon which is thousands of km away is elevated by an angle of 9.3976°.

For example, I have calculated two rays at this angle which differ in elevation at the observer by 3 ten-millionths of a degree. Fifty degrees around the planet from the observer, the lower ray is just returning to the surface, while the upper ray is already 200 km above the surface and is rising rapidly. Any opacity in the atmosphere would lower the visible horizon quite appreciably. For example, if a point 200 km away is the farthest we could see then the horizon would appear elevated by only 5°.

A particularly interesting set of phenomena accompany the motion of the sun.

Here again we must imagine that we are able to see in wavelengths of electromagnetic radiation to which the entire atmosphere is transparent. Suppose that we are standing on the equator at the time of the equinox so that the sun is directly overhead at noon. First of all, the sun travels from west to east because of the retrograde rotation of the planet. Second, the sun moves very slowly, only 3° per twenty-four hour day, but then Venusians are accustomed to much longer days. Around noon we see nothing unusual; even at 3:00 pm when we know the sun is 45° above the horizontal it appears to be at 46.2°. But along about 5:00 pm the sun begins to move very much more slowly and changes its shape instead. In fact, its vertical dimension gets smaller and smaller at just the same rate that its velocity of descent decreases. At 6:00 pm when it should be along the horizontal, it is still up 10.4°. After all, the horizon is up 9.4° Its distortion is now severe; its horizontal dimension is approximately normal but its vertical dimension has been reduced by a factor of 4. By 7:00 pm this vertical reduction has reached 250 times and by 8:00 pm it has reached 30 000 times. As midnight approaches its vertical dimension continues to get smaller and smaller. At some time – it is very difficult to calculate just when – the horizontal dimension of the little sliver that is the sun begins to increase very rapidly and just at midnight it reaches all the way around, defining by a very narrow bright line the distant horizon. Then the line breaks in the east, becomes shorter and shorter and the sun gradually reassembles itself at the western horizon, gradually regaining its normal circularity as it rises. Of course if we happen to be situated more than about ⅜ of a degree off of the latitude of the sun, we will not be treated to the sight of the uniformly illuminated midnight horizon. Instead, the little sliver of the sun will simply crawl like a worm across the horizon during the night from the point where it has set to the point where it is planning to rise.

A TWO-PARAMETER THEORY FOR VENUS SPECTRA

JOSEPH W. CHAMBERLAIN

Planetary Sciences Division, Kitt Peak National Observatory,* Tucson, Ariz., U.S.A.

Abstract. An approximate, analytic theory has been developed for the formation of spectral absorption lines in a hazy atmosphere that scatters isotropically and has a homogeneous mixture of scattering and absorbing matter. The behavior – i.e., the curve of growth and the dependence of equivalent width on incident and emergent angles – has been examined for various possible situations, with emphasis on the physical reasons for a particular behavior. It is emphasized that two ratios – essentially the line and continuum absorption coefficients relative to the scattering coefficient – are important in any quantitative theory of the curve of growth and phase variation of the absorption spectrum.

In a semi-infinite atmosphere, categories of absorption lines are fixed by *two* fundamental parameters, whose relative importance has not previously been clearly distinguished: The probability of photon absorption in a line determines whether the line is classified *weak* or *strong*. This probability is characterized by the parameters $w = \alpha_0/(\sigma+\kappa)$; here σ is the continuum scattering coefficient, κ is the continuum absorption coefficient, and α_0 is the gaseous absorption coefficient at the center of a line. The quantity w determines whether the core of a line will be nearly saturated or not.

A second parameter, $q = a_0/\kappa$, may also be formed from the same three fundamental coefficients. The role of the parameter q – or even the basic fact that the theory contains two independent, fundamental parameters – does not seem to have been fully appreciated in the literature. It is predominantly q, the ratio of line absorption to continuum absorption, that determines, for weak lines as well as strong ones, the slope of the curve of growth – i.e., how the equivalent width varies with the line-absorption coefficient. The competition between α_0 and κ for dominance in fixing the character of the curve of growth has no direct analogue in the simply reflecting model (i.e., an optically thin atmosphere with a diffusely reflecting ground). A physical explanation of the linear and square-root absorption laws, which emerge from the theory, may be obtained from a simplified discussion with random-walk theory of the mean number of scatterings.

The behavior of strong pressure-broadened lines is different in the central core, the outer wings, and the broad intermediate or *transition* region. The latter portion has no exact analogue in saturated lines in a simply reflecting atmosphere. All three regions of these strong lines follow the square-root absorption law, but for different reasons. The core and outer wings taken together are somewhat analogous to a pressure-broadened line in a simply reflecting atmosphere; but the extensive transition region is a unique result of multiple scattering in line formation. Numerical examples of

---

* Operated by the Association of Universities for Research in Astronomy, Inc., under contract with the National Science Foundation.

curves of growth have been compared with the linear and square-root asymptotes for various continuum albedos.

In strong lines, a broad transition region, between the saturated core and the outer wings, absorbs roughly in proportion to $1/\Delta\nu$ – or as the square root of the line absorption coefficient. This region has no analogue in ordinary, saturated Lorentz profiles. For weak continuum absorption the curve of growth continues to follow the square-root absorption law as w increases into the region of strong lines (where the central core is saturated), but the transition region dominates the equivalent width, except in the thin-crescent phase of the planet.

Line formation in finite but thick atmospheres may be treated with asymptotic formulae that express the emergent intensity in terms of Chandrasekhar's H functions times a correction factor of the order of unity. A major problem in interpreting planetary spectra from a finite hazy atmosphere involves distinguishing between continuous absorption throughout the atmosphere and absorption at the ground. Intuitively one would expect the qualitative differences to be slight if the atmosphere is optically thick (say, $\tau \gtrsim 3$). Indeed, it is possible to construct a finite atmosphere of thickness τ_0 and with continuum albedo $\tilde{\omega}_c = 1$ that closely mimics the continuous and line absorption of a semi-infinite atmosphere with a lower continuum albedo, $\tilde{\omega}'_c$. As we might have anticipated, the required finite thickness is $\tau_0 \sim [3\,(1 - \tilde{\omega}'_c)]^{-1/2}$, the same as the mean depth of continuum absorption in the semi-infinite atmosphere. Another interesting result, which also could have been anticipated from the invariance principles, is that there is a ground albedo that will cause *any* thick but finite atmosphere to reflect the same intensity as a semi-infinite one having the same scattering albedo.

Phase variations of equivalent widths constitute one of the more important types of data for investigating hazy planets. These phase variations have been derived for several limiting cases and approximations. The use of a single point on the disk to represent the entire planet and thereby avoid a complicated averaging works moderately well.

The available information on the various CO_2 absorption bands in Venus' spectra has been examined to ascertain whether a scattering model for the atmosphere can fit all the data without additional *ad hoc* assumptions. Some previous work has indicated apparent inconsistencies between theory and observations; however, a consistent model can be obtained, provided that the scattering albedo, $\tilde{\omega}_c$, is close to unity around 1.71 μ. Scattering albedos, for isotropic scattering, may be converted to equivalent values for other scattering functions with van de Hulst's *similarity relations*. A self-consistent scattering model is definitely indicated. The gaseous abundance 'above the cloud tops', although frequently mentioned in the literature, is a concept without meaning.

The visual albedos show that the atmosphere has an optical thickness of $\tau \gtrsim 20$, which is virtually infinite. The profile of the 1.05 μ band gives a CO_2 specific abundance (the amount in a column of unit cross section and a length of one photon mean free path) that is accurate to about a factor of two. The phase variation of the 8689 Å band is consistent with what would be expected for this specific abundance and the measured continuum albedo in this region. However, the phase variation of the strong

1.6 μ bands requires that the single-scattering albedo in the continuum, $\tilde{\omega}_c$, be *very* close to unity. Direct measurements of the planetary albedo indicate a somewhat smaller $\tilde{\omega}_c$, but the discrepancy is probably not serious. A third means of estimating the albedo in this region is provided by a moderately strong isotopic band at 1.71 μ; the strength of this band also seems to require a $\tilde{\omega}_c$ of essentially unity.

The detailed arguments leading to these conclusions are published elsewhere [1, 2].

References

[1] Chamberlain, J. W.: 1970, 'Behavior of Absorption Lines in a Hazy Planetary Atmosphere', *Astrophys. J.* **159**, 137–158.
[2] Chamberlain, J. W. and Smith, G. R.: 1970, 'Interpretation of the Venus CO_2 Absorption Bands', *Astrophys. J.* **160**, 755–765.

A NEW SHORT-WAVELENGTH CARBON DIOXIDE BAND IN THE SPECTRUM OF VENUS

T. OWEN and H. P. MASON

IIT Research Institute, Chicago, Ill., U.S.A.

In attempting to resolve the CO_2 band at 7159 Å identified in the spectrum of Venus by Spinrad (1962), we discovered a new band at 7105 Å. Both bands are fully resolved on our spectra and tentative J numbers have been assigned to the rotational lines. In both cases, the P branch appears to be anomalously weak. The 7159 Å band has been observed in the laboratory (unresolved) by Herzberg and Herzberg (1953) at path lengths on the order of 55 km atm. To our knowledge, the 7105 Å band has not been recorded previously. Following the predictions of Herman (1948) the two bands are given the assignments $5\nu_3 + 2\nu_2 + \nu_1$ (7159 Å) and $5\nu_3 + 2\nu_1$ (7105 Å). An unsuccessful attempt was made to find the third member of this triad, $5\nu_3 + 4\nu_2$. In view of the very large amount of carbon dioxide apparently required to produce these absorptions, further study in the laboratory and at the telescope is encouraged.

References

Herman, R. C.: 1948, *Astrophys. J.* **107**, 386.
Herzberg, G. and Herzberg, L.: 1953, *J. Opt. Soc. Am.* **43**, 1037.
Spinrad, H.: 1962, *Astrophys. J.* **135**, 651.

CALCULATIONS OF CO_2 ENERGY LEVELS: THE $\tilde{A}^1B_2$ STATE

W. S. BENEDICT

Institute for Molecular Physics, University of Maryland, College Park, Md., U.S.A.

The $\tilde{A}^1B_2$ state of CO_2, identified by Dixon (1963) as the upper level of the 'carbon monoxide flame bands', must be of importance in the upper atmospheres of Venus and Mars. New calculations of the high vibrational levels of the ground ($\tilde{X}^1\Sigma_g^+$) state, which lead to improved fits of the observed vibration-rotation bands, confirm Dixon's analysis, except that the v_2'' numbering must be lowered by two, and fix the energy of the $v=0$, $K=0$ level of $^1B_2$ at 45210 ± 10 cm^{-1} = 5.605 eV.

Reference

Dixon, R. N.: 1963, *Proc. Roy. Soc.* **A275**, 431.

HIGH SPECTRAL RESOLUTION INTERFEROMETRIC PLANETARY OBSERVATIONS IN THE 7–25 μ REGION

R. A. HANEL, V. G. KUNDE, T. MEILLEUR, and G. STAMBACH

Goddard Space Flight Center, Greenbelt, Md., U.S.A.

Abstract. The thermal emission spectra of Venus, Mars, Jupiter, and the moon were observed at the coudé focus of the McDonald Observatory 107-inch telescope in the 400–1400 cm^{-1} spectral range with spectral resolutions of 0.3–0.7 cm^{-1}. A preliminary interpretation of the Venus/lunar ratio spectrum allows identification of four upper state CO_2 bands in the Venusian atmosphere at 791, 828, 865, and 961 cm^{-1} and confirms previous observations of the broad absorption-like depression around 890 cm^{-1}. The rotational structure of the 791 and 961 cm^{-1} bands is well developed at this spectral resolution.

During three 10 day periods, between March and June 1969, the thermal emission spectra of Venus, Mars, Jupiter, and the moon were observed at the coudé focus of the recently completed 107-inch telescope of the McDonald Observatory. Spectra were recorded from 400–1400 cm^{-1} with spectral resolutions in the range of 0.3–0.7 cm^{-1}. The spectral range covers numerous intervals where the earth atmosphere is sufficiently transparent to permit precise measurements of planetary and lunar intensities.

A double beam Fourier transform spectrometer of the type described by Hanel *et al.* (1969) was modified for these observations. First, the spectral range was extended to the long wavelength limit of the beam splitter substrate (potassium bromide) by the use of copper-doped germanium detectors. Second, the spectral resolving capability of the instrument was increased by a new drive system which permitted movement of the Michelson mirror up to a distance of 5 cm. Third, the motion of the auxiliary mirrors used for calibration purposes was automated.

The observational procedures of this series of measurements followed the procedures established for the 1967 observations from the Harvard Observatory, Hanel *et al.* (1968). During each daily sequence of observations, measurements were obtained of the radiation from (1) blackbodies at room temperature, and about 25 °C warmer, (2) from the sky with and without the telescope, and (3) from the planetary and lunar object. The sky adjacent to the planet was measured in order to eliminate thermal emission from the telescope as well as from the earth's atmosphere. Lunar spectra were taken primarily to be able to remove the effect of absorption within the telescope and within the earth's atmosphere, at least in weak and moderately strong absorbing spectral regions.

Data reduction has progressed to a point where some spectra of Venus are available for inspection. Presented here are the initial results of the thermal emission spectra of Venus in the 10–13 μ region. Calibrated spectra reduced to absolute levels outside the earth's atmosphere and spectra of Mars and Jupiter are being prepared for publication at a later date together with a more detailed description of the observational procedures and instrumentation used.

The spectrum shown in Figure 1 was taken on April 20, 1969, at an instrumental

Sagan et al. (eds.), Planetary Atmospheres, 44–47.

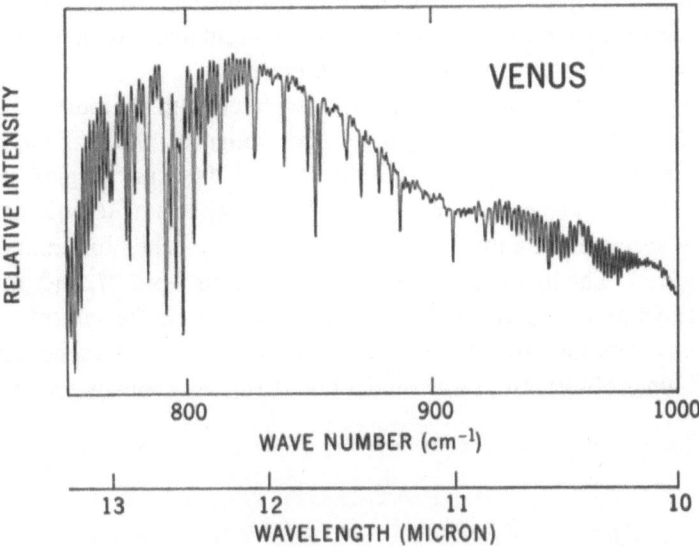

Fig. 1. Relative intensity spectrum for Venus. The spectral resolution is approximately 0.67 cm^{-1}. The spectrum is the average calculated from 32 interferograms and represents approximately 19.5 min of observing time. Telluric water vapor, carbon dioxide, and ozone lines appear in absorption.

resolution of 0.67 cm^{-1}. Displayed is the difference of the spectra obtained while viewing the planet and while viewing the adjacent sky almost simultaneously. The spectrum has been corrected for the responsivity of the interferometer. The spectrum of Figure 1 represents the emission from Venus multiplied by the transmission functions of the earth's atmosphere and the telescope. The planetary diameter was almost

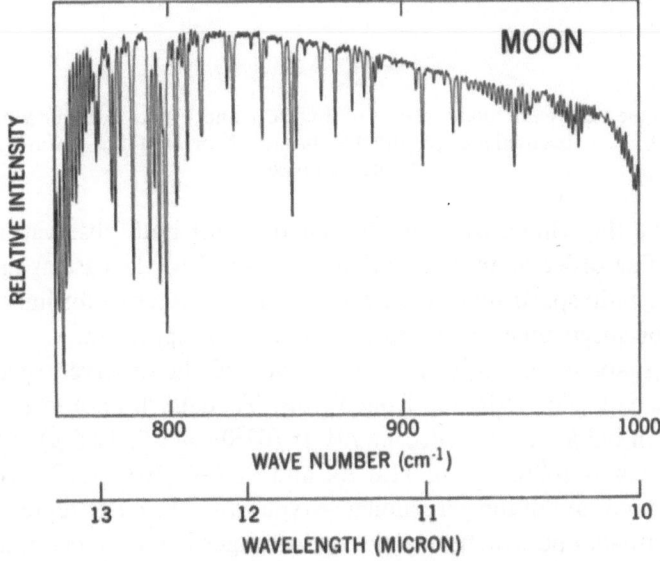

Fig. 2. Relative intensity spectrum for the moon recorded at the same spectral resolution as the Venus spectrum shown in Figure 1.

56 sec of arc while the circular field of view of the instrument covered only 30 sec of arc. The telescope was set to track the center of the apparent disk. At that time Venus was completely dark except for a small illuminated crescent.

Most of the observed spectral features are associated with the pure rotational lines of water vapor in the earth's atmosphere. The wings of the 667 cm^{-1} CO$_2$ and 1050 cm^{-1} O$_3$ telluric bands are evident in the 750 and 1000 cm^{-1} regions respectively.

To correct the Venus spectrum of Figure 1 for the effects of telluric and telescope transmission, a spectrum of a maria area near the center of the illuminated lunar disk was taken, Figure 2. The lunar spectrum was recorded on April 27, and, therefore, the transmission function of the earth's atmosphere may not be the same as it was while observing Venus. The amount of water vapor in the earth's atmosphere during both the Venus and lunar observations was similar and in this very transparent portion of the

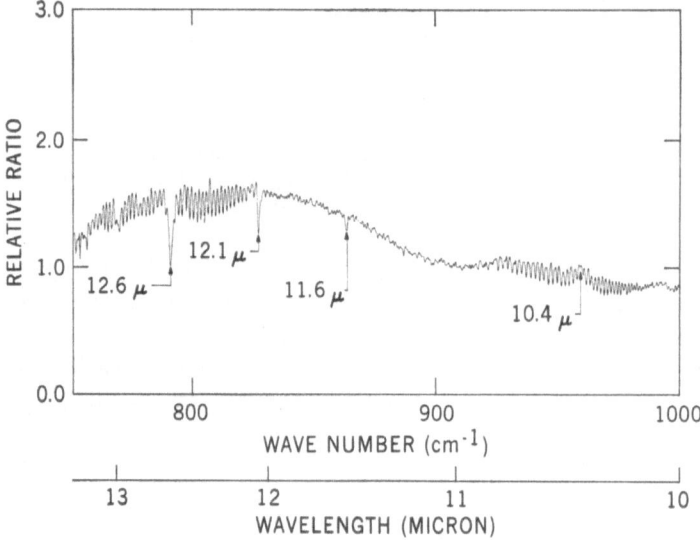

Fig. 3. Ratio of the Venus and lunar spectra of Figure 1 and Figure 2. Four vibration-rotation bands of ^{12}C^{16}O$_2$ have been identified and the broad absorption-like feature near 900 cm^{-1} should be noted.

telluric spectrum the telluric transmission functions for both observations should be the same, to a first order of approximation. This is indeed the case as may be seen in the Venus lunar ratio spectrum of Figure 3. The strong water vapor lines visible in the Venus and moon spectrum have disappeared in the ratio spectrum.

The general shape of the ratio spectrum curve and the observed spectral features are due to the emergent radiation from Venus. To date, four upper state ^{12}C^{16}O$_2$ bands have been definitely identified at 791.45 (02^{2}0 →11^{1}0, 12.6 μ), 828.18 (03^{1}0→ 12^{2}0, 12.1 μ), 864.51 (03^{1}0→20^{0}0, 11.6 μ), and 960.96 (10^{0}0→00^{0}1, 10.4 μ) cm^{-1}. The first three bands are of the perpendicular type while the latter represents a parallel band. The rotational line structure of the two stronger bands at 791 and 961 cm^{-1} is well-developed at this spectral resolution. The decline of the ratio spectrum in the 750 cm^{-1} region is associated with absorption in the Venus atmosphere from the R branch of the $^{12}C^{16}O_2$ ν_2 fundamental. The appearance of a line structure of the CO_2 bands indicates that the emission comes from a region in the Venus atmosphere where a negative temperature lapse rate exists.

Also apparent in the ratio spectrum is the broad absorption-like feature near 900 cm^{-1} that has previously been seen in the spectra of Sinton and Strong (1960), Hanel *et al.* (1968), and Gillett *et al.* (1968). It has generally been assumed that this depression is due to particles of the cloud layer. Further analysis based on model atmosphere calculations is being pursued.

References

Gillett, F. C., Low, F. J., and Stein, W. A.: 1968, *J. Atmospheric Sci.* **25**, 594.

Hanel, R. A., Forman, M., Stambach, G., and Meilleur, T.: 1968, *J. Atmospheric Sci.* **25**, 586.

Hanel, R. A., Forman, M., Meilleur, T., Wescott, R., and Pritchard, J.: 1969, *J. Appl. Opt.* **8**, 2059.

Sinton, W. M. and Strong, J.: 1960, *Astrophys. J.* **131**, 470.

LIMB DARKENING OBSERVATIONS OF VENUS
FROM 5 μ TO 18 μ

J. A. WESTPHAL

Mt. Wilson and Palomar Observatories, Carnegie Institution of Washington, California Institute of Technology
and
Division of Geological Sciences, California Institute of Technology

Abstract. Limb darkening measurements of the thermal radiation from Venus at 5, 9, 11, 13 and 18 μ are presented. This data, produced by deconvolving the observations with a measured instrumental response function, indicates a complex atmospheric structure which will require careful atmospheric modeling to fully interpret.

1. Introduction

In recent years several models for the atmosphere of Venus above the cloud layer have been proposed (see Samuelson, 1968, for a review). The direct, in situ, measurement of the temperature, pressure, and chemistry by the Venera probes (Avduevsky *et al.*, 1968, Vinogradov *et al.*, 1970) and the temperature-pressure measurements made by Mariner 5 (Kliore *et al.*, 1967) have greatly increased the knowledge of this part of the atmosphere of Venus. Unfortunately the only data bearing on the variation of opacity in the upper atmosphere comes from limb darkening measurements which with one exception have been made from the earth.

Previous limb darkening measurements have been made by Sinton and Strong (1960), Murray *et al.* (1963), Chase *et al.* (1963), Westphal *et al.* (1965), and Westphal (1966). All of these measurements were made in the region from 8–14 μ. Only those of Westphal (1966) were corrected for instrumental and atmospheric effects near the limb and extend to local zenith angles larger than about 60°.

Atmospheric models utilizing the Venera and Mariner data along with the broad band infrared (8–14 μ) measurements have shown the need for both higher spectral resolution, wider wavelength coverage, and more dependable data near the limb (see for example, Samuelson, 1968).

2. Observations

On 24 April 1969, a series of scans of Venus were made with the 200-inch Hale telescope to determine limb darkening profiles at several wavelengths. These measurements were designed to yield the best possible determination of the profiles very near the limb.

Previous experience with this type of measurement (Westphal, 1966) indicated that a very high signal to noise ratio was desirable and that careful attention to the determination of the instrumental profile was essential. Improvements in detectors and photometers during the four years since the previous measurements allowed the necessary

Sagan et al. (eds.), Planetary Atmospheres, 48–54.

signal to noise ratio to be obtained in narrow wavelength regions ($\approx$1 μ wide) centered at 5, 9, 11, 13, and 18 μ.

Several changes were made in procedure as well as in the equipment used for the 1964 measurements. Instead of a small round measuring aperture, a very narrow (0.75 arc sec) slit, 7.5 arc sec long was used. This slit was oriented accurately normal to a scan diameter so that the instrumental response function due to the slit alone was less than one arc second halfwidth. The added flux through this slit along with an improved doped germanium detector allow the measurements to be made in approximately one micron wide band passes. Table I lists the details of the filters. The errors

TABLE I

Filter μ	Wavelength 50% transmission
5	4.56–5.02
9	8.41–9.03
11	10.45–11.9
13	12.5–13.5
18	16.5–19.5[a]

[a] Long wavelength cutoff of 18 μ filter determined by atmospheric water vapor.

associated with the 1964 measurements were almost entirely due to uncertainties in the instrumental response function used to deconvolve the data. To reduce this uncertainty substantially, measurements of a bright star (β Pegasus) were made during the observations of Venus. The star was scanned normal to the slit used for the limb measurements. All the data were recorded on magnetic tape and reduced with an IBM 360-75 digital computer.

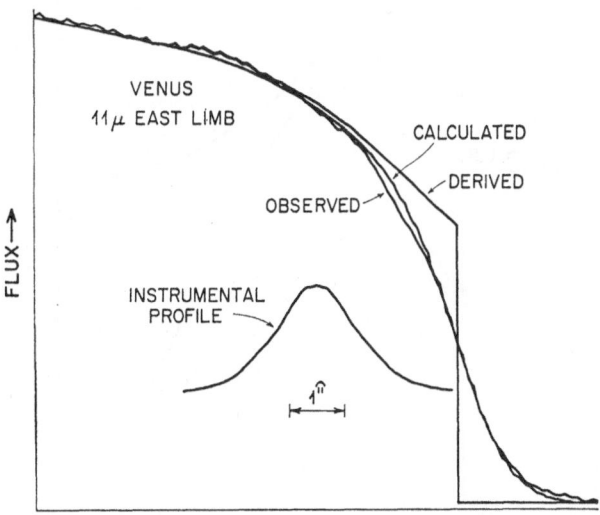

Fig. 1.

Figure 1 illustrates the data collected along part of the east limb of Venus at 11 μ. The data were reduced by the technique described in the earlier paper (Westphal, 1966) and the calculated, observed, and derived limb curves are shown along with the

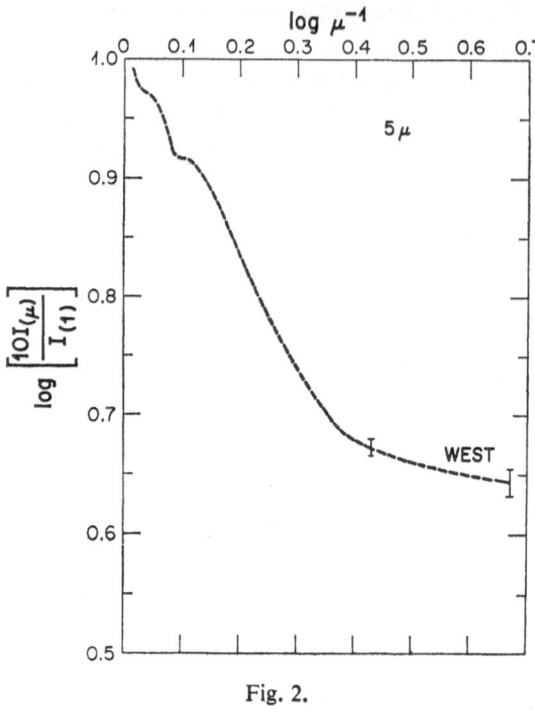

Fig. 2.

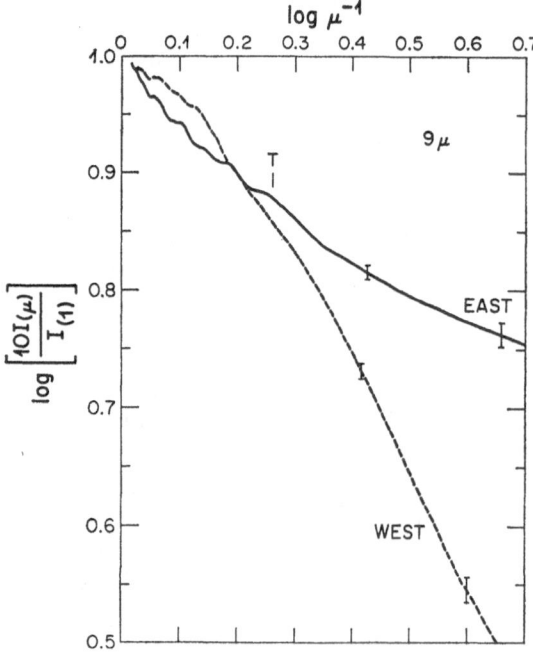

Fig. 3.

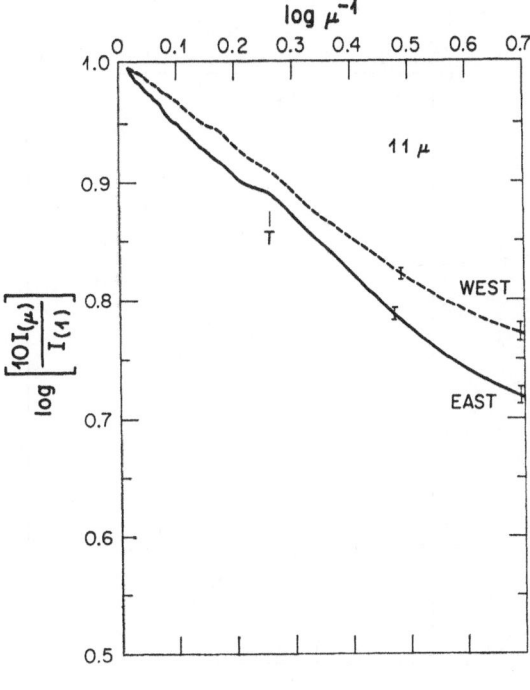

Fig. 4.

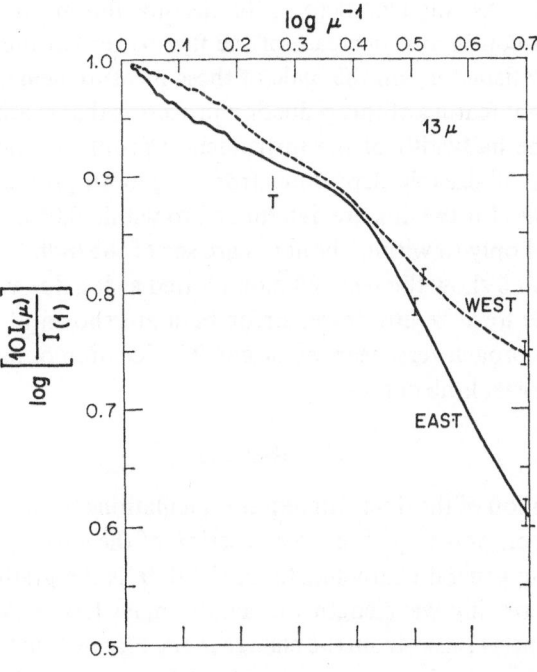

Fig. 5.

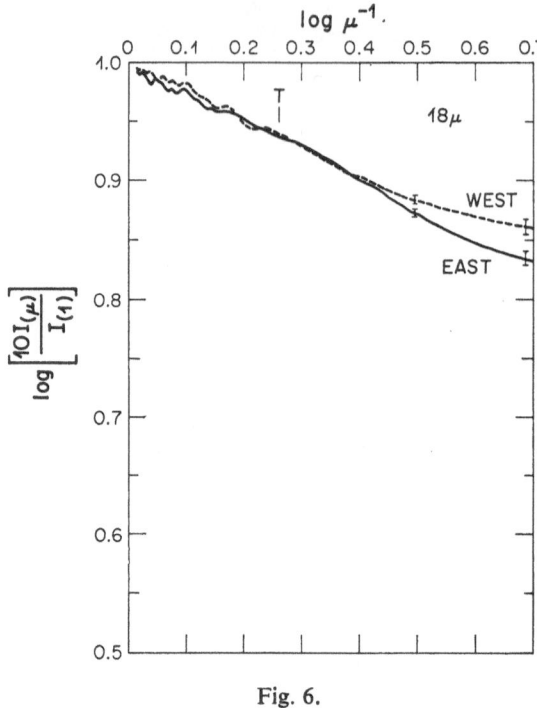

Fig. 6.

instrumental profile. The sensitivity of this deconvolution process is very high and again the errors in the final derived profile are almost entirely due to the uncertainty in the instrumental profile. As was true before, the deconvolution process depends on the assumption of a discontinuous decrease of the flux to zero at the limb. This assumption is very likely satisfactory on the scale of these measurements.

Another important feature of this reduction process is that spatial variations smaller than about $\frac{1}{3}$ of the halfwidth of the instrumental profile are not recoverable. This precludes detection of sizeable departures from a smooth profile very near the edge. Thus although values for the flux are determined to within 0.05 arc sec of the limb, the reliable data extend only to within about 0.6 arc sec of the limb.

Figures 2 through 6 show the reduced data plotted as log flux vs log μ^{-1}, where μ^{-1} is the secant of the local zenith angle. Error bars are shown at two places on each curve, the errors approach zero near the origin. The location of the sunrise terminator is shown on all the east limb curves.

3. Discussion

Detailed interpretation of the data will require calculations based on quite complicated models. In this paper, only a qualitative discussion of the curves will be attempted.

The most striking general characteristic of the data is the gradual decrease of limb darkening with increasing wavelength for zenith angles less than 65° (log $\mu^{-1} < 0.4$). This results, in a general sense, from the change in the Planck function with temperature and wavelength. The data closer than log $\mu^{-1} = 0.4$ to the limb show very complicated details and large differences between the east and west sides. As an example, the $13\,\mu$ data from the east limb show a very large increase in limb darkening for $\log \mu^{-1} > 0.4$. This is a spectral region of high CO_2 opacity and suggests the possibility of horizontal inhomogeneity in either temperature or opacity or both, related to the presence of sunlight.

The $9\,\mu$ data from the west limb also shows very strong limb darkening out to the limit of measurement, while the east limb has only moderate darkening. Since this is a region free from CO_2 absorption the differences must reflect changes in the opacity or temperature of the particulate material in the atmosphere.

The $11\,\mu$ east limb curve shows a definite increase in flux just at the terminator. Samuelson (1969) has suggested that such an effect could be due to the difference in thermal time constant between the particulate material and the surrounding CO_2 gas as the planet rotates into the sunlight. One would expect that a careful study might determine some of the thermal properties of the particles from this data.

The 5, 11, and $18\,\mu$ data from the west limb and the 9, 11, and $18\,\mu$ data from the east limb show a marked decrease in limb darkening for values of $\log \mu^{-1} > 0.4$. Since such an effect can be caused in several ways, only careful modeling can choose the most likely source of this flattening.

Five micron data is available only for the west limb, since reflected sunlight eastward from the terminator on the east limb profile precludes a unique determination of the thermal flux.

Another feature of the curves is that, at all wavelengths except $18\,\mu$, the west limb is brighter than the east limb for zenith angles less than about $50°$. This observation is consistent with the observation that the antisolar point is emitting more flux than the subsolar point in the $8–14\,\mu$ wideband measurements previously reported (Westphal *et al.*, 1965) and indicates that the upper part of the atmosphere has essentially a uniform longitudinal temperature structure.

Further high spatial resolution studies should include polarization measurements, particularly at 5 and $9\,\mu$ where most of the opacity should be due to the particulate material in the atmosphere.

It should also be possible to make limb darkening measurements at much higher spectral resolution (resolution ≈ 50) in regions of special interest.

4. Summary

The results reported in this paper, along with the in situ measurements from probes and the occultation measurements should allow a rather complete model of the upper atmosphere of Venus to be developed. The complexity of this region is greater than previously known and the likelihood of spatial and temporal inhomogeneities will require careful modeling.

References

Avduevsky, V. S., Marov, M. Ya., and Rozhoestvensky, M. K.: 1968, 'Model of the Atmosphere of the Planet Venus Based on Results of Measurements Made by the Soviet Automatic Interplanetary Station Venera 4', *J. Atmospheric Sci.* **25**, 537.

Chase, S. C., Kaplan, L. D., and Neugebauer, G.: 1963, 'The Mariner 2 Infrared Experiment', *J. Geophys. Res.* **68**, 6157.

Kliore, A., Levy, G. S., Cain, D. L., Fjeldbo, G., and Rasool, S. I.: 1967, 'Atmosphere and Ionosphere of Venus from the Mariner V S-band Radio Occultation Measurement', *Science* **158**, 1683.

Murray, B. C., Wildey, R. L., and Westphal, J. A.: 1963, 'Infrared Photometric Mapping of Venus Through the 8- to 14-Micron Atmospheric Window', *J. Geophys. Res.* **68**, 4813.

Samuelson, R. E.: 1968, 'The Particulate Medium in the Atmosphere of Venus', *J. Atmospheric Sci.* **25**, 634.

Samuelson, R. E.: 1969, private communication.

Sinton, W. M. and Strong, J.: 1960, 'Radiometric Observations of Venus', *Astrophys. J.* **131**, 470.

Vinogradov, A. P., Surkov, Yu. A., Andreichikov, B. M., Kalinkina, O. M., and Grechischeva, I. M.: 1970, this volume, p. 3.

Westphal, J. A., Wildey, R. L., and Murray, B. C.: 1965, 'The 8–14 Micron Appearance of Venus Before the 1964 Conjunction', *Astrophys. J.* **142**, 799.

Westphal, J. A.: 1966, 'The 10-Micron Limb Darkening of Venus', *J. Geophys. Res.* **71**, 2693.

WATER ON VENUS?

W. F. LIBBY and P. CORNEIL

Dept. of Chemistry and Institute of Geophysics and Planetary Physics, University of California,
Los Angeles, Calif. U.S.A.

Abstract. It is proposed that Venus may have polar seas which are acidic and thus cannot precipitate calcium carbonate. This leaves the carbon dioxide in the atmosphere. The argument is that the great equatorial land masses always have been too hot for liquid water and thus could not be weathered to give the sea salts necessary to form the precipitate. The action of steam on rocks is to liberate acids which are volatile and would dissolve in the polar seas. The volcanic vapors issuing in the early times consisting mainly of water and carbon dioxide would have begun polar seas at once since the expected equatorial (black body) surface temperature of the bare planet is too high (464 K) due to proximity of the sun. The accumulation of carbon dioxide in the atmosphere would have ensured the continued increase of the temperature due to the greenhouse effect. On earth, on the contrary, condensation over most of the planetary surface probably was possible from the beginning. Liquid water, ice-weathering, and river transport of salts to the seas all probably occurred from the beginning.

As the pressure at the surface probably approximates 100 atm (Venera 5 and 6) we can expect the polar seas to be below the boiling point although possibly hot. An isothermal layer of some thickness is naturally established over liquid water heated by infrared from above. Evaporation and condensation to form rain constitutes an efficient heat transport mechanism. Such a layer naturally would move toward lower latitude carrying moisture which then will rise and eventually move poleward in the high atmosphere causing rain and possibly the planet wide cloud cover. The atmosphere containing volatiles such as hydrochloric and hydrofluoric and sulfurous and sulfuric acids as well as carbon dioxide will form clouds which might be expected to consist of concentrated acid solutions. The main rain over the poles probably falls from altitudes well below the cloud top seen from earth. It is possible that the Venus clouds seen from earth are non aqueous just as our stratosphere carries dust clouds apparently of ammonium sulfate. At the moment it is very difficult to decide between these alternatives.

In the more polar regions the seas might conceivably be as cool as 50 °C.

1. Introduction

It has been known for quite some time that the atmosphere of Venus has water vapor. Photographs of the spectrum of Venus in the region of the 8200 Å water vapor band have indicated the presence of water (Dollfus, 1963; Bottema *et al.*, 1964; Belton and Hunten, 1966; Spinrad and Shawl, 1966; Owen, 1967). Owen (1967) estimates a maximum of about 16 μg of ppt H_2O per cm^2. Convincing proof for the existence of water vapor has, however, come from the measurements of the space probes, Venera 4, 5, 6 and Mariner V, in the atmosphere of Venus. Venera 4 indicated 1 to 8 mg liter^{-1} of water (Vinogradov *et al.*, 1968) whereas Venera 5 and 6 ('Venera 5 and 6', 1969) gave 4 to 11 at 0.6 atm total pressure and 25 °C. Observations of Mariner V also indicated small amounts.

Based on these small amounts of water, Sagan (1967) and Rasool (1968) have suggested that water may have been continuously lost from Venus at approximately the same rate as it is outgassed from the interior. They suggest that the water is dissociated by solar ultraviolet radiations of $\lambda < 1800$ Å in the stratosphere at the rate of 10^{11}–10^{12} mole cm^{-2} sec^{-1} and that the hydrogen produced by dissociation diffuses rapidly upwards with a velocity of 110 cm sec^{-1} at the 100 km level and at much greater speeds at higher altitudes and escapes from the planet with an escaping flux of about 10^{11} mole cm^{-2} sec^{-1} at the exosphere level. The atomic oxygen released in the photodissociation process is rapidly used up by oxidation.

This proposal, though applicable for the loss of some water from Venus, probably cannot account for the loss of water equivalent to the amount present as excess volatile on earth (Rubey, 1964). This has already been pointed out (Jastrow, 1968). In this paper we wish to re-examine the possibility of the existence of large quantities of condensed water in the relatively cooler regions of the planet. The principle of chemical similarity between sister planets induces us to consider such a possibility.

Libby (1968) proposed earlier that extensive ice caps might be present in the polar regions of Venus. At the time this model was proposed the high temperatures and pressures now known for the planet were not as firmly established and based on the then available data of 270 °C and 20 atm (from original reports of Venera 4) for the equator, it appeared to be possible that the poles could be cool enough to sustain ice caps as on earth. However, the present data from Venera 5 and 6 now on hand confirm the Mariner V and the microwave conclusions of the American School (see e.g., Jastrow, 1968) that the surface temperatures and pressures in the equatorial regions may well be about 800 K and 75 to 100 atm. Assuming a mean temperature gradient of about 3 K/degree latitude (Fabian *et al.*, 1968), a polar temperature of about 500 K might be possible. This temperature agrees well with the temperature at the poles derived from radioastronomical data (Clark and Kuzmin, 1965) − ⩾420 K; Pollack and Sagan (1965) −470 ± 95 K) and from theoretical calculations (Fabian *et al.* (1968) ∼500 K). Even at the originally reported equatorial temperatures the difficulty of preserving the ice caps at the poles was considerable and led to serious questions (Weertman, 1968; Owen, 1968; Bussinger and Holton, 1968). The higher temperatures at the poles now make the ice caps untenable, but appear to us not to necessitate the complete elimination of water. The boiling point of water at 100 atm is 314 °C (586 K).

2. The Concept of Polar Seas: Their Acidity and Failure to Precipitate CO_2

A careful examination of the factors that can influence the existence of liquid water on Venus indicates that indeed oceans of water may be preserved at Venus' poles, and since it is logical that the poles were relatively cooler than the equator in the early history of the planet liquid water might have condensed at the poles from the very beginning and have existed throughout the life span of the planet.

If we assume that the water on Venus is equal in amount to the water on earth the polar seas would have to be about 9 km deep and extend down to 55° latitudes. However, some water might have been lost from the planet through photodissociation, as suggested by Sagan and Rasool. Also Fricker and Reynolds (1968) conclude that more water may have stayed in the Venus crust than on earth because the temperature being higher would raise the H_2O partial pressure in the Venus atmosphere higher and keep the water in rocks higher. Such considerations would reduce the total amount of water in the oceans to a certain extent and make the polar seas narrower and shallower, but in our judgment there seems to be little possibility of avoiding oceans unless the planets are truly dissimilar chemically.

We assume that the polar seas on Venus would be nearly as old as the oceans on earth and since the oceans on earth are found to be at least 3 billion years old, oceans might have been formed at the poles of Venus about 3 billion years ago. The primitive oceans, developed by the condensation of accumulated moisture in the atmosphere, would have dissolved all the other soluble volatile gases, like HCl, HF, SO_3, CO_2, NH_3, etc. and would have been highly acidic. Unfortunately it is not possible to estimate precisely the amounts of water and of these dissolved gases developed 3 billion years ago. However, since the acidic volatiles predominate, the primitive oceans probably would have been highly acidic. Introduction of all the chlorine present on earth as excess volatile into the Venus' polar seas in the form of HCl, would yield a concentration of about 0.5 mole/l. Sulphur probably would have been oxidized to SO_3 and H_2SO_4 in the Venus' atmosphere (Corneil and Heber, 1968; Corneil and Budgor, 1969) to give amount equivalent to about 0.05 mole/l^{-1}. In addition to these two major acidic gases, CO_2 dissolves to the extent of about 0.5 mole/l^{-1} at an atmospheric pressure of 80 bar and with water temperature of about 200 °C. This acidity would appear to explain the failure of the CO_2 to precipitate $CaCO_3$ as happened on earth.

Our studies (Raman et al., 1969) of the corrosion of igneous rocks and minerals by such highly acidic solutions show that considerable attack of the sea bed would occur introducing metal cations; however the Venus oceans would appear to remain highly acidic even though substantial extraction occurs. Harvey (1945) estimates that when $CaCO_3$ precipitates, the concentration of Ca^{2+} and $[CO_3^{2-}]$ would be 0.15 and 8×10^{-6} mole/l^{-1} respectively at a pH of 5.1. Below this level of pH the carbonate precipitation does not occur. Since the pH of the sea water on Venus probably would not have risen to this level, no carbonates would have precipitated, and CO_2 is preserved in the atmosphere.

It is obvious that although the temperature of the water probably will be well below the boiling point in the polar regions some evaporation will occur from the surface and some of the water vapor, as described below, would be carried to lower latitudes by itinerant winds. However, since the temperature on the equatorial part of Venus apparently is much higher than the boiling point of water at the 100 atm pressure (314 °C) it cannot rain in these hotter regions. As a result the weathering of planetary rocks cannot take place on Venus and no rivers can flow and so no salts will be carried into the seas. This makes the sea bed the only source for salts and limits the accumulation of basic constituents in the sea water and the precipitation of carbonates probably would not occur. Under these circumstances the salt content of the seas may be fairly low, as compared to the sea water on earth.

3. Preservation of Polar Seas: Temperatures

Can such polar seas be preserved in the hot CO_2 atmosphere? We have studied the evaporation characteristics of similar acidic aqueous solutions in the presence of CO_2

at high pressure (Raman *et al.*, 1969). The experimental results are interesting and favor the preservation of Venus' polar seas. We find that with increased pressures of CO_2 increased amounts of thermal energy are required to heat the water. Only the sunlight can penetrate and heat the water at depths below the very surface, for the hot equatorial air descending over the polar seas radiates infrared, which is completely absorbed by the underlying moisture laden atmosphere and the top water layers. Our experiments have confirmed that the main deep water cannot be heated by this infrared and that only the visible sunlight and, of course, the outward flowing heat from the ocean floor can reach it. An approximately isothermal layer is established in the air over the water surface the thickness of which can only be estimated but which may be very substantial. At the top of the layer rain forms and the heat of condensation is released. This constitutes a kind of nearly isothermal heat transport device. In this way we estimate that much of the polar seas area may be as cool as 100 °C (373 K).

4. Wind Pattern at the Poles

Mintz (1961) found evidence for atmospheric circulation and proposed a model for it. In his model, however, no pattern is suggested for the winds at the poles. Goody and Robinson (1966) have also discussed the wind patterns. Their results indicate that winds are prevalent on Venus. The data of Venera 4 also indicated a steady vertical wind of $\leqslant 0.5$ m sec^{-1} during its descent into the atmosphere of Venus (Avduevsky *et al.*, 1968). We suggest a wind pattern at the poles, which would be compatible with the findings of other investigators.

Horizontal winds just above the surface of the polar ocean would be directed toward lower latitudes and these winds would carry moisture from the evaporating ocean surface and would be relatively cool. At a later stage the relatively cool horizontal winds would interact with the hotter winds at lower latitudes and vertical tropical winds would be set up, mixing the moisture upward to higher altitudes at relatively high speeds. We suggest that such vertical winds would prevail at around 50° latitude level. At the higher altitudes the winds would blow toward the poles where the moisture content would increase steadily until the relative humidity exceeded a certain critical value, when condensation would occur and clouds would form. These clouds would be at much lower altitudes as compared to the clouds observed from earth (Dollfus, 1955). Rain then would bring the water back to the oceans and deplete the moisture content of the upper altitudes above the poles.

Evaporation of water from the ocean and back condensation and rain result in efficient removal of heat brought to the surface of water by the itinerant winds. In other words, the process may be similar to an efficient 'heat pipe'. The heat liberated in the condensation process heats the upper zone and is radiated back into space ultimately. The wind conditions might vary with time, but apparently never could lead to drying of the oceans. The ocean protects itself by evaporation and formation of a cloud.

The suggested wind pattern does not allow much moisture to accumulate in the major portions of the atmosphere of the planet, especially near the equator, where observations have been made of the water content. The mixing vertical winds at the 50° latitude level may carry some water and other acidic volatiles aloft to form the observed cloud cover.

The suggested model would facilitate the greenhouse effect. (Pollack, 1968). We feel that a 'closed system' for water such as the óne proposed to exist at the poles, retains most of the water in a relatively narrow region and prevents the detection of water in major amounts in the cool upper atmosphere through earth based observations.

5. The Planetary Clouds

As stated previously some of the moisture will move into the equatorial belt and not all may be efficiently returned poleward to give low level rains. Some of this may eventually reach the greatest altitudes and condense to form acid laden droplets. We suggest that these may constitute the clouds shrouding the planet as a whole.

It is possible, however, that the great cloud cover consists mainly of a non-aqueous system as suggested by Kuiper since there is so little water in the very high layers visible from the earth (about 55 km at 10 °C/km and with the observed temperature of -30 °C). In our own stratosphere the moisture content is as low as that of Venus and we have clouds of solid particles said to be in part ammonium and metallic sulfates (Junge, 1963). The Venus clouds may have a similar constitution and at some lower level over the polar regions only true water clouds may occur. However, Venera 5 and 6 report 4 to 11 mg of water per liter at 0.6 atm and 25 °C. Since 23 mg would correspond to saturation, such air would have to be cooled to form droplets, but this would appear to be possible at a few kilometers higher altitudes. So it is not clear whether the observed clouds are aqueous or not even if there is abundant water on the planetary surface, as seems to us to be likely.

Acknowledgment

This research was supported in part by the United States Air Force Office of Scientific Research, Grant No. AFOSR 1255-67, and the National Aeronautics and Space Admin., Grant No. NSG 05-007-003.

References

Avduevsky, V. S., Marov, M. Ya., and Rozhdestvensky, M. K.: 1968, 'Model of the Atmosphere of the Planet Venus', *J. Atmospheric Sci.* **25**, 537–545.

Belton, M. J. and Hunten, D. M.: 1966, 'Water Vapor in the Atmosphere of Venus', *Astrophys. J.* **146**, 307.

Berkner, L. V. and Marshall, L. C.: 1964, 'The History and Growth of Oxygen in the Earth's Atmosphere', in *The Origin and Evolution of Atmospheres and Oceans* (ed. by P. J. Bracazio and A. G. W. Cameron), John Wiley and Sons, Inc., New York.

Bottema, M., Plummer, W., Strong, J., and Zander, R.: 1964, 'Composition of the Clouds of Venus', *Astrophys. J.* **140**, 1640–1641.

Bottema, M., Plummer, W., and Strong, J.: 1965, 'A Quantitative Measurement of Water Vapor in the Atmosphere of Venus', *Ann. Astrophys.* **28**, 225–228.

Clark, B. G. and Kuzmin, A. D.: 1965, 'The Measurement of the Polarization and Brightness Distribution of Venus at 10.6 cm Wave Length', *Astrophys. J.* **142**, 23.

Connes, P., Connes, J., Benedict, W. S., and Kaplan, L. D.: 1967, 'Traces of HCl and HF in the Atmosphere of Venus', *Astrophys. J.* **147**, 1230–1237.

Corneil, P. and Heber, D.: 1968, 'Sulphuric Acid and the Nature of the Lower Atmosphere of Venus', unpublished report, Chemistry Department, UCLA.

Corneil, P. and Budgor, A.: 1969, This report calculates that sulfur in a CO_2 atmosphere is found as $H_2SO_4 + SO_3$ at 590 K.

Dollfus, A.: 1955, 'Etude visuelle et photographique de l'atmosphere de Venus', *L'Astronomie* **69**, 413–425.

Dollfus, A.: 1964, 'Mesure de la vapeur d'eau dans les atmospheres de Mars et de Venus', *Mem. Soc. Roy. Sci., Liege, Serie 5*, **9**, 392–395.

Fabian, P. and Libby, W. F.: 1969, 'Ozone in the Atmosphere of Venus and Its Contribution to the Heat Budget', *Z. Geophys.* **35**, 1.

Fabian, P., Sasamori, T., and Kasahara, A.: 1958, 'Radiative-Convective Equilibrium Temperature Calculation of the Venus Atmosphere', unpublished report, National Center for Atmospheric Research, Boulder, Colo.

Fricker, P. E. and Reynolds, R. I.: 1968, *Icarus* **9**, 221.

Goody, R. M. and Robinson, A. R.: 1966, 'A Discussion of the Deep Circulation of the Atmosphere of Venus', *Astrophys. J.* **146**, 339–355.

Harvey, H. W.: 1945, *Recent Advances in the Chemistry and Biology of Sea Water*, Cambridge Univ. Press, pp. 61–68.

Jastrow, R.: 1968, 'The Planet Venus', *Science* **160**, 1403–1410.

Junge, C. E.: 1963, *Air Chemistry and Radioactivity*, Academic Press.

Kliore, A. and Cain, D. L.: 1968, 'Mariner 5 and the Radius of Venus', *J. Atmospheric Sci.* **25**, 549–554.

Kuiper, G.: 1969, private communication.

Libby, W. F.: 1968, 'Ice Caps on Venus', *Science* **159**, 1097; **160**, 1474; **161**, 916.

Mintz, Y.: 1961, 'Temperature and Calculation of the Venus Atmosphere', *Planetary Space Sci.* **5**, 141–152.

Mueller, R. F.: 1968, 'Sources of HCl and HF in the Atmosphere of Venus', *Nature* **220**, 55.

Ohring, G. and Mariano, J.: 1964, 'The Effect of Cloudiness on the Greenhouse Model of the Venus Atmosphere', *J. Geophys. Res.* **69**, 165–175.

Owen, T.: 1967, 'Water Vapor on Venus – A Dissent and a Clarification', *Astrophys. J. Letters* **150**, L121.

Owen, T., Bussinger, J. A. and Holton, J. R.: 1968, 'Ice on Venus: Can It Exist?' *Science* **161**, 916.

Pollack, J. B. and Sagan, C.: 1965, 'The Microwave Phase Effect of Venus', *Icarus* **4**, 62–103.

Pollack, J. B.: 1969, 'A Non-Gray CO_2–H_2O Greenhouse Model of Venus, *Icarus* **10**, 314.

Raman, A. and Johnson, R. G.: 1969, 'Corrosion of Igneous Rocks and Minerals by Aqueous Acidic Solutions in the Presence of CO_2', unpublished report, Department of Chemistry, UCLA.

Raman, A. and Nemes, P.: 1969, 'Evaporation of Water and Aqueous, Acidic Solutions in the Presence of CO_2 at High Pressures', unpublished report, Department of Chemistry, UCLA.

Raman, R.: 1969, 'pH Variations in Carbonate-Acid Reactions in the Presence of CO_2 at High Pressures', unpublished report, Department of Chemistry, UCLA.

Rasool, S. I.: 1968, 'Loss of Water from Venus', *J. Atmospheric Sci.* **25**, 663.

Rubey, W. W.: 1964, 'Geologic History of Sea Water', in *The Origin and Evolution of Atmospheres and Oceans* (ed. by P. J. Bracazio and A. G. W. Cameron), John Wiley and Sons, Inc., New York, pp. 1–63.

Sagan, C.: 1967, 'Origins of Atmospheres, of Earth and Planets', *International Dictionary of Geophysics*, Oxford, Pergamon Press, pp. 97–106.

Spinrad, H. and Shawl, S. J.: 1966, 'Water Vapor on Venus – A Confirmation', *Astrophys. J.* **146**, 328.

'Venera 5 and 6' *Pravda* No. 155 (18568) of 4 June 1969, Moscow.

Vinogradov, A. P., Surkov, U. A., and Florensky, C. P.: 1968, 'The Chemical Composition of the Venus Atmosphere Based on the Data of the Interplanetary Station, Venera 4', *J. Atmospheric Sci.* **25**, 535–536.

Wertman, J. W.: 1968, 'Venus: Ice Sheets', *Science* **160**, 1673.

Discussion

Morrison: The most recent Greenbank interferometry of Venus by Sinclair and coworkers shows no sign of cold poles. How does this affect your hypothesis?

Libby: If there is no special absorber in the equatorial region, we'll have to reconsider.

Golitsyn: The similarity theory also shows a very small temperature difference between equator and pole.

Sagan: I just want to support Dr. Morrison's remark. The early interferometric data which implied cold poles and the early phase data which implied a cold antisolar point have both been supplemented by more recent data suggesting a much more isothermal surface.

VEGETATIVE LIFE ON VENUS? OR INVESTIGATIONS WITH ALGAE WHICH GROW UNDER PURE CO₂ IN HOT ACID MEDIA AND AT ELEVATED PRESSURES*

JOSEPH SECKBACH and W. F. LIBBY

Institute of Geophysics and Planetary Physics and Dept. of Chemistry, University of California at Los Angeles, Calif., U.S.A.

Abstract. Experiments are described with algae grown in a new environment of pure CO_2 under pressure and in an acidic nutrient medium at elevated temperatures. One species found in hot-springs was observed to grow. If the planet Venus has acidic polar seas as we suggest, they may harbor photosynthetic life.

1. Introduction

A. THE POSSIBILITIES OF LIFE ON VENUS

Studies on the survival and growth of terrestrial life forms in simulated extraterrestrial environments have been attempted in several laboratories. However, to our knowledge little, if any, research has been reported concerning an environment close to that existing on Venus. In size and density, Venus shows more similarity to the Earth than to any other planet in the solar system [1, 2, 2a]. The analyses of the data on Venus obtained recently from Mariner V and the three Venera missions indicate that the Cytherean atmosphere consists in the main of some 50 to 100 atm of CO_2 and that the equatorial surface is extremely hot near 500 °C. At first sight it appears that the odds against any kind of life on the Earth's sister planet are quite high. But the information on the Venusian atmosphere was obtained mainly from the equatorial regions and not the poles which may be relatively cooler. In an attempt to preserve the chemical similarity principle, Libby suggested [1] that the Venus polar zones might have ice caps equivalent to our seas but in view of the latest evidence it seems that the temperature is too high for that, so we now think of warm acidic polar seas, the acid preventing the precipitation of the atmospheric carbon dioxide. Argument for these polar seas has been given recently [2]. It depends principally on the volatility of acids as compared to bases.

Plummer and Strong [5] suggested that Venus may not be unfavorable to life. Based on older data they [5] also proposed a model of Venus which contains ice caps at the polar regions and has lakes or seas which may have escaped detection. Sagan commented [6] in a paper concerning life on the Cytherean surface that the "strictures against life at the very poles may be slightly relaxed." In an earlier illuminating article on Venus, Sagan [7] hypothesized on "microbiological planetary engineering" for the composition and meteorology of the Cytherean atmosphere "for comfortable human habitation." He contemplated seeding the upper Venus atmosphere with

* This research was supported by NASA 05-007-003 and AFOSR 1255-67A.

Sagan et al. (eds.), Planetary Atmospheres, 62–83.

blue-green algae which should finally reduce the temperature and permit surface photosynthesis [7]. More recently Morowitz and Sagan assumed [8] further theoretical postulations about life-forms in Venus cloud levels. In addition, Kvashin and Miroshnichenko [9] summarized the Cytherean life possibilities and suggested that photosynthetic organisms may exist below the clouds and obtain their water from rain and mineral particles ejected from the Venusian surface. Further exploration over the Venus polar regions both by earth based optical and radar observation and space probes is necessary to answer questions about their conditions and possibly to answer finally the question of life on this 'morning star'.

B. THE EXTREME RANGES OF LOWER FORMS OF LIFE

Several microorganisms show optimal adaptation to relatively extreme conditions [10, 11] such as acid media or high temperatures (i.e., in sulfuric hot springs). They can grow under elevated hydrostatic pressure (i.e., Barophilic microbes in the depths of the ocean). Also many can tolerate anaerobic conditions (i.e., within animal intestine, sewage plants, H_2S, etc.). However, algae are the only microorganism which photosynthesize and evolve molecular oxygen. In addition some thermophilic algae can tolerate extremely acidic media as for example $1 N H_2SO_4$ [12]. Other extreme conditions for microbial growth were recently reviewed by Brock [10] and Vallentyne [11].

1. Life Under Pure CO_2

The first and main question pertinent to the possibility of Cytherean life seems to be whether plants can live in a CO_2 atmosphere. The inhibitory effects of higher concentrations of CO_2 on plants is reviewed in several sources [13, 14]. Rabinowitch suggested [13] that the inhibitory effect of higher doses of CO_2 is due to narcotic poisoning and acidification of the cellular content. The photosynthetic bacteria reduce CO_2 through an anaerobic mechanism [15] but without any evolution of O_2. Higher concentration of CO_2 inhibits or even prevents mold growth [16] while filamentous *Phycomycetes* show yeast-like development which is induced with a high concentration of CO_2 [17]. Also, the existence of water molds adapted to an anaerobic environment rich in CO_2 has been recently reported [18]. The fungus *Blastocladia pringsheimii* grows well under tank CO_2 which is used for the formation of resistant sporangia [19]. Among the higher green plants such high levels of CO_2 are toxic and observations in our laboratory have confirmed this with some hydrophytes (*lemna, Saqitaria*), desert plants (cacti, *Atriplex*) and seed germination tests (rice, *Saqitaria*). Ewart reported in the last century [20] that in an atmosphere of pure CO_2 moss plants (*Bryophyta*) may remain living in darkness for weeks; our studies with the bryophyte *Marchantia polymorpha* under light could not confirm this observation. Experiments with plant roots by Jacobson *et al.* show that above 95% CO_2 both oxygen and K^+ uptake decrease rapidly and the protoplasmic streaming is stopped [21]. Anion uptake is also severely inhibited by the presence of CO_2 [22]. Algae produce O_2 during photosynthesis and are therefore not usually associated with environments of low O_2. Nevertheless several

6—P.A.

algae are capable of performing highly specialized anaerobic metabolism, such as assimilation of CO_2 with H_2, H_2S or organic H donor [23].

2. HIGHER PRESSURE TREATMENT OF MICROORGANISMS

Effects of elevated pressures on microbial systems were recently reviewed [11, 24, 25]. Hydrostatic pressure is an important environmental parameter in life which exists in the depths of the oceans and subterranean areas. Because of the limit of light in that zone most of the species involved are bacteria or other heterotrophic organisms rather than photosynthetic plants. Most of the higher pressure experiments resulted in a negative effect on the organism tested. ZoBell [10, 11, 24, 46] isolated very few pure cultures of obligate barophiles from enrichment cultures of deep-sea samples. Biological materials exposed to increasing pressures slow down their growth and reproduction [10, 24, 25], and the ability to undergo bacterial cell-division is often lost which results in retardation of activity or death [26]. In contrast, a sudden compression and decompression of the hydrostatic pressure may not injure the microorganism [24]. When a high pressure of gas is suddenly released the bursting of over 90% of the tested bacteria cells was reported by Fraser [27]. Furthermore, the most severe effects were obtained with ca. 33 atm of CO_2 gas [27]. In a review on "environmental biophysics and microbial ubiquity" Vallentyne (11) emphasizes that "scientists are inclined to study single factors taken one at a time." When two or more environmental factors show antagonistic effects... "one expects to find an increased tolerance to each factor using combined action." ZoBell [11] reported that the maximum temperature for thermophiles is increased by compression. Heden's studies [25] suggest that nucleic acids are probably the target of hydrostatic pressure. A fairly elevated pressure increases the melting temperature of DNA and protects the biological activity against thermal inactivation [25]. Similar effects were observed earlier with enzymes and recently reviewed by Mortia [24]. Regnard [26, 28] observed many years ago that botanical material continues to function (evolution of O_2) under fairly high compression. Under hydrostatic pressure of 400 atmospheres, O_2 production proceeds slowly and after several days the tested algae cells were decomposed [28]. More recently Gross [29] observed severe color mutation in a green alga culture after being subjected to higher pressure. These *Euglena* cells show a higher percentage of mutants permanently lacking chlorophyll and having altered carotenoids when the treatment took place in darkness [29].

We now report on experiments with unicellular algae which thrive in conditions close to those possibly existing in the Venus polar regions. Pure cultures and mixtures of microorganisms (from greenhouse scrapes, soil, or ponds) were initially assayed in order to find the selected CO_2 adapted alga. A newly isolated green alga (*Scenedesums* sp.) was also examined. After selecting the algae which can thrive in a CO_2-rich environment we tested their tolerance to higher temperature and pressures as well as towards acidic media. Results obtained with the thermophilic-acidophilic alga – *Cyanidium caldarium* – may indicate some clues for supporting assumptions about Venusian life. Primary reports of this study were recently written and presented [30 to 32].

2. Methods and Materials

A. CULTURE AND MEDIA

The condition of growth and pigment determination for *Cyanidium caldarium* were recently described [30]. Algal samples were collected at acid sulphate springs of Yellowstone National Park (Wyoming) and were cultured in a double strength, Allen [12], medium pH 2–3, gassed with pure CO_2. The growth flasks were always agitated inside water bath (40 to 50 °C) under continuous fluorescent illumination (ca. 450 ft-c). For each CO_2 flask an air control was grown, otherwise both were treated similarly. Green algae were grown in mineral nutrition media [33, but mostly as 33a]. The algal flasks were agitated on rotating or wrist action shakers at 25–27 °C, and their CO_2 treatment followed basically that of *Cyanidium*.

B. ALGAE CELLS AND CHLOROPHYLL DATA

Growth values were determined both by turbidity of the cells (O.D. 580 mμ) and the volume of packed cells measured in graduated Hopkins centrifuge tubes. Pigments were extracted from *Cyanidium* with dimethylformamide (DMFA): the specific absorption coefficient for chlorophyll at 665 mμ in this solvent is 13.86 μg/ml [34]. The green algal chlorophyll was determined on a methanol extract using the constants obtained by Mackinney [35].

1. *Dry Algae Powder and Chemical Analysis*

The algae were harvested by centrifugation and washed twice with distilled water to remove all soluble materials. The concentrated cell-precipitate was frozen in acetone-dry ice and lyophilized at -50 °C in a vacuum of 5 μ overnight or longer. Then the dry algal powder was placed in the oven at 40–50 °C or at room temperature inside a vacuumed desiccator. The determination of C and H followed the classical micro Pregel method.

2. *Morphology*

Light, fluorescent, and electron microscopical examinations were made. The cell diameter was measured and the standard deviation was calculated from prints taken with a Zeiss Ultraphoto II light microscope. For the fine structural studies the cells were fixed with unbuffered $KMnO_4$, embedded in polyester resin (Westopal W) and the blocks were sectioned with a LKB Ultratome and the tissue reviewed in a Hu 11 A Hitachi electron microscope.

3. *High Pressure Experiment*

Cyanidium caldarium cells were transferred to plexiglass cylinders (Figure 1) placed in room temperature or in 45–50 °C (water bath) and connected to CO_2 and argon tanks. All containers were exposed to relatively low light intensity of ca. 50–100 foot candles obtained from fluorescent desk lamps. The algal suspension was subjected to increasing pressures of CO_2 and Ar at a rate illustrated in Figure 2. When the final pressure was

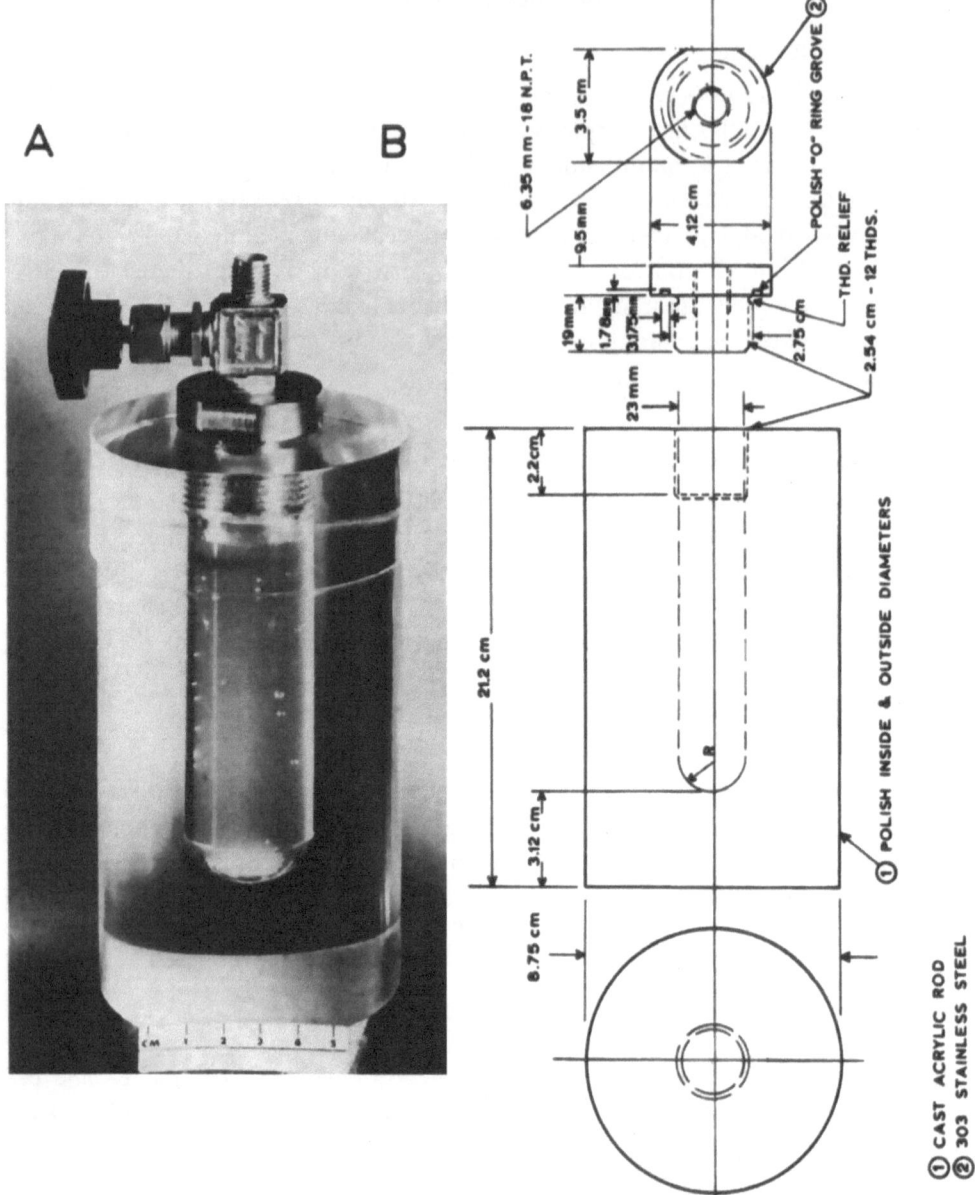

Fig. 1. The high pressure vessel for growing algae at various temperatures and under illumination while subjected to the gaseous pressure. These containers were designed to fit into a Cary 14 recording spectrophotometer. Each plexiglass pressure cylinder was tested at 200 atm at room temperature prior to the experiment. (A) a general view, (B) cross-section of the cylindrical pressure chamber and its dimensions.

reached the algae were kept for periods up to 15 days under high pressure (50 atm). At the end of this treatment the algae cells were examined for their number, pigment spectra and appearance in the microscope. Subcultures were transferred into 1 atm of CO_2 for determining the restored growth after the high pressure treatment.

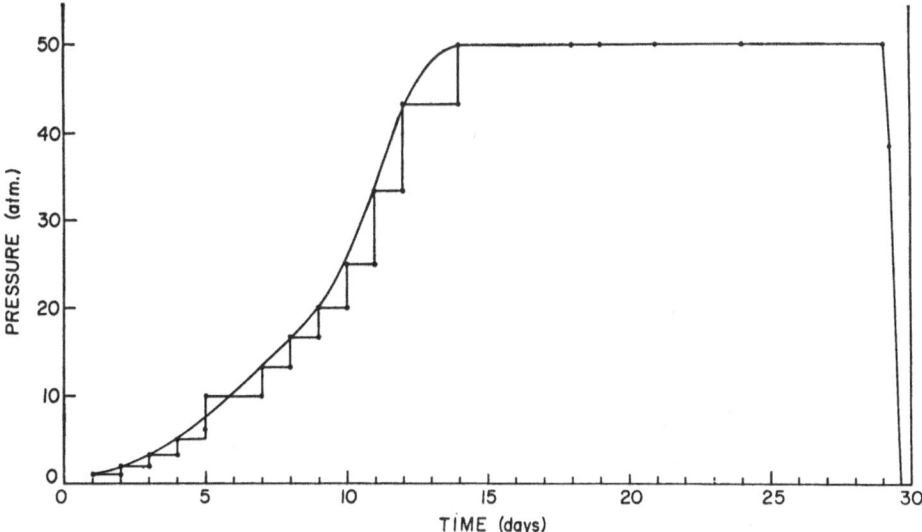

Fig. 2.   Pressurization rate of alga cells under CO_2 or/and Ar and the decompression during the experimental period.

3. Results and Discussions

A. GROWTH RATES

Results obtained with a mixture of microorganisms (algae and bacteria) treated for 40 days with a higher concentration of CO_2 (Figure 3A) show a lag period for ca. two weeks with almost no obvious increase of cellular material. After this adaptation period, a rapid promotion of cellular volume takes place for the following three weeks (Figure 3A). Subcultures of the previous CO_2 grown cells in new nutrient media [33] gassed with CO_2 or air result in a prompt acceleration of all the cells for the first five days, regardless of the gas source (Figure 3B). The cultures grown under CO_2 show lower growth than the aerated controls; this decline is more pronounced with the duration of this treatment (Figure 3B). Identification* of the predominant algae surviving after the pure CO_2 treatment suggested a few species from the *Chlorococcalean* unicellular members (*Chlorella, Chlorococcum, Scenedesmus,* and *Cocomyxa*). Thus, it was established that at room temperature a steady growth occurs with several green algae in an agitated culture supplied with a constant stream of CO_2.

The next step was to begin with a pure *Chlorococcalean* culture and examine the response to pure CO_2 application. A newly isolated pure green algal culture†

* We acknowledge the assistance of Professor H. C. Bold (University of Texas, Austin): R. A. Lewis (University of California, San Diego, Scripps Institution of Oceanography, La Jolla). H. S. Forest (State University College, Geneseo, New York), and to F. R. Trainor (University of Connecticut, Storrs) who identified the algal cultures.
† This alga was isolated by Dr. G. A. Zavarzin (Institute of Microbiology Academy of Sciences, Moscow, U.S.S.R.) from rain water accumulated at his country house near Moscow. He innoculated it into mineral medium under H_2. Further purifications by Prof. R. A. Lewin (Scripps Institution of Oceanography, La Jolla, Calif.), resulted in a bacteria-free culture which was kindly given to us and labelled as 'Zav-9'. Additional taxonomical examinations suggested (see previous acknowledgement) that this alga is probably *Scenedesmus* sp. We wish to thank all who assisted us with this organism.

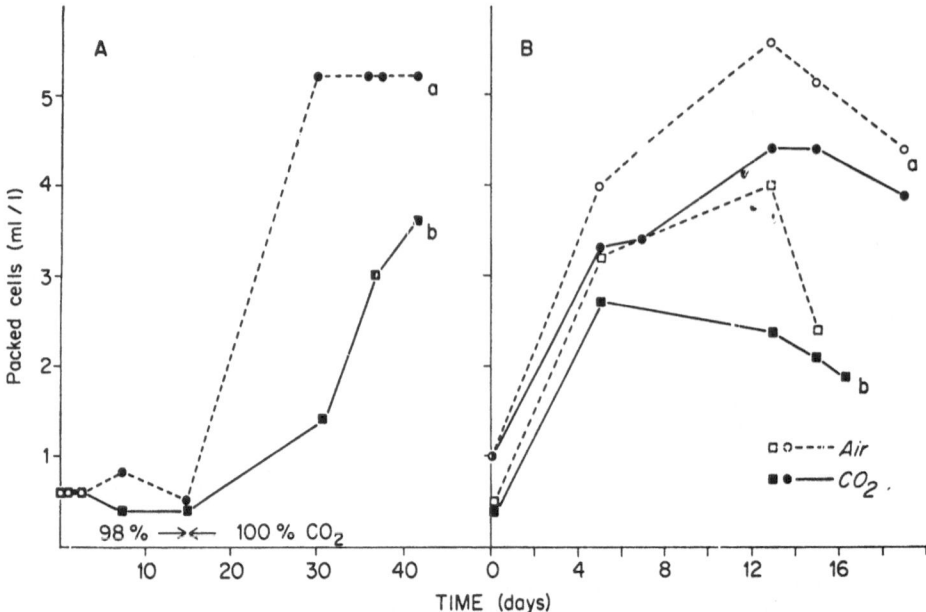

Fig. 3.   Growth rates of two cultures labeled a and b containing a mixture of green algae under higher concentrations of CO_2 (A), and the subculturing of these CO_2 grown cells in new media gassed with CO_2 or air (B).

(*Scenedesmus* sp.) was treated with CO_2 for a few weeks and the cellular appearance and growth rates are shown in Figures 4 and 5 respectively. The CO_2 treated alga is larger in size and appears more spherical than the 'lens-like' aerated cells (Figure 4). Morphological modifications were previously observed in plants or fungi indicating an increase in size when subjected to abnormal conditions of growth. Some molds show dimorphism in cellular development under 100% CO_2 [17]. Working with higher plants, Madsen [36] determined an additional fresh weight and thickness of leaves with the increasing CO_2 concentration. The growth of this alga (Zav-9) under CO_2 is almost as good – if not better – than the aerated control as illustrated (Figure 5) for the volume packed cells and chlorophyll content. When the pH of the growing media was monitored during the first 15 days in a reproducible experiment, the CO_2 culture showed a decrease from 7 to 5.5 while the aerated cells increased the pH to 10. Table I represents the photosynthetic activity of this green alga as compared with *Cyanidium caldarium* under CO_2 or air treatments. The technique used for total O_2 determination was previously described [30]. It is clear that the photoysnthetic values are quite similar for both algae incubated in different conditions (Table I). The CO_2 grown green algal cell has ca. triple content of chlorophyll and produces 1.5 times more O_2 per packed cell than the control. Since the chlorophyll is lower in the aerated cultures, they show higher O_2 per pigment unit than the CO_2 cells (Table I). Thus, this alga (*Scenedesmus*) seems to have a larger pH range as well as the ability to thrive under 100% CO_2 and shows higher physiological activity.

Pigment absorption spectra of this alga grown for 33 days on CO_2 or air seems to be

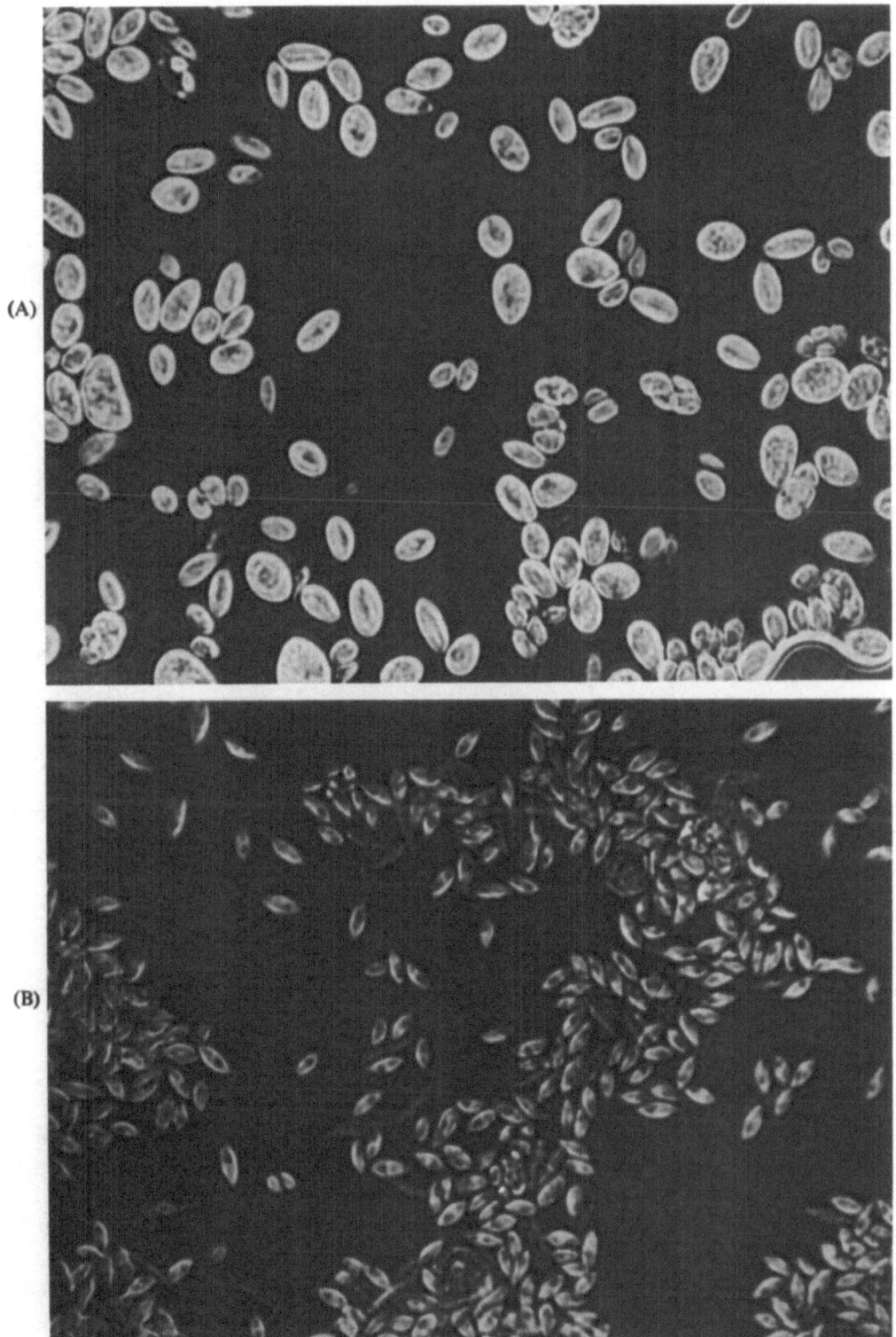

Fig. 4.   The appearance of newly isolated green alga (*Scenedesmus* sp.) cells grown under pure CO_2 for 13 days at room temperature (A), as compared with the aerated control cells (B). The CO_2 cell is larger than the control and shows higher activity. $\times 500$.

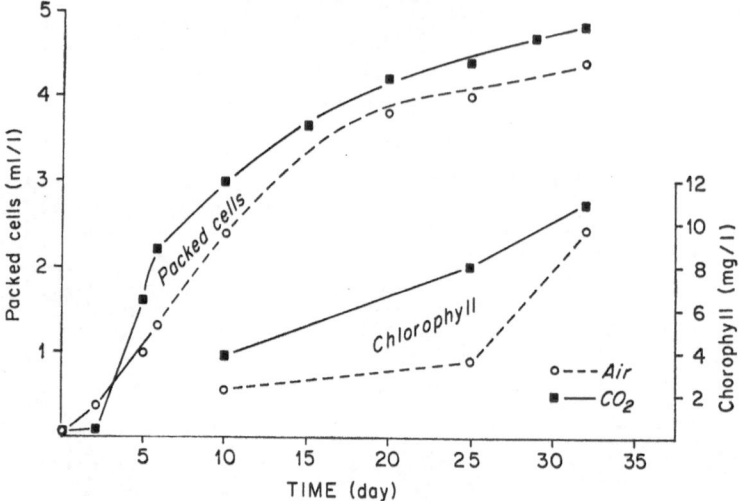

Fig. 5.   Growth rates and chlorophyll content of Zav-9 (*Scenedesmus* sp.) as a function of CO_2 (■ —) or air (○ – –). Cells were grown on a rotating shaker at 26 °C illuminated with ca. 400 ft-c.

TABLE I

Photosynthetic activity of algae cells grown under CO_2 and air at different conditions

Expt.	Treatment	Chlorophyll packed cells	Photosynthetic rates per minute, total O_2 per	
			Chlorophyll	Packed cells
		g/l	ml/g	ml/l
Zav-9 (*Scenedesmus* sp.) 27 days, 26°	CO_2	3.65	106	387
	Air	1.19	215	256
	Ratio CO_2/Air	(3.0)	(0.5)	(1.5)
Cyanidium caldarium 6 days, 45°	CO_2	2.9	127	369
	Air	1.4	170	236
	Ratio CO_2/Air	(2.1)	(0.7)	(1.5)

similar for both cells as illustrated in Figure 6. All major peaks and the 'shoulder' of the methanol extract are located at the same wavelengths.

Greater attention was given to the cultures of *Cyanidium caldarium*, the hot-spring acidophilic alga. The growth curves of *Cyanidium caldarium* cells grown at 45° in 100% CO_2 are presented in Figure 7. The growth values are expressed as: (a) absorbancy of the turbid algae suspension (units of O.D. at 580 mμ which also gives an estimate of cell mass); (b) volume packed cells; and (c) chlorophyll *a* content extracted with DMFA. In general the CO_2 treatment results in higher values than the control, as can be judged from these curves (Figure 7), each one of which was composed from three different experiments. Table II shows the daily growth rates from each value as

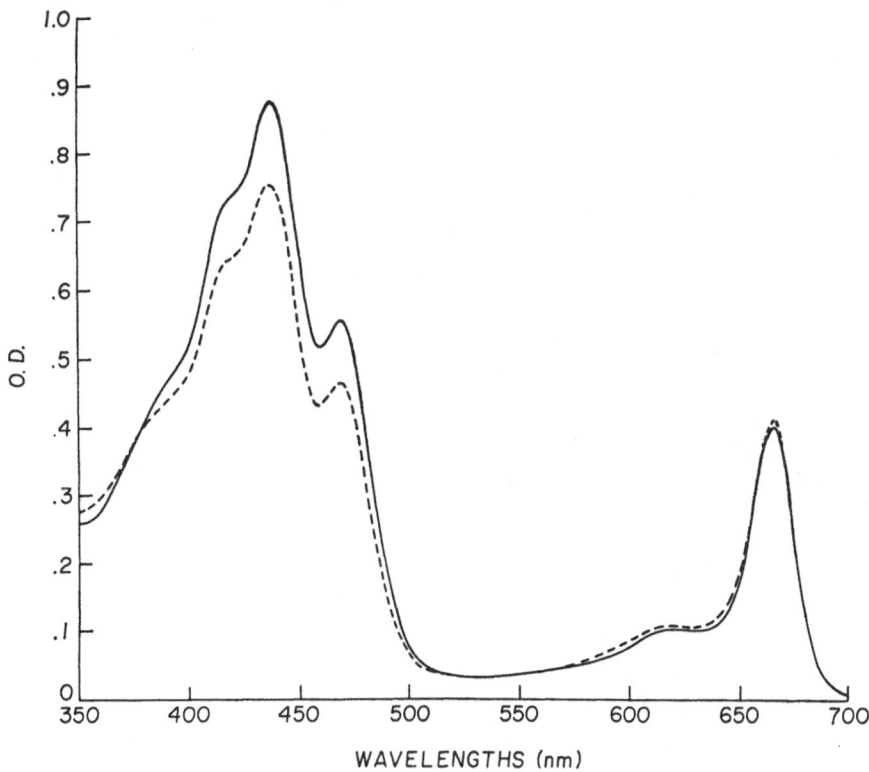

Fig. 6. Pigments absorption spectra in methanol of Zav-9 grown for 33 days in CO_2 (– – –) or air (—). These curves are composed from extracts of 5 ml algal suspensions containing 0.815 μl (CO_2 cells) and 1.75 μl (air control) packed cells per ml of methanol.

calculated from the alga production during their 10th to 20th day of treatment. It is clear from Table II that the cell volume and chlorophyll production rate is about fourfold higher for the CO_2 grown culture. However, a similar concentration of chlorophyll is present in both cultures per cell volume or number (Table III), after 18 days of treatment. The *Cyanidium* pigment absorption is similar for both cultures, which have the major peaks of the DMFA extract located at the same wavelength (Figure 8). Total photosynthetic O_2 evolution (measured O_2 evolution plus dark O_2 uptake for the same time) and the chlorophyll amounts are shown in Tables I and III for 6 and 18 days old cells respectively. When the chlorophyll level per cell volume is similar for both treatments (Table III), CO_2 grown cells are more photosynthetically active since their O_2 production is triple that of the control. This ratio decreases when the chlorophyll content is double within the CO_2 cell (Table I) of the younger culture.

B. LIGHT INTENSITY STUDIES

Earlier studies have reported that *Cyanidium caldarium* is extremely dependent upon light intensity [38]. The recorded optimal light is at relatively low intensities [12, 37,

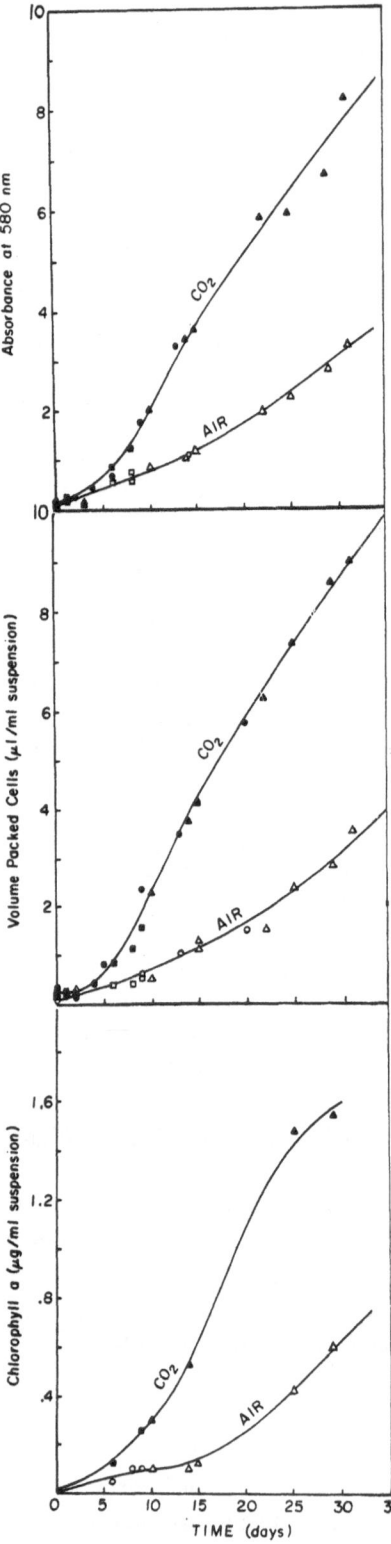

Fig. 7. Growth of *Cyanidium caldarium* as a function of CO_2 (solid figures) or air (open figures) at 45 °C in agitating containers for 35 days. Growth values are expressed in (a) units of O.D. at 580 mμ (model B Beckman Spectrophotometer); (b) volume of packed cells (μl/ml suspension), and (c) chlorophyll content (μg/ml). All data are from three experiments.

TABLE II

Daily growth rates of *Cyanidium caldarium* per ml suspension at 45°

Treatment	a	b	c
	Cell number $\times 10^6$	mμl packed cells	mμg Chlorophyll *a*
CO_2 (1 atm)	3.80	365	92
Air (1 atm)	1.04	90	24
Ratio	3.64	3.95	3.82

TABLE III

Total photosynthetic O_2 evolution and chlorophyll content of *Cyanidium caldarium* grown for 18 days at 45°

Treatment	Chlorophyll/Cell g/l	O_2/Chlorophyll ml/g/min	O_2/cell volume ml/l/min
CO_2	1.85	234	435.0
Air	1.88	71	133.5
Ratio	1.0	3.3	3.2

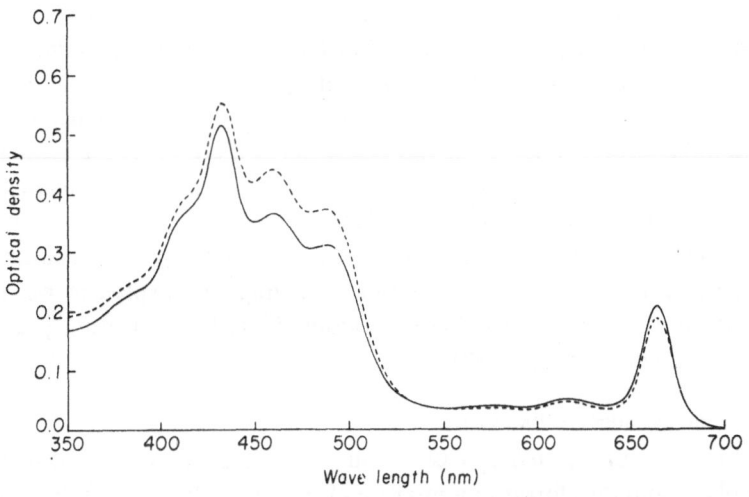

Fig. 8. Pigments absorption spectra in dimethylformamide from *Cyanidium* grown for 10 days at 40 °C in CO_2 (– – –) or in air (——). The CO_2 extract was diluted with two volumes of DMFA for a closer comparison with the control.

38] of ca. 400 ft-c. Allen reported [12] that high light intensities ($9–15 \times 10^3$ lux) causes loss of chlorophyll *a* and phycocyanin and the *Cyanidium caldarium* cell appears light yellow green. In contrast to this negative effect of higher light levels, other investigators [30, 39] have reported increasing photosynthetic rates with increasing

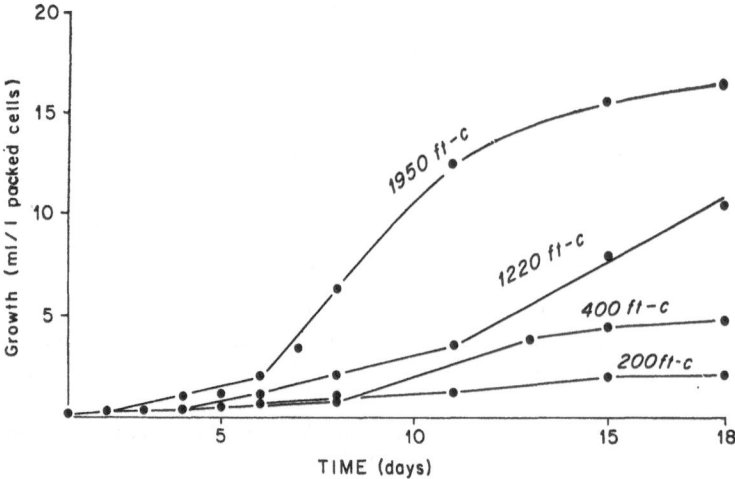

Fig. 9.   Growth of *Cyanidium caldarium* as a function of light intensity at ca. 40 °C under pure CO_2.

intensities of light and concentrations of CO_2 for this alga. We studied the effect of light intensity on the growth of *Cyanidium* [31] and observed that the higher the light level the better the growth, as illustrated in Figure 9. These cells were cultured under pure CO_2 for 18 days and show a progressing increase in cellular volume after the first week of treatment with increasing light intensities. An increase in the light from three to five times the control intensity (400 ft-c) caused a corresponding promotion of the volume packed cells of two to threefold, as observed during the last determinations (Figure 9). Additional experiments show that *Cyanidium* under high light grows much faster than under low light whether gassed with pure CO_2 or air. The culture gassed with CO_2 exposed to ca. 2000 ft-c was growing fastest of all. *Cyanidium* rate of growth was found to be proportional to the light intensity (Figure 9). These results of increasing growth at elevated light intensities are not surprising since *Cyanidium caldarium* culture was isolated in Yellowstone Hot Springs where the light level can reach 8000 ft-c in full sunlight [40]. The results obtained by others with *Cyanidium caldarium* which suggested low light optima for this alga might be explained on a basis of a different strain which was isolated from Sonoma County, California [12] or a variation in the nutrient media [37, 38].

C. THIN LAYER PIGMENT SEPARATION

Methanol extracts of *Cyanidium* CO_2 or air grown cells were applied to sucrose and silica gel plates and developed in a mixture of organic solvents [31]. Results from this study do not indicate any new band for the pure CO_2 grown cells, both cultures have similarly the four major pigment bands. Results of the quantitative studies are in preparation [31].

D. CHEMICAL COMPOSITION

The results from three sources of chemical determinations [47] are presented in Figure 10. All the results are calculated as % of the freeze-dried (lyophilized) cells, excluding the C and H which are based on the ash-free fraction. The ash content from the CO_2 culture is only half the control, while the C+H levels ($52\% + 7.5\%$ respectively) are similar for both treatments. Some minerals are detected only in air grown culture (Mo, Na, Va), or are more abundant within these cells (Mn, Ca, Fe, S, Si), while other elements are within a close range (Cu, Zn, Mg, P, N, C, H). The ash-rich control fraction contains mainly cations belonging to the trace nutrients (Mn, Nz, Cu, Mo, Va, Ca, Fe) and less from the macronutrients (Na, K). The fact that the chlorophyll content per cell volume is similar in both cultures (Figure 7 and Table III) is confirmed here by its components (H, C, N, and Mg) detected at close levels for both treatments.

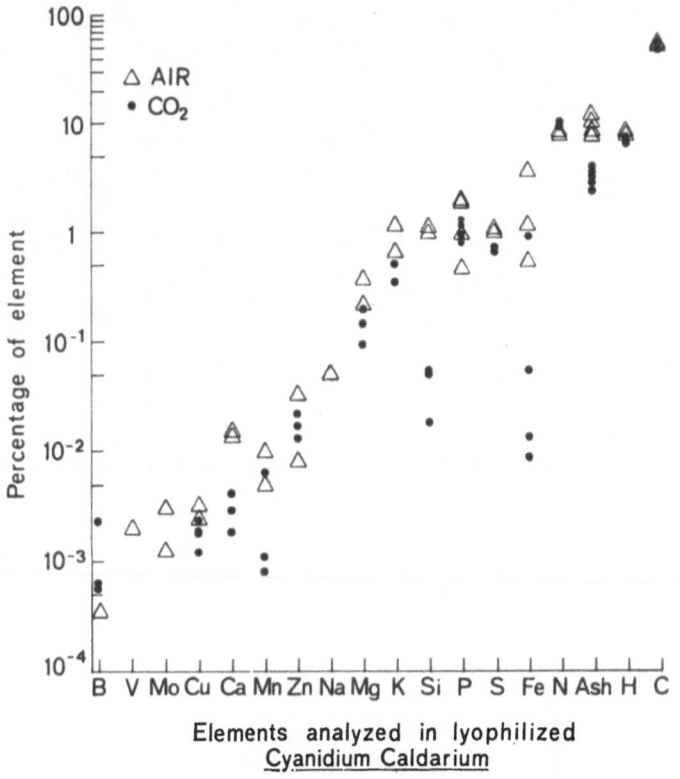

Fig. 10. The percentage of chemical elements analyzed from the lyophilized *Cyanidium* cells. The values for C and H are calculated for an ash-free weight while the rest are expressed per lyophilized powder. The chemicals were detected by three different sources.

E. CELLULAR MORPHOLOGY

The general appearance of *Cyanidium* grown at 50° (Figure 11) shows that cells vary in size and extend from 3.5 to 8 μ in diameter. The histograms of the cell distribution (Figure 12) illustrates that the CO_2 grown cell reaches a larger diameter range than the air control. The average diameter of the CO_2 treated cell is 4.78 ± 0.86 μ whereas the control cell reaches a size of 3.88 ± 0.77 μ. Additional observations with the fluorescent microscope demonstrated a bright red color from both cultures which

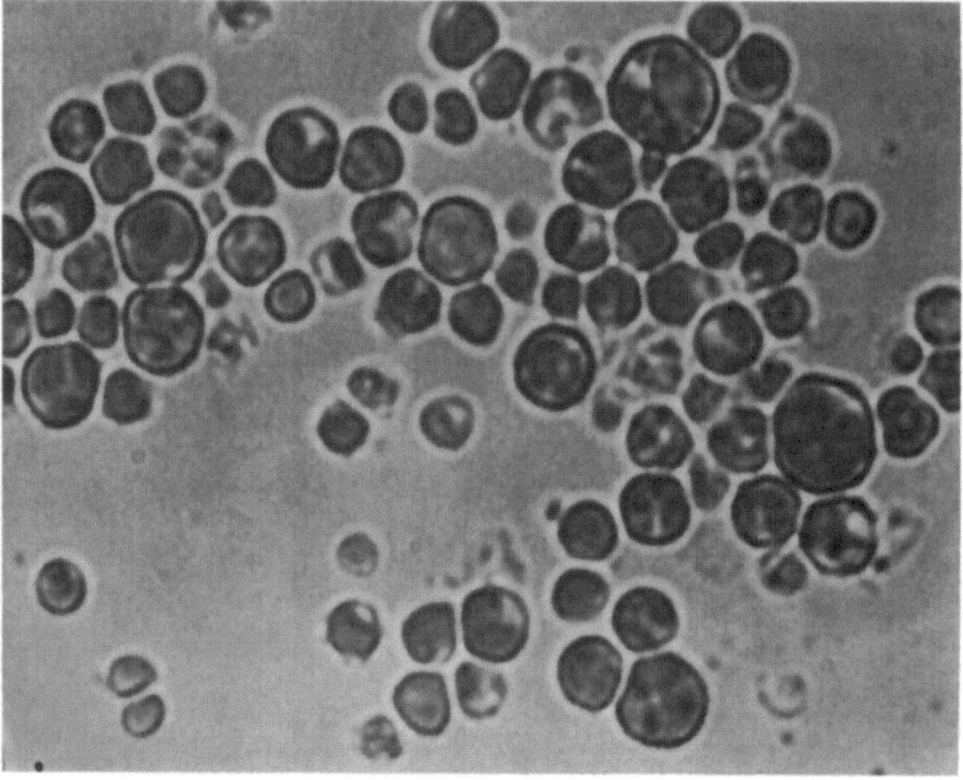

Fig. 11. The appearance of *Cyanidium caldarium* cells grown in CO_2 for nine days at 50 °C. Several of these cells are larger in diameter than the air control cells. These cells are not uniform in size or shape. × 1.250.

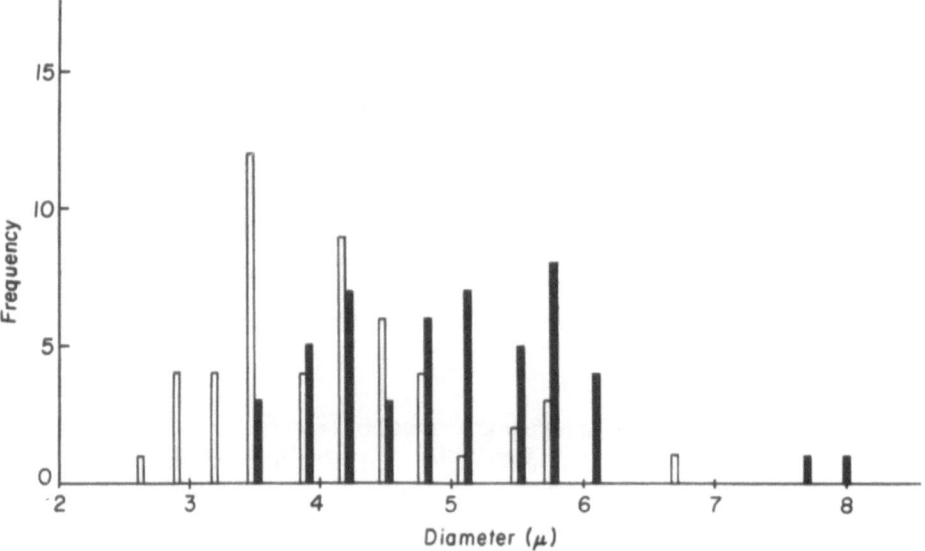

Fig. 12. Distribution of the diameter size from *Cyanidium caldarium* cells grown under CO_2 (solid histograms) or on air (open histograms), same growth conditions as for the previous figure.

indicates active chlorophyll. When these cells grow at room temperature (25°) the CO_2 culture reveals also larger cells than the control.

The fine structure of the CO_2 grown cell is shown in Figure 13. One cell is in the division stage and three out of the four autospores are visible. One endospore has a mitochondrium (Mit) which gives further evidence of its presence within this genus [41]. Several electronmicrographs from both cultures (unpublished observations) illustrate clearly mitochondria, relatively small vacuoles, and conspicuously large plastids. Several chloroplasts as shown here (Figure 13) may occupy almost all the entire

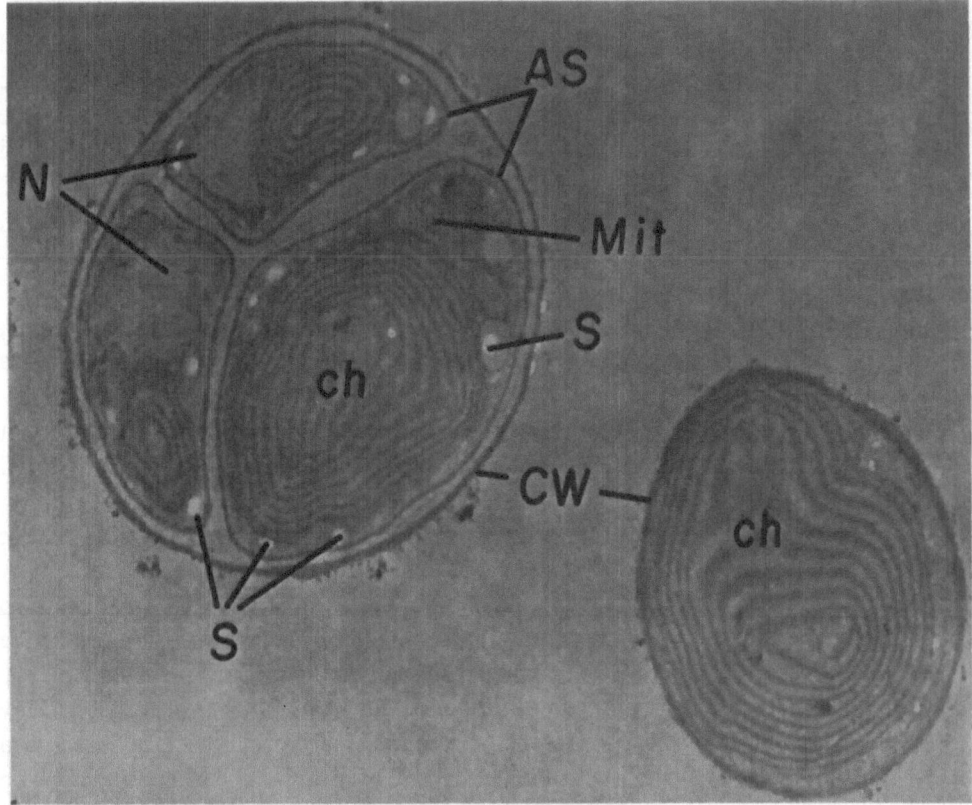

Fig. 13.   The fine structure of *Cyanidium caldarium* cell grown on CO_2 at 40°C for nine days. The cell at the right contains a large chloroplast (ch) which has eight lamillar bands. The left cell is dividing into autospores (AS), which illustrate the cellular organelles [i.e. the nucleus (N), chloroplast (ch), mitochondrium (Mit), and storage granules (S)]. All the endospores are still within the mother cell wall (CW). × 53 000, unstained.

cytoplasmic volume. On the other hand, other publications of this genus' ultrastructure [42, 43] show relatively large vacuoles and small parietal plastids. No striking differences in the fine structure were observed between the two treatments, however the CO_2 group seems to contain a larger number of storage granules. Confirmations of earlier studies indicate that the starch content of leaves from certain species is found to be higher with enriched CO_2 concentrations than when they were placed in atmospheric air [36]. Also the total content of carbohydrates measured in plants increased in proportion to the concentration of CO_2 in the surrounding air [36]. Starch accumulation was observed in *Oedogonium* after CO_2 application. The higher storage material observed in the electronmicrographs of the CO_2 treated cells (Figure 13) could possibly serve as a mechanism to control the mineral uptake via a starch-glucose osmotic pressure regulator. The presence of starch should reduce the uptake, and indeed the CO_2 cell is lower in chemicals (Figure 10). The presence of many lamillar bands within these chloroplasts and their rich orientation suggests active photosynthetic functions.

F. HIGH PRESSURE STUDIES

We undertook the growing of *Cyanidium* under pure carbon dioxide at elevated temperatures for supporting assumptions about the Cytherean life theory. We have found that this acidic hot-spring alga thrives better under CO_2 than in air, also the rate of photosynthesis (O_2 evolution), cell's size and number exceed the aerated control.

Initial studies with a few blue-green hot-spring algae (*Mastiqocladus laminosum*, *Synechococcus lividus* clones Y52 and 53), under CO_2 at 45° [33a] resulted in a discoloration and bleaching of these anaerobic treated cultures within the first two weeks. Similar results were obtained with fresh water blue-green algae (*Nostoc* sp. 588, *Anabaena cylindrica* Lem B 629, supplied from the collection of algae, Indiana University, Bloomington, Ind.) and with some marine species (*Dunaliella* sp., *Prymnesium* and *Chlamydomonas* sp.) which were cultured under high doses of CO_2 at 26° in mineral enriched sea-water. Accelerated growth always occurred in the aerated controls.

In order to approach further the extraterrestrial conditions of Venus' poles we have tested *Cyanidium* in elevated pressures of CO_2, acidic media and at 25 to 50 °C. Several experiments were conducted with this hot-spring alga at higher pressures (up to 50 atm). The rates of pressurization and decompression is shown in Figure 2. We found that when the pressure is built up within a very short period of time it kills the culture. Furthermore, Frasser [27] developed a system of bursting microorganisms by a rapid release of the high CO_2 pressures. Higher pressure of air (49 atm plus 1 atm CO_2) causes a faster damage to *Cyanidium* than 50 atm of CO_2. This super-oxidation effect was not obtained with lower O_2 levels (3.5 atm O_2 to cells previously gassed with CO_2). As judged from the coloration of a series of algae which were subjected to increasing CO_2 pressures at 25°, it seems that additional pressure from 1 to 10 atm does not damage this alga. Variation in the 'in vivo' absorption spectra was observed in cells which were subjected to from 20 to 50 atm CO_2 for ca. 10 days. The phycocyanin absorption peak (620 mμ) is very reduced or completely removed while the chlorophyll peak (675 mμ) is detectable. When the cells from the high pressure treatment were returned to atmospheric pressure of CO_2 (or air) they resumed a slow growth and have normal absorption spectra patterns (Figure 14). Thus, it seems that the high CO_2 pressure arrests the algae growth and reduces the cellular diameter to ca. 70% as compared with their initial size. It was noticed that higher light intensities causes a discoloration of the algal suspensions under the high CO_2 pressure while this

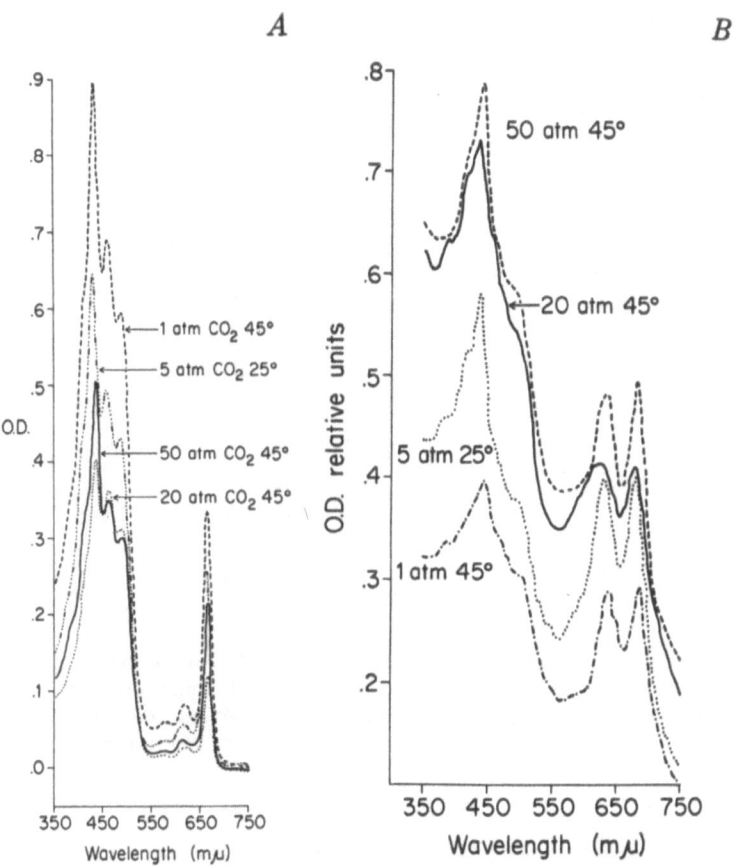

Fig. 14. Pigments absorption spectra from *Cyanidium caldarium* cells which were subjected to higher pressures of CO_2 for a few days and then transferred to 1 atm for an additional period of a few weeks. All the peaks are similar to the control culture (grown under 1 atm CO_2) which indicates normal pigmentation as determined with a DMFA extract (A), or from the 'in vivo' spectra of the algal suspensions (B).

was delayed with lower levels of illumination. Later, these treatments took place under relatively low light intensities of ca. 50 to 100 ft-c. The high pressure of CO_2 alters not only the 'in vivo' absorption spectra of *Cyanidium* cells, but also a shift in the pigment absorption peak from 432 mμ to 413 mμ was observed within DMF extracts (Figure 15). The absorption at these wavelengths is related to the carotenoids and chlorophyll *a* contribution. It was shown [44] that bacteria cells may alter their carotenoid content due to higher concentrations of CO_2. Also Gross [29] reported on pigment mutation of the green alga *Euglena* under higher hydrostatic pressures.

The question was asked whether the suppression effect of 50 atm of CO_2 is due to the CO_2 elevated dose or to high pressure 'per se'. Since higher pressures of O_2 have been shown to be toxic to this alga, we treated *Cyanidium* with 49 atm argon and 1 atm CO_2. Results of this are presented in Figure 16 and Table IV. There is a continuation of growth under high pressures of Ar which provides a hydrostatic pressure due to its

7—P.A.

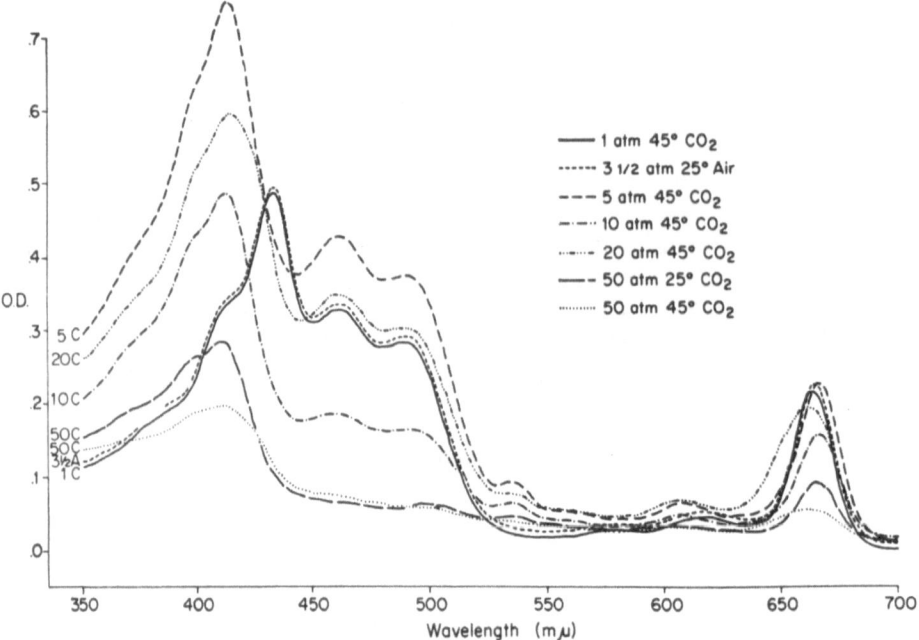

Fig. 15.   The DMFA extract shows a shift from the peak at 430 mμ to lower wavelengths which is caused by the high CO_2 pressure treatments. The control cells contain a peak at 432 mμ which also resembles the low air pressure cells. Chlorophyll *a* peak at 665 mμ and most of the other peaks do not alter from the control cells grown under 1 atm of CO_2.

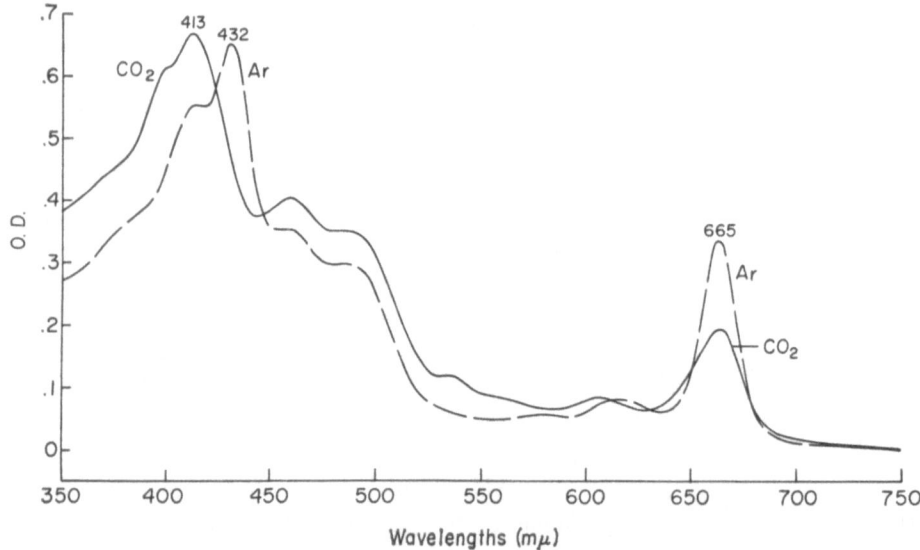

Fig. 16.   Higher pressures of argon (49 atm + 1 atm CO_2) at 45 °C do not arrest the *Cyanidium* growth. The absorption spectra of the Ar treatment (– – –) is similar to control cells (1 atm), in contrast to cells subjected to 50 atm CO_2 (——). The main difference between the 50 atm of CO_2 and Ar absorption peaks is in the blue region of the spectrum.

inertness. On the other hand, 50 atm of CO_2 cause arrest of growth and 82% of the initial cells survive. There is an increase of ca. 25% of the algae cells under the high pressure Ar as determined by number and volume packed cells (Table IV). Chlorophyll examinations of these cells show higher values for both, 93% of the original level remained in the CO_2 culture while the Ar culture had about tripled its pigment content. The chlorophyll per cell is promoted because of the lower light intensity used, twice as much was detected in the Ar treated cells than the CO_2 treated cultures (Table IV). The shift of the 432 mμ peak was not detected with Ar, while it was demonstrated with 50 atm of CO_2 (Figure 16).

TABLE IV

The growth yield and survival of *Cyanidium caldarium* cells after being subjected to 50 atm[a] of CO_2 or/and argon at 45° in acidic media for 15 days.

Pressure	50 CO_2	49 Ar + 1 CO_2
Algal cell/ml suspension	4.7×10^7 (0.82)[b]	7.3×10^7 (1.28)
Packed cells (ml/1)	3.35 (0.82)	5.10 (1.24)
Chlorophyll *a*/suspension (mg/l)	6.2 (0.93)	18.0 (2.7)
Chlorophyll *a*/packed cells (g/l)	1.85 (1.11)	3.54 (2.2)

[a] See high pressure application rate in Figure 2.
[b] The ratios of the final cells and chlorophyll over the initial values (prior to the high pressure treatment) are parenthesized.

In studying higher pressures of gas we have also to consider the physical and chemical properties involved. The linear function of the CO_2 solubility in higher pressures is well known. Also the higher the temperature the less gas dissolves in the solvent, and the higher the pressure the more gas gets in the solution. Since *Cyanidium* cells were incubated in mineral media of ca. pH 2, there was not expected a decline in the H^+ ions concentration caused by higher pressures of CO_2. Electronmicroscopical observations (unpublished data) indicate that the *Cyanidium* cell may possess most of its constituents after being under 50 atm of CO_2.

These primary high pressure experiments – although not sufficient enough to draw any generalizations – may provide increasing evidence that algae thrive in this new unexplored ecological niche which may resemble Venus' poles. We feel that as long as this question about life on Venus is not 100% clear examinations on this subject are worthwhile.

G. FINAL COMMENTS

The search for life outside the Earth intensifies with the capacity of man to travel and enter into deep space. Venus with its extreme conditions might provide milder ecological environments for lower vegetation at its cooler poles. Our studies with *Cyanidium caldarium* show that this acidic hot-spring alga thrives in pure CO_2 and under higher pressures. This raises the question of the adaptation range for photosynthetic microorganisms and suggests the study of their biological behavior under the extreme conditions on Venus. It also emphasizes the danger of contamination (or colonization) of other planets with terrestrial microbes.

According to Rubey [45] the primitive Earth was very acidic and abundantly rich with CO_2 (ca. 90%) and the pressure at the initial stage was >14 atm. A gradual increase of terrestrial CO_2 may cause the organisms to "adapt themselves by generations of evolutionary changes... [so] the effect would be much less disastrous" [45]. Thus, there may have been organisms suitable to Venus in the earliest stages of life on earth.

Acknowledgments

We are indebted to: Professor S. M. Siegel for his active interest and valuable suggestions toward this project; Professor P. D. Voth for the *Marchantia* plants; Professor R. W. Castenholz for the thermophilic blue green algae; Professors F. A. Eiserling and J. M. Christie for light and electron microscopy, respectively; Miss H. King for C and H determination; and Mr. H. Kappel and Mr. J. Vanek for technical assistance with the high pressure setup.

References

[1] Libby, W. F.: 1968, *Science* **159**, 1097; 1969, *Umsch. Wissensch. Tech.* **13**, 420; *Sky Telescope* **35**, 296.
[2] Raman, A., Corneil, P., and Libby, W. F.: 1969, unpublished report.
[2a] Fabian, P. and Libby, W.F.: 1969, *Z. Geophys.* **35**, 1.
[3] Tass: 1967, *Pravda*, No. 295 (17977), 6 (October 22); 1967, in *Izvestia*, No. 257 (15651, October 31); Reese, D. E. and Swan, P. R.: 1968, *Science* **159**, 1228; Vinogradov, A. P. Surkov, Yu. A., Florenskiy, K. P., and Andreychikov, B. M.: 1968, *Dokl. Akad. SSSR* **179**, 37; 1968: *J. Atmospheric Sci.* **25**, 535.
[4] Tass: 1969, *Pravda*, No. 155 (18568), June 4; Brichant, A. L.: 'Analysis of Essential Data from Venera 5 and 6', NASA, Goddard Space Flight Center, Contract No. NAS-512487; Hindley, K.: 1969, *New Scientist* **42**, 700.
[5] Plummer, W. T. and Strong, J.: 1965, *Astron. Acta* **11**, 375.
[6] Sagan, C.: 1967, *Nature* **216**, 1198.
[7] Sagan, C.: 1961, *Science* **133**, 849.
[8] Morowitz, H. and Sagan, C.: 1967, *Nature* **215**, 1259.
[9] Kvashin, A.N. and Miroshnickenko, L. I.: 1968, *Priroda-(Nature)* No. 11, 77.
[10] Brock, T. D.: 1969, *Symp. Soc. Gen. Microbiol. (Microbial Growth)* **19**, 15.
[11] Vallentyne, J. R.: 1963, *N.Y. Acad. Sci.* **108**, 342.
[12] Allen, M. B.: 1959, *Arch. Mikrobiol.* **32**, 270.
[13] Rabinowitch, E. I.: 1945, *Photosynthesis and Related Processes*, Vol. I, p. 330; 1951, Vol. II (pt. 1), p. 903. Interscience Publishers Inc., New York.

[14] Osterlind, S.: 1949, *Symp. Bot. Upsol.* **10**, 123; Gessner, F.: 1959, *Hydrobotanik* Deut. Verlag. der Wissensch. Berlin **2**, 232; Stalfelt, M. G.: 1960, in *Encyclop. Plant Physiol.* **5** (pt. 2), 5, 81, Springer Verlag, Berlin.

[15] Gest, H. and Kamen, M. D.: 1960, *Encycloped. Plant Physiol.* (*ibid.*) **5**, (pt. 2), 568.

[16] Golding, N. S.: 1940, *J. Dairy Sci.* **23**, 891.

[17] Bartnicki-Garcia, S.: 1963, *Bacteriol. Rev.* **27**, 293; Bartnicki-Garcia, S. and Nickerson, W. J.: 1962, *J. Bacteriol.* **84**, 829.

[18] Held, A. A., Emerson, R., Fuller, M. S., and Gleason, F. H.: 1969, *Science* **165**, 706.

[19] Emerson, R. and Cantino, E. C.: 1948, *Am. J. Botany* **35**, 157.

[20] Ewart, A. J.: 1896, *J. Linnean Soc. London, Bot.* **31**, 404.

[21] Jacobson, L., Schaedle, M., Cooper, B., and Young, L. C. T.: 1968, *Physiol. Plant.* **21**, 119.

[22] Jacobson, L., Schaedle, M., Cooper, B., and Young, L. C. T.: 1967, in *Use of Isotopes in Plant Nutrition and Physiology*, IAEA/FAO. Vienna, p. 303.

[23] Spruit, C. J. P.: 1962, in *Physiol. and Biochem. of Algae* (ed. by R. A. Lewin), Academic Press, New York and London, p. 47.

[24] Morita, R. Y.: 1967, *Oceanogr. Mar. Ann. Rev.* **5**, 187.

[25] Heden, C-G.: 1964, *Bacteriol. Rev.* **28**, 14.

[26] McKeen, C.: 1936, *Biol. Rev. Cambridge Phill. Soc.* **11**, 441.

[27] Frasser, D.: 1951, *Nature* **167**, 33.

[28] Regnard, P.: 1884, *Compt. Rend. Soc. Biol., Paris* **36**, 164.

[29] Gross, J. A.: 1965, *Science* **147**, 741.

[30] Seckbach, J., Shugarman, P. M., and Baker, F. A.: 1969, unpublished report.

[31] Seckbach, J., Nathan, M. B., and Gross, H.: 1969, unpublished report.

[31a] Seckbach, J. and Kaplan, T. R.: 1969, unpublished report.

[32] Seckbach, J.: 1969, *Plant Physiol. Sup.* **44**; Seckbach, J. and Libby, W. F., *ibid.*; Seckbach, J. and Libby, W. F., XI International Botanical Cong., Seattle, Wash., Aug.-Sept. 1969.

[33] Green Algae Nutrient Solution (*Chlamydomonas*) described by Levine, R. P. and Ebersold, W. T.: 1958, *Vererbung* **89**, 631.

[33a] Blue-green algae medium described by Castenholz, R. W.: 1967, *Nature* **215**, 1285, was used one half strength of green algae.

[34] Volk, S. L. and Bishop, N. I.: 1968, *Photochem. Photobiol.* **8**, 213.

[35] Mackinney, G.: 1941, *J. Biol. Chem.* **140**, 315.

[36] Madsen, E.: 1968, *Physiol. Plantarum* **21**, 168.

[37] Halldal, P. and French, C. S.: 1958, *Plant Physiol.* **33**, 249.

[38] Brown, T. E. and Richardson, F. L.: 1968, *J. Phycol.* **4**, 38.

[39] Fukuda, I.: 1958, *Botan. Maq. Tokyo* **71**, 70.

[40] Brock, T. C. and Brock, M. L.: 1969, *Limnol. Oceanogr.* **14**, 334.

[41] Klein, R. M. and Conquist, A.: 1967, *Quart. Rev. Biol.* **42**, 219.

[42] Rosen, W. G. and Seigesmund, K. A.: 1961, *J. Biophys. Biochem. Cytol.* **9**, 910.

[43] Mercer, F. V., Bogorad, L., and Mullens, R.: 1962, *J. Cell Biol.* **13**, 393.

[44] Cost, H. R. and Gray, E.: 1968, *Bacter. Proc.* p. 47.

[45] Rubey, W. W.: 1964, in *The Origin and Evolution of Atmospheres and Oceans* (ed. by P. J. Brancazio, and A. G. W. Cameron), J. Wiley and Sons, Inc., pp. 1–63.

[46] ZoBell, C. E.: 1952, *Science* **115**, 507; ZoBell, C. E. and Johnson, F. H.: 1949, *J. Bacter.* **57**, 179; ZoBell, C. E. and Morita, R. Y.: 1957, *J. Bacter.* **73**, 563.

[47] The chemicals were analyzed by Elec. Microanalytical Laboratories at Torrance, Calif., and by Pacific Spectrochemical Laboratory, Inc., Los Angeles, Calif.

Discussion

Sagan: The polar temperature you think possible in your previous paper is about 200 °C or higher. Have you grown organisms at 100 atm pressure and at these temperatures?

Libby: No, our high pressure experiments were done at about 50 °C or less.

VENUS CLOUD CONTRASTS

DAVID L. COFFEEN

Lunar and Planetary Lab., University of Arizona, Tucson, Ariz., U.S.A.

Abstract. Contrast of the Venus cloud patterns is measured by photoelectric scans using a sequence of narrow-band interference filters. The contrast appears constant in the range 3100–3600 Å, drops to the limit of detectability at about 4100 Å, and is absent at longer wavelengths.

Photoelectric scans of Venus were made parallel to the terminator (Figure 1) using the two-channel scanner of Hall (1968), similar to the scans of Hall and Riley (1968). A 300 μ (2.0 arcsec) circular focal plane aperture was used with a 1 sec sweep time and a 6 mm sweep length. Eight Thin Films interference filters were used (Figure 2). The λ 3557 filter was placed permanently in Channel 1 to monitor the seeing. The other filters were placed successively in Channel 2. For each filter four scan integrations were made, each of 15 sec duration. For two of these the Wollaston prism was set at approximately 45° to the direction of polarization of Venus; for the other two the prism was turned by exactly 90°. Thus the averaged scans are insensitive to the presence of polarization.

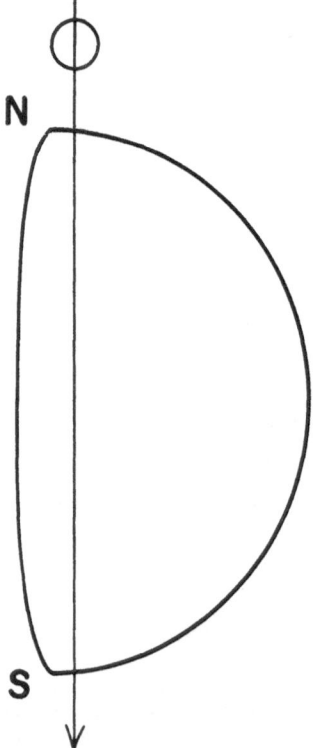

Fig. 1.   Geometry for Venus scans at 83°.0 phase angle.

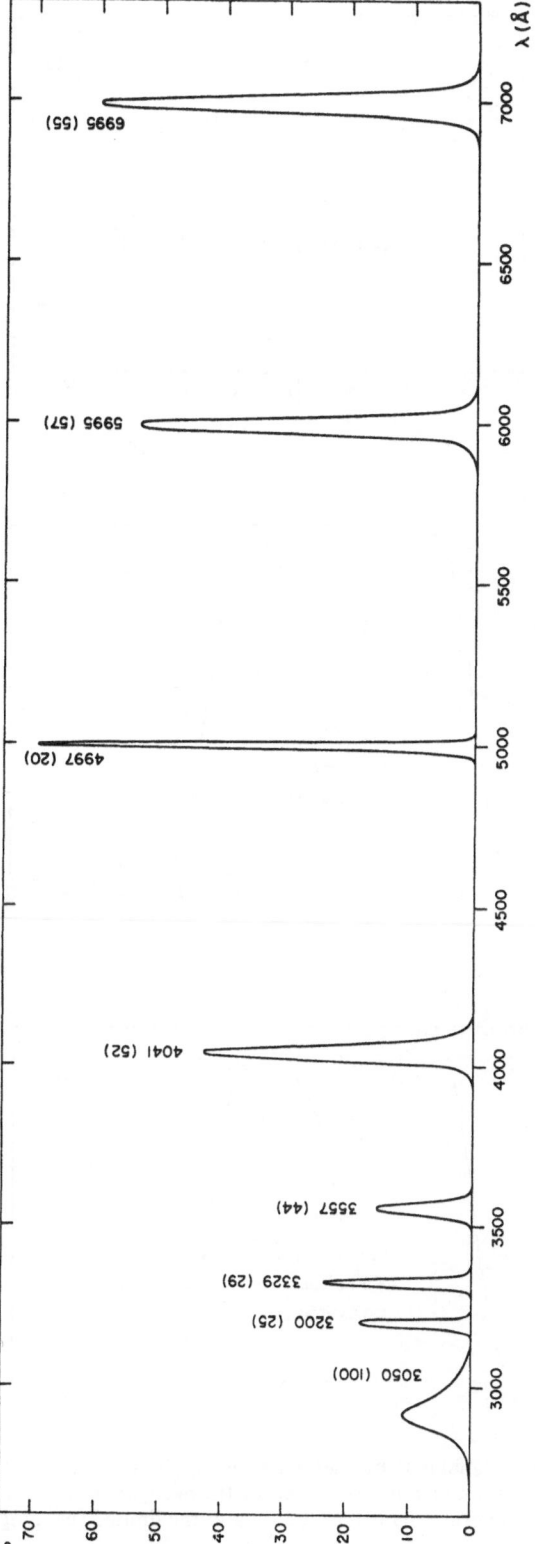

Fig. 2. Transmission curves of interference filters. Each curve is labeled with the effective wavelength and halfwidth (full width at half-transmission points) for Venus observations.

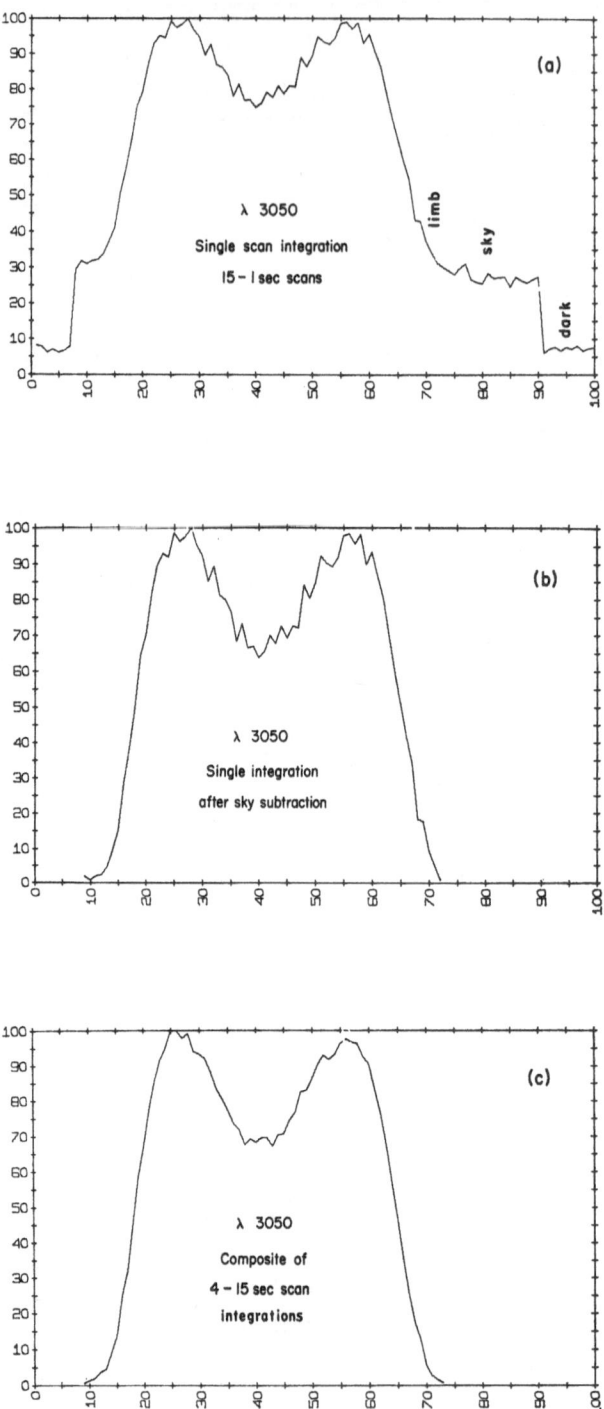

Fig. 3.   Data reduction. The signal is somewhat noisy due to photon statistics. In spite of the large telescope diameter (72″), the photon count is limited by the narrowness of the filters ($\Delta\lambda \sim 50$ Å), the focal plane aperture diameter (300 μ), and the limited scanning time.

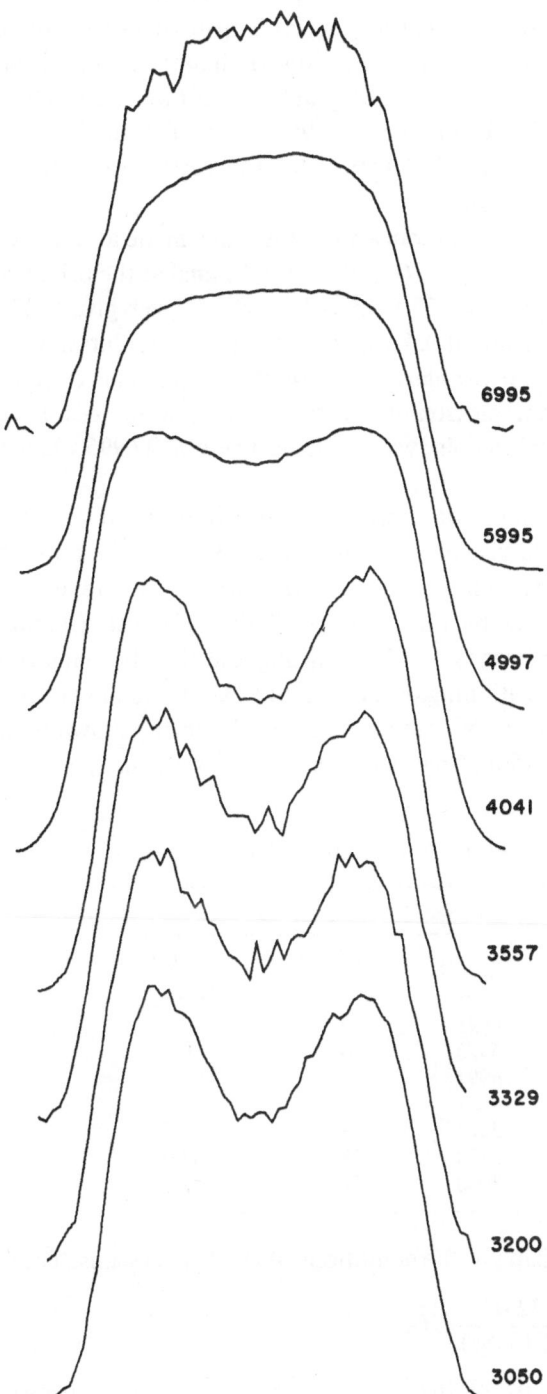

6995

5995

4997

4041

3557

3329

3200

3050

Fig. 4. Composite North-South scans of Venus at eight wavelengths.

No depolarizer was used. The two photomultipliers were never interchanged, but the filters are sufficiently narrow that any spectral mismatch of the tubes is unimportant. One 16/1 prescalar was used between the discriminator and multichannel analyzer for some of the filters to maintain an acceptable count rate. These daytime observations were made 2230–2400 UT, 17 January 1969 on the Perkins 72″, at a phase angle of 83°0. Seeing excursions were 3–4 arcsec; the rms seeing was considerably better. Sky was clear but for a thin haze.

Figure 3 shows the steps in reduction of the scans made at one wavelength. The sky background varied from 0% to 76% of the total signal at the brightest region, depending on wavelength and time. The multichannel analyzer gives 100 points along the scan, or a resolution of about 0.4 arcsec. But the true resolution is determined by the seeing plus instrumental profiles. Figure 4 shows the final composite scans for all eight filters. On intercomparison of the scans there is no noticeable variation of seeing with wavelength. Signal to noise was particularly low at λ 6995 where the S-13 cathodes have low sensitivity.

Venus showed on this date a strong central dark region in the ultraviolet. The feature was observable visually through the λ 4041 filter in spite of the faintness of the image. The contrast will be defined as the intensity difference of the brightest and darkest regions, divided by the intensity of the brightest. Channel 1 was used to monitor the scan appearance at λ 3557 throughout the observing run. Changes in the seeing and/or in the positioning of the scan line caused small variations in the contrast measured on the sequence of composite Channel 1 scans, shown in the first column of Table I. The corresponding Channel 2 contrasts C (column 3) were corrected for these

TABLE I

Cloud contrast $C = (I_B - I_D)/I_B$

Channel 1 C (3557) %	Channel 2		
	λ	$C(\lambda)$ %	$C'(\lambda)$ %
—	6995	0	0
30.0	5995	0	0
29.5	4997	0	0
31.5	4041	8.5	8.6
28.0	3329	29.0	33.1
29.0	3200	29.0	32.0
32.0	3050	32.0	32.0

variations by normalizing to the conditions of the λ 3050 scans. Thus for each filter:

$$C'(\lambda) = \frac{32.0\%}{C(3557)} C(\lambda).$$

The corrected values are plotted in Figure 5, showing the contrast as a function of wavelength for this date. For λ 4997 and longer wavelengths no central dip is seen, giving zero contrast. The central feature is detectable at λ 4041, and strengthens rapidly to λ 3557. The peak contrast of 32% is sensibly constant from λ 3050 to λ 3557.

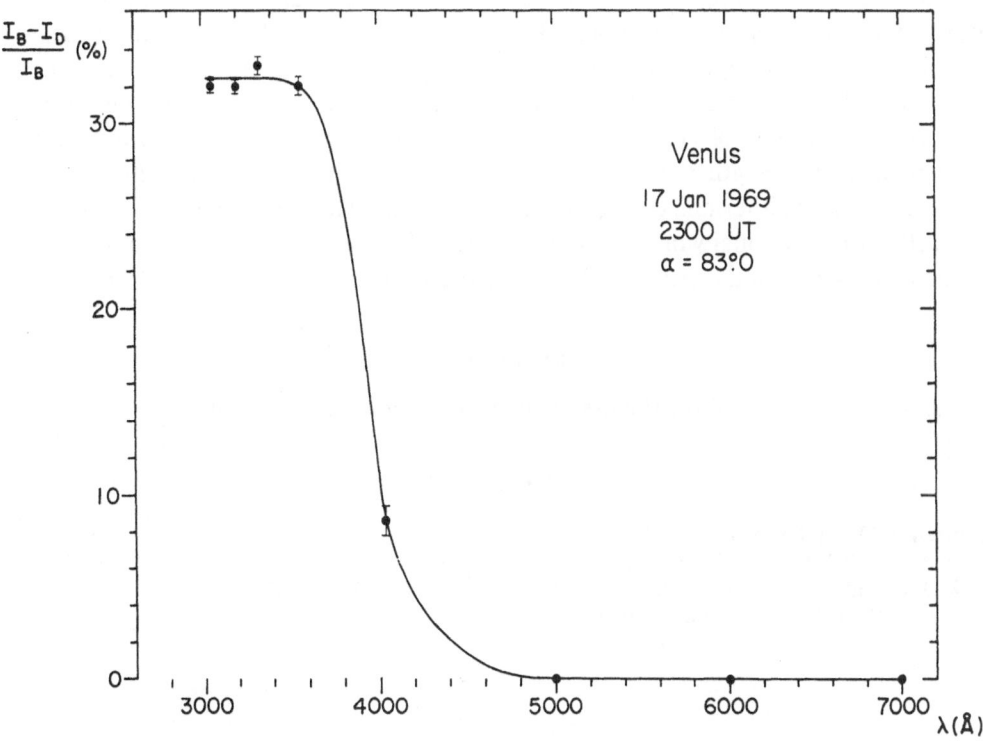

Fig. 5. Venus clouds as a function of wavelength. The measurements are taken from the scans of Figure 4, and are corrected slightly by the reference channel measurements as shown in Table I. The error bars represent one standard deviation from the mean of the four scan integrations made at each wavelength. The actual contrast values were taken from the composite scans of Figure 4. Those values could be read with a resolution of 0.5% contrast.

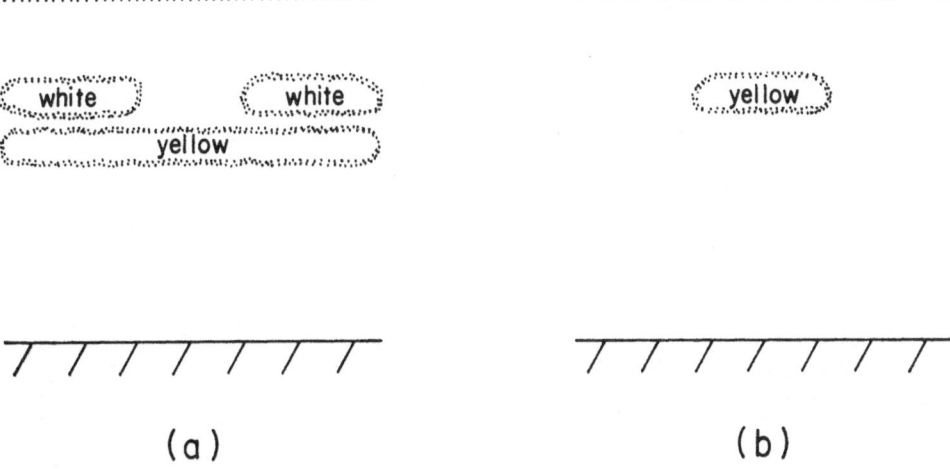

Fig. 6. Two possible explanations of the ultraviolet cloud contrasts. The central dark area is caused by absorbing clouds. The bright areas may be a white 'UV haze' cloud layer (a), or may simply be the light return from the deep molecular atmosphere (b).

The relative smoothness of the contrast curve gives no suggestion of an origin as molecular absorption lines. The probable origin is absorption by the cloud particles. The total molecular optical depth of the atmosphere is >100 at λ 3050, assuming 100 atm surface pressure of pure CO_2, giving a Bond albedo of $>97\%$ (Kahle, 1968) at λ 3050, regardless of surface reflectivity. The cloud particles must be absorbing over a wide wavelength range, with more absorption at shorter wavelengths. This could occur with either one or two cloud layers, as shown in Figure 6.

Further observational work should be done on the *absolute* reflectivities of different cloud regions and on the wavelength dependence of the limb-darkening of Venus.

Acknowledgment

I wish to thank Dr. Hall for the use of the Lowell Observatory facilities.

References

Coffeen, D. L.: 1969, *Astron. J.* **74**, 446.
Hall, J. S.: 1968, *Lowell Observatory Bulletin* 143, **7**, 61.
Hall, J. S. and Riley, L. A.: 1968, *Lowell Observatory Bulletin* 145, **7**, 83.
Kahle, A. B.: 1968, *Astrophys. J.* **151**, 637.

ON THE NATURE OF THE VENUS CLOUDS

GERARD P. KUIPER

University of Arizona, Tucson, Ariz., U.S.A.

Abstract. A comparison of a wide range of Venus observations with the ultraviolet, visible, and infrared reflection spectra and thermodynamic properties of a variety of candidate materials indicates that the principal constituent of the Venus clouds is partially hydrated $FeCl_2$.

1. Introduction

Hypotheses on the composition of the Venus clouds have proliferated in recent years. The Proceedings of the CalTech-JPL Lunar and Planetary Conference [1] describe the following interpretations, held by different authors: small water droplets, small ice crystals, hail stones (at depth); derivatives of methane, ethane, and benzene; and dust. Other recent publications have considered snow flakes (at depth), NH_4Cl, an ice-HCl solution, volcanic products, a carboniferous swamp, a planetary oil field, and the 'global-seltzer ocean' theory. A useful critical view of this curious assortment is made in the NASA Handbook of Venus [2]. Of necessity, many questions had to be left open for lack of data.

With the very low, but finite, mixing ratio of H_2O/CO_2 now established from the NASA 990 Jet observations [3, 4] and with the reinterpretation of the $2\,\mu$ region of the Johns Hopkins balloon spectrum of Venus [5], coupled with the new polarization data of Venus by Coffeen [6], the problem of water or ice clouds can be dealt with definitively. One of the arguments used in past years in support of the ice hypothesis, the low reflectivity of the planet between $3-4\,\mu$, is reinterpreted on the basis of new spectrophotometric data and found to lead to a very different conclusion.

2. Clouds Visible in Ultraviolet

Ross [7] in his extensive photographic coverage of Venus in 1927, using the 60-inch Mt. Wilson telescope, established that the planet shows a variable cloud pattern in UV light (3600 Å $\pm$); but that in yellow or red light no clear markings appear except possibly on rare occasions. Wright [8] had previously taken a few photographs at the Lick Observatory with the same general result. The cloud pattern may show changes even in a few hours. Recently the French observers (Boyer and Camichel [9]; Boyer [10]; Boyer and Guérin [11] noted that the cloud pattern at times moves across the disk in retrograde motion roughly 90° per day, corresponding to a pseudo-rotation period of 4–5 days. No strict periodicity exists, however, either in this motion of the upper atmospheric layers nor in the details of the cloud forms. The observed motion could obviously either be real or a group velocity caused by atmospheric waves. The conclusions by Boyer *et al.* have been corroborated by observations at the McDonald and Catalina Observatories, a selection of which is published in *Comm. LPL* No. 102.

Sagan et al. (eds.), Planetary Atmospheres, 91–109.

Ross [7] found that the UV clouds on Venus become gradually invisible at wavelengths substantially longer than 4000 Å. Recent observations at the Catalina Observatory by J. Fountain and S. Larson have shown that the cloud contrast continues to increase shortward of 3600 Å, down to at least 3200 Å. Undoubtedly, this contrast is related to the steep decrease of the planetary albedo (Figure 1). It would be of distinct interest to observe the cloud contrast telescopically from above the terrestrial ozone layer, between 2000–3000 Å.

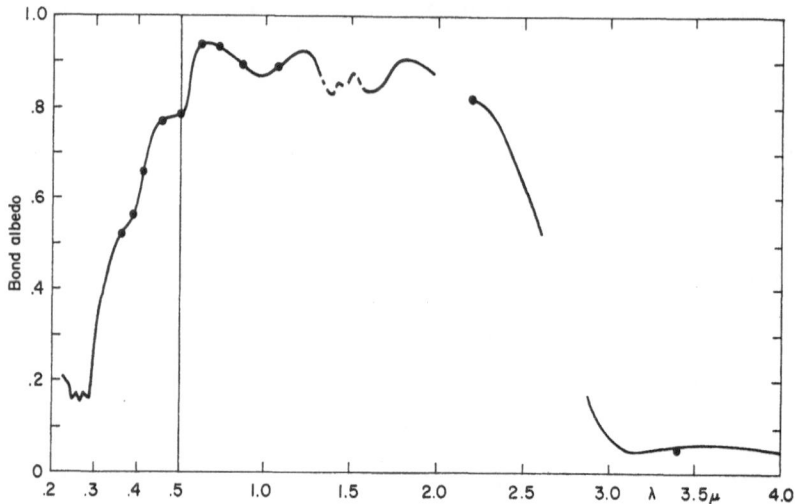

Fig. 1.   Bond albedo vs. wavelength 0.2–4.0 μ. Details near 1.5 μ somewhat uncertain.

3. The Yellow Haze Layer

Observations of the planet in yellow or red light normally show only indistinct markings. This is probably related to the excessively high albedo of the planet in yellow and red light (Figure 1) which must be attributed to a coefficient of scattering of the cloud particles very close to unity. As a result the cloud cover may be somewhat likened to a terrestrial snow surface without shadows. Yet it will be very important to ascertain whether breaks and irregularities may be seen with image resolution 10 or 100 times greater than achieved with earth-based telescopes.

The sources for Figure 1, which summarizes the data on the bond albedo of Venus for the wavelength interval 0.2–4.0 μ, have been described in a more extensive publication [12]. The curve has been corrected for absorptions by CO_2, as derived from records with higher resolution. For the two gaps the CO_2 absorptions were so large that allowance for them was not feasible. The region near 1.5 μ is somewhat uncertain.

The albedo curve is clearly very informative on the nature of the Venus cloud cover. The broad absorptions have the appearance of being caused by a solid or a liquid. The identification of these absorptions has proceeded on the basis of laboratory studies. I am indebted to Rev. Godfrey Sill (O. Carm.) for major assistance in these laboratory comparisons and for helpful comments on the chemistry of the Venus atmosphere. This laboratory program is considered in Section 6.

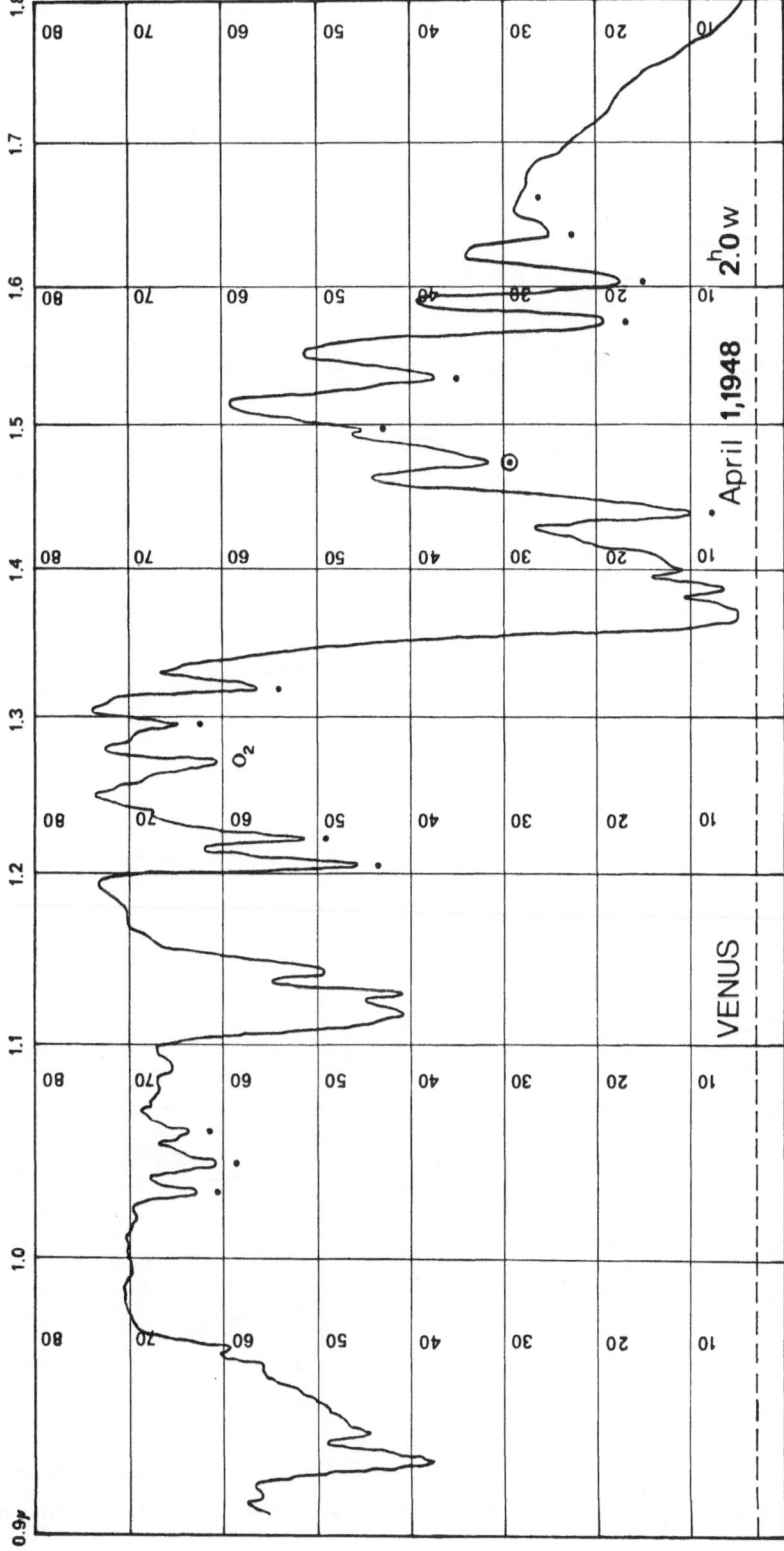

Fig. 2. Venus spectrum, 0.9–1.8 μ resolution 250. Dots CO_2 bands. Encircled dot shows isotopic band. Unmarked absorptions, telluric H_2O.

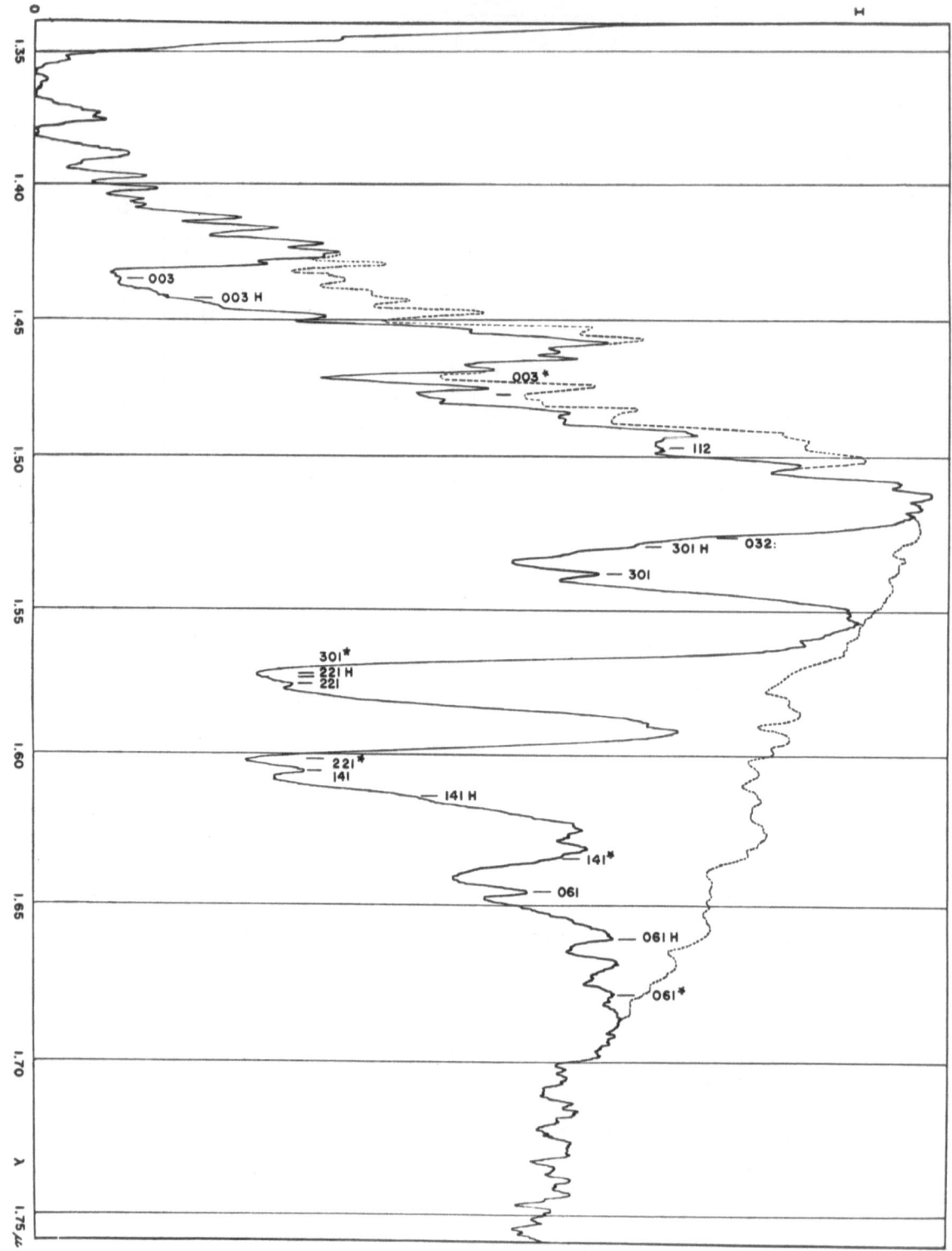

Fig. 3.   Venus spectrum, **KPNO**–36″, June 17, 1962, 1.35–1.75 μ. Upper curve is solar spectrum.

4. Spectroscopy, Composition, and Temperature Profile of the Venus Atmosphere

Substantial information has recently been derived on the atmosphere from spectroscopy in the near-infrared, 1–4 μ. These observations have all been made with the use of lead-sulfide cells, first developed during World War II. Reference is made to a spectrum obtained by the author with the 82-inch telescope of the McDonald Observatory, 0.9–1.8 μ, on April 1, 1948 (Figure 2), showing the now well-known strong CO_2 bands, including the $C^{13}O_2$ isotopic band at 1.47 μ. This band was shown separated from the telluric H_2O in spectra obtained in 1962 (Figure 3), allowing the first determination of C^{13}/C^{12} for Venus (same as for the earth), and a similar determination of the O^{18}/O^{16} ratio from the 2.15 μ band.

During the last three years enormous progress in the IR spectroscopy of Venus and other planets has been achieved by the Connes [13] which led to the discovery of HCl and HF on Venus and set upper limits for a number of other gases. It also definitely established the presence of CO (made very probable previously by the work of Moroz [14]). The very-high resolution observations by the Connes, made at the Haute Provence (el. 650 m), have been supplemented by the author with comparatively low-resolution interferometer observations from the NASA CV 990 at elevations 12–13 km [3, 4]. These observations have definitely established the presence of water vapor in the Venus atmosphere [4]. They also showed the absence of ice absorptions near 2 μ [3].

To illustrate the nature of the high-altitude interferometer spectra, we reproduce in Figure 4 the spectrum of the moon, used as comparison for the elimination of both telluric and solar absorptions, and one of the Venus spectra, Figure 5 (obtained Nov. 27–28, 1967). The ratio spectrum between the two will eliminate solar and remaining telluric absorptions and is given in Figure 6, together with identifications.

The utility of the high-altitude spectra may be judged by reference to corresponding ground-based spectra taken at the Catalina Observatory (el. 2580 m). Figure 7 of the moon, Figure 8 the ratio spectrum Venus/Moon. They may be compared with Figures 4 and 6.

Table I collects the composition data on the Venus atmosphere presently available. A critical compilation of available data on the Venus temperature profile [4] is reproduced in Figure 9.

5. Planetary Diameter and Cloud Level

The published measures of the optical diameter of the planet are reviewed in the NASA Handbook [2]. The adopted mean value is that derived by de Vaucouleurs [15], 12 240 $\pm$ 15 km ($R = 6120 \pm 8$ km). This value coincides with the level of the tropopause in our Figure 9 and should be, within the uncertainties, equivalent to the upper boundary of the dense haze layer expected to form the visible limb of the planet. Subsequent photographic measures by the writer on the Venus crescent near inferior conjunction have given $R = 6100$ for red and near-infrared light; and $R = 6145$ km

8—P.A.

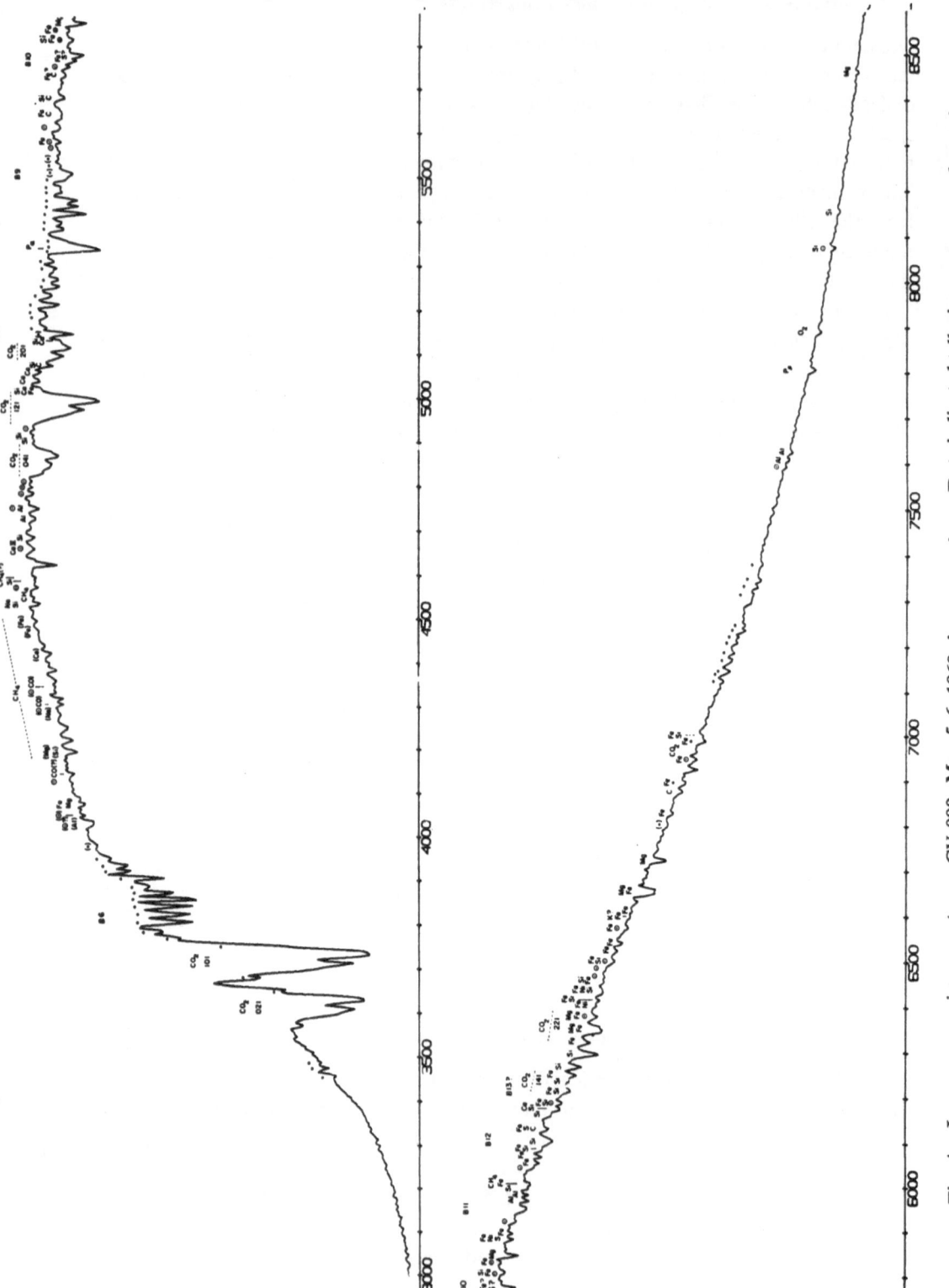

Fig. 4. Lunar comparison spectrum. CV 990, May 5–6, 1968, in two sections. Dots indicated telluric water-vapor absorptions. Attention is called to the solar Brackett lines, B6–B13. *Abscissae*: wave numbers.

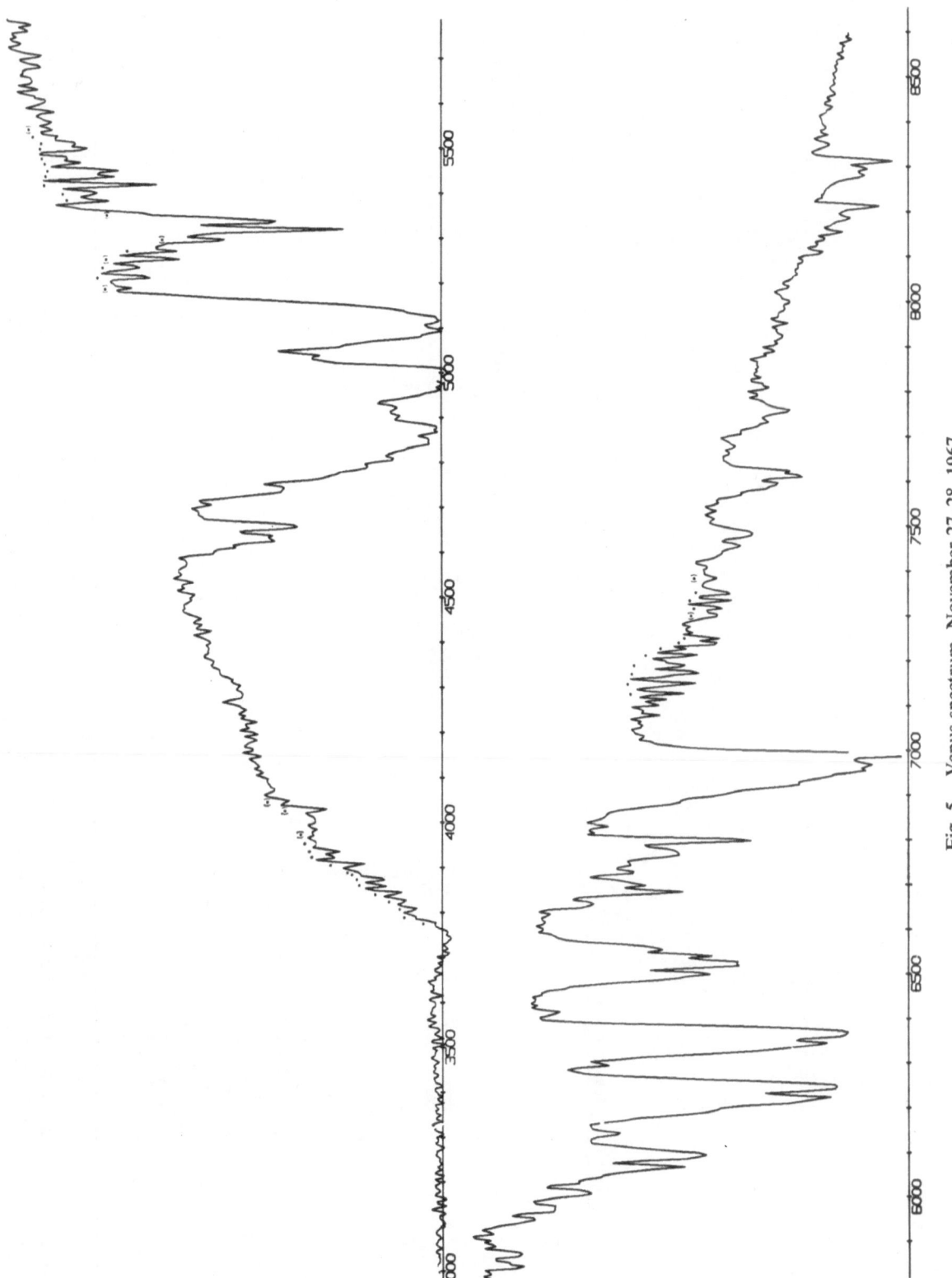

Fig. 5. Venus spectrum, November 27–28, 1967.

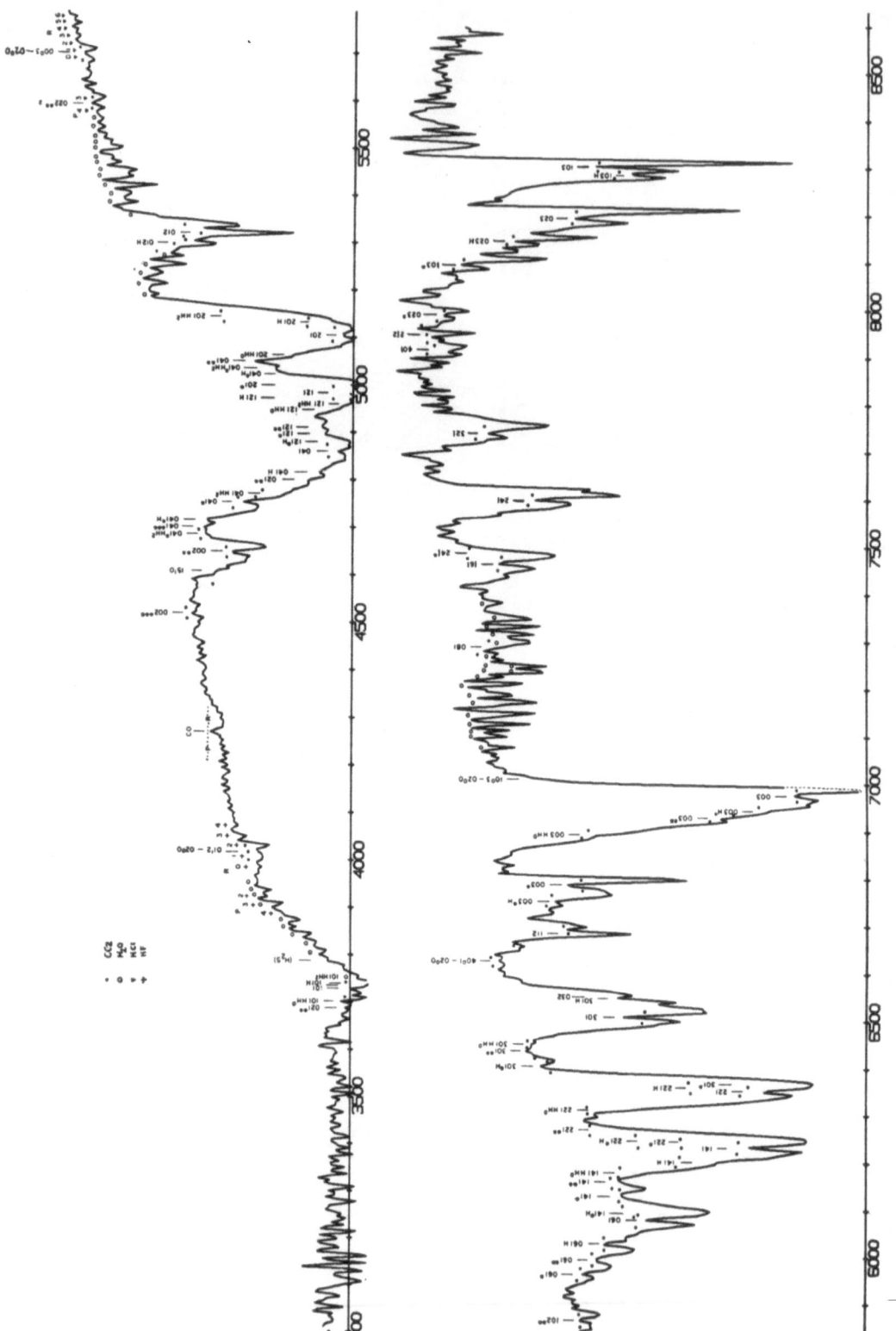

Fig. 6. Ratio spectrum, Venus/Moon, based on spectra of Figures 4 and 5, with identifications. CO_2 designations as in *LPL Comm.* No. 15, Table 1. Venus H_2O absorptions marginally recorded. HCl, HF absorptions are open circles. Identifications $\lambda < 3680$ cm^{-1} uncertain (omitted).

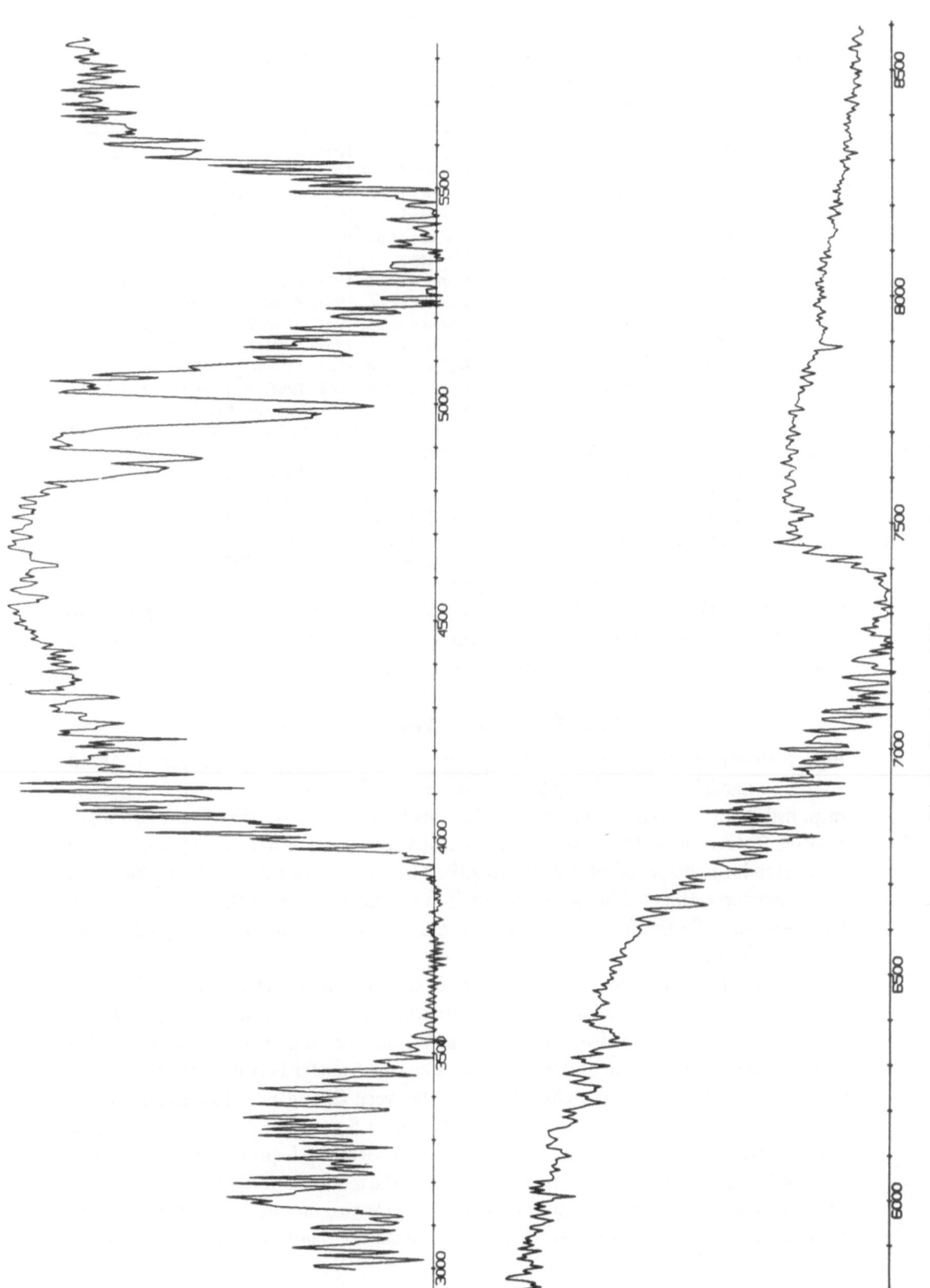

Fig. 7.   Earth-based Moon comparison.

TABLE 1

Composition of Venus atmosphere

Gas	Mixing ratio	Source
CO_2	0.95 ± 0.02	Venera 4, 5, 6 (*Pravda*, June 4, 1969).
N_2	0.035 ± 0.015	Venera 4, 5, 6 (*Pravda*, June 4, 1969).
CO	$10^{-4.34}$	Connes *et al.*, 1968, *Astrophys. J.* **152**, 741.
HCl	$10^{-6.2}$	Connes *et al.*, 1967, *Astrophys. J.* **147**, 1230–1237.
HF	$10^{-8.2}$	Connes *et al.*, 1967, *Astrophys. J.* **147**, 1230–1237.
CH_4	$< 10^{-6}$	Connes *et al.*, 1967, *Astrophys. J.* **147**, 1235.
CH_3Cl	$< 10^{-6}$	Connes *et al.*, 1967, *Astrophys. J.* **147**, 1235.
CH_3F	$< 10^{-6}$	Connes *et al.*, 1967, *Astrophys. J.* **147**, 1235.
C_2H_2	$< 10^{-6}$	Connes *et al.*, 1967, *Astrophys. J.* **147**, 1235.
HCN	$< 10^{-6}$	Connes *et al.*, 1967, *Astrophys. J.* **147**, 1235.
H_2O	$10^{-6.0}$	Kuiper, G. P. *et al.*, *Comm. LPL* No. 100.
O_2	$< 10^{-5.0}$	Belton and Hunten, 1968, *Astrophys. J.* **153**, 970.
O_3	$< 10^{-8}$	Jenkins *et al.*, 1969, *Contrib. KPNO* No. 421.
SO_2	$< 10^{-7.5}$	Cruikshank, D. P. and Kuiper, G. P.: *Comm. LPL* No. 97.
COS	$< 10^{-6}$	Cruikshank, D. P., *Comm. LPL* No. 98.
COS	$< 10^{-8}$	Based on *Comm. LPL* No. 100.
C_3O_2	$< 10^{-6.3}$	Based on *Comm. LPL* No. 100.
H_2S	$< 10^{-3.7}$	Cruikshank, D. P., *Comm. LPL* No. 98.
NH_3	$< 10^{-7.5}$	Based on *Comm. LPL* No. 100.

for the near UV (3600 Å). This difference, of 0.7% is regarded real and appears to establish the reality of the separation of the UV haze layer from the top of the dense yellow haze, already inferred from its rapid horizontal motions.

6. Identification of the Venus Haze Layer

Figure 10 shows laboratory reflection curves of six substances powdered to about 50–100 μ particle size, investigated to determine whether *silicate dust* might be responsible for the yellow haze layer. The conclusion is clearly negative. (The slight positive features near 1.95 and 1.5 μ are caused by minute water absorptions by the white standard composed of pulverized LiF. The rise shortward of 0.3 μ in the silica curve and some others is likewise attributed to an imperfection of the white standard. Otherwise all reflection curves, including those of Figure 11, are accurate and in any case comparable.)

In Figure 11 *sulfur* and a selected number of *carbonates* and *metallic halides* are shown. These curves show enormous differences and only one substance, *partially-hydrated FeCl$_2$*, corresponds closely with the Venus reflection curve of Figure 1. This material has an absorption at 0.24 μ due to a charge transfer between OH and Fe; at 0.29 μ an absorption due to a charge transfer between Cl and Fe; absorptions due to *d–d* transfers of Fe^{2+} at 0.35, 0.5, 1.0, and around 1.5 μ; and a deep absorption due to the hydrate centered on 3.0 μ. Each of these 7 absorptions is present, approximately in the expected amount, on Venus (Figure 1), and the identification is regarded certain. The only questions remaining pertain to the particle size (about 2.5 μ for Venus and 50–100 μ for the laboratory samples) which affects the *depth*, not the *wavelength* of

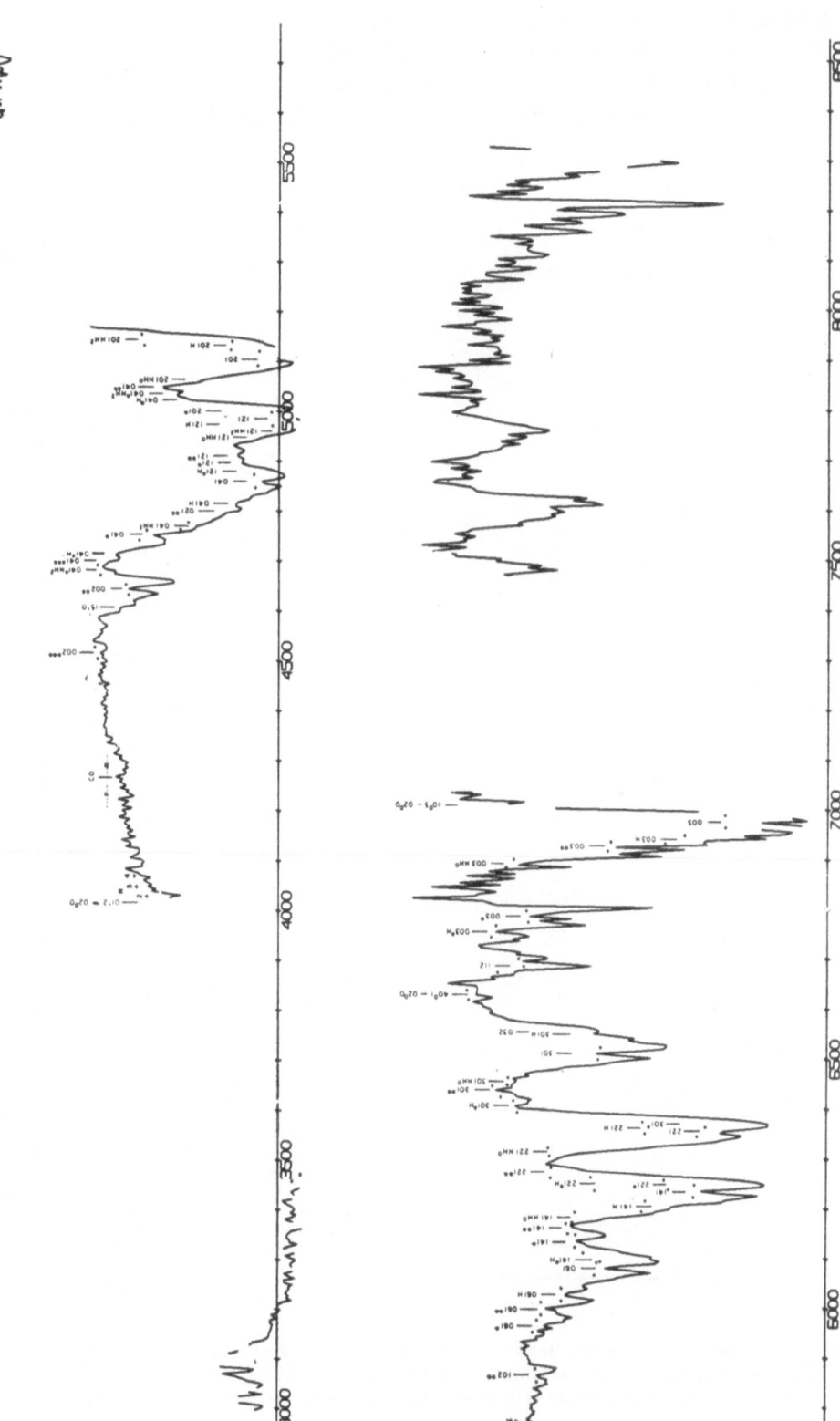

Fig. 8.   Ratio spectrum, Venus/Moon, November 7, 1967 (ground-based).

absorptions; and to the rise at 3.5–4.0 μ in Figure 11c (q), not present in Venus. It is possible that this difference is due to the *small particles* of the Venus cloud layer (smaller than the wavelength) or else due to an admixture of a *sulfate* or a *hydrate* with an extended absorption beyond 3 μ (cf. Figure 12). The refractive index of $FeCl_2 \cdot 2H_2O$

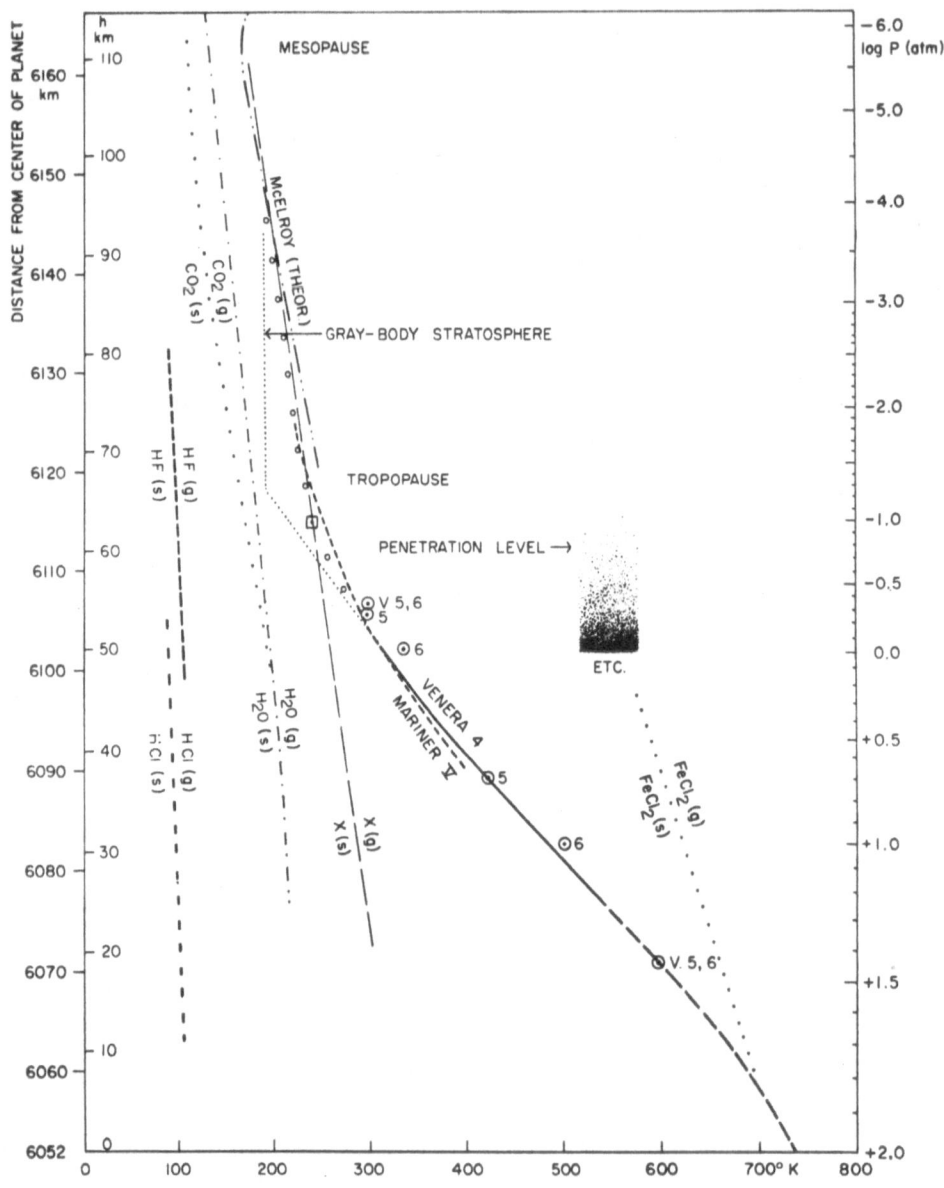

Fig. 9. Compilation of atmospheric pressure, temperature, and altitude relationships for planet Venus; dot in square, rotational temperature and pressure derived from CO (Connes *et al.*, 1968); small open circles: relationship adopted in computation of pressures.

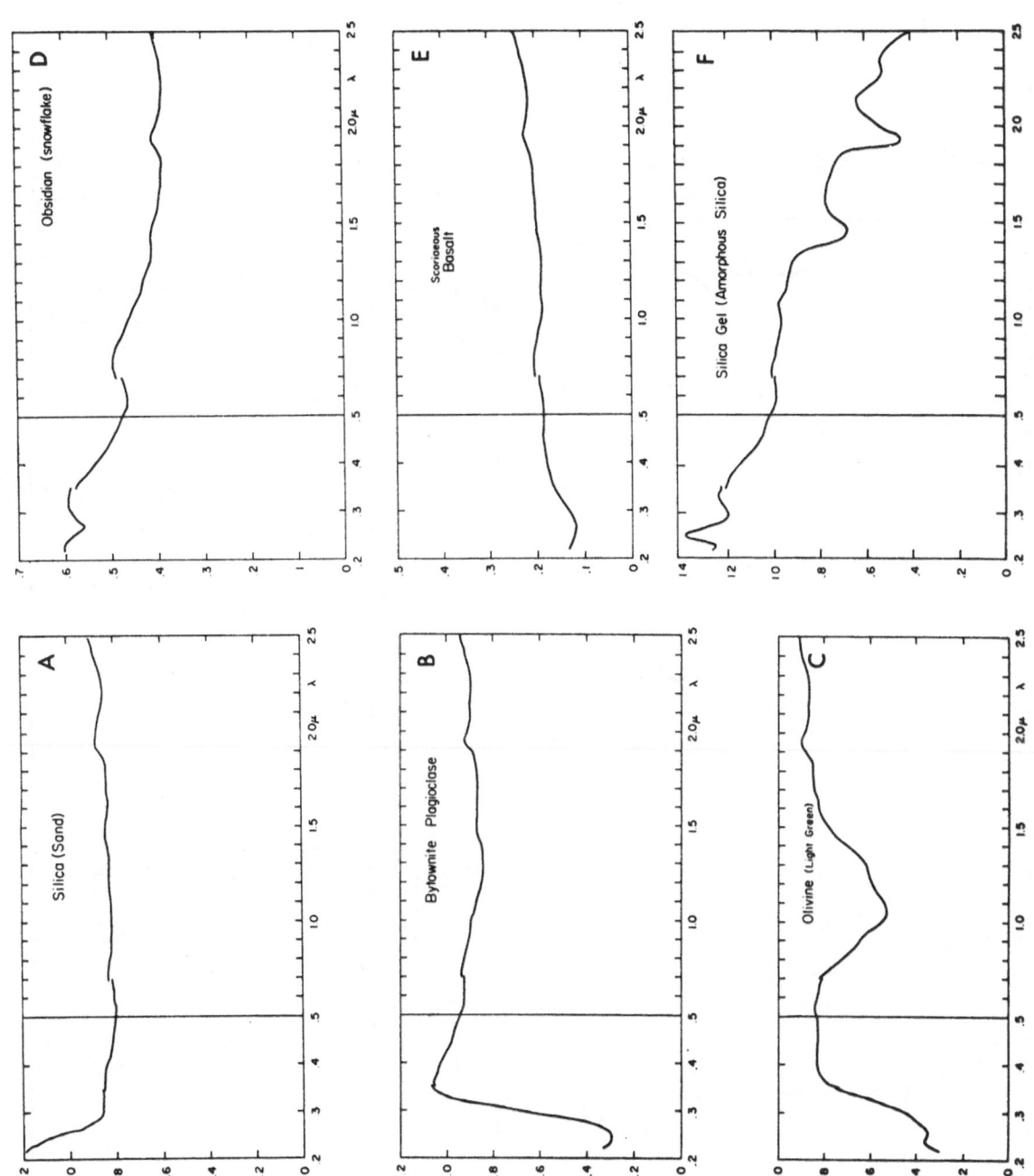

Fig. 10. Reflectance measurements of possible solid constituents in Venus atmosphere made by G. Sill (O. Carm.) with a Zeiss Reflectance Spectrophotometer, using a Li F (white) standard.

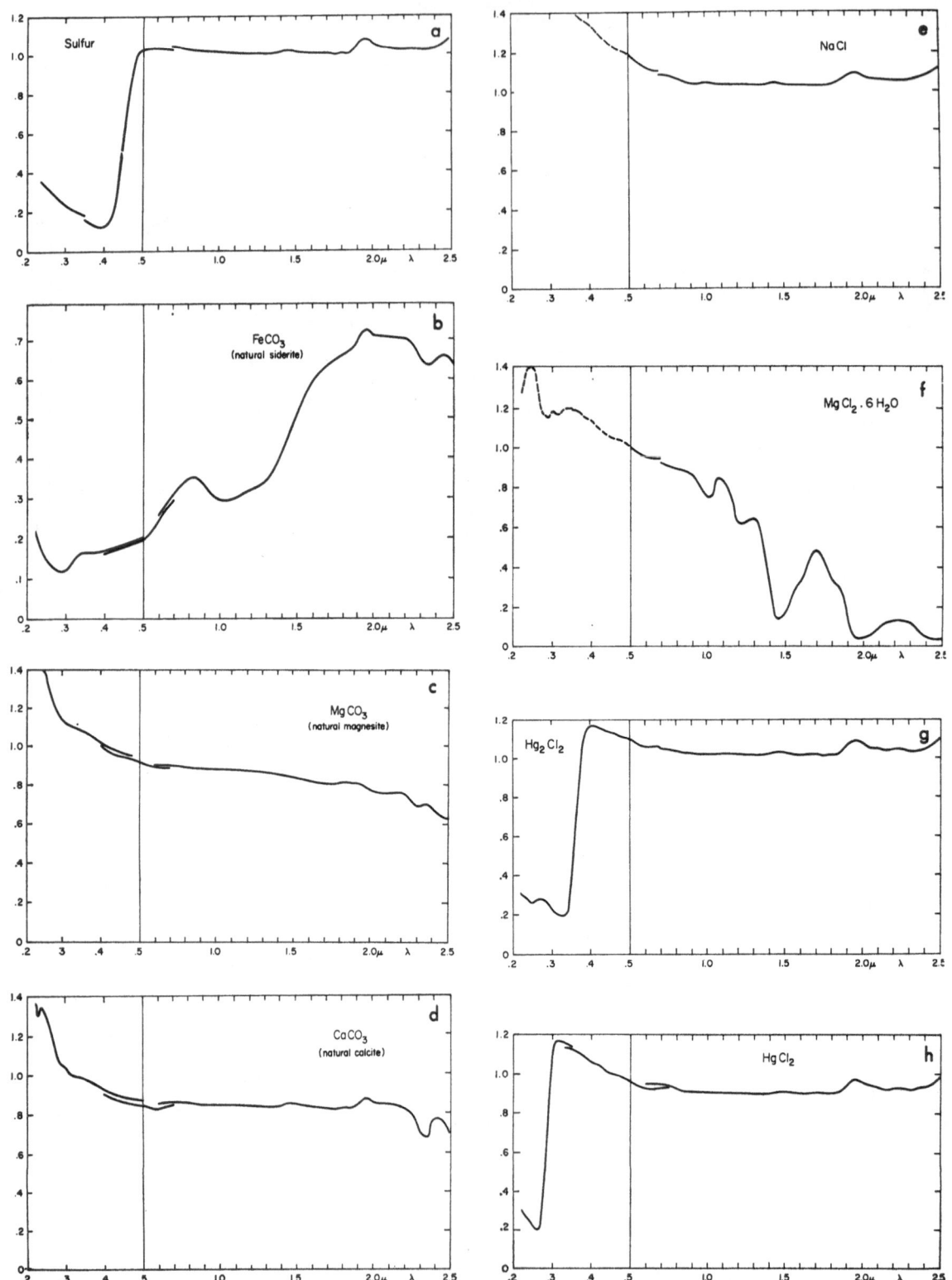

Fig. 11a.

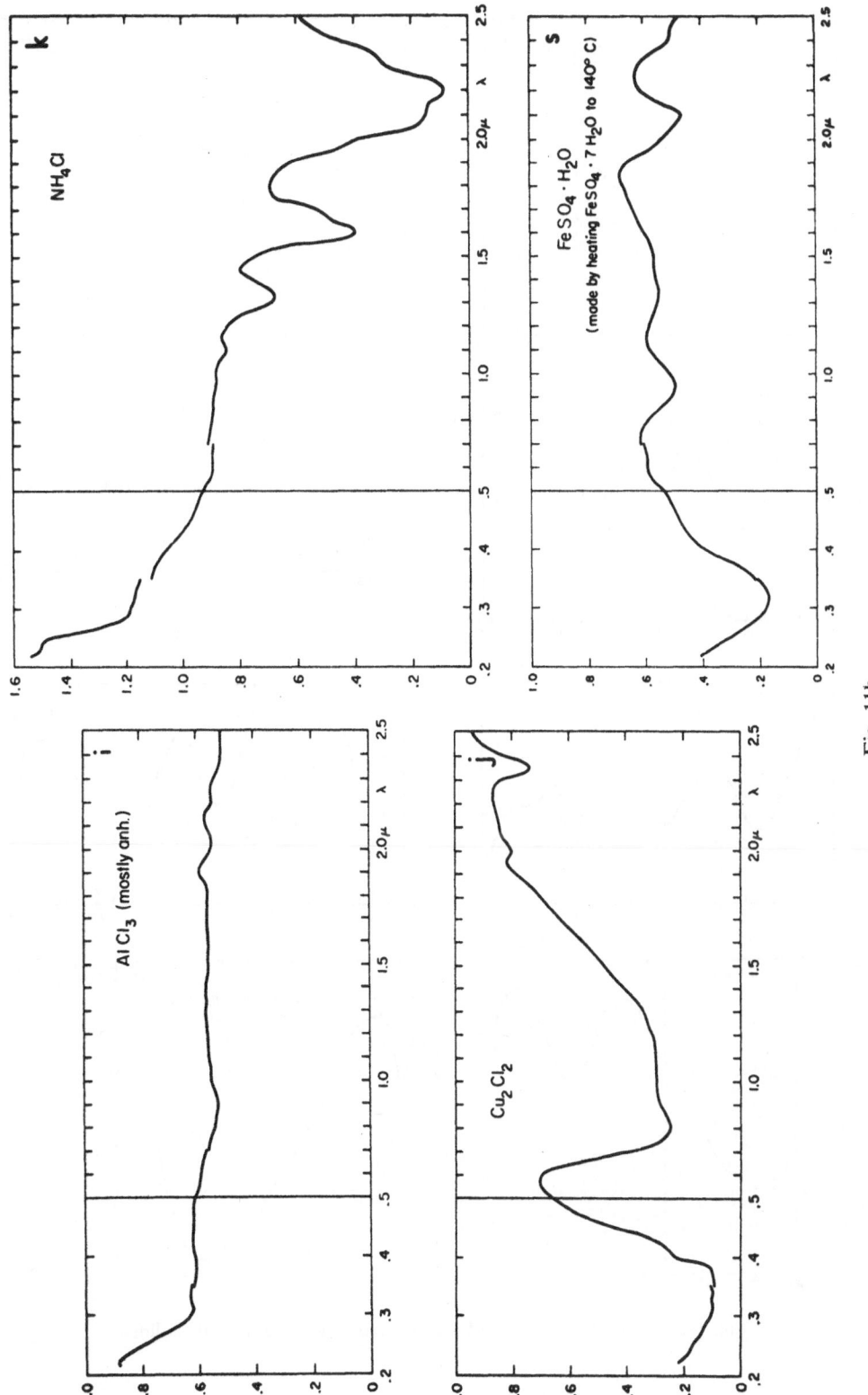

Fig. 11b.

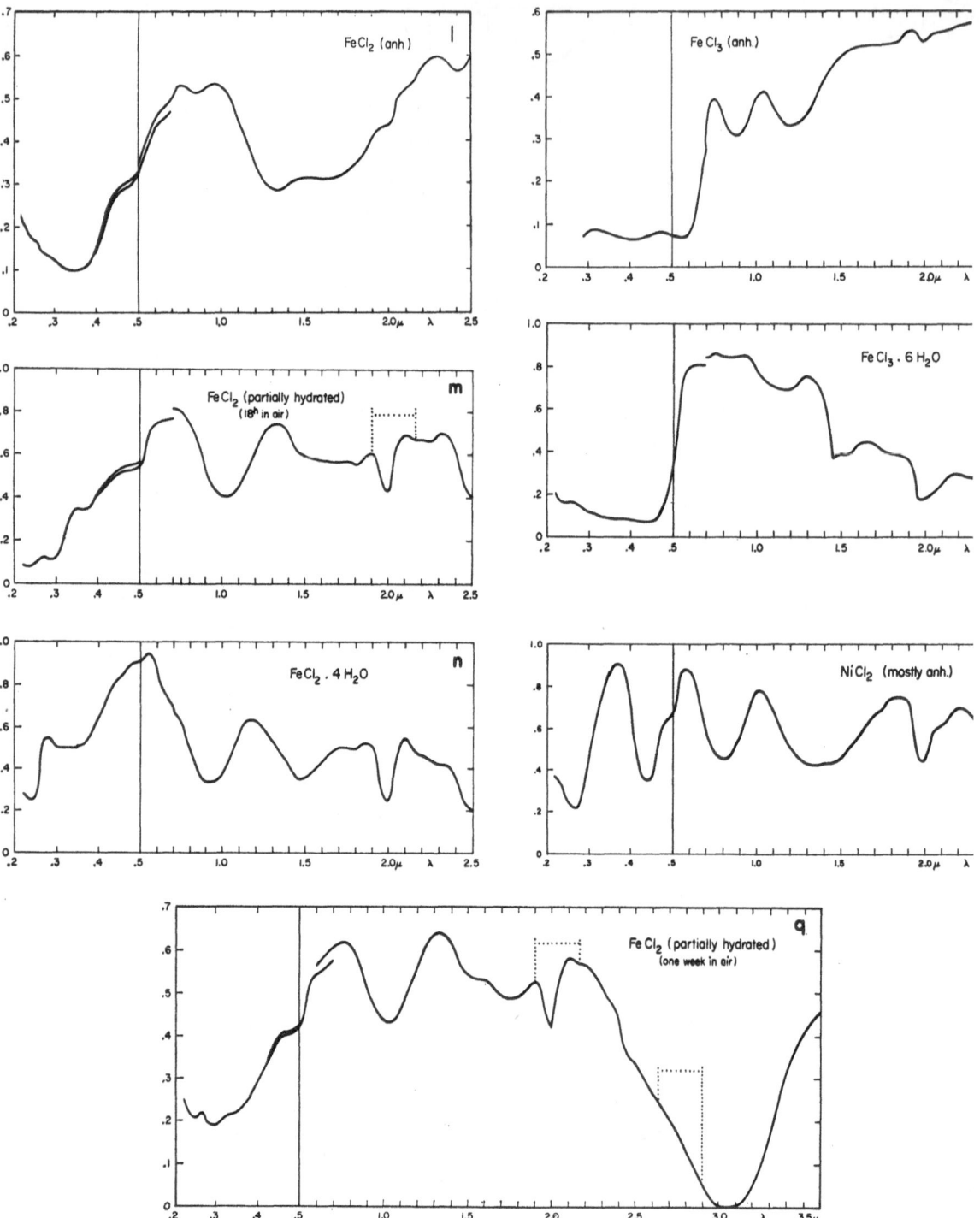

Fig. 11c.   Reflectance curves of sulfur, carbonates and metallic halides.

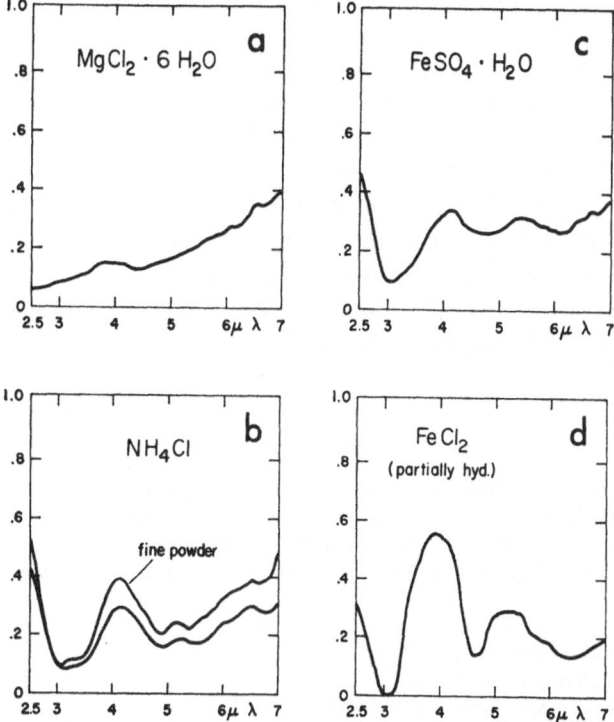

Fig. 12.   Reflectance curves, 2.5–7 μ for four substances.

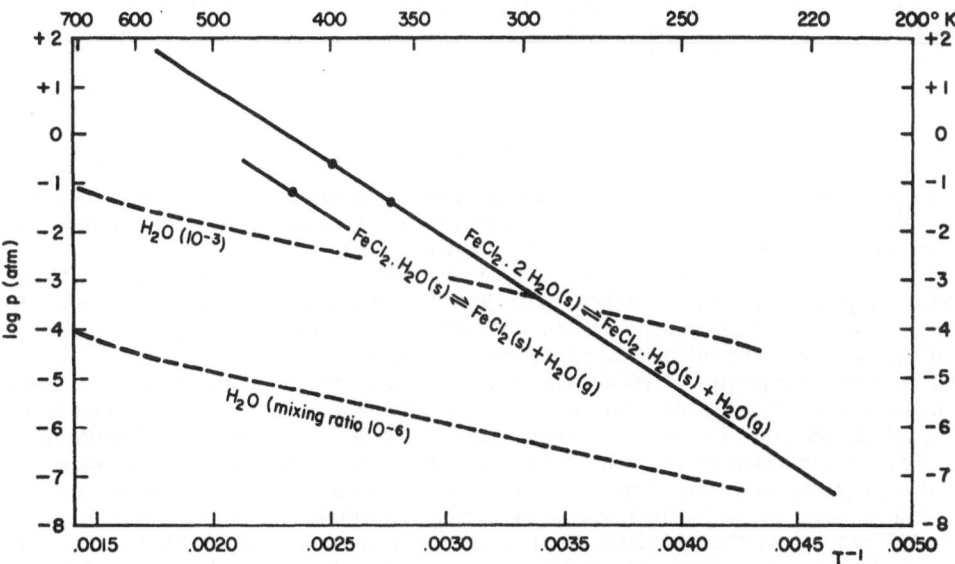

Fig. 13.   Dissociation pressures for hydrated ferrous chloride in p, T diagram, also showing two H_2O mixing ratios for Venus atmosphere.

about 1.55, is in agreement with the results obtained by Coffeen for the Venus cloud particles [6].

A clinching argument favoring the identification of partially hydrated $FeCl_2$ is found from the remarkable agreement between the *dissociation pressure of H_2O* for the cloud material (cf. Figure 13) and the actual water-vapor pressure in the observable layers of the Venus atmosphere. If the atmosphere water-vapor content were, e.g., 100 times larger, the clouds would be tetrahydrate (Figure 11c (*n*)); if the vapor pressure were 10 times smaller, one would expect $FeCl_2$, anhydrous (Figure 11c (*l*)). Clearly, neither one satisfies the Venus reflection curve.

There are some reasons to suspect that the UV cloud layer is composed of $NH_4Cl(s)$ but present evidence is insufficient to establish this [4].

References

[1] CalTech-JPL Lunar and Planetary Conference, Sept. 13–18, 1965.
[2] *Handbook of the Physical Properties of the Planet Venus*, NASA, SP 3029, 1967.
[3] Kuiper, G. and Forbes, F.: 1967, *Comm. LPL* No. 95, **6**, 177.
[4] Kuiper, G., Forbes, F., Steinmetz, D., and Mitchell, R.: 1968–69, *Comm. LPL* No. 100, **6**, 209.
[5] *Comm. LPL* No. 95, p. 187.
[6] Coffeen, D.: 1968, *Astron. J.* **74**, 446.
[7] Ross, F.: 1928, *Astrophys. J.* **67**, 57.
[8] Wright, W.: 1927, *Publ. Astron. Soc. Pacific* **39**, 220.
[9] Boyer, C. and Camichel, H.: 1967, *Ann. Astrophys.* **24**, 531; 1965, *Compt. Rend. Acad. Sci. Paris* **260**, 809.
[10] Boyer, C.: 1965, *Astronomie* **79**, 223.
[11] Boyer, C. and Guérin, P.: 1966, *Compt. Rend. Acad. Sci. Paris* **263**, 253.
[12] Kuiper, G.: 1968–69, *Comm. LPL* No. 101, **6**, 229.
[13] Connes, *et al.*: 1967, *Astrophys. J.* **147**, 1230–1237; Connes, J., Connes, P., and Maillard, J.: 1969, 'Near Infrared Spectra of Venus, Mars, Jupiter, and Saturn', CNRS, Paris.
[14] Moroz, V.: 1965, *Soviet Astron.-A.J.* **8**, 566.
[15] de Vaucouleurs, G.: 1969, *Icarus* **3**, 187.

Discussion

Morrison: There have been new measurements from three spacecraft of water vapor in the lower atmosphere. We suggest that some tenths of a percent water is compatible with some of the radio observations. Would you object to the presence of water clouds?

Kuiper: What I would object to is this: that one would make hypotheses believing that Venus must be a replica of the Earth. Certainly Jupiter has a very different composition from the Earth. Venus may have originated somewhat later than the Earth and therefore collected fewer volatiles. I would say that the identification of the visible clouds is definite.

However the vapor pressure of ferrous chloride at $T = 240$ K is about 10^{-23} atm. So there is a transportation problem for the particles. And therefore, what is found in the top layer must have a bearing on what is further down. The presence of water clouds below the visible layer is very improbable.

L. Young: I want to make a comment on the 7820 Å carbon dioxide band that Spinrad measured, and that you asked about. One of the reasons he thought he found a very high temperature is that he was assuming it was on the linear part of the curve of growth. Now we've made measurements from 1967 on; we have about 24 plates. And in general the temperature is about 230 K, and we did not find any high temperatures. I believe 256 K was the highest temperature that we found, during that year. The phase angles varied from 26° to 116°. I really believe that Spinrad's high temperatures are a result of the assumption about the linear part of the curve of growth.

Kuiper: Well, I'm very happy to know that this puzzle has been solved.

Boyce: Did you in your diameter measurements take any account of the difference in size of the seeing disk between the red and blue images?

Kuiper: You see, we didn't use a disk. We used an extremely narrow crescent. I would never dare use a disk.

Suess: Did you do this experiment with a large excess of CO_2 for the laboratory absorption?

Kuiper: No, we did not.

GEOCHEMICAL PROBLEMS IN THE PRODUCTION
OF THE VENUS CLOUDS

GODFREY T. SILL

Lunar and Planetary Laboratory, University of Arizona, Tucson, Ariz., U.S.A.

Abstract. Spectroscopic evidence is strong that the clouds of Venus are composed of ferrous chloride hydrate. At the surface of the planet it appears as though the vapor pressure of ferrous chloride may be quite low (10^{-6} atm), especially if such vapor is allowed to come to equilibrium with the other constituents of the Venus atmosphere. The formation of magnetite (Fe_3O_4) may be favored, and the spectrum of magnetite is shown in relation to the spectrum of the Venus cloud cover. The presence of ferrous chloride in the upper atmosphere may be due to a non-equilibrium condition.

The spectrum of volatiles, such as the halides of mercury, are considered vs the albedo of Venus. The presence of these volatiles seems incompatible with the spectroscopic evidence. It is suggested that the planet has been depleted of its volatiles, as it has been depleted of its water.

1. Spectroscopic evidence in the region of $0.2\,\mu$ to $4.0\,\mu$ indicates that the clouds of Venus are composed of hydrated ferrous chloride (Kuiper, 1969, *Comm. LPL*, no. 101; cf. also this symposium). The clouds of Venus accessible to observation in the near UV, visible, and near IR are those of the upper atmosphere at a height of 60 km and a temperature of about 250 K. Any cloud material must be influenced by production rates of the material near the surface of the planet, just as the water vapor clouds of earth are dependent on evaporation rates near the surface.

Anhydrous ferrous chloride is not known for its high volatility, even at the surface temperature of Venus, 700 K. Extrapolation along the modified Clausius-Clapeyron equation of vapor pressure of $FeCl_2$ yields a vapor pressure of approximately 10^{-5} atm at 700 K. Using more recent vapor pressure and thermodynamical data, Brewer *et al.* (1963) calculate that the vapor pressure of $FeCl_2$ would be 10^{-6} atm at 700 K. This low vapor pressure causes problems in creating an abundant Venus cloud cover, especially if the vapor of $FeCl_2$ is allowed to come to equilibrium with the surface rocks of Venus. Using the thermodynamic data of Brewer *et al.* (1963) and Lewis and Randall (1961), the author has calculated the equilibrium pressure of $FeCl_2$ as a very low $10^{-7.5}$ atm in the reaction:

$$\tfrac{1}{3}\,Fe_3O_4(s) + 2\,HCl(g) + \tfrac{1}{3}\,CO(g) = FeCl_2(g) + H_2O(g) + \tfrac{1}{3}\,CO_2(g).$$

where $Fe_3O_4(s)$ would be magnetite on the surface of Venus, or perhaps granular sized in the atmosphere, and where $FeCl_2(g)$ would be the quantity of ferrous chloride produced by the constituents of the Venus atmosphere. The abundances used in the calculations were:

$$P_{H_2O} = 10^{-4}\ \text{atm (from mixing ratio of } 10^{-6})$$
$$P_{HCl} = 10^{-4.2}\ \text{atm}$$
$$P_{CO} = 10^{-2.3}\ \text{atm}$$
$$P_{CO_2} = 10^2\ \text{atm}$$

Sagan et al. (eds.), Planetary Atmospheres, 110–115.

With the same abundances, the equilibrium vapor pressure of $FeCl_2(g)$ in the following reaction yielded a pressure of 10^{-6} atm.

$$FeO(s) + 2\,HCl(g) = FeCl_2(g) + H_2O(g).$$

In any case, if $FeCl_2$ is produced through the reaction of HCl on the surface rocks of Venus, the quantity of vapor so produced is quite small, and seems to be unable to account for a Venus cloud or haze layer at the observed level of 60 km.

The thermodynamic data used is not completely certain; the calculated $FeCl_2$ vapor could differ from the above by, perhaps, an order of magnitude or so. In any case the vapors of $FeCl_2$ may not necessarily be produced by reaction with surface rock. If the vapor is liberated into the Venus atmosphere the above reactions may not specify the true state of affairs. If the oxide is produced by the reaction of ferrous chloride vapor and water vapor, then the oxide should initially be treated as a vapor and not as a solid, or perhaps microcrystalline aggregates of oxide analogous to a vapor. This state of the oxide would undoubtedly favor the production of $FeCl_2$ vapor, instead of decreasing it. In any case the author intends to investigate this problem in the laboratory.

As quoted by Kuiper (1969) $FeCl_3$ is a common volcanic emanation, especially in the Hawaiian group. Any $FeCl_3$, if not already decomposed into $FeCl_2$ and chlorine, would soon be reduced in the Venus atmosphere to $FeCl_2$. The high vapor pressure of $FeCl_3$ at 700°, converted to $FeCl_2$ would yield virtually all solid $FeCl_2$ with a minor amount present as vapor. The solid $FeCl_2$ would be quite stable in the Venus atmosphere and with any vertical transport mechanism the $FeCl_2$ particles should be easily raised to higher altitudes. Admittedly the explanation is contrived, but does not seem unreasonable. At the higher elevations and cooler temperatures, the $FeCl_2$ would become hydrated, and may perhaps be the dominant influence in controlling the pressure of $H_2O(g)$ in the Venus atmosphere.

If any oxides of iron are produced in the Venus atmosphere, and are of sufficiently small particle size to be buoyed up to the upper atmosphere, they do not seem to show themselves spectroscopically. The spectral reflectances of several oxides of iron are shown in Figure 1. These do not match the Bond albedo of Venus, and Fe_3O_4 as a matter of fact, would serve to uniformly decrease the albedo of the planet. (FeO spectral reflectivity was not measured, but being a black solid like Fe_3O_4 is presumed to have a similar reflectivity.)

2. It has been suggested by Lewis (1969) that the cloud layers of Venus could be composed of various compounds of mercury and elemental mercury itself, each of the cloud materials condensing at a level where their concentration exceeds their vapor pressure. The lowest level is formed from HgS, then succeeded at higher levels by Hg_2Br_2, Hg_2I_2, Hg, and the uppermost haze level of 240 K composed of Hg_2Cl_2 overlying a deep cloud of liquid mercury droplets. Figure 2, g and h, shows the spectral reflectivity of the chlorides of mercury and Figure 3 the bromides and iodides of mercury. They all show the same type of reflectivity – low in the near UV (and blue for some), with a sharp rise to around 100% reflectivity, which then continues out to

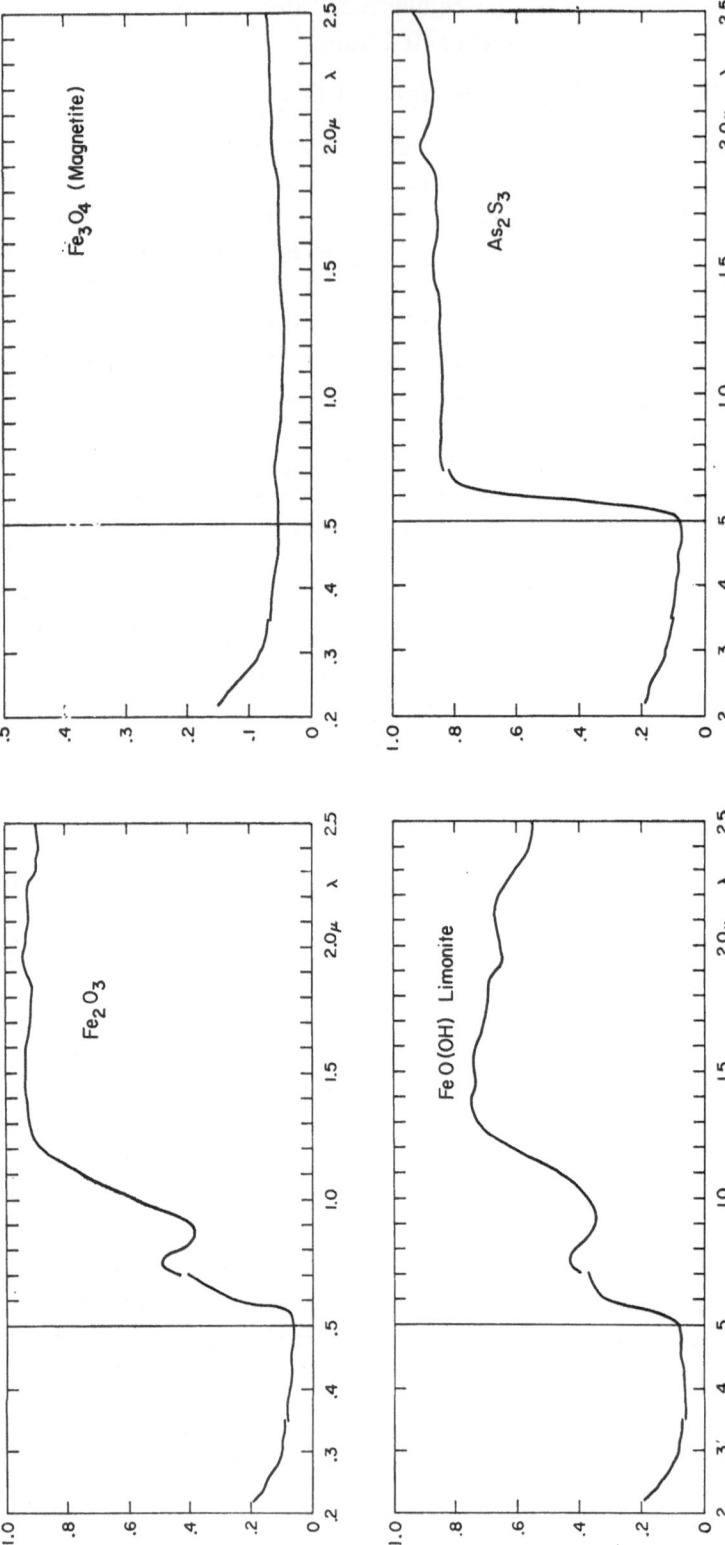

Fig. 1. Spectral reflectivities of oxides of iron, from 0.2 to 2.5 μ. All measurements are vs a white standard, LiF.

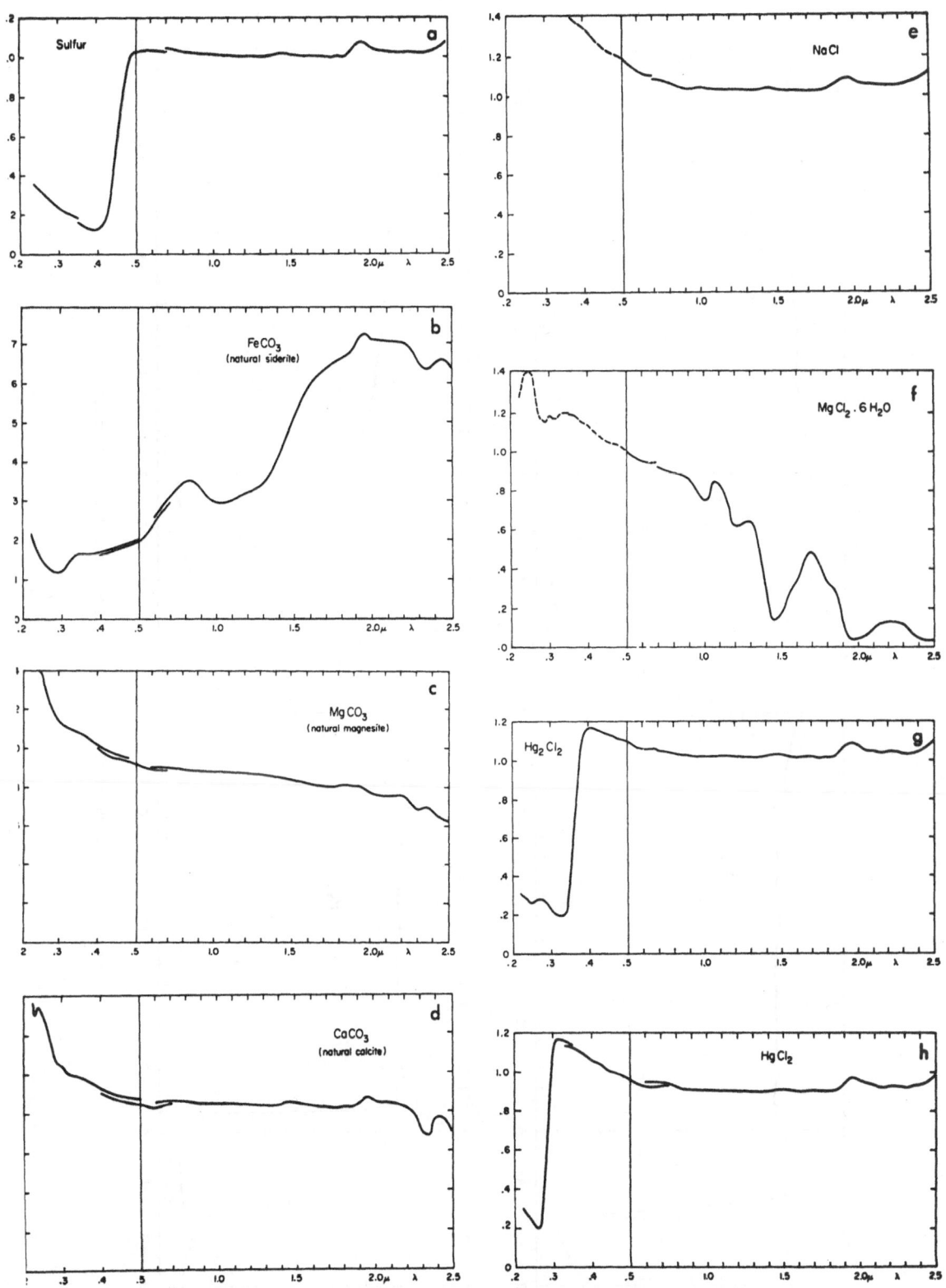

Fig. 2. Spectral reflectivities of the chlorides of Mercury, g and h.

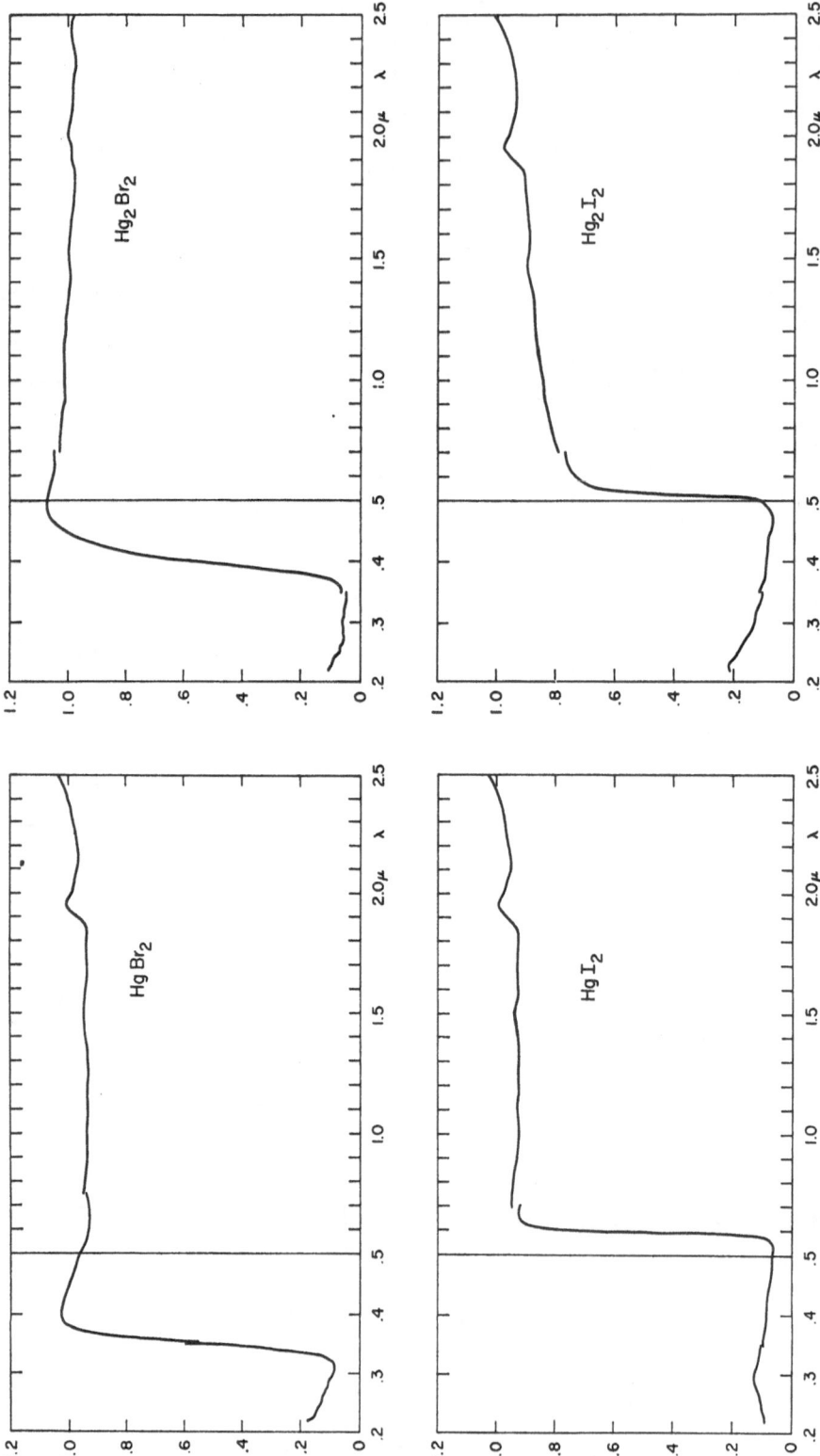

Fig. 3. Spectral reflecivities of the bromides and iodides of Mercury.

2.5 μ in the near IR. None of these has the characteristic features of the Venus Bond albedo, and it would seem that their presence is not proven observationally.

As Anders (1968) points out, Venus might have passed through a thermal history quite different from the earth. If, as he notes, Venus in its accreting stages did not fall below 400 K until late in its formation interval, then the planet would have been depleted of many volatiles, such as the H_2O he mentions. The same could quite easily be applied to the element mercury. Therefore it may not be valid to take the abundance of Hg from the earth's crust and apply it to Venus. There may not be clouds of Hg compounds in the atmosphere of Venus simply because there is too little Hg available.

In conclusion, there do seem to be difficulties in producing clouds of ferrous chloride hydrate in the atmosphere of Venus, but these problems can be eased if vertical mixing is fairly rapid in the atmosphere of Venus, and if the whole atmosphere is not in equilibrium with the planetary surface. Other more volatile compounds seem at first sight, to be more favorable constituents of the clouds of Venus, but elemental abundances and depletion factors must be taken into account, as well as the primary evidence of spectroscopic observation of the planet.

Note Added in Proof. A reaction similar to the first one described in this article was tested in the laboratory. Ferrous oxide was treated with vapors of hydrogen chloride at 700 K to see whether ferrous chloride would be produced according to the reaction:

$$FeO(s) + 2HCl(g) = FeCl_2(g) + H_2O(g).$$

Nitrogen gas was bubbled through concentrated hydrochloric acid at a rate of 10 cc/min, and passed over FeO in a furnace at 700 K. The cool glass outlet tubing, where it left the furnace, began to show traces of green $FeCl_2 \cdot 4H_2O$ within one hour's time. Despite the unfavorable thermodynamics of the reaction, the reaction did go, undoubtedly due to the fact that equilibrium conditions did not prevail: evidently the $FeCl_2$ sublimed out of the reaction vessel as soon as it was formed. However, this may also be the case on Venus with a wind blowing across the surface. Equilibrium conditions may not prevail between the surface of the planet and a moving atmosphere.

References

Anders, Edward: 1968, 'Chemical Processes in the Early Solar System, as Inferred from Meteorites', *Act. Chem. Res.* **1**, 289–298.

Brewer, Leo, Somayajulu, G. R., and Brackett, E.: 1963, 'Thermodynamic Properties of Gaseous Metal Dihalides', *Chem. Rev.* **63**, 111–121.

Kuiper, G. P.: 1969, 'Identification of the Venus Cloud Layers', no. 101, *Commun. Lunar Planetary Lab.* **6**, 4.

Lewis, G. N. and Randall, M.: 1961, *Thermodynamics*, 2nd ed., McGraw-Hill.

Lewis, John S.: 1969, 'Geochemistry of the Volatile Elements on Venus', *Icarus*, in press.

THE TROUBLE WITH VENUS

CARL SAGAN

Laboratory for Planetary Studies, Cornell University, Ithaca, N.Y., U.S.A.

Venus is the closest planet. Its surface has never been seen at optical frequencies; nevertheless we now know with at least fair reliability, and in some cases with remarkable accuracy, its surface temperature and pressure, its atmospheric structure, its period of rotation, the obliquity of its rotation axis, the mean surface dielectric constant, its ionospheric structure, and even a little about its surface topography. And yet the clouds of Venus, visible to the naked eye and known to be clouds since the time of Lomonsov, continue to elude our efforts to understand them comprehensively. Not only do we disagree on the chemical composition of the clouds, but it is not even settled whether they are condensation clouds or non-condensable aerosols. And yet there is a very wide variety of relevant data on the clouds. Indeed, the ratio of potentially diagnostic data points to mutually exclusive hypotheses is of the order unity.

This is not the first time that studies of Venus have been beset with troubles, with a simultaneous multiplicity of hypotheses and data. The debate on the nature of the Venus microwave emission bears some interesting parallels, and might conceivably serve as a methodological guide, to studies of the Venus clouds. The natural first explanation of the high microwave brightness temperature of Venus, discovered by Mayer *et al.* (1958) in 1956, was that we are observing thermal emission from a hot surface. But this straightforward hypothesis was greeted by considerable skepticism – some of which, I believe, has psychological rather than scientific roots. If Venus really had a thermometric temperature in excess of 600 K, then a variety of pleasant possibilities – a habitable, ocean-covered Venus, for example – would be removed from the field of reasonable discourse. As new observations of Venus were performed, new varieties of non-thermal explanations for the microwave emission emerged. It is instructive here to consider the evolution of just one of the non-thermal hypotheses, the ionospheric model.

In the ionospheric model, free-free emission in a dense Cytherean ionosphere is invoked to explain the microwave spectrum. The free-free optical depth is given by $\tau \propto \lambda^2 T_e^{-3/2} \int N_e^2\, dz$, where λ is the wavelength of the emitted radiation, T_e the electron temperature, N_e the mean ionospheric electron density, and z the vertical dimension. The proportionality factors are known. The microwave spectrum, as it seemed to be some years ago, could be explained on this model; setting $\tau = 1$ at $\lambda \simeq 1$ cm, we find that at wavelengths much longer than 1 cm we are observing a constant brightness temperature at the electron temperature, T_e, which is specified at about 600 K. Shortward of 1 cm we look through the now optically thin ionosphere to the lower atmosphere and surface, which to match the microwave spectrum must have low and possibly even congenial temperatures. This was the attractive feature of the model in the original formulation by Jones (1961). But when this model is compared with the

Sagan et al. (eds.), Planetary Atmospheres, 116–128.

range of information known about Venus before the space vehicle successes of mid-October 1967, it runs into an impressive variety of independent embarrassments. First, the value of $\int N^2_e \, dz$ is specified by the requirement for a fit to the microwave spectrum. The resulting values of the electron density are 10^9 or 10^{10} electrons cm^{-3}, some three or four orders of magnitude greater than the peak ionospheric electron density on the Earth. Even under the most generous assumptions, it is difficult to justify such high electron densities on Venus. Secondly, an optical depth of unity at 1 cm wavelength implies an optical depth at 68 cm wavelength, say, of $(68)^2 = 4624$. But, in fact, radar reflectivities of 10 or 15% were known at 68 cm wavelength. A third and weaker piece of conflicting evidence is the fact that the interferometric radius of Venus is less than the optical radius of Venus, implying that the emitting region lay below the clouds – consistent with the hot surface model – rather than above the clouds, as is evidently required by an ionospheric model.

Finally, the requirement of optical depth unity near 1 cm implies that Venus should exhibit limb brightening at wavelengths near 1 cm – because, as we scan from the center of the disk towards the limb, we are looking through longer paths of the emitting layer. On the other hand, if we were to believe the hot surface model, the opacity at 1 cm must be attributed to the atmosphere or clouds – this was realized more than a decade ago – and Venus at 1 cm should exhibit limb darkening and not limb brightening. Interferometric studies of Venus at such short wavelengths were impossible in the 1960's and remain impossible today. The question of limb darkening or limb brightening at 1 cm wavelength could, therefore, be most expeditiously settled by flying a small microwave radiometer to the vicinity of Venus and crudely scanning across the disk – an objective of the Mariner II mission to Venus. This is, incidentally, an excellent example of a critical experiment, performed by space vehicle, that could not be performed from the surface of the Earth. Despite time-constant problems and calibration difficulties, particularly on the 13.5 mm channel of the Mariner II radiometer, the 19 mm channel showed distinct limb darkening consistent with the hot surface model and inconsistent with the ionospheric model (Barath *et al.*, 1964).

With such an array of data opposed to the ionospheric model, it would seem that the model would have had no adherents by the mid-1960's. This, however, was not the case. It is always possible to make *ad hoc* revisions of a simple theory, in order to keep up with the accretion of embarrassing data. Thus, it was proposed that ion-electron recombination in the Venus ionosphere might be sufficiently rapid for there to be a hole in the center of the night hemisphere, large enough for the first radar Fresnel zone to penetrate the ionosphere at inferior conjunction and be reflected by the surface but small enough not to compromise the passive microwave emission coming primarily from the ionosphere. The limb darkening difficulty could be explained away by postulating a sufficiently ingenious array of ionized layers, strategically placed. And, in one attempt to come to grips with the high electron density implied for free-free emission, it was suggested that the bulk of the microwave emission comes, not from the interaction of charged particle with charged particle, but from the interaction of charged particle with the instantaneous Coulomb field of molecules in the neutral atmosphere. This idea requires very frequent collisions to explain the observed micro-

wave intensity. The required density of neutral atmosphere would place the Venus
ionosphere well below the visible cloud tops, roughly at the 1 bar pressure level. But at
such depths adequate ionization either from solar electro-magnetic or solar particulate
radiation is not to be expected. Accordingly, it was seriously proposed in one paper
that the required high ionization rate deep in the clouds of Venus may be due either
to a significantly greater primary cosmic ray flux at Venus than at Earth, or to many
orders of magnitude more radioactive materials in the atmosphere of Venus than
of Earth – evidently the result of a Cytherean nuclear war. These examples illustrate
an understandable and very human tendency towards the selective rejection of dis-
quieting data, and the lengths to which a sufficiently desperate theoretician may be
driven. The failure of the ionospheric model may not be merely of academic interest:
the epic-making Soviet entry vehicle, Venera 4, appears to have been crushed by the
weight of the overlying atmosphere at the 550 K, 20 atmosphere level. Might this be
due to a too confident reliance on the ionospheric model – which implied, because of
the low surface temperatures, pressures of at most a few atmospheres?

Before the Venera 4, and Mariner V successes, it was possible to construct a truth
table which compared the range of observations with the range of models. The
observables in question were the microwave spectrum, the phase effect (about which,
more in a moment), interferometric observations, the Mariner II limb darkening
observations, the radar spectrum and the radar diameter. Tested against these observations were models of synchrotron or cyclotron emission, the free-free emission model
just discussed, a glow discharge model, a droplet electrical discharge model, and the
hot surface model. Only the hot surface model survived analysis by such a truth table
(Sagan, 1967, 1969).

Another interesting implication of the debate on the Venus microwave emission is
the possible existence of false positives. The reported variation of the microwave
brightness temperature with phase angle at 3.15 and 10 cm wavelength had precisely
the form anticipated from the solution of the one-dimensional equation of heat
conduction for an impressed sinusoidal thermal wave. The time-independent components of the brightness temperature at the two wavelengths were the same, as they
should be; the amplitude of the time-dependent term declined with depth and the
phase lagged with depth, again as the theory of heat conduction predicts. This seemed
to be so clearly what is expected for a hot surface heated by sunlight that a detailed
analysis of the data appeared warranted (Pollack and Sagan, 1965); the analysis gave
values for the sense of rotation of Venus, the obliquity of its rotation axis, the dielectric constant, and the ratio of electrical to thermal skin depths of the surface
material which were entirely consistent with other information – primarily radar
data.

Nevertheless, and in spite of great care taken by the observers to insure the reality
of the reported phase effect, the observations appear to be in error. Observations at
wavelengths of 4.15 cm and 2 cm, respectively (Dickel et al., 1968; Morrison, 1969),
and a more recent set of 3 cm observations at the Naval Research Laboratory (Mayer,
1969) fail to show any statistically significant phase effect whatever. Indeed, because
of the large heat capacity and infrared opacity of the massive atmosphere known to

exist on Venus, a detectable microwave phase effect is not to be expected – a point which also emerges in a novel context in Golitsyn's (1970) similarity theory of the Venus atmospheric circulation. And the possibility that both the early observations, which show a phase effect, and the more recent observations, which do not, are correct – entailing very special emissivity distributions over the disk, or the outgassing of many tens of atmospheres in less than a decade, or an extremely prolate and nutating planet – seem too desperate even for Venus. We seem to be faced with the possibility that Nature and radioastronomical techniques have contrived a noise spectrum which simulates the solution of the one-dimensional equation of heat conduction.

It seems to me that a related procedure, also using a truth table, and bearing in mind the possibility of false positives, might be very useful in studies of the vexing questions of the nature and composition of the Venus clouds. Even though the relevant observables about Venus are not in as good an observational shape as could be desired, they can be used to test the various models. Among the observations and consistency checks which seem relevant are the following: (1) the absorption in the blue and ultra-violet, which gives Venus a naked eye color which is approximately a pale lemon yellow; (2) the very high visible albedo of Venus; (3) the near infrared spectrum of Venus, which is approximately flat between $0.7\,\mu$ and about $2\,\mu$ (although there is some dispute about the reality of features reported at 1.5 and $2\,\mu$), and a precipitous decline longward of $2\,\mu$ to an extremely low albedo which appears to remain constant between 3 and $4\,\mu$; (4) the microwave spectrum between a few millimeters and some meters, which now appears to have a peak in the vicinity of 6 cm; (5) the radar spectrum; (6) direct chemical investigations by Veneras 4, 5, and 6; (7) possible attenuation layers in the Venus atmosphere which may have been observed by the Mariner V S-band occultation experiment; (8) consistency of proposed compositions with the pressures and temperatures known near the clouds; (9) the dependence of the polarization of scattered light on phase angle and on wavelength; (10) the dependence of the geometric albedo on phase angle and on wavelength, as in the halo effect; (11) the density of scatterers in the clouds, as determined from near infrared spectroscopy; and (12) the question of support of aerosol particles, in the case of non-condensable clouds. There are also questions of (13) the geochemical plausibility and (14) the cosmic abundance of proposed cloud constituents.

To be run against these observables in the truth table are at least the following proposed cloud constituents: (a) 'dust' – that is, various geochemically common silicates and oxides; (b) hydrocarbons; (c) carbon suboxide and its polymers; (d) ammonium chloride; (e) various mercurous halides; (f) ferrous chloride dihydrate; (g) polywater; (h) water ice. All these materials have been seriously proposed as principal constituents of the Venus clouds. While, for the truth table analysis of the microwave emission, it did not turn out that mixed hypotheses were necessary – that a significant contribution to the microwave emission came from two conceptually distinct sources – it is nevertheless possible that 'the' clouds are not composed entirely of one material, or that there is more than one cloud layer, different layers contributing in different proportions to the parameters (1) through (14) above. I think the time is rapidly approaching when a systematic truth table analysis, involving (1) through (14) and (a) through (h), can be drawn up. I do not propose to make such a truth table here, but I cannot resist the temptation to discuss how ice clouds would fare in such an analysis, and to make a few critical comments on some of the other proposed cloud constituents:

In order to support ice crystal clouds at approximately the 230 K temperature level (corresponding to a few tenths of a bar pressure), it is easy to show that the present Venus atmosphere below the clouds and far from saturation must consist of a few tenths of a percent water vapor mixing ratio by volume. This is just the amount announced by independent observations aboard Veneras 4, 5, and 6. While there is some cause for skepticism, particularly about the Venera 4 oxygen measurements, this does not seem to be the case for the water vapor measurements, as discussions at conferences in Marfa, Wood's Hole, and elsewhere clearly indicate. I think we must take seriously the Venera water vapor results. Next, it has long been known that a few tenths of a percent water vapor mixing ratio is the sort of value needed, along with CO_2, to explain the high surface temperature of Venus via the greenhouse effect (Sagan, 1960); and this conclusion has recently been confirmed in detail by Pollack (1969) and by Ohring (1969). As we have heard at this meeting (Pollack and Morrison, 1970), the microwave spectrum of Venus in the vicinity of 13.5 mm is definitely consistent with, although not uniquely indicative of, a water vapor mixing ratio of a few tenths of a percent. In addition, the radar observations require an additional opacity source, consistent with a few tenths of a percent water vapor (Kroupenio, 1970). Note that the Venera measurements, the greenhouse model calculations, the microwave resonance line and the radar spectra are the only means now available for probing the water vapor content of the lower atmosphere of Venus. They all give results consistent with ice clouds on Venus.

The infrared spectrum of the Venus clouds is very close to that of ice crystals with particle sizes in the few micron range; in particular, the deep and flat absorption longward of 3 μ strongly indicates the presence of condensed rather than bound water (Pollack and Sagan, 1968). Plummer (1970b) argues that the 2 μ ice band is present also, implying particle dimensions of several microns. The high visual albedo is just in the range expected for multiple scattering in ice clouds with particle dimensions of a few microns, and the cloud density once advertised as much too diffuse for consistency with terrestrial ice clouds now emerges – after allowance for anisotropic multiple scattering is made – as precisely in the range of terrestrial cirrus clouds (Potter, 1969a; Hansen, 1969). A Venus halo effect, characteristic of hexagonal ice crystals, has been sought for by O'Leary (1967) near phase angle $180° - 22° = 158°$. O'Leary found a marginal positive result from his earlier measurements, and a somewhat more definite identification in his more recent efforts (O'Leary, 1970). He finds that a larger halo effect would have been observed if $>5\%$ of the multiply scattering clouds are composed of pure hexagonal ice crystals of radii $>$ a few microns, where diffraction is small. But since the infrared spectra imply a particle size distribution function peaked at a few microns, and since the clouds are surely not without contaminants (see below), O'Leary's observations seem to imply ice crystals as a major cloud constituent. The increased reflectivity at a given phase angle depends not only on the real part of the refractive index, but also on the geometry of the scattering crystals. Veverka (1971) has searched unsuccessfully for a polarimetric halo effect at 158°; but his sensitivity was such that his negative findings are not in contradiction to O'Leary's results. There is no inconsistency with water ice clouds in the radar spectrum or in the occultation measurements.

The ultraviolet albedo of water clouds is much higher than that of Venus; the real part of the refractive index as extracted from polarization measurements by Coffeen (1968) is > 1.43 compared to about 1.3 for pure ice. However, both these observations can be very trivially made consistent with ice clouds if we include an admixture of dust particles, particularly those containing iron which gives a strong ultraviolet absorption and a large refractive index. Indeed, the rings of Saturn, which now appear clearly as largely water ice, also exhibit absorption in the blue and ultraviolet, probably for similar reasons. The photometric properties of the Venus clouds in the visible are consistent with ice clouds with particle diameters of a few microns (see, e.g., Arking and Potter, 1968; Potter, 1969b).

One of the major arguments against water ice as a principal constituent of the Venus clouds has for many years been the apparent inconsistency between the spectroscopic abundances and the equilibrium vapor pressures at the presumed cloud temperatures (Sagan and Kellogg, 1963; Chamberlain, 1965; many subsequent papers). As the vapor pressure goes exponentially as the negative reciprocal temperature, a variation of a few tens of degrees in estimates of rotational temperature implies many orders of magnitude variation in the derived water vapor abundance. Considering the apparent time-variability (Schorn *et al.*, 1969) in the near-infrared spectroscopic water vapor abundances (possibly connected with the four day retrograde rotation in the clouds?); the range of reported cloud temperatures (210–250 K); possible departures from saturation above the clouds; and the effects of inhomogeneity and anisotropy in the clouds (see, e.g., Sagan and Regas, 1970), I am not convinced that the infrared spectroscopic measurements provide a crushing argument against the case for ice clouds on Venus – although they represent certainly the strongest adverse argument. At one time it appeared that ice clouds were inconsistent – at least at thermodynamic equilibrium – with the quantities of HCl and HF determined from the Connes interferometric spectra. But more recent calculations, performed with the water vapor abundance as an unknown, shows this apparent inconsistency to be almost removed (Lewis, 1969a).

All of the other proposed principal cloud constituents seem to run into one or another difficulty: 'dust' introduces problems about the ability of the slow Venus circulation to raise and maintain dense aerosols, as well as difficulties with the absence of huge quantities of SiF_4 (Lewis, 1969b) and with the infrared spectrum; NH_4Cl runs into problems with the amount of gaseous ammonia needed in equilibrium with it; there are doubts on the very reality of polywater, and about the interpretation of occultation attenuation data in terms of clouds rather than scintillation eddies in the Venus atmosphere (Fjeldbo, 1969); carbon suboxide and hydrocarbons do not appear to match the Venus infrared spectrum (Plummer, 1970a, 1969), and the latter are particularly unlikely because of the absence of evidence for simple gaseous hydrocarbons in equilibrium with putative hydrocarbon clouds (Sagan, 1961). As for $FeCl_2 \cdot 2H_2O$, it is unstable in the presence of the water abundances quoted by the Venera 4, 5, and 6 experimenters; and, while it shows an absorption at $3\,\mu$, the feature has shallow wings and does not have the saturation character of the infrared observations. The $3\,\mu$ absorption in $FeCl_2 \cdot 2H_2O$ is in fact due to water, but in the bound state. A stronger absorption can be obtained with more bound water per $FeCl_2$ moiety, but the vapor pressure data seem to be inconsistent with the tetrahydrate or higher hydrates (Kuiper, 1970). John Lewis (private communication, 1970) points out that $FeCl_2 \cdot 2H_2O$ is so involatile it should condense out at > 500 K, and at this symposium both Kuiper and Sill have mentioned attendant serious transport problems, and the necessity to postulate strong departures from equilibrium.

Perhaps none of these objections will be considered individually decisive. The question of the composition of the Venus clouds is by no means solved, at least to the satisfaction of all investigators in the field at the present time. But I will be surprised if we are more than a few years from such a solution.

There are a range of other troubles with Venus. The structure of the atmosphere seems reasonably well-determined (Figure 1), although the variation of this structure with position on the disk and with time of day needs to be studied further. Another question is whether the clouds of Venus ever exhibit breaks. The impression of almost

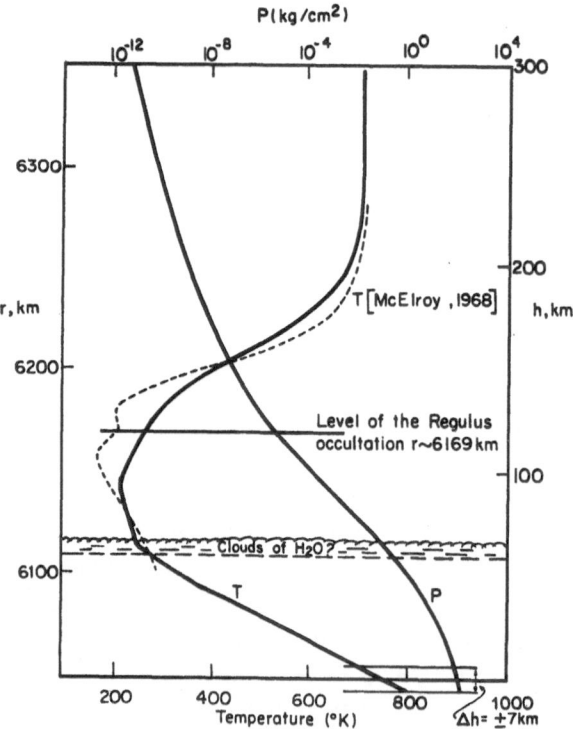

Fig. 1.   Structure of the Venus atmosphere as determined from Veneras 5 and 6. This diagram is redrawn from one in the paper by Avduevsky *et al.*, 1970 (and with the addition of the question mark in the clouds).

all observers is that the Venus cloud deck is uniform and without breaks. But on rare occasion there have been reports of breaks in the clouds. One such event is recorded in the drawing by J. H. Focas, reproduced in Figure 2. We can compare this unusual representation of Venus with the usual representation of the earth in Figure 3 at

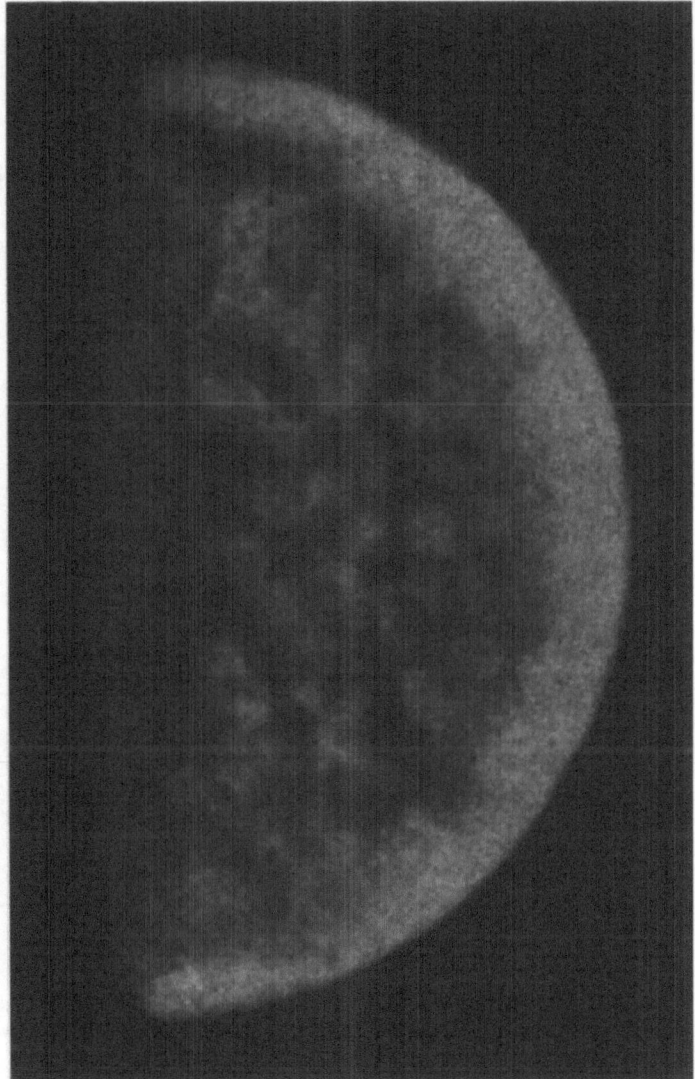

Fig. 2.   An unusual drawing of Venus by J. H. Focas showing apparent resolution of the clouds into individual elements [taken from *Atlas des Planètes* (ed. by V. de Callataÿ and A. Dollfus), Gauthier-Villars, Paris, 1968].

approximately the same phase angle. The search for such breaks requires high resolution and is a natural experiment for a Venus orbiter. (Such vehicles will also be of very great importance in mapping temperatures and water vapor abundances over the disk of the planet.) The radar altimeters on the Venera 5 and 6 spacecraft give a variation

Fig. 3.   Apollo photograph of the Earth taken at very roughly the same phase angle
as the drawing of Venus in Figure 2. The Earth shows about 50% cloud cover.

in altitude of the respective landing sites of many kilometers. This appears to be
inconsistent with the time delay radar measurements of topography performed at
Lincoln Laboratory (Smith *et al.*, 1970), but the lateral resolution of the ground-based
radar is some hundreds of kilometers; perhaps at a much smaller lateral resolution
scale the topography is rougher. Why Venus should be rotating in a retrograde
direction, why it should be locked or almost locked in rotation at moments to inferior
conjunction with the earth (see, e.g., Gold and Soter, 1969), and why the clouds of
Venus show a four day retrograde rotation are clearly important problems which are

far from solved at the present moment. The radar topography of Venus suggests mountain ranges and large circular basins (impact craters?), but definite knowledge of the geomorphology of Venus lies in the future. A similar remark applies to meteorological questions of the circulation of the Venus atmosphere, both in the stratosphere (see, e.g., Gierasch, 1970), and in the deep atmosphere. And the explanation of the high surface temperature of Venus is certainly one of the major intellectual problems about the planet. There is a misimpression occasionally encountered that the large optical depths required for ice clouds to explain the high visible and near infrared albedos imply that very little sunlight penetrates through the clouds to the lower atmosphere and surface. This is incorrect. Clouds of water ice need not produce a 'dirty' greenhouse, and 10 to 20% of the incident sunlight would emerge out of the bottom of reasonable ice clouds (Sagan and Pollack, 1967; Pollack, 1969). Although the case for the greenhouse model is, I believe, better today than it ever was, there is still no unanimity on the subject.

Even if the Venus atmosphere is composed of as much as a few tenths of a percent of water vapor by volume, there is a large discrepancy between the water abundance in the atmosphere of Venus and that in the atmosphere and hydrosphere of the earth – a discrepancy of perhaps a factor of a thousand. Here again the variety of explanations put forth seriously, is, at the very least, sobering. Venus may have started out with much less water than the earth because it was closer to the sun – although this explanation tends to put the problem out of reach, depending on initial and inaccessible conditions. Serpentinization of olivine may be responsible, although why this should be more effective on hot Venus than on cool earth is far from clear (Sagan, 1960); nevertheless, Rubey (1969) states that this may be a promising possibility. Perhaps all the water is locked up as polywater (Donahoe, 1970), although again why this should occur more readily on Venus than on Earth is far from clear – even if we admit the reality of anomalous water. Finally, perhaps the two planets started out with comparable complements of water, but photodissociation of water, selective escape of hydrogen, and preferential oxidation of the crust of Venus by huge quantities of oxygen occurred (see, e.g., Sagan, 1968; Eck et al., 1967). Calculations suggest that this is barely possible, but it requires very special circumstances and is not an idea marked by a striking economy of hypothesis.

The microwave, Venera, and Mariner observations and the Golitsyn (1970) similarity theory all seem to imply that there is nowhere on the Venus surface where liquid water could exist – even with a surface pressure of 100 atm. Accordingly life, particularly life based on familiar chemistries, seems implausible on the Venus surface (cf. Sagan, 1970). This leaves the clouds. Especially if the clouds are composed of condensed water – but even if they are not – life in the clouds is not by any means out of the question. There is water vapor, there is carbon dioxide, there is sunlight, and very likely there are small quantities of minerals stirred up from the surface. These are all the prerequisites necessary for photoautotrophs in the clouds. In addition the conditions are approximately S.T.P. The only serious problem that immediately comes to mind is the possibility that downdrafts will carry our hypothetical organisms down to the hot, deeper atmosphere and fry them faster than they reproduce. To circumvent this difficulty, and to show that organisms might exist in the Venus clouds based purely on terrestrial biochemical principles, Harold Morowitz and I (1967) devised a hypothetical Venus organism in the form of an isopycnic balloon, which filled itself with photosynthetic hydrogen and maintained a constant pressure level to avoid downdrafts. We calculated that, if the organism had a wall thickness comparable to the unit membrane thickness of terrestrial organisms, its minimum diameter would be a few centimeters. This heuristic argument had at least one salutary consequence: The *Saturday Evening Post* ran a cartoon showing a ping pong player (dressed in Florida sports shirt and Bermuda shorts) about to serve, and interrupted by the cry from his ping pong ball, "Stop! I am a friendly visitor from another planet!"

While it is not out of the question that life exists in the Venus clouds, it seems quite unlikely to have arisen there. If we wish to take seriously a possible exobiological interest in Venus, we must postulate some earlier epoch (perhaps before outgassing produced a large atmospheric infrared opacity, a strong greenhouse effect, and high surface temperatures) when the surface conditions were much more clement. Thus the runaway greenhouse scenario (Sagan, 1960) seems connected with the question of life on Venus.

The probability that the clouds of Venus are such a pleasant environment – the most earth-like extraterrestrial environment in the solar system, so far as we know – opens up the perhaps amusing prospect of astronauts floating in somewhat larger balloons, ballasted and valved for the pressures of the lower Venus clouds, and clad in shirtsleeves and 19th Century oxygen masks. We can imagine them peering wistfully down at the inaccessible surface of Venus below – wistfully, because Rayleigh scattering in an atmosphere with 100 bar surface pressure will prevent them from seeing any images of surface features, even if there are breaks in the clouds (although, in the vicinity of $1\ \mu$, there might just conceivably be an imaging window between the extinction due to Rayleigh scattering at shorter wavelengths and the extinction due to molecular absorption at longer wavelengths).

Rather than such a manned venture, most of us would much prefer to see a series of small orbiters and entry probes (cf. Hunten and Goody, 1969) and perhaps unmanned buoyant stations to more thoroughly characterize our still enigmatic nearest planetary neighbor. Until then, I suspect we shall continue to have some trouble with Venus. But perhaps, in paraphrase of *Julius Caesar*, Act I, Scene 2, "the trouble lies not in our stars but in ourselves".

Acknowledgment

This work was supported by the Atmospheric Sciences Section, National Science Foundation under grant GA 10836.

References

Arking, A. and Potter, J.: 1968, *J. Atmospheric Sci.* **25**, 617.
Avduevsky, V. S., Marov, M. Y., and Rozhdestvensky, M. K.: 1970, *Radio Sci.*, in press.
Barath, F. T., Barrett, A. H., Copeland, J., Jones, D., and Lilley, A. E.: 1964, *Astron. J.* **69**, 49.

Chamberlain, J. W.: 1965, *Astrophys. J.* **141**, 1184.
Coffeen, D. L.: 1968, *J. Atmospheric Sci.* **25**, 643.
Dickel, J. R., Medd, W. J., and Warnock, W. W.: 1968, *Nature* **220**, 1183.
Donahoe, F. J.: 1970, *Icarus* **12**, 424.
Eck, R., Dayhoff, M. O., Lippincott, E. R., and Sagan, C.: 1967, *Science* **155**, 556.
Fjeldbo, G.: 1969, Remarks at URSI Conference on Planetary Atmospheres and Surfaces, Wood's Hole, Mass., August.
Gierasch, P.: 1970, *Icarus* **13**, 25.
Gold, T. and Soter, S.: 1969, *Icarus* **11**, 356.
Golitsyn, G.: 1970, *Icarus* **13**, 1.
Hansen, J. E.: 1969, *Astrophys. J.* **158**, 337.
Lewis, J.: 1969a, private communication.
Lewis, J.: 1969b, *Icarus* **11**, 367.
Hunten, D. H. and Goody, R. N.: 1969, *Science* **165**, 1317.
Jones, D. E.: 1961, *Planetary Space Sci.* **5**, 166.
Kroupenio, N. N.: 1970, this volume, p. 32.
Kuiper, G. P.: 1970, this volume, p. 91.
Mayer, C. H.: 1969, private communication.
Mayer, C. H., McCullough, T. P., and Sloanaker, R. M.: 1958, *Astrophys. J.* **127**, 1.
Morowitz, H. and Sagan, C.: 1967, *Nature* **215**, 1259.
Morrison, D.: 1969, *Science* **163**, 815.
Ohring, G.: 1969, *Icarus* **11**, 171.
O'Leary, B. T.: 1967, *Astrophys. J.* **146**, 754.
O'Leary, B. T.: 1970, *Icarus* **13**, in press.
Plummer, W. T.: 1969, *Science* **163**, 1191.
Plummer, W. T.: 1970a, in press.
Plummer, W. T.: 1970b, *Icarus* **12**, 233.
Pollack, J. B.: 1969, *Icarus* **10**, 314.
Pollack, J. B. and Morrison, D.: 1970, this volume, p. 29, also *Icarus* **12**, 376.
Pollack, J. B. and Sagan, C.: 1967, *J. Geophys. Res.* **73**, 5943.
Pollack, J. B. and Sagan, C.: 1965, *Icarus* **4**, 63.
Potter, J.: 1969a, *J. Atmospheric Sci.* **26**, 511.
Potter, J.: 1969b, *Bull. Am. Astron. Soc.* **1**, 258.
Rubey, W. W.: 1969, private communication.
Sagan, C.: 1960, Jet Propulsion Lab. Tech. Rept. 32–34.
Sagan, C.: 1961, *Science* **133**, 849.
Sagan, C.: 1967, *Nature* **216**, 1191.
Sagan, C.: 1968, *International Dictionary of Geophysics* (ed. by S. K. Runcorn), Pergamon Press, London, p. 2049.
Sagan, C.: 1969, *Comments Astrophys. Space Phys.* **1**, 94.
Sagan, C.: 1970, 'Life', *Encyclopedia Brittannica*.
Sagan, C. and Kellogg, W. W.: 1963, *Ann. Rev. Astron. Astrophys.* **1**, 235.
Sagan, C. and Pollack, J. B.: 1967, *J. Geophys. Res.* **72**, 469.
Sagan, C. and Regas, J.: 1970, *Comments Astrophys. Space Phys.*, in press.
Schorn, R. A., Barber, E. S., Gray, L. D., and Moore, R. C.: 1969, *Icarus* **10**, 98–104.
Smith, W. B., Ingalls, R. P., Shapiro, I. I., and Ash, M. E.: 1970, *Radio Sci.*, in press.
Veverka, J.: 1971, *Icarus* **14**, to be published.

PART II

MARS

A. OPTICAL PROPERTIES

COLORIMETRY OF MARTIAN FEATURES BY MEANS OF AREA SCANNING

PETER B. BOYCE*

Lowell Observatory, Flagstaff, Ariz., U.S.A.

1. Introduction

The contrast between light and dark areas on Mars has been the subject of two photo-electric studies by Younkin (1966) and McCord (1969), both of which were somewhat limited in scope. Younkin's data are based on one night's observations. McCord had a more suitable instrument for making a comparison between two areas on Mars, but still used only six nights. The agreement between these two studies is excellent. However, there is ample evidence from photographs that there are short-term changes in the contrast and brightness of Martian surface features, especially in the case of blue clearing. Visual observations are unsuitable for investigating these changes, due to the limited wavelength range of the eye's sensitivity, as well as the poor accuracy of the eye as a photometric instrument. Even photography is of limited usefulness when high accuracy is desired. Consequently, a program of continuous photoelectric observation of Mars was set up for the 1969 opposition. The technique of area scanning was chosen as the best method of recording the contrast of Martian surface features.

2. Instrumentation

Area scanning is a technique whereby a small aperture is swept rapidly and repeatedly across the disk of the planet. Each successive sweep is added in phase with the previous sweeps, this process continuing until the signal-to-noise ratio of the accumulated scan reaches the desired value. Area scanning was pioneered by Rakos (1965) and has proven to be an extremely powerful method of obtaining accurate data under conditions in which the smearing introduced by the seeing in the Earth's atmosphere is an important factor.

A new dual-beam area scanner was designed and constructed at Lowell Observatory for the purpose of making these contrast measurements. The design embodies a dichroic filter which provides two beams. One beam is used for reference and includes an interference filter 75 Å wide centered at 6000 Å. The other beam passes through a filter wheel with eight filters ranging in wavelength from 3400 Å to 5300 Å. The filter bandwidths range from 150 Å in the ultraviolet to 75 Å in the green.

The scanning is accomplished by physically moving the entrance slit at a rate of two complete sweeps per second. A round aperture 1.2 sec of arc in diameter was used for all the Mars observations. However, in order to calibrate the sensitivity of the photo-meter, as well as to measure the amount of smearing due to atmospheric seeing, the

---

* Visiting Astronomer, Cerro Tololo Inter-American Observatory.

aperture was replaced with a slit, and standard stars were observed each night. In order to provide a measure of the quality of the seeing during the night, a Gaussian curve was fitted to the standard star profile. The standard deviation of the Gaussian provides a quantitative measure of the smearing introduced by the atmospheric seeing.

The dual-beam area scanning method has the advantage that it provides simultaneous observations in two colors, thereby eliminating any possibility that errors in positioning or changes in seeing will affect the measured contrast. For instance, on any particular night the continual repetition of the reference beam scan allows us to state that the variation in contrast ratio to be expected due to changes in the seeing is somewhat less than 2%.

Because the final area scan is composed of from 100 to 300 repeated scans, the area scanner produces accurate data even under conditions of changing transparency. This is most important in the case of an extended program because many more nights are usable than would be the case for conventional photometry.

The observations were made at the $f/75$ focus of the Lowell-Tololo reflector at the Cerro Tololo Inter-American Observatory in Chile. The telescope was shared between this program and the Lowell-NASA International Planetary Patrol Program. Consequently, there was an hourly photographic record in four colors which has proved to be extremely useful in interpreting the photoelectric scans. Observations were made on every possible night from May 28 to July 8, 1969.

3. Blue Clearing

On the night of June 10, 1969, there was a partial blue clearing over Mare Cimmerium. Figure 1 shows the appearance of Mars in red, blue, and green. The southern two-thirds of Mare Cimmerium is visible with extreme clarity, even on the blue photographs. This is not the usual case and represents a strong blue clearing over that portion of the planet. Notice that the Trivium Charontis-Cerberus complex near the center of the disk, although visible in the red, is not visible in the blue photograph, indicating that the blue clearing was present only over the southern part of the planet. Figure 2 shows several typical area scans made at approximately the same time as the photographs in Figure 1. The strong dip to the left is Mare Cimmerium; the shallower dip in the 6000 Å scan is Trivium Charontis-Cerberus. For each of the scans in the left side of the figure a monitor scan at 6000 Å was obtained simultaneously. The differences between the monitor scans are very small; a representative one is shown in the right-hand side, along with a typical instrumental function which includes the smearing due to the finite aperture size combined with that due to the Earth's atmosphere. Defining the contrast as the ratio of the intensity of the dark area to the intensity of the neighboring bright area, we see that it is possible to measure the contrast of the Trivium Charontis region at wavelengths as short as 4600 Å. However, the Mare Cimmerium is visible on the scans, and its contrast can be measured even at 4000 Å.

Figure 3 is a plot of the contrast of Mare Cimmerium and Trivium Charontis-Cerberus for June 10 as a function of wavelength. Also included in the figure is the

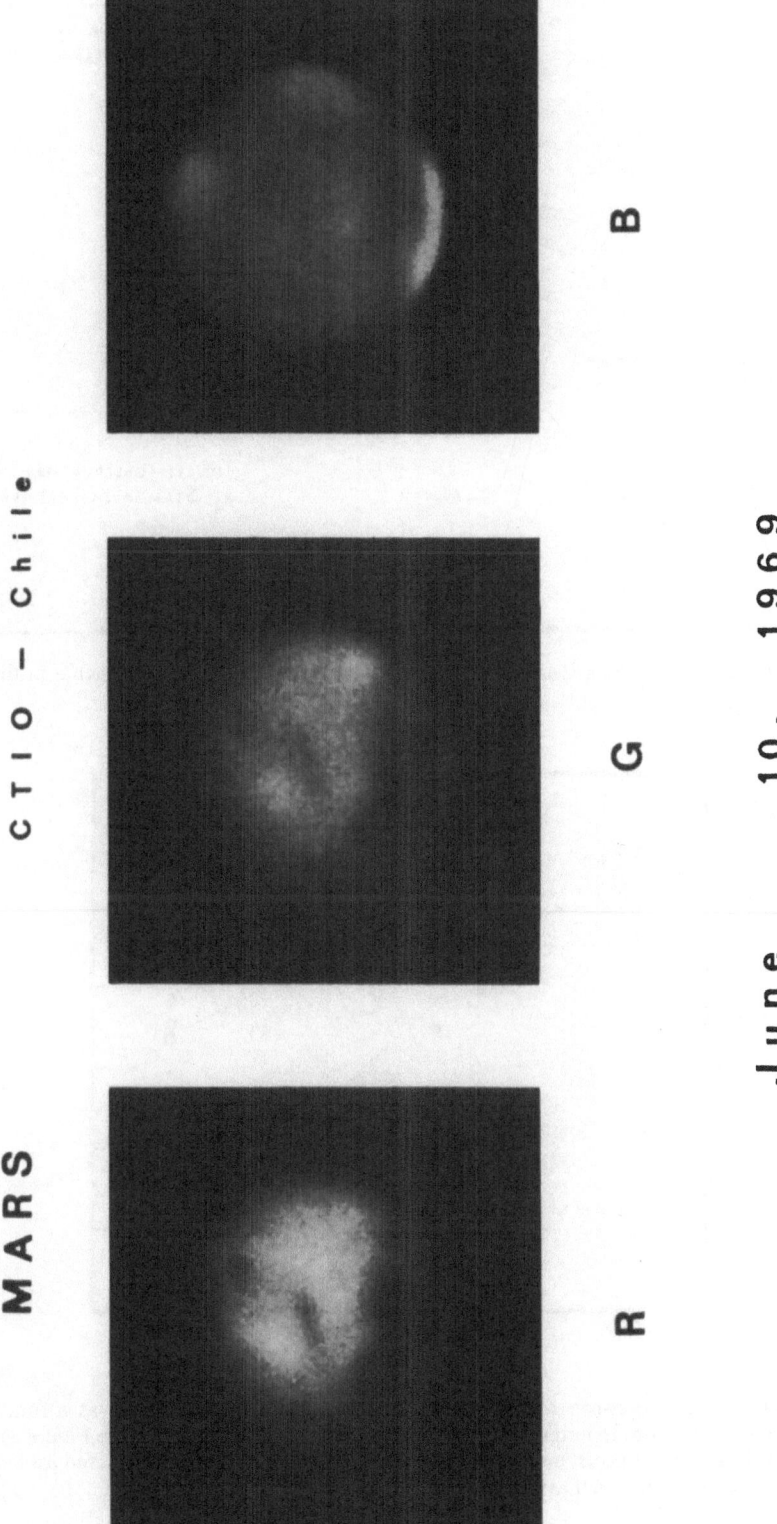

Fig. 1.   Photographs of Mars in three colors, showing strong blue clearing over part of the disk.

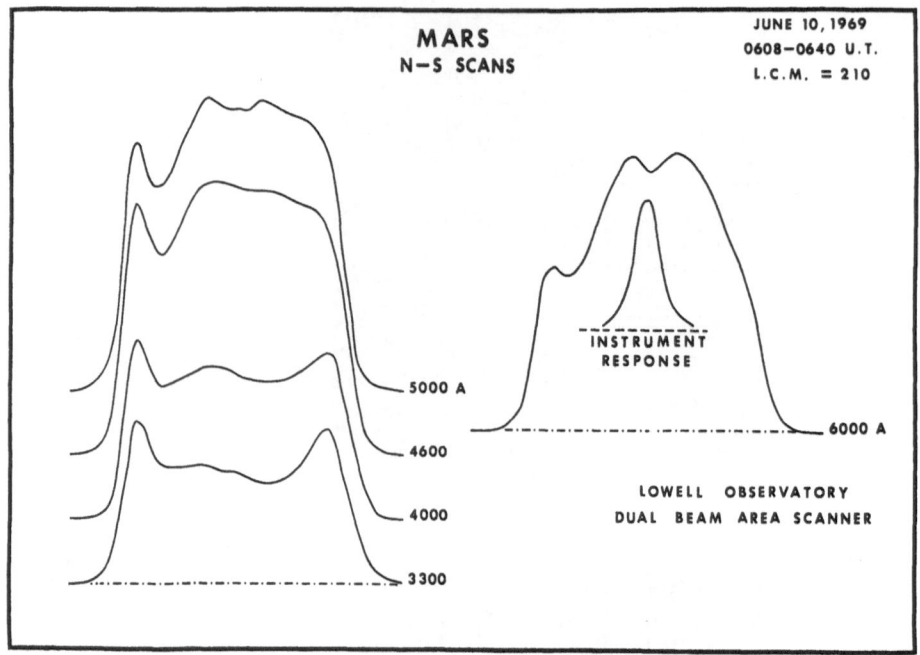

Fig. 2. N–S intensity profiles of Mars through Trivium Charontis-Cerberus. North is at the right.

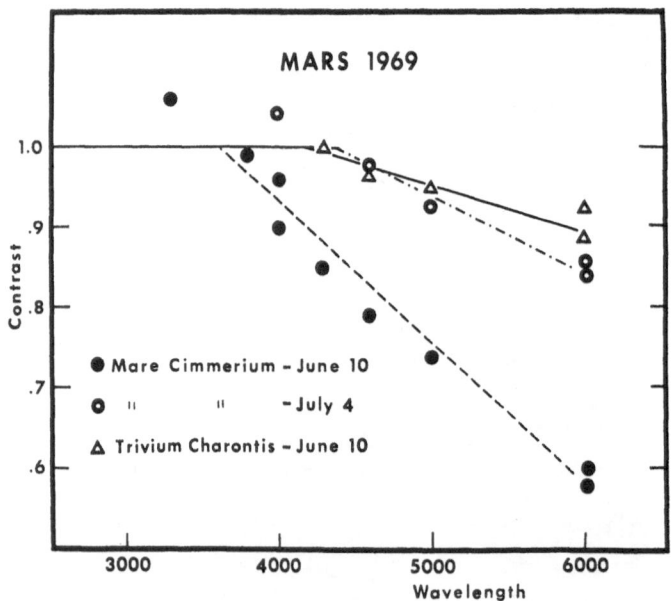

Fig. 3. Contrast ratios between Martian dark areas and adjacent light areas as a function of wavelength. The Mare Cimmerium exhibited strong blue clearing on June 10 (see Figure 1) which is apparent as shift of the contrast curve toward the blue. These ratios are uncorrected for smearing by the terrestrial atmosphere and finite scanning aperture width.

contrast measured for Mare Cimmerium on July 4, when there was no visible blue clearing. Two things are immediately obvious: (1) The contrast seems to increase approximately linearly with wavelength. Reduction of more scans has shown that the normal curve for Mare Cimmerium is typical of the contrast curve for most other dark areas on the disk of Mars. (2) Although these contrast measurements are uncorrected for the effects of seeing, there is a large change in the contrast between the blue-cleared and non-blue-cleared condition. Since the width of the atmospheric smearing profile was approximately the same on both days, this change cannot be due to variations in the conditions of the Earth's atmosphere, but must be intrinsic to Mars.

The area of blue clearing on June 10 seems to have no relationship to the light or dark areas since the line between the clear and non-clear areas of the disk runs down the middle of the dark Mare Cimmerium. In the blue photograph, an east-west bright band is present between the equator and latitude $-25°$. It is also quite evident on the photoelectric scans at 4000 Å. This brightening has the appearance of a meteorological phenomenon because of its long east-west extent and transient character. Such cloud-like phenomena have been noted before by Slipher (1962). The rapidity with which the blue clearing can appear and disappear also argues for this being a meteorological phenomenon.

4. Changes in the Polar Caps

The polar caps were visible on all north-south scans taken at wavelengths below 5000 Å. By taking the ratio of the brightness of the polar caps to an average area on the disk and normalizing to the albedo of the whole Martian disk as determined by Tull (1966), it was possible to determine an approximate albedo for the polar caps as a function of wavelength. Figure 4 shows albedos observed on several days. Again, these measurements are uncorrected for the smearing effects due to seeing in the Earth's atmosphere, which is extremely severe for an area on the limb of the planet, such as the polar caps. However, it is immediately evident that the caps have a relatively flat albedo below 4500 Å and that they change in brightness from day to day. The north cap, which at this time of Martian season should more properly be called a 'hood', shows variation in extent and brightness from day to day. This is very typical behavior for the fall hood. At the time of these observations, Mars was just past the fall equinox in the northern hemisphere. Notice that at its brightest the north cap is very similar to the stable southern cap; however, the north cap did show a change in intensity by a large factor during the course of one day from July 4 to July 5. Again, this is a strong indication of some type of meteorological phenomenon, perhaps deposition and sublimation of solid carbon dioxide on the surface.

5. Hellas

This large bright area to the south of Syrtis Major has been known to undergo changes in brightness at the time of the northern autumnal equinox. At this time Hellas will sometimes, but not always, become darker than the desert areas. From a survey of plates available at the Planetary Research Center of the Lowell Observatory,

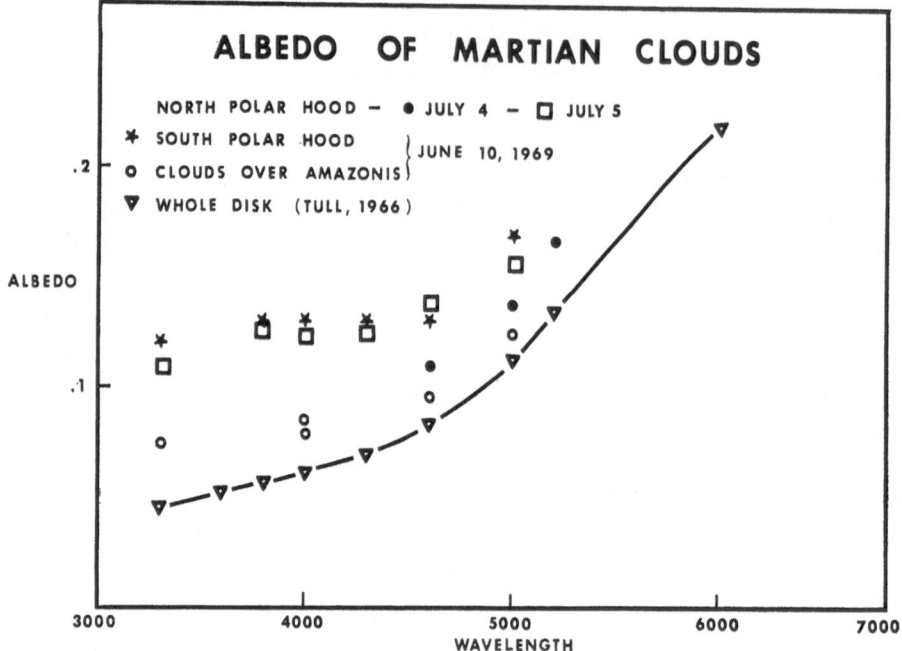

Fig. 4. Albedos of polar caps and 'W clouds' on Mars. Uncorrected for smearing.

it is apparent that Hellas showed this darkening during the oppositions of 1907, 1937, and 1969, but remained quite bright during the opposition of 1922. The darkening, when it does occur, apparently has no relationship to phase angle.

During the present opposition, Hellas did become quite dark for a period of at least forty days. Figure 5 shows two sets of north-south scans across the Hellas and Syrtis Major areas. The top set of scans was made when Hellas was faint. It is apparent that Hellas is not present on these scans at any wavelength. The brightening on the limb at 4000 Å is not at the position of Hellas and is due to the south polar cap. The bottom set of scans was taken one month later after Hellas had once again reached its normal brightness. It is extremely bright at 6000 Å, becoming less visible at shorter wavelengths. Although these scans have not yet been calibrated in absolute intensity, it is apparent that there is approximately a 30% change in the brightness of Hellas.

Examination of the Planetary Patrol photographs shows that Hellas was not always uniformly dark, undergoing changes in appearance from day to day during the month of June. The most violent fluctuations in the appearance of Hellas occurred during the period of June 14 to 22, when Hellas was recovering its normal bright appearance. At this time, daily changes in brightness and appearance occurred. By June 17, Hellas had become quite bright, but on the 18th it had once again reverted to its dark appearance. On the 20th of June, the eastern half of Hellas was bright, but the western half was still dark. However, by June 22 the whole area of Hellas had become bright. The rapidity and reversibility of these changes seems to indicate a phenomenon that is connected in some way with Martian meteorology.

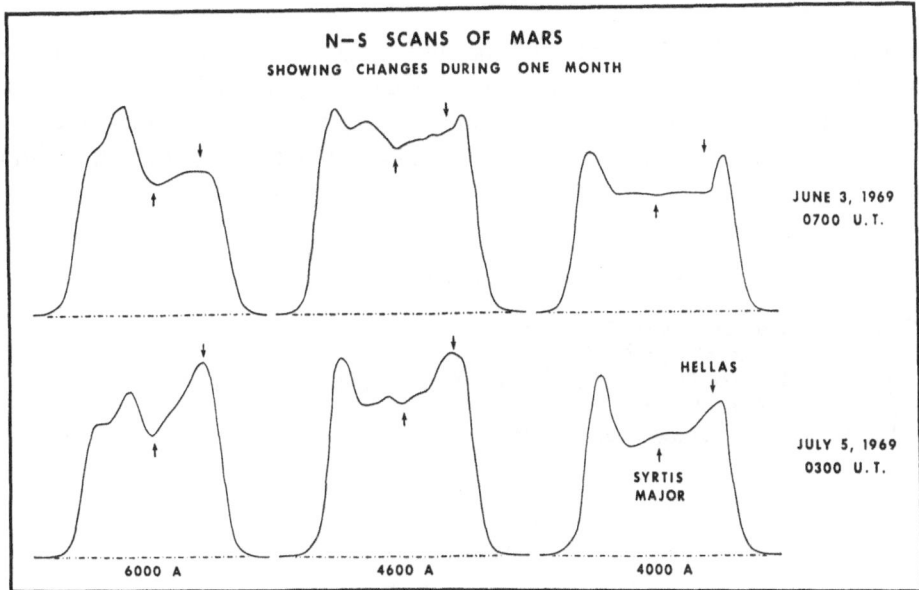

Fig. 5.   N–S intensity profiles of Mars through Hellas and Syrtis Major. North is at the left.

This conclusion is reinforced by the diurnal brightening of Hellas which was very apparent as Hellas began to regain its normal appearance. Hellas usually showed a strong brightening in the Martian afternoon, a type of behavior very similar to that of the Martian 'clouds', which are known to be transient phenomena. A detailed photometric investigation of Hellas is sufficient to show the rapidity with which large areas on the surface of Mars can change.

6. Conclusion

It is apparent from photoelectric observations that the Martian blue clearing is indeed due to a change in the contrast of the Martian surface markings. Such changes can be extremely rapid and often cover large areas of the Martian surface. The autumnal polar hood of Mars has likewise been shown to change drastically in intensity over a very short time span. Finally, the area Hellas has also been shown to vary in brightness with a rather short time scale. In this respect, Hellas is one of the most peculiar areas on Mars. It is tempting to think that the lack of craters in the Mariner VII photographs of Hellas is connected in some way with the remarkable brightness changes that Hellas underwent in the month preceding the Mariner encounter.

In any event, it is evident that the surface of Mars undergoes many large-scale, rapid variations of intensity of a type that is very difficult to explain without invoking some sort of meteorological phenomena. It is too early at this time to do more than speculate upon mechanisms which could be responsible for such changes, but it is necessary to be aware that such changes exist and will have to be taken into account by any theory of the Martian surface and lower atmosphere.

PETER B. BOYCE

Acknowledgments

I wish to thank Dr. V. M. Blanco and the staff of the Cerro Tololo Inter-American Observatory for their help and cooperation during my stay in Chile. This research was supported in part by the Air Force Cambridge Research Laboratories, Office of Aerospace Research, under Contract AF 19(628)4070 and in part by NASA grant NGR-03-003-001.

References

McCord, T. B.: 1969, *Astrophys. J.* **157**, 79.
Rakos, K. D.: 1965, *Appl. Opt.* **4**, 1453.
Slipher, E. C.: 1962, *The Photographic Story of Mars*, Northland Press, Flagstaff.
Tull, R. G.: 1966, *Icarus* **5**, 505.
Younkin, R. L.: 1966, *Astrophys. J.* **144**, 809.

LONGITUDINAL VARIATIONS, THE OPPOSITION EFFECT AND MONOCHROMATIC ALBEDOS FOR MARS*

WILLIAM M. IRVINE and JAMES C. HIGDON

Dept. of Physics and Astronomy, University of Massachusetts, Amherst, Mass., U.S.A.

and

SUSAN J. EHRLICH

Smith College, Northhampton, Mass., U.S.A.

Abstract. Observations of Mars previously reported in 10 narrow bands between 3150 Å and 1.06 μ and in UBV are analyzed for brightness variations which correlate with longitude of the central meridian. Such an effect is found for $\lambda \geqslant 5000$ Å, with some evidence for such a correlation at $\lambda = 4570$ Å. The data are then corrected to the mean (over longitude) brightness and a linear phase curve fitted to those observations with phase angle $i \geqslant 15°$. An opposition effect (anomalous brightening at small phase angles) is found for wavelengths $\lambda \leqslant 5500$ Å, in contrast to a result previously reported. The magnitude at zero phase, phase coefficient, and monochromatic albedo are computed for Mars as a function of wavelength.

1. Introduction

Multicolor photoelectric photometry of Mars between 1963 and 1965 has been reported in two previous papers (Irvine *et al.*, 1968a, Paper I; Irvine *et al.*, 1968b, Paper II). The observations were made using 10 narrow bands isolated by interference filters between 3150 Å and 1.06 μ and also in UBV. The narrow bands were labeled *v-u-s-p-m-l-k-h-g-e* as shown in Table I of Hopkins and Irvine (1969). The observations were conducted from two sites, one in South Africa and one in France; for the present paper these results are combined.

2. Longitudinal Variations

From the data presented in Paper I and Paper II we selected those observations which were not denoted by an asterisk; that is, we selected observations made under superior observing conditions. We then further selected those observations corresponding to phase angles $i \geqslant 15°$. A linear least squares fit to this data was made, and the residuals R were plotted versus longitude of the central meridian on Mars ω. The correlations found are shown in Figure 1. The longitudinal effect is easily observed. Its magnitude increases with wavelength out to the long wavelength limit of our observations (1.06 μ). It is clearly visible for wavelengths as short as 5000 Å, and there is some evidence for the effect in the band at 4570 Å. The solid line in Figure 1 is a least squares fit using a 6th-order polynomial, with the obvious constraint that the curve and its first derivative be periodic with period 2π. Note that the planetocentric declination of the Earth was $D_E \simeq 20°$ during the periods of observation.

* Contribution from the Four College Observatories, #64.

Sagan et al. (eds.), Planetary Atmospheres, 141–155.

WILLIAM M. IRVINE ET AL.

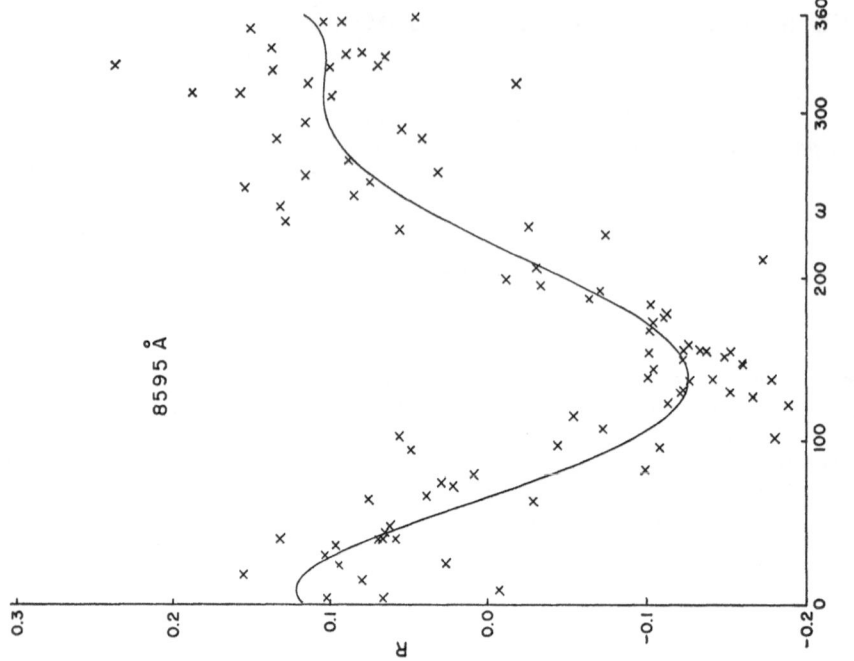

Fig. 1b.

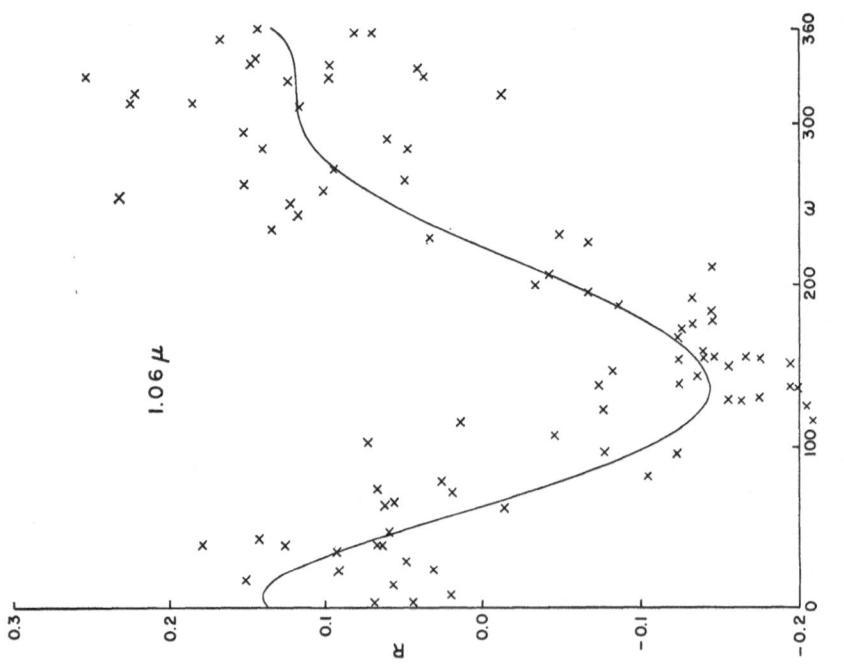

Fig. 1a–g. Longitudinal brightness variations for Mars. Longitude of central meridian denoted by ω, residual from linear phase curve by R. Only points with phase angles $i \geqslant 15°$ plotted.

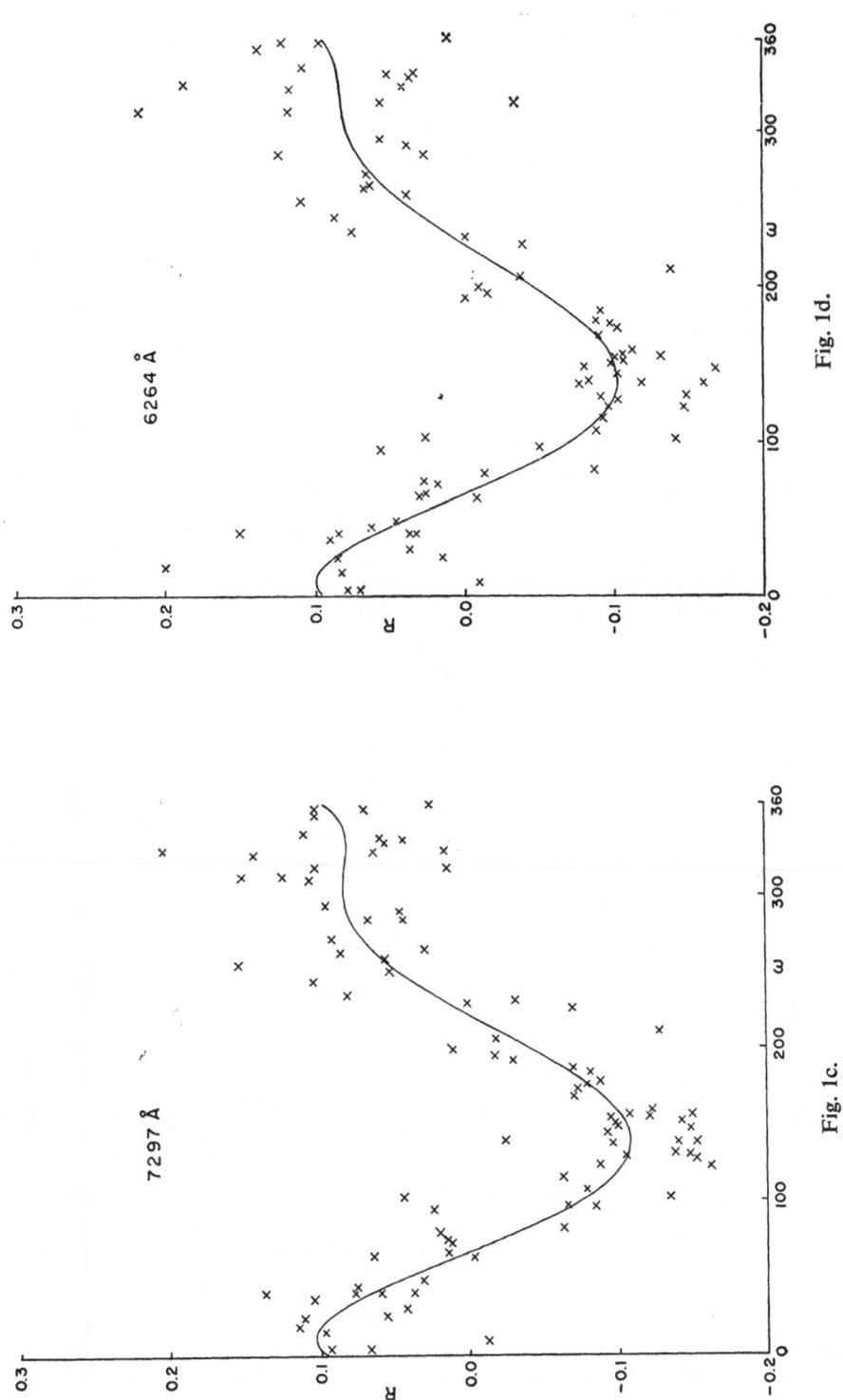

Fig. 1d.

Fig. 1c.

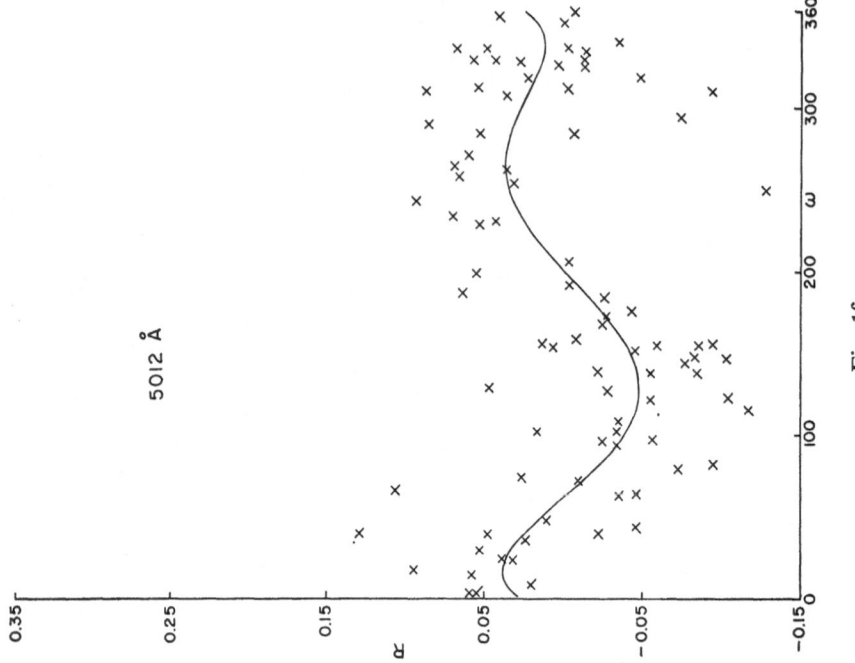

5012 Å

Fig. 1f.

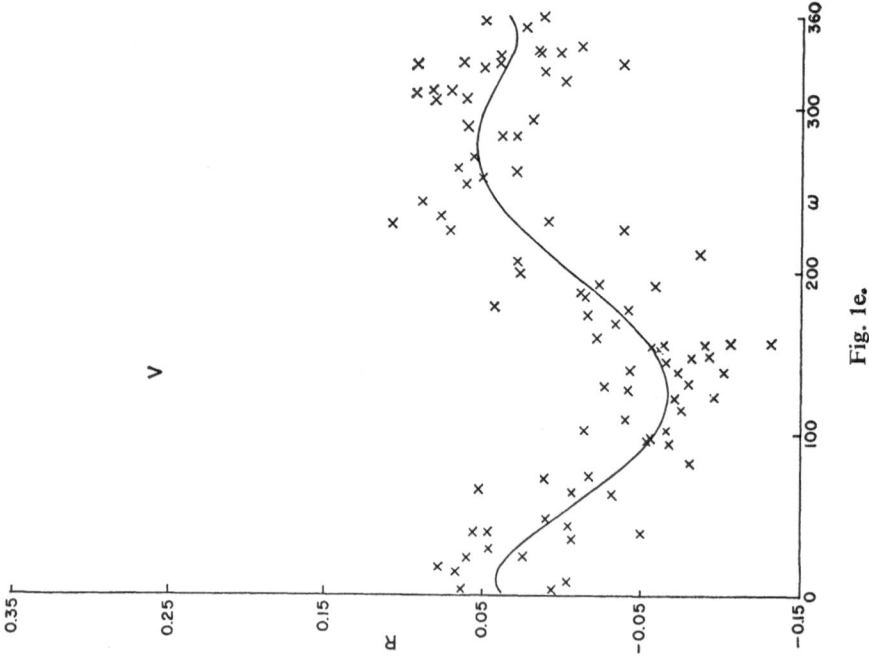

V

Fig. 1e.

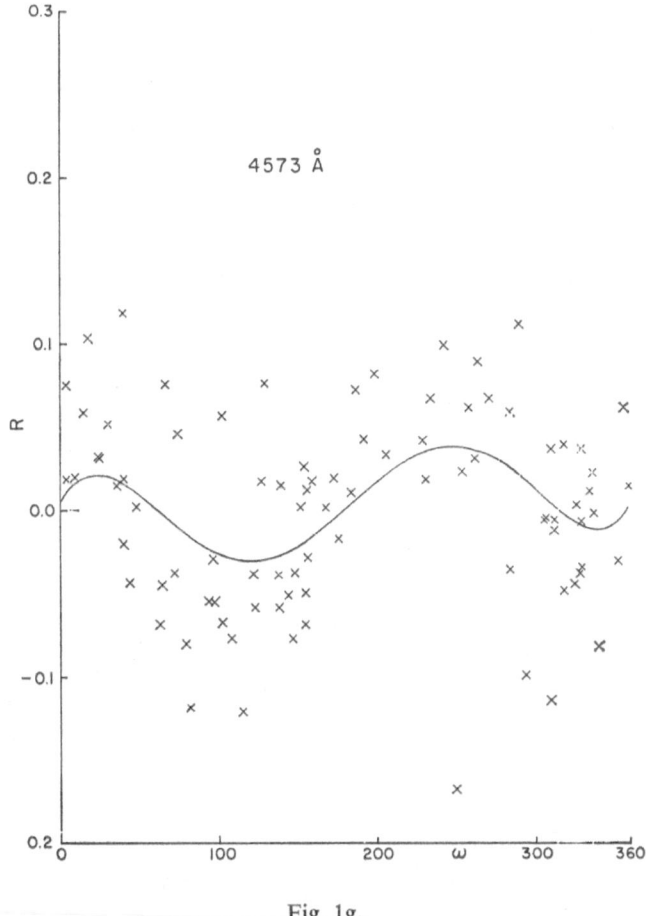

Fig. 1g.

For the shorter wavelengths the correlation observed no longer seems related to surface features and is apparently not statistically significant (see for example Figure 2). This is to be expected from the well known loss of observable surface detail on Mars at wavelengths $\lambda \lesssim 4550$ Å.

3. Phase Curves and the Opposition Effect

The observations in filters m-l-k-h-g-e and $V (\lambda \geqslant 4570$ Å) were then corrected to a mean longitudinal brightness using the least squares fit illustrated in Figure 1. The resulting data, and also the corresponding observations for wavelengths $\lambda \leqslant 4500$ Å, were then plotted versus phase angle and a linear least squares fit was made (remember that this data includes only observations for $i \geqslant 15°$). The resulting straight line is the full line shown in Figure 3, and the corresponding magnitudes at zero phase $m(1, 0)$ (all the data have been reduced to unit distance) and phase coefficient a are given in Table I, columns 2 and 3. We note that the narrow band color of the sun is zero on our magnitude system. Standard errors for $m(1, 0)$ are typically ~ 0.015 m. The observational data for phase angles $i \leqslant 15°$ were then corrected for the longitudinal effect and

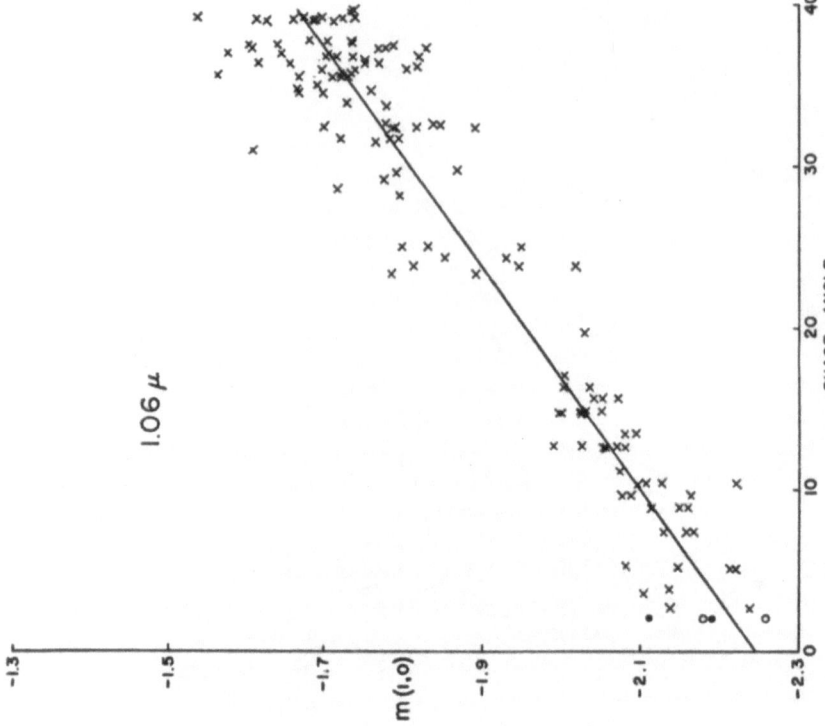

Fig. 3a–m.   Phase curves for Mars. Full line fitted to points with $i \geqslant 15°$, dashed line fitted to all points. Circles are observations under poorer conditions at small phase angles not used in least squares fits (open circles are 'starred' observations from Paper I, filled circles from Paper II).

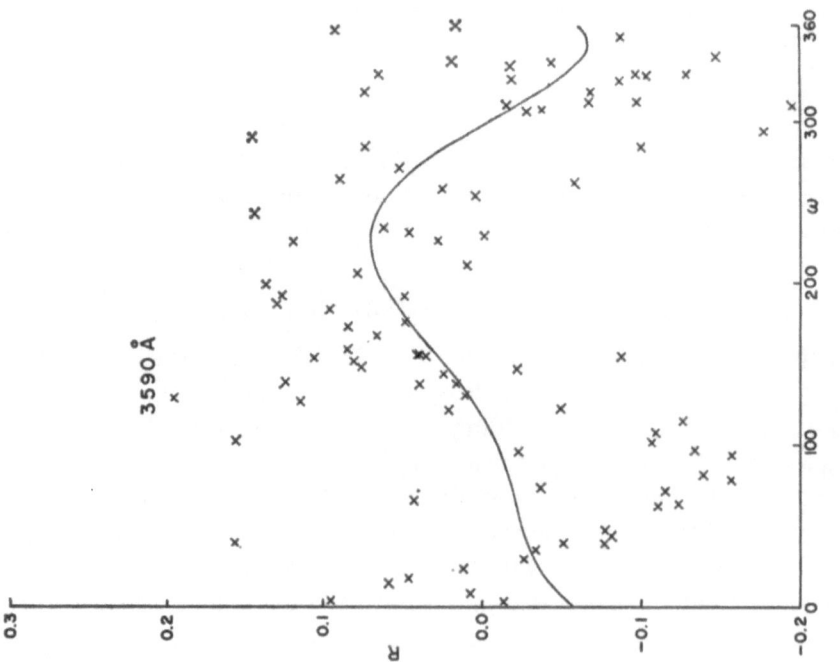

Fig. 2.   Same as Figure 1, but shorter wavelength.

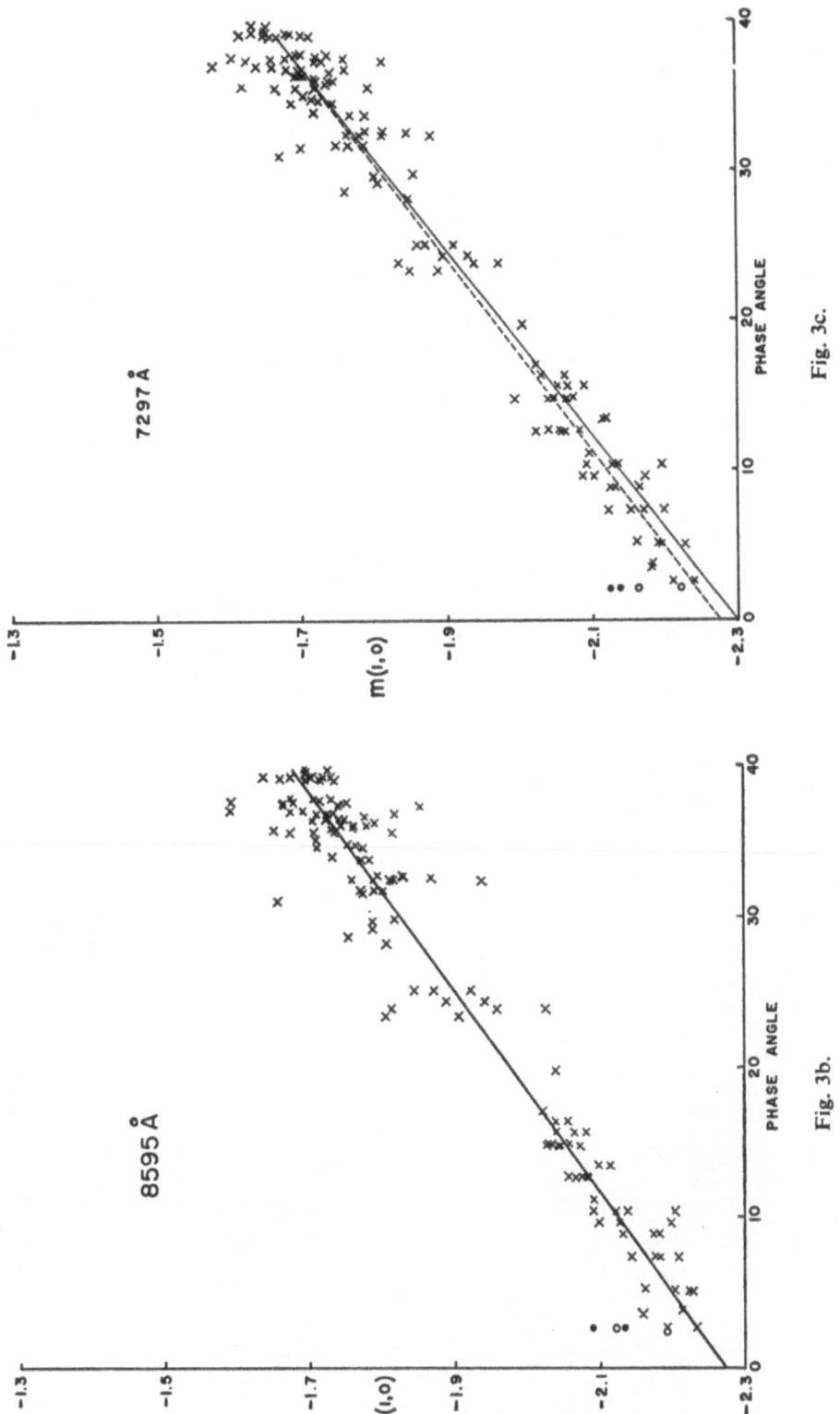

Fig. 3c.

Fig. 3b.

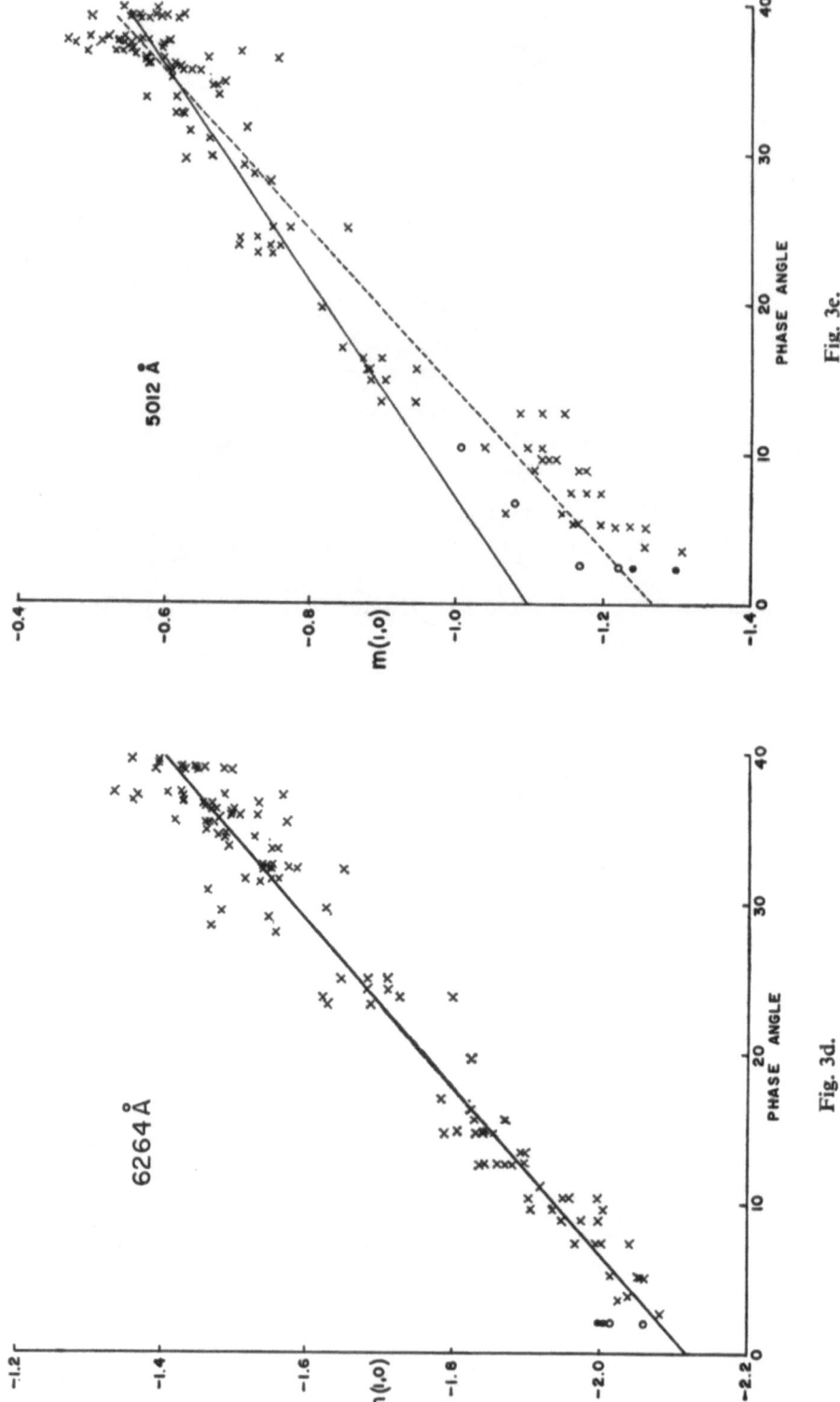

Fig. 3e.

Fig. 3d.

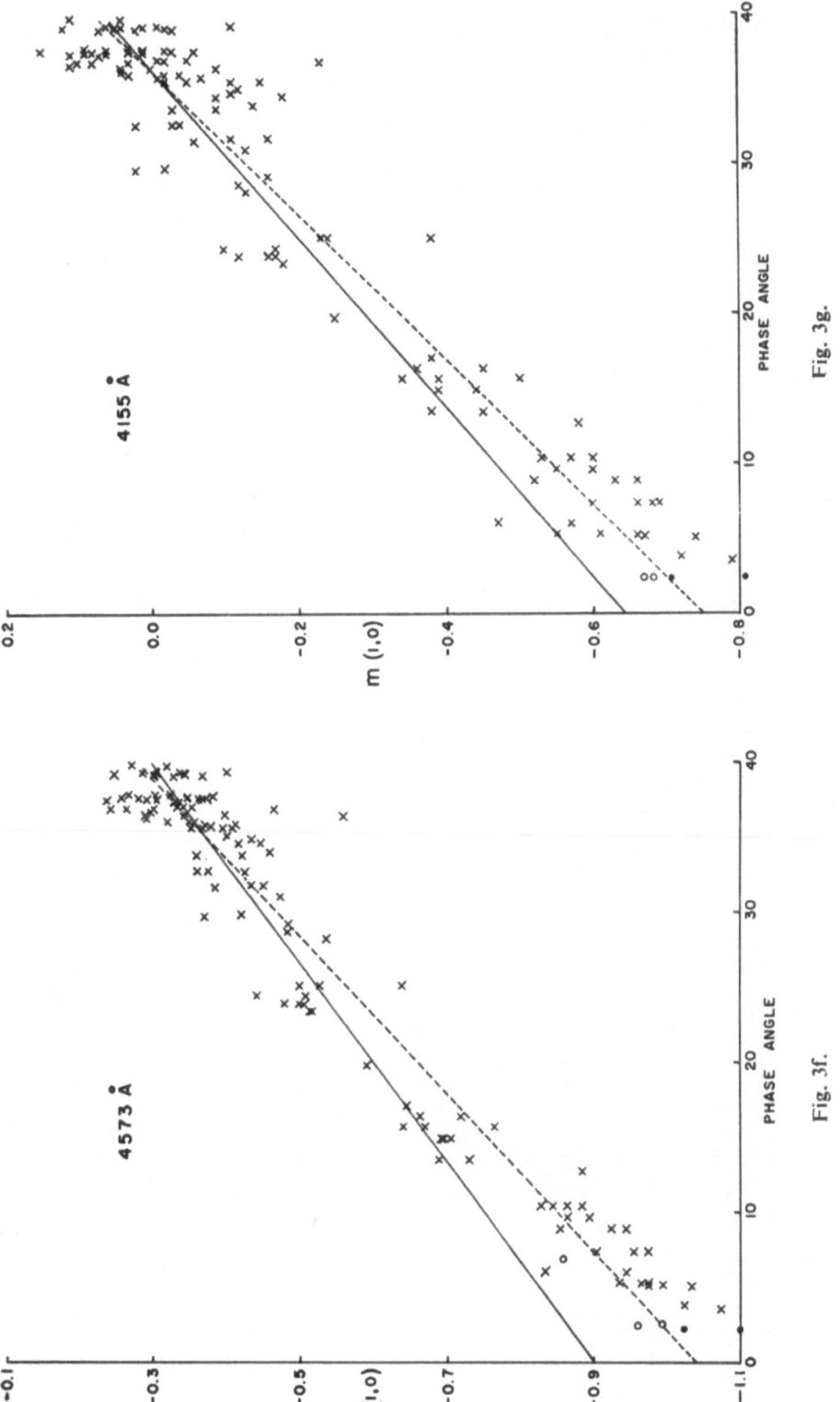

Fig. 3g.

Fig. 3f.

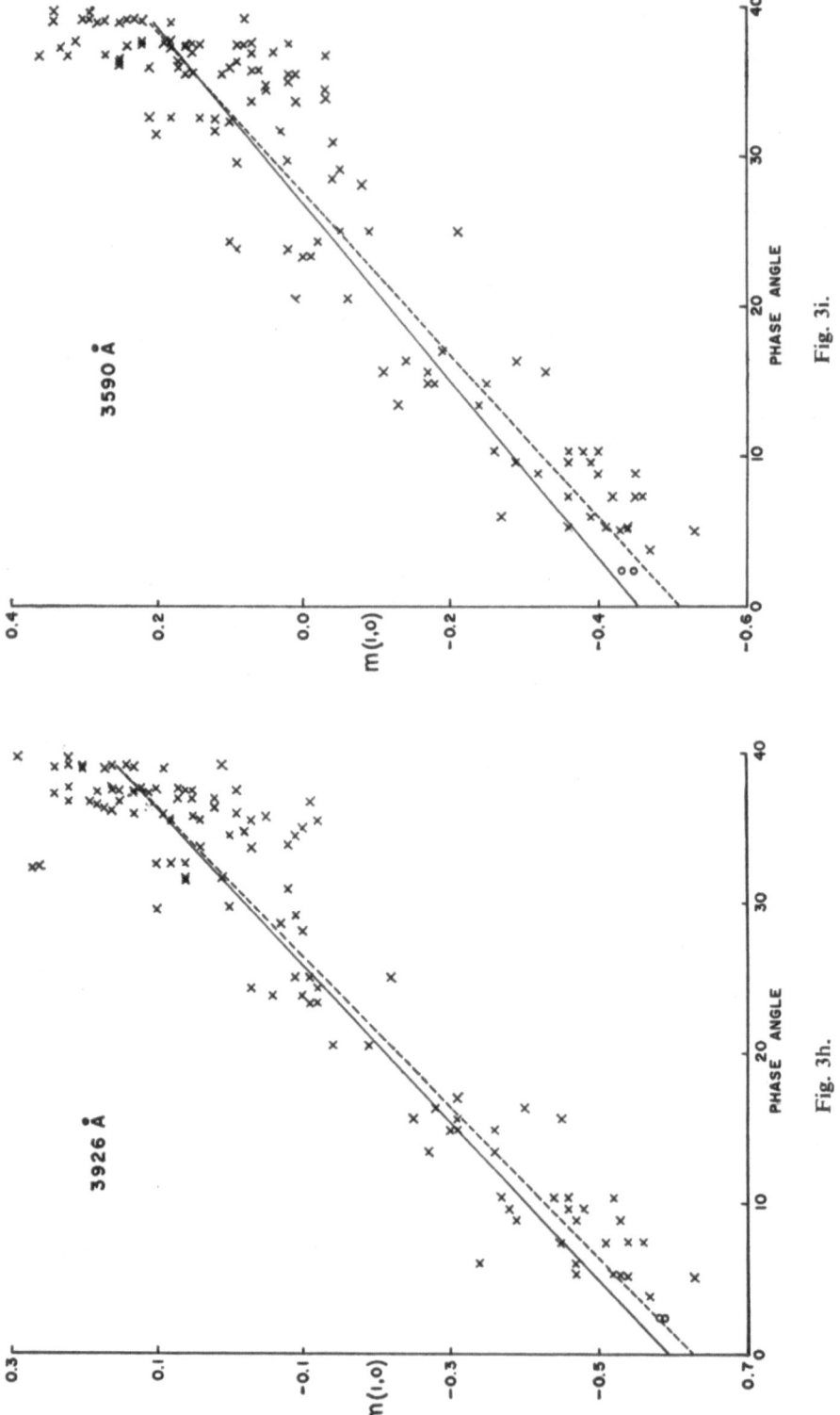

3926 Å

3590 Å

Fig. 3h.

Fig. 3i.

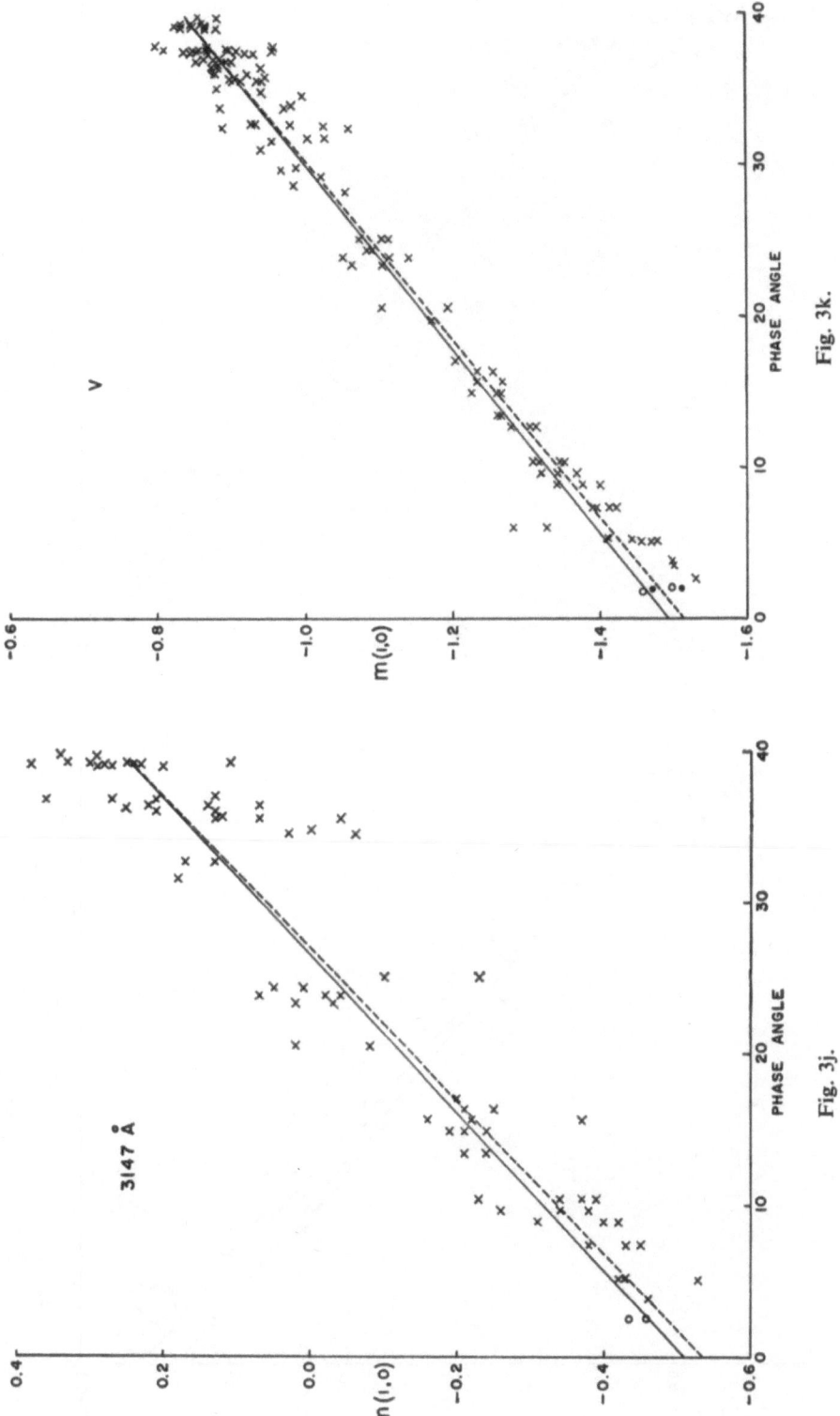

Fig. 3k.

Fig. 3j.

WILLIAM M. IRVINE ET AL.

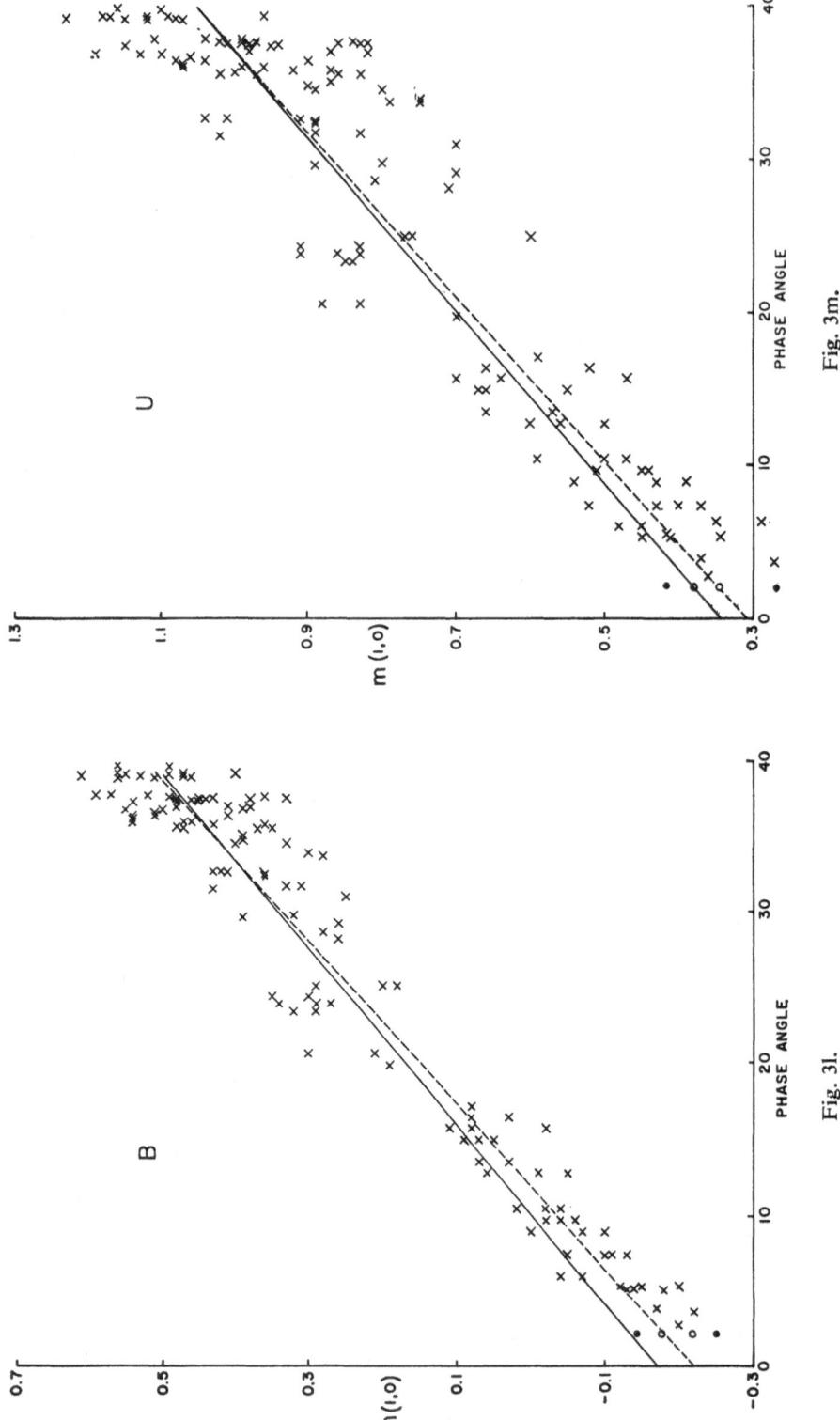

Fig. 3m.

Fig. 3l.

TABLE I

Spectral reflectivity of Mars

λ	$m(1, 0)^a$	a^a	$m(1, 0)^b$	a^b	$m(1, 0)^c$	ΔM_0^d	A
3147 Å	−0.51	0.019	−0.54	0.020	−0.60 ± 0.05	0.09 ± 0.05	0.052
3590	−0.45	0.017	−0.51	0.019	−0.58 ± 0.05	0.13 ± 0.05	0.053
3926	−0.60	0.019	−0.63	0.020	−0.70 ± 0.04	0.10 ± 0.04	0.057
4155	−0.65	0.018	−0.75	0.021	−0.81 ± 0.06	0.16 ± 0.06	0.006
4573	−0.90	0.015	−1.04	0.019	−1.13 ± 0.08	0.23 ± 0.10	0.086
5012	−1.10	0.014	−1.27	0.018	−1.38 ± 0.09	0.28 ± 0.15	0.112
6264	−2.12	0.018	−2.12	0.018	−2.12 ± 0.01	0 ± 0.02	0.244
7297	−2.30	0.016	−2.27	0.016	−2.27 ± 0.02	0 ± 0.02	0.308
8595	−2.27	0.015	−2.27	0.015	−2.27 ± 0.01	0 ± 0.02	0.322
1.06 μ	−2.25	0.015	−2.24	0.015	−2.24 ± 0.01	0 ± 0.02	0.314
U	0.34	0.018	0.31	0.019	0.22 ± 0.07	0.12 ± 0.07	0.052
B	−0.17	0.017	−0.22	0.019	−0.32 ± 0.04	0.15 ± 0.04	0.074
V	−1.49	0.016	−1.52	0.016	−1.58 ± 0.04	0.09 ± 0.04	0.154

[a] Linear fit to data with $i \geqslant 15°$.
[b] Linear fit to data at all i.
[c] Including estimated opposition effect.
[d] Column 2 minus column 6.

added to the plots, the least squares fit performed for all the data (dashed line in Figure 3), and the resultant intercept and slope listed in columns 4 and 5 of Table I. No significant change in the mean curve was found for filters *k-h-g-* or *e* ($\lambda \geqslant 6250$ Å). An opposition effect (anomalous brightening for small phase angles) was, however, observed for filter *V* and shorter wavelengths. This finding is in contradiction to the result previously reported (Irvine *et al.*, 1968b) for this data, although anomalous brightening at the oppositions of 1967 and 1969 has been reported by Bugaenko *et al.* (1967), O'Leary (1967), and Murphy (1969). A (necessarily rough) extrapolation of our results to zero phase results in the values of $m(1, 0)$ shown in column 6 of Table I, where the errors listed are 'eyeball' estimates. We also list the difference ΔM_0 between the $m(1, 0)$ in columns 2 and 6 of Table I (i.e., the 'magnitude' of the opposition effect).

A comment on the internal consistency of our results is in order here. The phase coefficients *a* listed in column 3 of Table I appear anomalously small at λ 4573 and λ 5012 and rather large at λ 6264, both compared to the other narrow band data and also to the broad band (*B* and *V*) results. The wavelength range $5000 \leqslant \lambda \leqslant 4500$ Å will, of course, be most subject to changes in the 'blue haze', and our curves may be weighted by unusual atmospheric conditions. We also note from Paper II that the observations near opposition in bands λ 4155, λ 4573, and λ 5012 may be anomalously bright because of uncertainties in transformation to the standard magnitude system. The combination of these effects makes the value of ΔM_0 in column 9 of Table I particularly uncertain for λ 4573 and λ 5012.

Our results for the opposition effect may be compared with those of O'Leary (1967) and O'Leary and Rea (1968). We do not confirm the existence of an opposition effect at wavelengths $\lambda > 6000$ Å, as those authors report. Rather it seems that at least part of the apparent effect in their data may be due to the selection of an asymptotic phase coefficient a (for $i > 16°$, derived ultimately from Wooley *et al.* (1955)) which is too small, and, in the case of band $R(0.7\ \mu)$, their choice of an $m(1,0)$ from the linear extrapolation (column 2 of Table I) which is fainter by about 0.06 m than is indicated by our data.

At wavelengths $\lambda \leqslant 5500$ Å the opposition effect which we observe is significantly less than that reported by O'Leary and Rea, and does not show the strong wavelength dependence which they report. In fact our results could be read as indicating no wavelength dependence of the effect for 5500 Å $\geqslant \lambda \geqslant 3150$ Å. At U this difference is in part the result of our finding a 'no-opposition-effect' $m(1, 0)$ of 0.34, considerably brighter than used by O'Leary and Rea and derived from de Vaucouleurs (1964). For B and V the difference may be partly due to O'Leary's observations extending to smaller phase angles, and conservatism on our part in the extrapolation of our results.

On the other hand, our observations were made during a different opposition, and parameters such as atmospheric aerosol content may play an important role in determining both a and m. This discussion points out the difficulty of determining the magnitude of the opposition effect on a planet like Mars, for which atmospheric and surface conditions change both during an apparition and from apparition to apparition.

4. Albedos

Values of the geometric albedo including the opposition effect could be obtained from the values of $m(1, 0)$ in column 6 of Table I using the standard formula (e.g., Paper II). The relatively large uncertainties in ΔM_0 make this appear unprofitable, however. Rather we shall use the values in columns 4 and 5 to determine p, and Russell's Rule (Paper II) to find the phase integral q and the spherical albedo $A = pq$; note that the first-order inclusion of the opposition effect increases p and decreases q by the same factor, so that A is left unchanged. Values of A calculated in the manner described, using parameters given in Paper II for the semi-diameter of Mars and the magnitude of the Sun ($V = -26.81$), are listed in the last column of Table I. They fall, not surprisingly, between the values previously quoted in Papers I and II.

Acknowledgments

The observations of Mars discussed here were obtained under a grant to Professor Donald H. Menzel at the Harvard College Observatory. The present research was supported in part by NASA grant NGR 22-010-023 and NSF grant GP 7793 to the University of Massachusetts. Computations were performed principally at the Amherst College Computing Center, and the assistance of their staff is gratefully acknowledged. We are also grateful to Mr. Neil Hopkins for his aid.

References

Bugaenko, L. A., Koval', I. K., and Morozhenko, A. V.: 1967, *Trans. IAU (Commission 16)*, Prague.

de Vaucouleurs, G.: 1964, *Icarus* 3, 187.

Hopkins, N. B. and Irvine, W. M.: 1969, in this volume, p. 349.

Irvine, W. M., Simon, T., Menzel, D. H., Charon, J., Lecomte, G., Griboval, P., and Young, A. T.: 1968a, *Astron. J.* 73, 251 (Paper I).

Irvine, W. M., Simon, T., Menzel, D. H., Pikoos, C., and Young, A. T.: 1968b, *Astron. J.* 73, 807 (Paper II).

Murphy, R. E.: 1969, paper at Albany, N.Y., meeting of A.A.S.

O'Leary, B. T. and Rea, D. G.: 1968, *Icarus* 9, 405.

Wooley, R. v.d. R., Gottlieb, K., Heinz, W., and de Vaucouleurs, A.: 1955, *Monthly Notices Roy. Astron. Soc.* 115, 57.

MIE SCATTERING AND THE MARTIAN ATMOSPHERE

WALTER G. EGAN and KENNETH M. FOREMAN

Research Dept., Grumman Aerospace Corporation, Bethpage, N.Y., U.S.A.

Abstract. It has been suggested that the discrepancy between radio occultation determinations of the Martian atmospheric surface pressure (3.8 to 7 mb) and those deduced from optical polarization measurements and a simple Rayleigh atmosphere model (about 10 mb) are the results of sub-micron sized aerosols in the Martian atmosphere. Based on observed viewing angle dependence of the polarization of the Martian disk in the visual range, a Mie scattering analysis has been made utilizing the measured complex index of refraction of limonite and bulk solid CO_2. The results of this study indicate that limonite aerosols alone are unsatisfactory to explain the viewing angle observations, whereas solid CO_2 (and H_2O ice) aerosol spheres, having a dominant particle radius range between 0.28 and 0.35 μ, could bring planetary and laboratory observations into compatibility. It is suggested, further, that solid CO_2 aerosols could explain limb brightening in the blue spectral range. Various distributions of solid CO_2 and H_2O Mie particles with radii up to 0.35 μ show an opposition effect. However, the role of these aerosols in explaining the Mars opposition observations is very dependent on the optical properties of the underlying Mars surface material.

Measurements of the Martian atmospheric surface pressure by the radio frequency occultation determinations of Mariners IV, VI, and VII yield values between 3.8 and 7 mb (Kliore *et al.*, 1966; Kliore, 1969). The spectroscopic data have been in reasonably good agreement, yielding a value of the order of 10 mb (Kaplan *et al.*, 1964). Still another technique, polarization measurement, has been yielding pressure values that are too large in comparison to the previously noted methods (Dollfus and Focas, 1966). Part of the problem with polarization lies in the nature of the laboratory specimen used for surface simulation as well as in the planetary model used for interpretation of laboratory polarization data compared to earth-based observations of the planet. Polarization is an extremely sensitive indicator of atmospheric as well as surface optical effects (Dollfus and Focas, 1966).

It cannot be emphasized too strongly that average gross features are involved in both telescopic observations and surface simulations, and as large a scale as possible of laboratory surface simulation is necessary (Egan, 1967). Further, simultaneous consideration of many aspects of surface simulation, e.g., photometry and polarimetry and thermal effects, etc., is required (Egan, 1969).

The polarimetric results of Dollfus have been examined and refined analysis yields an average surface pressure of about 20 mb (O'Leary and Pollack, 1969), a factor of 3 or more greater than the 4 to 7 mb value range deduced from Mariner spacecraft data. Our analyses based on further refinements of our large scale surface simulation experimental data have indicated a total surface pressure of about 10 mb for a simple Rayleigh atmosphere model (Foreman, 1969). Thus, an aerosol contribution equivalent to the optical properties of the order of 5 mb of atmosphere appears indicated. Various investigators (e.g., Kuiper (1964), Rea and O'Leary (1965), Coulson (1969)) have calculated the Mie scattering effects of terrestrial, water ice, and silica aerosols.

O'Leary (1967) has questioned the significance of any surface simulation attempt in view of the numerous possibilities for aerosol effects. Based on Mars observations and various optical models, our opinion is that the best approach still lies in a laboratory simulation as close as possible to Mars data and with an allowance being made for a generally small atmospheric perturbing effect.

More extensive Mie scattering calculations for estimating the Mars aerosol effect have become possible this past year with the determination of the complex index of refraction of Venango County (Pennsylvania) limonite (Egan and Becker, 1969) and of bulk solid CO_2 (Egan and Spagnolo, 1969) in the spectral range between 0.35 and 1.0 μ. Both substances are potential candidate constituents of Mars aerosols. It is found that the refractive index of solid CO_2 is close to that of H_2O ice. Therefore, computations for dry ice may be considered as also approximately valid for H_2O ice particles.

This report summarizes a particular aspect of polarization that appears to permit an inference of the nature of aerosols in the Martian atmosphere; a Mie scattering computer program designed and run by Mrs. Jaylee Mead had supplied us with the single scatterer data with which this study has been concerned.

Polarization observations by Dollfus of the central light 'desert' regions of Mars, in the visual spectral range, exhibit a small but real variation of polarization with viewing angle (Figure 1) (Dollfus, 1966). This is of the order of 0.1% to 0.2% based

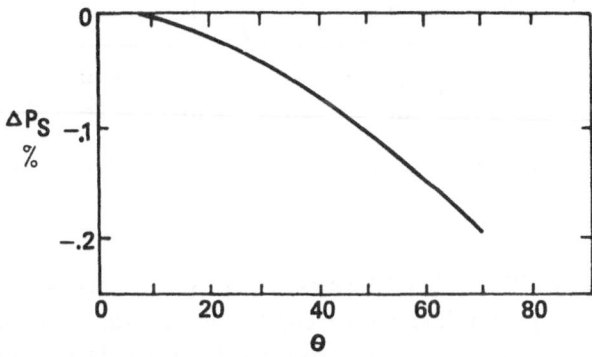

Fig. 1. Mars oblique polarization difference (Dollfus, 1965) $\Delta P_S = P_\theta^\circ - P_0^\circ$.

on the most updated information analyzed by Dollfus (Cann *et al.*, 1965). In the course of our simulation experiments, we also have discerned a comparable real variation of polarization with viewing angle (and wavelength dependent), however, of opposite sign from Dollfus' observations (Egan and Foreman, 1968). A possible explanation for this subtle paradox is the existence of Mie scatterers in the Martian atmosphere which could produce sufficient negative polarization to bring the particulate limonite surface simulation data into agreement with the planetary characteristics. Rayleigh scattering of atmospheric molecules will only produce positive polarization, and, thus, cannot be expected to explain the problem. However, any atmospheric Rayleigh scattering polarization would have to be compensated for additionally by Mie scattering.

The change of Martian polarization with viewing angle shown in Figure 1 amounts to a few tenths of a percent difference in polarization between the center of the disk and the limb, for a particular phase angle, with the limb having the lower polarization.

Our laboratory simulation for the limonite particulate specimen that most closely matches the polarization-brightness characteristics of the bright regions of Mars (i.e., Venango County, Pennsylvania limonite (Egan, 1969)) indicates an oblique polarization difference between 60 degrees and 0 degrees viewing angles for three wavelengths as shown by Figure 2. The data points for a range of phase angles between 5 degrees

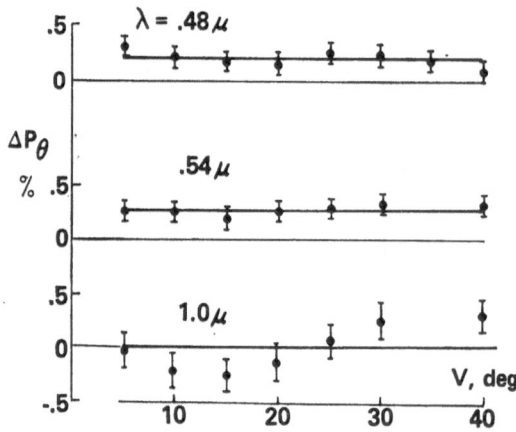

Fig. 2.   Oblique polarization difference Venango limonite $\Delta P_\theta = P_{60°} - P_{0°}$.

and 40 degrees fall on either side of the rms average values shown. The data accuracy is higher for the 0.48 μ and 0.54 μ wavelengths because the photomultiplier sensor signal to noise ratio is greater than for the one used at 1.0 μ. The approximate spectral variation of rms average values are shown in Figure 3. In the visual range (0.54 μ), the limonite specimen will be in agreement with the Martian data if the aerosol component produces a decreasing polarization (i.e., negative ΔP) with increasing viewing angle, at any particular phase angle. (This trend also should hold true for all phase angles.)

Mie scattering by individual particles is dependent on the phase angle, V, not viewing angle, θ, except when the scatterers exist in a layer above a planet's surface. In this latter situation, as the viewing angle increases, the atmospheric and aerosol layer optical path length increases; the polarization contribution of a widespread aerosol layer can become relatively large with viewing toward the limb.

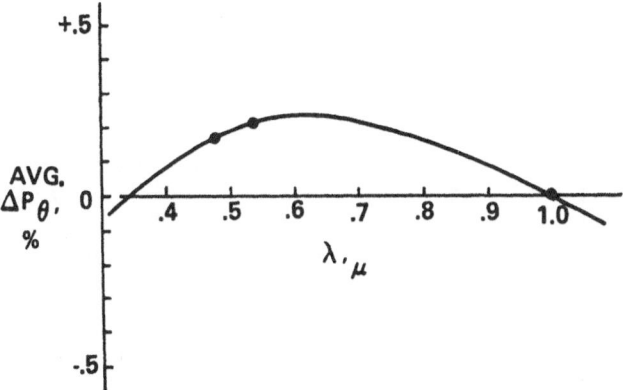

Fig. 3. Average oblique polarization difference – Venango limonite.

The results of the Mie scattering calculations* (simple spheres and single scattering) for Venango County limonite and bulk solid CO_2 at 0.54 μ and 1.0 μ are summarized in Figure 4. In this illustration is shown the variation of the $\sum_{r_0}^{r} B_r P_r$ product term with largest particle radius, r, for a flat (i.e., equal number) distribution of particle sizes between r_0 and r (r_0 is 0.005 μ radius and r can be examined up to a maximum of 0.5 μ). ($\sum_{r_0}^{r} B_r P_r$ represents the relative polarization of Mie aerosols, and is the numerically integrated product of relative brightness and polarization for integrating increments of particle radii, Δr, of 0.005 μ. The summation is for all particle sizes up to the maximum, r, considered.) This product term is proportional to the polarization contribution of Mie aerosols to the total polarization characteristic of the planet, which also includes the atmosphere and the surface contributions. The curves are shown for a phase angle of 40 degrees.

Additional results were computed for various non-flat distributions (e.g., Gaussian, step rise-exponential decay) and were found to produce virtually no variation in qualitative results although some detailed quantitative differences were evident.

It is apparent that the Venango County limonite produces a negative polarization term at 0.54 μ and zero polarization at 1.0 μ (as required by Figure 3) almost uniquely at a dominant particle radius of about 0.27 μ. However, additional data, not presented, but generated by our computer study, shows that at other phase angles between 10 degrees and 30 degrees, this dominant particle size yields positive polarization at $\lambda = 0.54$ μ. Therefore, we can conclude that limonite is not a good choice, in itself, for Martian aerosols, at least under Mie scattering theory constraints.

For solid CO_2, the situation appears different. In the particle size range between about 0.28 and 0.35 μ radius, the viewing angle variation of polarization difference, ΔP, is of the correct sign and magnitude at the two wavelengths. (It should be noted

---

* Provided by Mrs. Jaylee Mead in private communications, July–October, 1969.
   12—P.A.

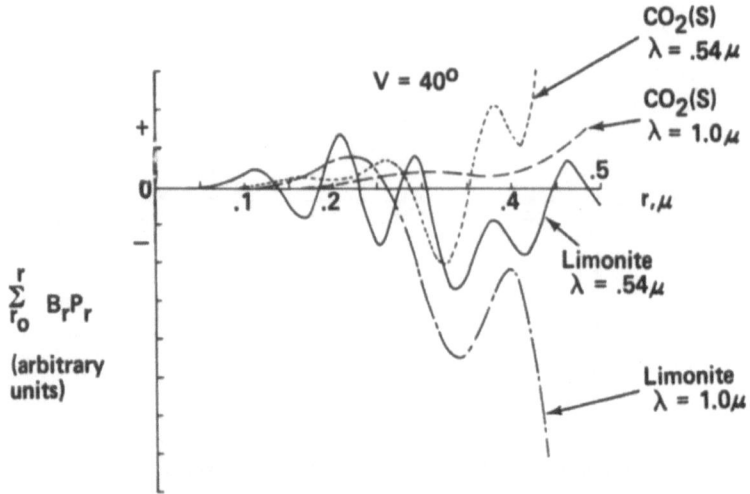

Fig. 4.   Integrated product of brightness and polarization vs. Mie scatterer size.

that for $\lambda = 1.0~\mu$, we must subtract the summation ordinate value for all particle radii sizes smaller than $0.28~\mu$ from the value at larger sizes and this results in a virtually zero polarization term for dominant particle radii between $0.28~\mu$ and $0.35~\mu$, as required.)

If we examine the photometric properties of solid CO_2 and limonite as a function of Mie particle radii at a 0 degrees phase angle, for instance, Figure 5, it is evident that the solid CO_2 particles are generally an order of magnitude brighter in the blue ($0.365~\mu$) than in the red ($0.740~\mu$), and increases in brightness with particle size (up to at least $r = 0.5~\mu$). Limonite aerosols, however, generally do not exhibit a strong color differentiation, and, in many size ranges are of the same order of brightness. Thus, if solid CO_2 aerosols of submicron size exist at the limbs of Mars, they could be expected to contribute to the commonly observed limb brightening in the blue portion of the spectrum.

In Figure 6, a further examination of solid CO_2 aerosols for other phase angles in the visible region shows that the particle size range between $0.28~\mu$ to $0.35~\mu$ remains compatible in sign, although varying, to some degree, in magnitude for the smaller phase angles.

Continuing to explore the photometric properties of solid CO_2 aerosols, in association with a Venango County limonite simulated surface, as a function of wavelength, one can see from Figure 7: (1) a 1–2 mm particulate limonite surface, at a wavelength of $0.54~\mu$* has an opposition effect close to that observed by O'Leary (1967) for Mars but above that inferred from de Vaucouleurs (1964); (2) solid CO_2 aerosols spheres, with a flat size distribution between $0.005~\mu$ radius and approximately $0.33~\mu$, show an opposition effect particularly at the shorter wavelengths. It should be noted that in

* Currently, data available are for this wavelength only.

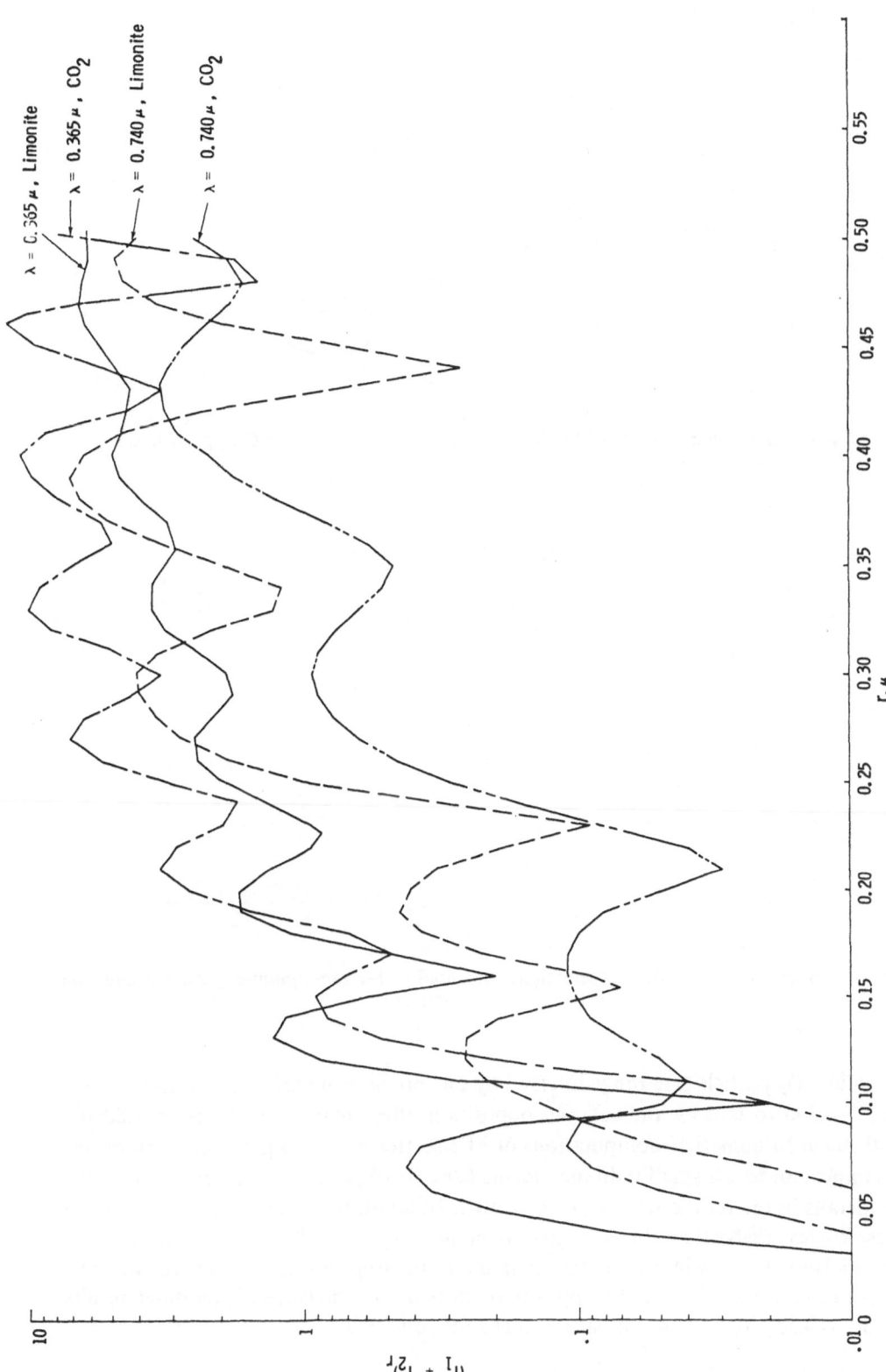

Fig. 5. Relative brightness of Mie particles ($V = 0°$).

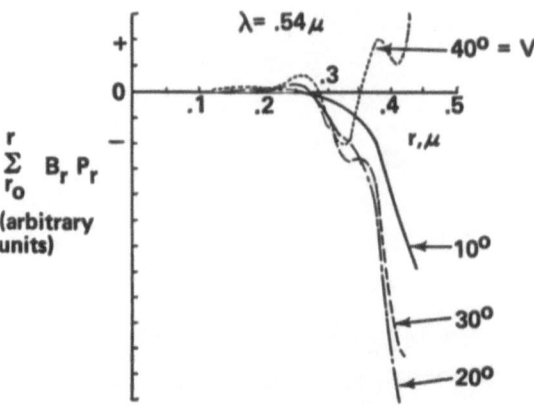

Fig. 6.   Integrated product of brightness and polarization vs solid CO_2 particle size and phase angle.

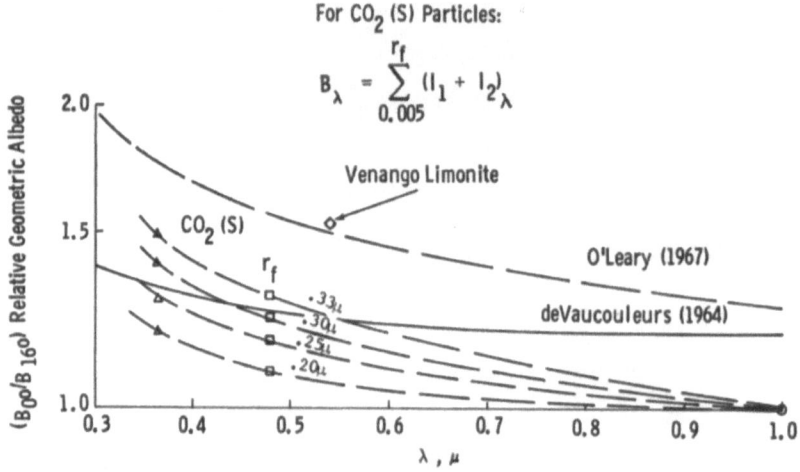

Fig. 7.   Opposition effect: Mars observations compared to 1–2 mm limonite specimens and Mie calculations for solid CO_2.

the solid CO_2 particle size range inferred by our previous polarization considerations (i.e., ~0.300 to 0.325 μ radius), the opposition effect may be as large as 1.80 for $\lambda = 0.365\ \mu$. In numerical computations of Mie scattering photometric characteristics, it is important to use small radii increments ($\varDelta r \gtrsim 0.005\ \mu$) for the particles so that the fine details in the results will not be obscured. In addition, it is advisable to examine phase angles greater than the 16 degrees reference angle, usually employed for opposition analyses, for possible secondary maxima in the opposition effect curve; we have found no such violation of the opposition effect in our analyses of computer results for the ~0.30 μ particle radius size deduced for solid CO_2.

It should be noted that if other wavelength determinations with limonite surface specimens similarly follow the trend (Figure 7) of the Mars opposition observations, then the need for aerosol particles to explain the opposition phenomena are considerably lessened. Additional studies of opposition effect are discussed in the other conference paper by Mead (1970). Her results are based on a particular and different set of assumptions, but her conclusions do not conflict with those of this paper which are based on a limonite surface simulation model that closely matches the Martian observations.

Figure 8 presents the variation of vapor pressure of solid CO_2 with temperature, and an indication of the equilibrium temperatures between the solid and a pure gaseous CO_2 atmosphere. For an atmospheric surface pressure of 7 mb, near the

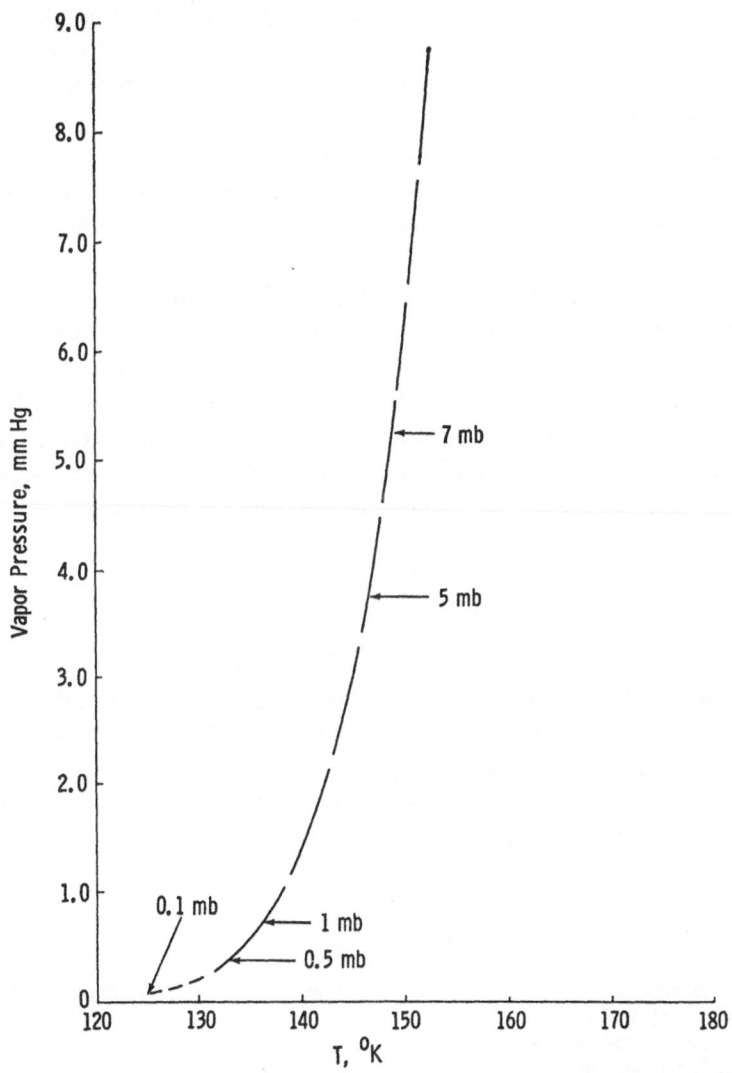

Fig. 8.   Vapor pressure of solid CO_2 (NBS Circular No. 564, revised).

Kliore *et al.* inferred surface pressure (Kliore *et al.*, 1966; Kliore, 1969), the equilibrium temperature is almost 149 K. Radiation-dominated models of the Martian atmospheric dynamics have been studied by Gierasch and Goody (1967/1968) and earth type convective models by Leovy and Mintz (1966).

Profile models of the Martian atmosphere have temperature distributions which depend upon the assumed condition of surface temperature (Gierasch and Goody, 1967/1968; Leovy and Mintz, 1966; Johnson, 1965). Except for polar and high winter latitudes, the near-surface temperature generally does not seem to approach a low enough value to result in dry ice production. However at altitudes where the atmospheric pressure is of the order of a magnitude lower or less (i.e., at 10 km or above) atmospheric temperatures may become low enough because of atmospheric radiative and dynamic processes to coincide with the temperature at which solid CO_2 will form, as indicated by the phase curve. The 1969 Mariner spacecraft photographs (Leighton *et al.*, 1969) show that clouds do occur at altitudes of approximately 15 to 40 km above the Martian surface and it is reasonable to consider these clouds as consisting of solid CO_2 as well as H_2O ice.

The mean aerosol particle size of the order of 0.30 μ radius that is inferred by our results, could be considered also to be representative of a distribution of aerosol particle sizes between, say, 0.28 to 0.35 μ. It might be noted that this size range is quite unlike our terrestrial experience (Junge, 1967).

However, the total environment of Mars apparently is quite different from that of the Earth, and perhaps these results are not as unreasonable as they first may appear. Certainly, the limitations of the Mie scattering theory constraints must be recognized. In addition, our assumptions about a simple quiescent aerosol volume distribution are rightly questionable. Nevertheless, this study has improved the perspective concerning Mars, and hopefully enables one to discern what remains significant in the abundant speculation about the Martian atmosphere that pervades the literature.

Acknowledgments

We gratefully thank Mrs. Jaylee Mead (NASA-Goddard Space Flight Center) for her considerable assistance in performing and providing us with her Mie scattering computer calculations. The authors express their appreciation to the Grumman Aerospace Corporation for its sponsorship and encouragement in the pursuit of this study, and to their colleagues Dr. G. McCoyd, Dr. N. Milford, and L. Smith for stimulating discussions on various aspects of the problem.

References

Cann, M. W. P., Davies, W. O., Greenspan, J. A., and Owen, T. C.: 1965, *NASA CR-298 Technical Report*, IIT Research Institute.
Coulson, K. L.: 1969, *Appl. Opt.* **8**, 1287.
de Vaucouleurs, G.: 1964, *Icarus* **3**, 187.
Dollfus, A.: 1966, *Compt. Rend. Acad. Soc. Paris* **262**, Serie b, 519.

Dollfus, A. and Focas, J. H.: 1966, *Polarimetric Study of the Planet Mars* (AFCRL 66-492 (AD635, 928), Air Force Cambridge Research Laboratories, through European Office of Aerospace Research).

Egan, W. G.: 1967, *J. Geophys. Res.* **72**, 3233.

Egan, W. G.: 1969, *Icarus* **10**, 223.

Egan, W. G. and Becker, J. F.: 1969, *Appl. Opt.* **8**, 720.

Egan, W. G. and Spagnolo, F. A.: 1969, *Appl. Opt.* **8** (in press).

Egan, W. G. and Foreman, K. M.: 1968, *Astron. J.* **78**, S92.

Foreman, K. M.: 1969, AIAA Paper No. 69–52 (presented at the 7th Aerospace Sciences Meeting, New York City, Jan. 20–22, 1969).

Gierasch, P. and Goody, R.: 1967/1968, *Planetary Space Sci.* **15**, 1465; **16**, 615.

Johnson, F. S.: 1965, *Science* **150**, 1445.

Junge, C.: 1967, NASA TTF-11, 091 (N67-36177).

Kaplan, L. D., Münch, G., and Spinrad, P.: 1964, *Astrophys. J.* **139**, 1.

Kliore, A.: 1969, presentation of Mariners VI and VII radio occultation experiment results, NASA, Washington, D.C.

Kliore, A., Cain, D. L., and Levy, G. S.: 1966, *Proc. 7th International COSPAR Symposium, Vienna, May 11–17, 1969*.

Kuiper, G. P.: 1964, *Comm. Lunar Planetary Lab.* **2**, 79.

Leighton, R. B., Horowitz, N. H., Murray, B. C., Sharp, R. P., Herriman, A. H., Young, A. T., Smith, B. A., Davies, M. E., and Leovy, C. B.: 1969, *Science* **166**, 49.

Leovy, C. B. and Mintz, Y.: 1966, Memorandum RM-5110-NASA Contract No. NASr-21-(07), Rand Corporation, Santa Monica.

Mead, J. M. : 1970, this volume, p. 166.

O'Leary, B. T.: 1967, *Space Science Laboratory Series* **8**, 103.

O'Leary, B. T. and Pollack, J. B.: 1969, *Icarus* **10**, 238.

Rea, D. G. and O'Leary, B. T.: 1965, *Nature* **206**, 1138.

THE CONTRIBUTION OF ATMOSPHERIC AEROSOLS TO THE MARTIAN OPPOSITION EFFECT*

JAYLEE M. MEAD

Laboratory for Space Physics, NASA Goddard Space Flight Center, Greenbelt, Md., U.S.A.

Abstract. Mie scattering calculations have been made for atmospheric aerosols having various indices of refraction to determine their possible contribution to a Martian opposition effect, such as that reported by O'Leary in 1967. Neither substances with a real index between 1.20 and 1.50, such as ice, water, or solid CO_2, nor highly absorbing materials, such as limonite, can produce the observed effect. Submicron-sized spherical particles with refractive indices of 1.55 to 2.00 do, on the other hand, exhibit a marked increase in reflectivity at small phase angles and might be responsible for the enhanced brightness at the shorter wavelengths.

Observations of Mars made by O'Leary (1967a, b) during the 1967 opposition show an 'opposition effect', i.e., a non-linear surge in brightness as the planet approaches 0° phase angle (the angle at the planet between the directions to the sun and to the observer). Figure 1, based on data taken from O'Leary and Rea (1968), shows the observed reflectivities, adjusted for the color of the sun, as a function of phase angle α, for five colors: U, B, V, R, and I. The observations were made at phase angles of 1.2° to 8° and are indicated by solid lines. These were fitted to the linear phase functions for $\alpha \geqslant 10°$, reported by de Vaucouleurs (1964).

As O'Leary and Rea pointed out, the opposition effect is much more pronounced at shorter wavelengths than at longer wavelengths, as evidenced by the fact that the U and B observations depart much more from the linear extrapolation than do the curves at R and I. The reflectivity, or albedo, on the other hand, is much greater at longer wavelengths than at shorter ones.

This increased opposition effect in the blue and ultraviolet could be primarily a surface effect in that the surface may have a much greater increase in reflectivity at shorter wavelengths; alternatively, it could be primarily due to light scattering in the atmosphere, as suggested by O'Leary (1967a).

Rayleigh scattering by molecules and by particles small compared to the wavelength of observation does not provide a sudden increase of brightness near 0° phase angle. Therefore, if the effect is primarily atmospheric, particles of larger size must be responsible.

Because of the low albedo in the U, a small brightness contribution by atmospheric aerosols will have a comparatively large effect; in the I, where the surface is much brighter, a small brightness contribution by aerosols will cause little or no change in the total brightness.

* The complete version of this paper appears in *Icarus* **13**, No. 1 (1970).

Sagan et al. (eds.), *Planetary Atmospheres*, 166–169.

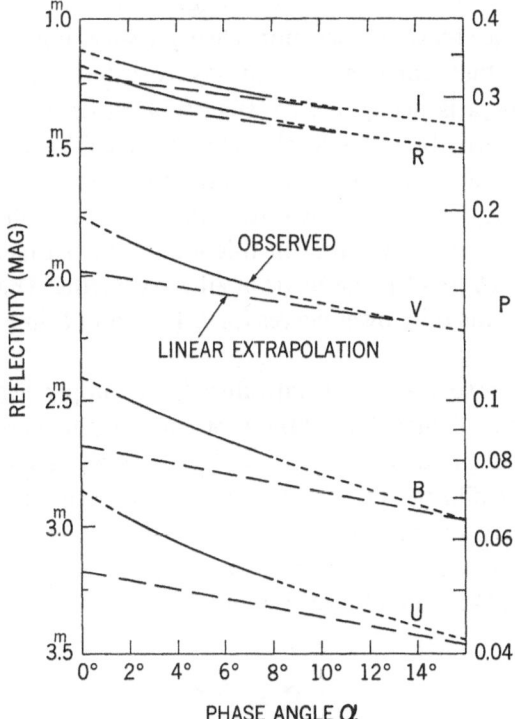

Fig. 1.   The Martian opposition effect in five colors, after O'Leary and Rea (1968), adjusted for the color of the sun. The observations (solid lines), which cover $1.2° \leqslant \alpha \leqslant 8°$, have been fitted to the linear phase functions for $\alpha \geqslant 10°$ reported by de Vaucouleurs (1964). Reflectivity is shown on a logarithmic scale on the right and on an equivalent magnitude scale on the left.

The purpose of this study is to investigate the contribution which atmospheric aerosols might make to the Martian opposition effect, under the assumption that the increased enhancement in the shorter wavelengths, where the surface albedo is low, is primarily an atmospheric effect rather than a surface effect.

The Mie scattering theory, which describes single scattering by spherical particles, was used to make light scattering calculations for substances such as ice, water, and solid CO_2, which have no significant absorption in the wavelength range under consideration. In addition, calculations were made for highly absorbing materials, such as limonite.

Various submicron particle size distributions were used to compute integrated scattering intensities for these aerosols as a function of phase angle and wavelength. The results indicated that neither substances having a refractive index between 1.20 and 1.50, which includes ice ($n=1.31$), water (1.33), and solid CO_2 (1.35: Egan and Spagnolo, 1969), nor highly absorbing materials, such as limonite (complex index $\tilde{n}=2.23-0.669i$ at $\lambda=0.365\mu$: Egan and Becker, 1969), can produce the opposition effect. Refractive indices of 1.55 to 2.00, on the other hand, were found to exhibit a definite increase in reflectivity at small phase angles.

A model was constructed to give the total brightness contribution by the surface plus atmospheric aerosols. At longer wavelengths, where the Martian albedo is higher and where surface markings are more clearly visible, it is reasonable to assume that the observed brightness comes almost entirely from the surface, and the brightness contribution by aerosols is negligible. As suggested by de Vaucouleurs (1968), we took the lunar photometric function developed by Hapke (1963) and modified it to fit the observed Martian brightness-phase curve at these longer wavelengths. We assumed that the phase curve for the surface would have the same shape in all colors; only the albedo would change. This means that in this model, the surface brightness would increase by 30% from 16° to 0° phase angle at all wavelengths. These assumed surface functions are shown as the thin lower curves for I, V, B, and U in Figure 2 (R has been omitted for simplicity).

The effect of the aerosols was then introduced. The upper heavy solid curves in Figure 2 represent the calculated brightness of surface plus aerosols for refractive index 1.75 and a skewed gaussian-type particle distribution peaked at 0.4 μ. At shorter wavelengths, where the albedo and surface contrast are greatly reduced, atmospheric aerosols are seen to play a significant role.

The calculated phase curves are in reasonable agreement with the observations, shown as dashed lines. One should bear in mind, however, that there was a good deal

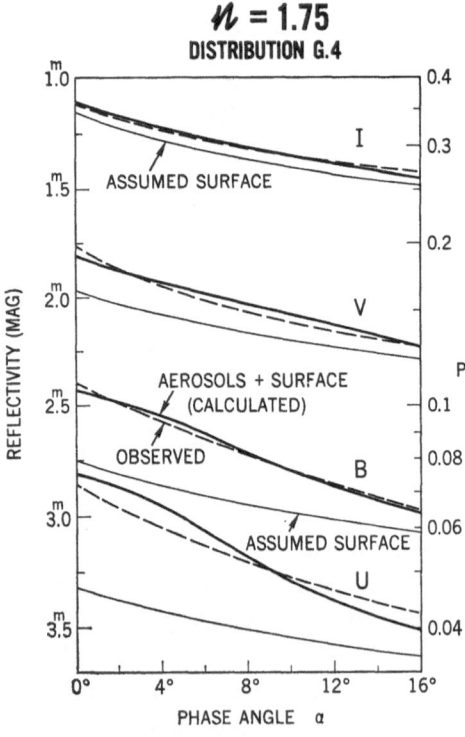

Fig. 2.   Comparison of model with Mars observations. For each wavelength, the thin lower curve is the assumed surface reflectivity, the heavy upper curve is the calculated brightness from the surface plus aerosols, and the broken curve is the Martian observational data.

of arbitrariness in obtaining this fit. It is by no means a unique solution to the problem. It does show, however, that the presence of a small amount of atmospheric aerosols, with the proper index of refraction, could provide the observed increased opposition effect for Mars in the ultraviolet, where the albedo is very low, but at the same time make a negligible contribution in the infrared, where the surface albedo is high.

References

de Vaucouleurs, G.: 1964, 'Geometric and Photometric Parameters of the Terrestrial Planets', *Icarus* **3**, 187–235.

de Vaucouleurs, G.: 1968, 'On the Opposition Effect of Mars', *Icarus* **9**, 598–599.

Egan, W. G. and Becker, J. F.: 1969, 'Determination of the Complex Index of Refraction of Rocks and Minerals', *Appl. Opt.* **8**, 720–721.

Egan, W. G. and Spagnolo, F. A.: 1969, 'Complex Index of Refraction of Bulk Solid Carbon Dioxide', *Appl. Opt.* **8**, 2359–2360.

Hapke, B. W.: 1963, 'A Theoretical Photometric Function for the Lunar Surface', *J. Geophys. Res.* **68**, 4571–4586.

O'Leary, B. T.: 1967a, 'The Opposition Effect of Mars', *Astrophys. J.* **149**, L147–L149.

O'Leary, B. T.: 1967b, 'Mars: Visible and Near Infrared Studies and the Composition of the Surface', Ph.D. thesis, University of California, Berkeley, Calif., 165 pp.

O'Leary, B. T. and Rea, D. G.: 1968, 'The Opposition Effect of Mars and its Implications', *Icarus* **9**, 405–428.

ULTRAVIOLET POLARIZATION MEASUREMENTS OF MARS AND THE OPACITY OF THE MARTIAN ATMOSPHERE*

ANDREW P. INGERSOLL

Division of Geological Sciences, California Institute of Technology, Pasadena, Calif., U.S.A.

Abstract. Ground-based polarimetric data taken near maximum elongation are presented. These data are analyzed assuming an optically thin, Rayleigh scattering atmosphere, and a surface whose polarization varies inversely as the surface albedo. The best fit to the data yields an optical depth for Rayleigh scattering corresponding to a surface pressure of 6 ± 1 mb, if carbon dioxide is the principal constituent. There is no need to postulate the existence of fine dust in the Martian atmosphere. This method is potentially capable of resolving elevation differences on the Martian surface.

1. Introduction

This paper is a report of polarization measurements of Mars at 200 Å resolution from 3200 Å to 7000 Å wavelength. The measurements were made near maximum elongation, when the solar phase angle θ was 42.5°. Two quantities were determined for each wavelength interval: the percentage polarization of reflected sunlight from a region of Mars, and the direction of polarization relative to the Sun-Mars line in the terrestrial sky. Combining these data with the photometric brightness measurements of Irvine *et al.* (1968) yields an estimate of the optical depth for Rayleigh scattering on Mars. The analysis is based on the assumption of an optically thin, Rayleigh scattering atmosphere, and a surface of low albedo which polarizes inversely as its albedo. The derived value of surface pressure is in accord with the Mariner occultation estimates, and the probable error is no larger than that obtained from the occultations. Thus there is no inconsistency between these polarization measurements and the model of a pure Rayleigh scattering atmosphere of carbon dioxide.

The significance of this conclusion is first, that previous polarization studies have been cited as evidence for high surface pressures (Pollack, 1967), or fine dust in the Martian atmosphere (Dollfus and Focas, 1966), but in fact, the polarization measurements are consistent with a simpler model in agreement with spacecraft results. The main difference between these measurements and previous measurements is the full wavelength coverage from 3200 to 7000 Å. Data at the shorter wavelengths are especially important because the atmosphere is the dominant source of polarization only at wavelengths below 4000 Å. The second aspect of these results is that the method may be used to measure topographic relief on Mars, as an alternative to the radar method (Pettengill *et al.*, 1969) and the spectroscopic method (Belton and Hunten, 1969). The method may also be used in searching for an atmosphere on Mercury.

* Contribution No. 1694 of the Division of Geological Sciences, California Institute of Technology.

2. Experimental Procedure

The measurements were made on August 12–13, 1969 from 03 30–05 30 UT, with the 100-inch telescope on Mt. Wilson. The instrument was operated at the Cassegrain focus, off the axis of the telescope, in such a way that the polarization introduced by the reflection at the Cassegrain flat was cancelled by reflection at a second mirror in the photometer (Figure 1). Thus, letting the axis of the telescope be the x-axis and the deflected beam be the y-axis, the direction after the second reflection must be the z-axis in order that the polarizations cancel. It is also necessary that both mirrors be of the same material (i.e., aluminum).

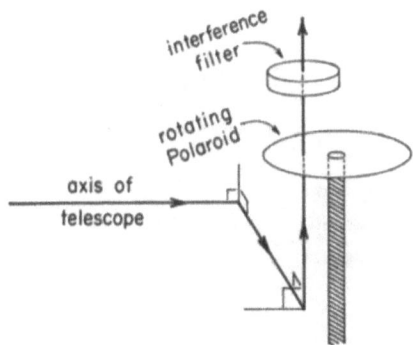

Fig. 1. Ray path in photoelectric polarimeter for use at East- or West-arm Cassegrain focus.

After the second right-angle reflection, the beam passed through a rotating Polaroid disk, then through an interference filter, and then into the detector, an S-20 photo-multiplier tube. The a.c. component of the signal was synchronously demodulated using a phase reference driven at twice the frequency of the Polaroid disk. (A 180° rotation of the Polaroid corresponds to a full cycle in the plane of polarization.) The d.c. component of the signal was amplified separately, and both outputs were simultaneously recorded on a d.c. strip chart recorder.

If we think of the incoming flux as the superposition of two linearly polarized fluxes $F^\perp$ and $F^\parallel$, then the output of the d.c. channel is proportional to $(F^\perp + F^\parallel)$, and the a.c. channel to $(F^\perp - F^\parallel) \times \cos \psi$, where ψ is the phase of the reference signal relative to the $\perp$ position of the Polaroid. This is equivalent to resolving the polarized flux $(F^\perp - F^\parallel)$ along a direction at an angle $\psi/2$ to the $\perp$ direction. Thus measurements at two values of ψ are sufficient to determine the plane of polarization and degree of polarization of the incident beam. The system was calibrated in advance using plane-polarized light in known orientations.

This system was tested using stars whose polarization has been measured or which have degrees of polarization below some threshold value (Hall, 1958; Appenzeller, 1968). The probable error determined in this way was several tenths of 1% in the degree of polarization, and several degrees of angle in the plane of polarization. Since the measured polarizations varied from 1.3% at 7000 Å, to 13% at 3200 Å, this was considered sufficient accuracy for the purposes of the experiment.

3. Analysis of Results

Measurements were made of a large bright area (Arabia) at Martian longitude 290° to 330°, which was located in the center of the disk at the time of observation. The solar phase angle was 42.5°. The plane of polarization was found to lie perpendicular to the Sun-Mars line at wavelengths less than 4500 Å, but appeared to deviate up to 10° at longer wavelengths. The observed degree of polarization vs wavelength is plotted in Figure 2.

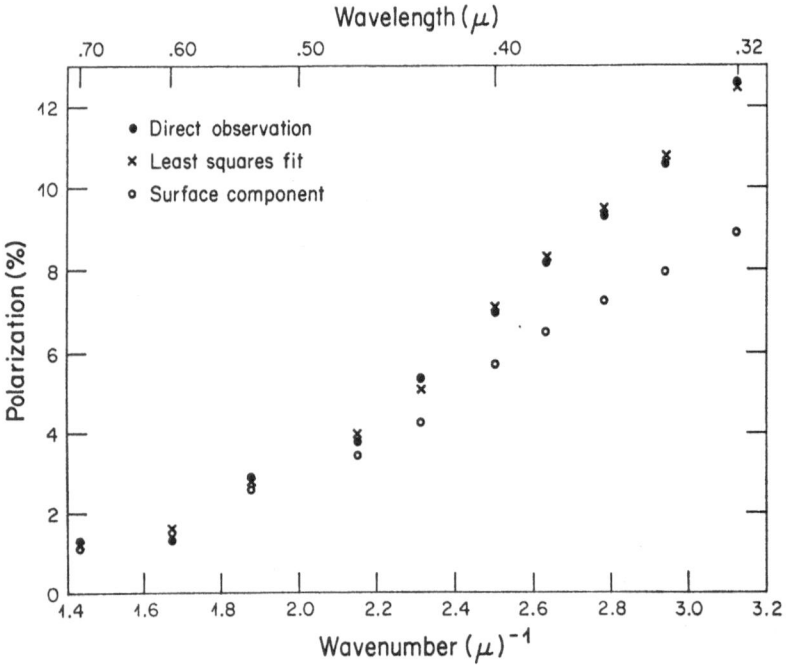

Fig. 2. Polarization of Arabia at the center of the Martian disk for solar phase angle 42.5°.

The most important part of the data analysis is separating effects of the surface from those of the atmosphere. To do this, we need a model which predicts how each component of the reflected light should vary with wavelength, and how these components should combine to produce the observed signal. The simplest model of the atmosphere is a layer of Rayleigh scatterers, i.e., particles small compared to the wavelength. This model includes the case of a pure, molecular atmosphere, or one with very fine suspended particles. We shall also assume that the atmosphere is optically thin, which follows from the Rayleigh scattering assumption and the fact that the planet has a low albedo. The attenuation of the incident solar beam as it passes through the atmosphere will be neglected; only first-order scattering, directly out of the incident beam, will be considered. Illumination of the atmosphere by the surface, and vice-versa, will be neglected because the albedo of the surface and the optical depth of the atmosphere are both small.

The polarization measurements described here refer to a point on the disk defined by μ_0 and μ, the cosines of the angles of incidence and reflection, respectively. In analysing the data, it is necessary to know the absolute value of the intensity from the point, as well as its direction and degree of polarization. Unfortunately, however, most of the absolute brightness studies refer to the integrated disk, at various wavelengths and phase angles. Therefore, we need a scattering law for the surface and atmosphere in order to obtain the brightness of a point of the disk from the average brightness of the whole planet.

For an optically thin, Rayleigh scattering atmosphere, the scattering law is known. We define the optical depth

$$\tau = \frac{8\pi^3}{3\lambda^4} \frac{(n^2 - 1)^2}{N_0^2} \int_0^\infty N \, dz, \tag{1}$$

where λ is the wavelength of the scattered light, n is the index of refraction of the medium at STP, N_0 is Loschmidt's number, z is the vertical coordinate, and N is the number density of molecules in the atmosphere. For a point (μ_0, μ) the intensity of light scattered by the atmosphere is

$$I_a^\perp = \frac{3F_\odot\tau}{16\pi\mu}, \qquad I_a^\parallel = \frac{3F_\odot\tau}{16\pi\mu} \cos^2 \theta, \tag{2}$$

in a specified wavelength interval, where $F_\odot$ is the flux of sunlight at Mars in the interval. The symbols $\perp$ and $\parallel$ refer to components resolved along directions perpendicular and parallel to the plane of scattering, and the subscripts 'a' refer to the atmosphere. Equation (2) is singular at the limb, but the singularity is integrable, and the equation may be integrated over the illuminated portion of the disk to yield the flux per unit solid angle emanating from the atmosphere:

$$F_a^\perp + F_a^\parallel = \frac{3F_\odot\tau}{8\pi} (\pi - \theta)(1 + \cos^2 \theta) \, r^2, \tag{3}$$

where r is the radius of the planet. Finding τ is our main task, but first we must develop a set of relations for the surface, similar to Equations (2)–(3).

We assume a surface scattering law of the form (Minnaert, 1941; Harris, 1961)

$$I_s^\perp + I_s^\parallel = F_\odot C \mu_0^k \mu^{k-1}. \tag{4}$$

Here k will be regarded as a parameter of the problem independent of wavelength; the analysis is insensitive to the value of k for $0.75 \leqslant k \leqslant 1.0$. The quantity C is related to the brightness of the surface, and is therefore wavelength dependent. Integrating (4) over the illuminated disk we obtain, for the flux per unit solid angle emanating from the surface:

$$F_s^\perp + F_s^\parallel = F_\odot r^2 C \int\limits_{\theta - (\pi/2)}^{\pi/2} [\cos(\phi - \theta) \cos \phi]^k \, d\phi \int\limits_{-1}^{+1} (1 - \mu^2)^k \, d\mu. \tag{5}$$

Then, combining Equations (3) and (5) and dividing by $r^2 F_\odot$, we obtain an expression for the geometric albedo of the planet at phase θ in terms of τ, k, and C. The geometric albedo of Mars $p(\lambda, \theta)$ has been measured at the wavelengths and phases of interest in this study (Irvine *et al.*, 1968) and therefore we may express C in terms of τ and k, for each wavelength of interest:

$$C = \frac{p(\lambda, \theta) - (3\tau/8\pi)(\pi - \theta)(1 + \cos^2 \theta)}{\int\limits_{\theta - (\pi/2)}^{\pi/2} [\cos(\phi - \theta) \cos \phi]^k \, d\phi \int\limits_{-1}^{1} (1 - \mu^2)^k \, d\mu}. \tag{6}$$

Finally, an important assumption of the analysis is that the degree of polarization of light scattered by the surface is inversely proportional to C, that is, inversely proportional to surface brightness. This law appears to hold for the Moon (Gehrels *et al.*, 1964), and for powders which have been observed in the laboratory (Dollfus, 1961; Coffeen, 1965). Thus we assume

$$I_s^\perp - I_s^\parallel = F_\odot \beta \mu_0^k \mu^{k-1}, \tag{7}$$

where β is a constant, like k, which is independent of wavelength. The degree of polarization from the surface δ_s is β/C, which follows from Equations (4) and (7).

The polarized component of intensity from a point on the disk is obtained by combining $I_a^\perp - I_a^\parallel$ and $I_s^\perp - I_s^\parallel$, Equations (2) and (7), respectively. Dividing this by the total intensity we obtain an expression for the degree of polarization of a point on the disk:

$$\delta = \frac{(3\tau/16\pi\mu)(1 - \cos^2 \theta) + \beta \mu_0^k \mu^{k-1}}{(3\tau/16\pi\mu)(1 + \cos^2 \theta) + C \mu_0^k \mu^{k-1}}. \tag{8}$$

Equation (8) involves the undetermined quantities τ, β, k, and C. However C can be eliminated from the problem by Equation (6), and k is treated as a parameter in the range $0.75 \leqslant k \leqslant 1.0$. τ varies as λ^{-4} and β is a constant independent of wavelength, so the problem reduces to determining the two constants $\tau\lambda^4$ and β.

The experimental data in Figure 2 refer to a large bright area centered at $\mu_0 = 0.91$, $\mu = 0.95$. Using these values, the degree of polarization δ was computed from Equation (8) for various $\tau\lambda^4$ and β. For each τ, values of C were determined from Equation (6), using determinations of $p(\lambda, \theta)$ from Irvine *et al.* (1968). It is necessary to assume that the area where the polarization measurements were made is typical of the planet as a whole; this is probably valid for bright areas, which comprise about 70% of the area of Mars. In this way, the values of $\tau\lambda^4$ and β which produce the best least-squares fit to the observational data in Figure 2 are

$$\beta = 0.9 \times 10^{-3}, \qquad \tau\lambda^4 = 2.1 \times 10^{-4} \, (\mu m)^4. \tag{9}$$

The uncertainty in each of the above quantities is about $\pm 15\%$, which includes the probable error from the least-squares fit for a given k, as well as the variation for different k in the range $0.75 \leqslant k \leqslant 1.0$. The points calculated in this way are also plotted in Figure 2.

4. Conclusions

We may interpret the estimate of $\tau \lambda^4$ in Equation (9) assuming the Rayleigh scattering opacity is due entirely to an atmosphere of pure carbon dioxide. This yields an estimate of the number of molecules in an atmospheric column, and therefore of the Martian surface pressure P_s on the region of Mars where the measurements were made:

$$P_s = 6 \pm 1 \text{ mb}. \tag{10}$$

From the derived values of β and C we may compute the polarization of light scattered by the surface alone, $\delta_s = \beta/C$. These values are also plotted in Figure 2.

As stated in the Introduction, the interpretation presented here is based on the simplest model one might propose for the Martian atmosphere. Other models could be made to fit the data equally well, but the agreement between our estimate of surface pressure Equation (10), and the generally accepted value, is significant. Our conclusion is that polarization measurements are consistent with a molecular atmosphere without suspended fine particles. On the other hand, thin clouds are observed in the Martian atmosphere (Leighton et al., 1969), but these do not necessarily contribute to the observed polarization. If the light scattered by these clouds is unpolarized, their presence will have no effect on the present interpretation.

The optical depth reported here is small even at the shortest wavelengths, and therefore the atmosphere is quite transparent. If this model is correct, then the disappearance of surface features in the blue is a property of the surface itself, and is not due to obscuration by the atmosphere. On the other hand, the light scattered by the atmosphere is a significant fraction of the total light from the planet at the shortest wavelengths, because the surface albedo is low. According to our model, the surface albedo decreases to 60% of the planetary albedo at 3200 Å, and so the surface must be extremely dark in the ultraviolet.

The results reported in this paper are preliminary. Future observations should include direct measurements of brightness and polarization across the disk, and should be analysed to include effects associated with observed surface features, clouds, and pressure variations associated with surface topography. The present study indicates the usefulness of polarization measurements in the ultraviolet, in investigations of thin planetary atmospheres.

Acknowledgments

I should like to thank James A. Westphal for much advice, encouragement and instruction in various aspects of observational astronomy, and Sol. L. Giles for help in building the equipment used in this study.

13—P.A.

References

Appenzeller, I.: 1968, *Astrophys. J.* **151**, 907.

Belton, M. J. S. and Hunten, D. M.: 1969, *Science* **166**, 225.

Coffeen, D. L.: 1965, *Astron. J.* **70**, 403.

Dollfus, A.: 1961, in *Planets and Satellites* (ed. by G. P. Kuiper and B. M. Middlehurst), University of Chicago Press, Chicago, p. 343.

Dollfus, A. and Focas, J.: 1966, *Compt. Rend. Acad. Sci. Paris* **B262**, 1024.

Gehrels, T. *et al.*: 1964, *Astron. J.* **69**, 826.

Hall, J. S.: 1958, *Publ. U.S. Naval Observatory, 2nd Ser.* **17**, 271.

Harris, D. L.: 1961, in *Planets and Satellites* (ed. by G. P. Kuiper and B. M. Middlehurst), University of Chicago Press, Chicago, p. 272.

Irvine, W. M. *et al.*: 1968, *Astron. J.* **73**, 807.

Leighton, R. B. *et al.*: 1969, *Science* **166**, 49.

Minnaert, M.: 1941, *Astrophys. J.* **93**, 403.

Pettengill, G. H. *et al.*: 1969, *Astron. J.* **74**, 461.

Pollack, J. B.: 1967, *Icarus* **7**, 42.

SOME PROBLEMS OF ANISOTROPIC SCATTERING IN PLANETARY ATMOSPHERES*

H. C. van de HULST

Sterrewacht Leiden, The Netherlands

Abstract. The similarity rules to compare atmospheres with anisotropic and isotropic scattering are reviewed. Omission of a narrow diffraction peak in the scattering pattern is permitted and corresponds to a special application of the similarity rules. It is shown that the extrapolation length for conservative scattering can be found with great precision from a formula involving only ω_2 and ω_3. An asymptotic expression for high-order scattering in a semi-infinite atmosphere is given and it is shown by some examples that this expression can be used to find by interpolation the terms of any order. Finally, the way in which the contrast between a dark and bright area near the center of the disk of Mars is affected by an overlying haze is computed for isotropic and for anisotropic scattering.

Much of the literature on the photometry and spectroscopy of cloudy atmospheres still deals with the cloud particles on the assumption of isotropic or linearly aniso-tropic light scattering. The reason for this is that calculations with a realistic phase function are complicated and time-consuming. In this paper we report on various approaches which we have recently tried out and which have in common that accurate quantitative answers to certain problems can be given without an excessive amount of computer time.

1. Similarity Relations

The question, first clearly posed during a symposium at Tucson 2 years ago, and answered in the Symposium Volume (Van de Hulst and Grossman, 1968) is the following. If a certain multiple scattering problem has been solved on the assumption of isotropic scattering, and we wish to change to an anisotropic law, can we then avoid a completely new computation and transform the total optical depth b and the single scattering albedo a in such a manner that the resulting intensity of reflected radiation is closely similar to the earlier result? The answer to this question is positive and the necessary relations can be summarized by two rules

transform a so that $y = (1 - a)/k = $ constant

transform b so that $ba(1 - g) = $ constant

Here k is the characteristic root, or inverse diffusion length and $g = \overline{\cos \alpha}$ is the anisotropy factor of the single scattering phase function. In the conventional repre-sentation of the scattering law by a series of Legendre functions,

$$\Phi (\cos \alpha) = \sum_{n=0}^{N} \omega_n P_n (\cos \alpha)$$

* Condensed version of paper presented at this symposium. Section 3 has been added after the symposium.

we have $\omega_0 = a$, $\omega_1 = 3ag$. Both rules were derived from the requirement that the diffusion through very thick layers should be made as closely similar as possible. Therefore, these rules should not be expected to give useful results for thin layers, or, generally, in any situation where single scattering, or light scattered only a few times, dominates the result.

Each of these rules degenerates into a simple limit: A semi-infinite layer ($b = \infty$) should remain semi-infinite and a conservative atmosphere ($a = 1$, $k = 0$) should remain conservative.

Examples given in the earlier paper and further tests by Hansen (1969a, b) showed that a difference of 2–5% between the exact results computed for 'similar' situations was not uncommon. This means that the relations are certainly good to 10–20% accuracy, but that they cannot be trusted if a 1% accuracy or better is required.

We now present two additional tests, which do not, however, change this conclusion.

A. ADDITION OF A FORWARD PEAK

This is a logical test. Strict forward scattering is no scattering at all. Hence, if inside an atmosphere formed by real scatterers and characterized by the values b', a', g', k' we sprinkle a well-mixed medium of fictitious conservative forward scatterers, characterized by the values b'', $a'' = 1$, $g'' = 1$, $k'' = 0$, nothing really changes. The combined atmosphere, characterized by the values b, a, g, k therefore gives reflection function, transmission function, and internal radiation field exactly as the atmosphere we started out with.

These identical situations should certainly be called similar. We shall now see if they indeed obey the similarity rules given above. The formal relations for the combined atmospheres are

$$\text{total extinction depth:} \quad b = b' + b''$$

which is the sum of

$$\text{total absorption depth:} \quad b(1 - a) = b'(1 - a')$$

and

$$\text{total scattering depth:} \quad ba = b'a' + b''$$

Taking only the forward component of the last relation we have

$$\text{scattering depth times } \overline{\cos \alpha}: \quad bag = b'a'g' + b''$$

Finally the attenuation suffered by the total flux in the diffusion domain (a description valid if the depth is very large) is

$$\text{diffusion depth:} \quad bk = b'k'$$

Combination of these various equations easily leads to the proportionalities

$$\frac{b}{b'} = \frac{1 - a'}{1 - a} = \frac{a'(1 - g')}{a(1 - g)} = \frac{k'}{k}$$

which agree with the rules stated in Section 1.

The preceding transformation is presented in the form of a logical test, but also has practical applications. First, the actual scattering pattern of particles large compared to the wavelength has a diffraction peak, which, though not infinitely sharp, is very strongly concentrated around the forward direction. The necessity to include this peak can be very bothersome in numerical calculations, because it greatly increases the number of Legendre functions which have to be retained in the expansion of the phase function. A practical consequence of the result just derived is that we can simply omit the diffraction peak, provided it is sharp, and provided we attach to the remaining phase function the appropriate values of a' and g'. The same conclusion has been reached by Hansen (1969a).

Secondly, if a result is sought, say for a phase function with a certain g' and a' and we wish to make use of tables (e.g. Van de Hulst, 1968a) available for a somewhat different anisotropy g, we can, by arbitrarily adding a fictitious forward peak, quite easily transform to this value g. Upon transforming a' to a and b' to b accordingly, we can then obtain the result from the existing tables by interpolation in a and b.

B. CONSERVATIVE, SEMI-INFINITE ATMOSPHERES

Under the particular assumption that $b = \infty$ (semi-infinite atmosphere) and $\omega_0 = a$ $= 1$ (conservative scattering) any phase function is 'similar'. The exact theory (Chandrasekhar, 1950; Busbridge and Orchard, 1968) shows that in this case the value of ω_1 is irrelevant. All quantities must, therefore, be slowly varying function of ω_2, ω_3, etc.

Numerical examples show that this is indeed true. Figure 1 shows by way of illustration the value of the extrapolation length $(1-g)q$. This has the well-known value 0.7104 for isotropic (or linearly anisotropic) scattering. The first two decimals remain the same throughout the figure; only the third and fourth are given. Values for $N=2$ for various ω_2 were taken from Horak and Janousek (1965). Values for $N=3$, representing the combinations $\omega_2=1$, $\omega_3=\frac{1}{2}$ and $\omega_2=1$, $\omega_3=1$ were computed from Busbridge and Orchard (1968). In addition, the figure shows three points for $N=\infty$ corresponding to the Henyey-Greenstein functions for $g=\frac{1}{4}$, $\frac{1}{3}$, and $\frac{1}{2}$ (Van de Hulst, 1968a and unpublished work). The fact that these points fit smoothly into the pattern, although here ω_4 etc. are non-zero, shows that the influence of those higher terms is very small indeed. The linear approximation

$$q(1 - g) = 0.7104 + 0.0020\omega_2 - 0.0010\omega_3$$

gives results good to 0.05% for all points shown in Figure 1 and even beyond that up to the Henyey-Greenstein function with $g=\frac{3}{4}$.

The extrapolation length discussed here forms a particularly favourable case. A less extreme example is the 'escape function', which is the solution of the Milne problem. Its empirical linear approximations in the domain of (ω_2, ω_3) corresponding to Figure 1 are in the perpendicular direction

$$K(1) = 1.259 + 0.012\omega_2 - 0.012\omega_3$$

and in grazing directions

$$K(0) = 0.433 - 0.028\omega_2 - 0.021\omega_3$$

which lead to errors of 1 or 2% at most.

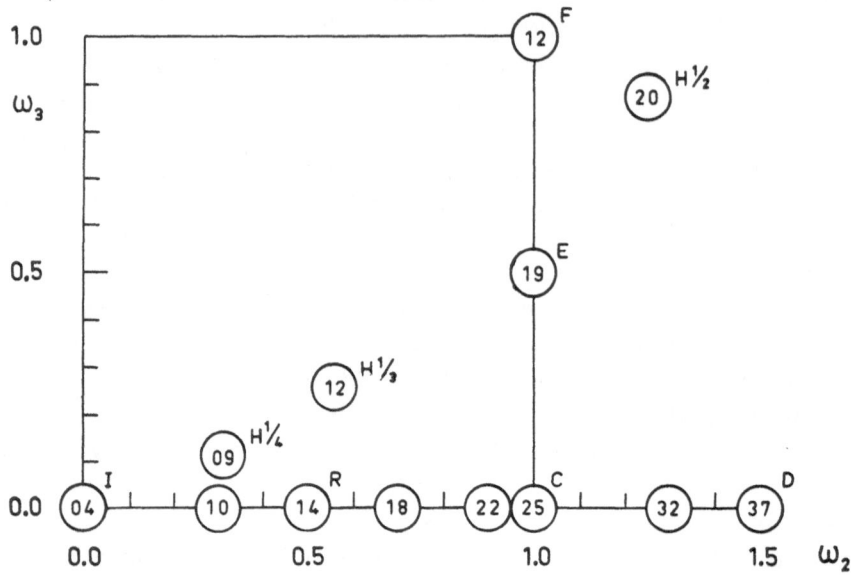

Fig. 1. Values of the extrapolation length for conservative atmospheres plotted against the coefficients ω_2 and ω_3 of the Legendre function. The third and fourth decimals following 0.71 are given.

2. High-Order Scattering

In certain problems it would be very attractive to fall back to the very simple description in which the diffusely reflected radiation is made up of the contribution of single scattering, two successive scatterings etc. If the form of the scattering pattern is fixed and only the albedo a is retained as variable we obviously can expand the reflected intensity (in any direction or integrated over a range of directions) in the form

$$f = f_1 a + f_2 a^2 + \cdots + f_n a^n + \cdots$$

in which f_n signifies the contribution given by quanta which have been scattered n times in succession. Such an expansion can be particularly handy if a has to be varied continuously as, e.g., inside an absorption line. Belton (1968) has shown that with known coefficients f_n a theory of the curve of growth follows directly.

It is not too much trouble to find the low-order terms, say f_1 to f_5, with reasonable accuracy. The principle simply is to take half-steps as follows:

Intensity of n-th order
↓ (integration over angles, using single scattering pattern)
source function of $(n + 1)$-th order
↓ (integration over optical depth)
intensity of $(n + 1)$-th order

However, for thick layers the convergence is extremely slow; this is the main reason why this 'successive-order' method has not found a wider use. The asymptotic behavior of f_n for $n \to \infty$ is as a geometric series as long as the optical thickness b is finite, but gets a different character for semi-infinite atmospheres. The dominant term in the asymptotic behavior for $b = \infty$ has been derived by Uesugi and Irvine (1969). Further terms can best be found (Van de Hulst, 1970) by starting from the expansion

$$f = G_0 + G_1 t + G_2 t^2 + G_3 t^3 + \cdots$$

which may in turn be derived from a similar expansion in terms of k (e.g. Van de Hulst, 1968b). Here $t = (1-a)^{1/2}$ and k again is the smallest characteristic root, or inverse diffusion length. We find

$$f_n = -(4\pi)^{-1/2} G_1 (n+c)^{-3/2} \{1 + O(n^{-2})\}$$

where

$$c = G_3/G_1 - \tfrac{1}{4}$$

Figure 2 shows by means of a few examples how this knowledge may be used to find the numerical values for all n. The abscissa is $(c+4)^{-2}$; the number 4 is chosen arbitrarily to match some average value of c. The ordinate is $f_n(n+c)^{3/2}$ with the correct value of c. Further specifications for the four examples are given in Table I.

TABLE I

Example	1	2	3	4
Quantity	URU	R(1, 1)	URU	R(1, 1)
anisotropy factor g	0	0	0.75	0.75
$-G_1$	2.31	3.66	4.62	7.33
c	1.15	3.76	4.00	6.75

Here *URU* is the fraction of the energy reflected if the incident radiation has uniform intensity. It also equals the Bond albedo of a planet covered by such an atmosphere. Further $R(1, 1)$ is the reflection function for perpendicular incidence and reflection.

Examples 1 and 2 in Figure 2 vary very little over the entire range; example 1 even starts out at $n=1$ with less than 1% difference from its asymptotic value. Examples 3 and 4 refer to the Henyey-Greenstein function for $g = \tfrac{3}{4}$, which is highly anisotropic. Here it was fully expected that the radiation emerging after only 2 or 3 scatterings should not yet conform to the asymptotic law. Nevertheless a smooth interpolation between $n=3$ and $n=\infty$ appears possible.

Having found from Figure 2 the values of f_n for any n we may use any summation method, numerical or analytic, to sum the power series and find the function f for an arbitrary value of a.

3. Contrast of Surface Markings

The blue haze phenomena on Mars have been discussed in this symposium. A number of models have been suggested to explain the visibility of contrast between dark and light surface features if the haze is partially dispersed. The time appears ripe to replace such qualitative suggestions by calculations based on precise models. This is indeed relatively simple.

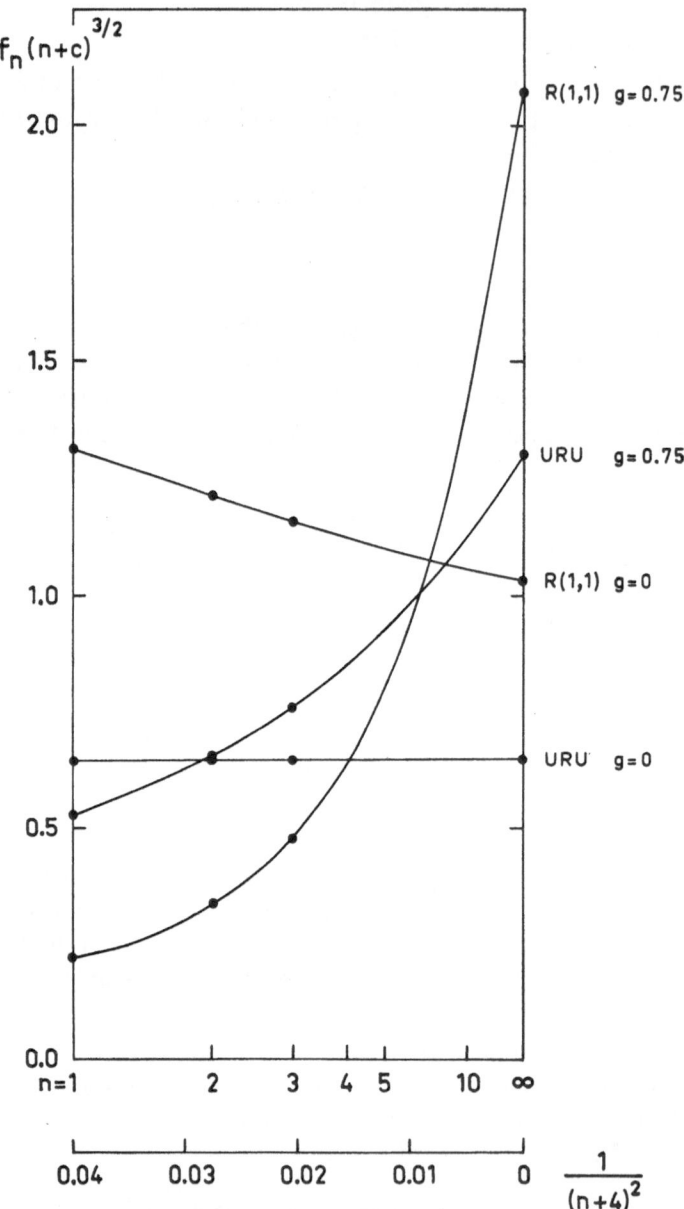

Fig. 2.   Graphical interpolation between the contributions of single, double, and triple scattering
($n=1, 2, 3$) and the asymptotic expression for $n\to\infty$, illustrated by four examples.

Let the atmosphere be homogeneous and the ground surface reflect with uniform intensity (i.e., by Lamberts' law) a fraction p of the incident radiation; further let μ_0 and μ be the cosines of the angles of incidence and reflection, both taken positive, and $\phi - \phi_0$ the azimuth-difference. Define

$$
\left.
\begin{aligned}
R'(\mu, \mu_0, \phi - \phi_0; a, b, g) &= \text{reflection function} \\
t_1(\mu_0; a, b, g) &= \text{transmitted flux} \\
r(a, b, g) &= \text{reflected flux with uniform} \\
&\quad\ \text{incidence (this was called} \\
&\quad\ URU \text{ in the last section)}
\end{aligned}
\right\}
\begin{aligned}
\text{of bare} \\
\text{atmosphere}
\end{aligned}
$$

$R(\mu, \mu_0, \phi - \phi_0; a, b, g; p) = $ reflection function of atmosphere backed by the ground surface

Thus we have the well-known relation, omitting the common variables:

$$
R(\mu, \mu_0, \phi - \phi_0) = R'(\mu, \mu_0, \phi - \phi_0) + \frac{p}{1 - pr}\, t_1(\mu)\, t_1(\mu_0)
$$

Throughout these definitions reflection from a white Lambert surface has been taken to define the unit reflection function.

The equation just given shows that the brightness p reflected from the bare ground surface may be either increased (by atmospheric scattering) or decreased (by atmospheric extinction) in the presence of an atmosphere. It seems rather difficult, by mere handwaving, to guess what happens to the contrast between a darker and brighter surface area on the planet.

In order to give one definite example, we have selected in Figure 3 two regions seen near the center of the disk at opposition so that $\mu = \mu_0 = 1$. The values of $R(1, 1)$, $t_1(1)$ and r were taken from the tables in a book in preparation. With a view to the actual situation of the blue haze on Mars, we have chosen regions with $p = 0.25$ (bright region; brightness plotted as abscissa) and $p = 0.05$ (dark region; brightness plotted as ordinate). The two diagrams correspond to two scattering laws, isotropic scattering, and Henyey-Greenstein function with $g = 0.5$.

Inside each diagram the single scattering albedo a and the optical depth b of the haze layer are varied. If the haze has completely cleared ($b = 0$), we see the full contrast ratio 5; if it is completely dense ($b = \infty$), the contrast ratio is 1. Somewhat surprisingly, we find that the contrast ratio at intermediate values depends strongly on b, but very little on a. Separate photometry of the dark and bright markings would in practice be necessary to determine a. Although the similarity rules cannot be expected to give a precise answer in this application, where the optical depths are small, the two diagrams are strikingly similar, with the curves of constant b and of constant a shifted approximately as predicted by the similarity rules.

Evidently calculations of this kind for different assumptions of the ground albedo and reflection law, and for different haze scattering patterns, and for different directions of incidence and reflection, are needed before such curves can be used as a firm basis for the interpretation of the observational data.

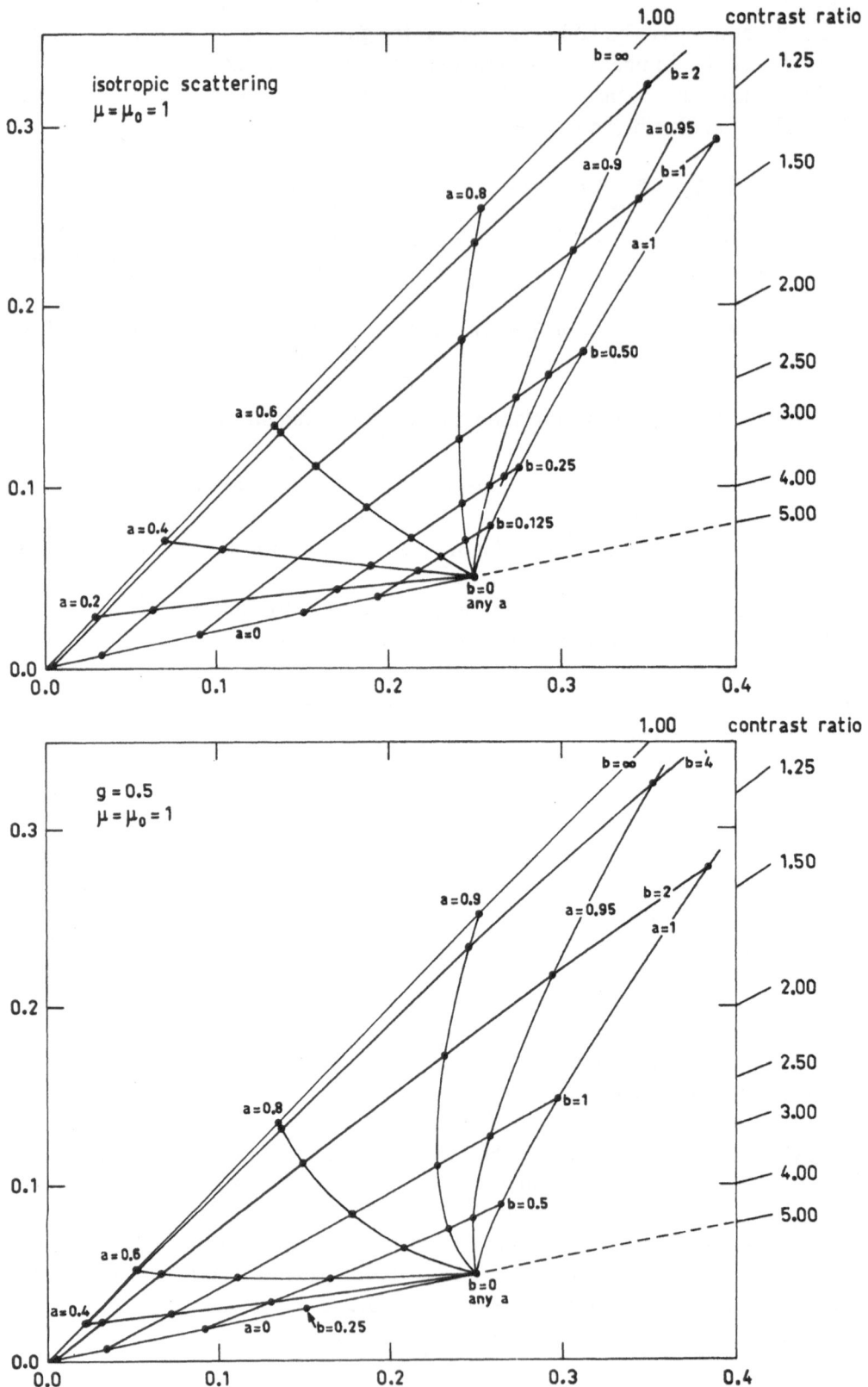

References

Belton, M. J.: 1968, *J. Atmospheric Sci.* **25**, 596.
Busbridge, I. W. and Orchard, S. E.: 1968, *Astrophys. J.* **154**, 729.
Chandrasekhar, S.: 1950, *Radiative Transfer*, Oxford.
Horak, H. G. and Janousek, A. L.: 1965, *Astrophys. J. Suppl.* **11**, 277.
Hansen, J. E.: 1969a, *J. Atmospheric Sci.* **26**, 478.
Hansen, J. E.: 1969b, *Astrophys. J.* **158**, 337.
Van de Hulst, H. C.: 1968a, *J. Computational Phys.* **3**, 291.
Van de Hulst, H. C.: 1968b, *Bull. Astron. Inst. Neth.* **20**, 77.
Van de Hulst, H. C.: 1970, *Astron. Astrophys.*, in press.
Van de Hulst, H. C. and Grossman, K.: 1968, in *The Atmospheres of Venus and Mars* (ed. by J. C. Brandt and M. B. McElroy), Gordon and Breach, New York, p. 35.
Uesugi, A. and Irvine, W. M.: 1969, *Astrophys. J.* **159**, 127.

←

Fig. 3.   Contrast between two areas near the center of the Mars disk, with brightness 0.25 and 0.05, as affected by an overlying haze; a = single scattering albedo, g = anisotropy factor, b = optical thickness of haze layer.

B. CO₂ ABSORPTION

AN INTERPRETATION OF THE MARS SPECTRUM TAKEN
BY THE CONNES*

L. D. KAPLAN† and L. D. GRAY YOUNG

Space Sciences Division, Jet Propulsion Laboratory,
California Institute of Technology, Pasadena, Calif., U.S.A.

Abstract. Lines of the 2-0 and 3-0 bands of carbon monoxide and (many) bands of carbon dioxide appear prominently in the Connes' Mars spectrum [1]. Five carbon dioxide bands were measured to construct a curve of growth for CO_2 lines formed in the Martian atmosphere [2]. A similar curve of growth was constructed for the 2-0 band of carbon monoxide. From these curves, we have computed the rotational temperature of the atmosphere, the surface pressure, and the abundance of CO and CO_2. The surface pressure is found to be approximately equal to the CO_2 partial pressure, i.e. $p_s \sim 5$ mb. The CO concentration by volume was found to be slightly less than one part per thousand.

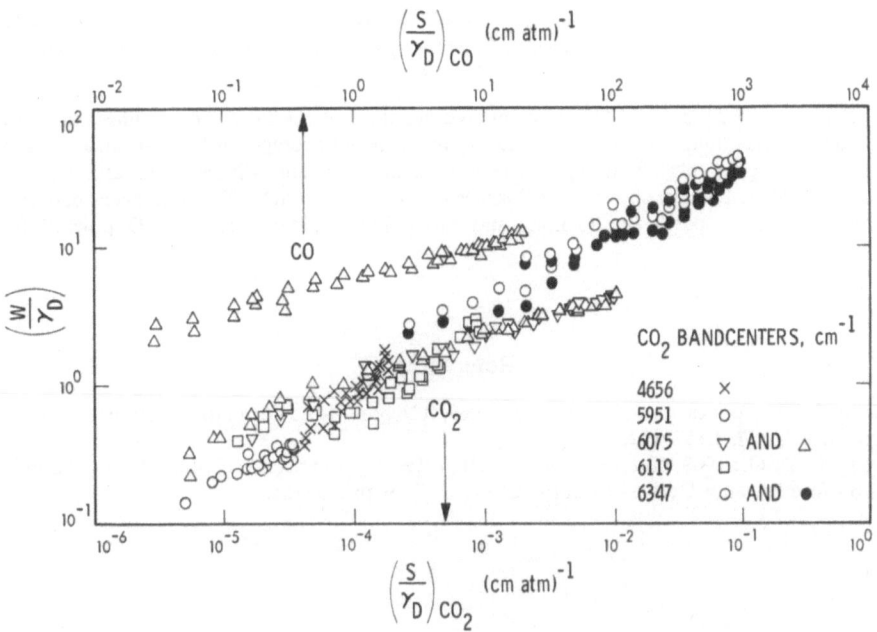

Fig. 1. Curves of growth for CO lines and CO_2 lines formed in the Martian atmosphere. The upper abscissa corresponds to CO lines and the lower abscissa to CO_2 lines. The open circles at the lower left of the figure refer to the CO_2 band at 5951 cm^{-1} while the open circles at the upper right refer to the 6347 cm^{-1} CO_2 band uncorrected for telluric absorption; the solid circles are for the corrected 6347 cm^{-1} CO_2 lines. The CO_2 bands have the following intensities in cm^{-1}/km atm: 4656, $Sv = 2.5$; 5951, $Sv = 0.47$; 6075, $Sv = 123$; 6119, $Sv = 7.8$; 6347, $Sv = 1150$. Lorentz half-widths of $\gamma_L(CO) = 0.07$ cm^{-1} and $\gamma_L(CO_2) = 0.10$ cm^{-1} (at stp) were used. The uncertainty in the half-widths is estimated to be ten percent.

* This paper presents the results of one phase of research carried out at the Jet Propulsion Laboratory, California Institute of Technology, under Contract Number NAS 7-100, sponsored by the National Aeronautics and Space Administration.
† Faculté des Sciences de Paris, Laboratoire de Spectroscopie Moléculaire, 9, Quai Saint-Bernard, Tour 13, 75-Paris (5E).

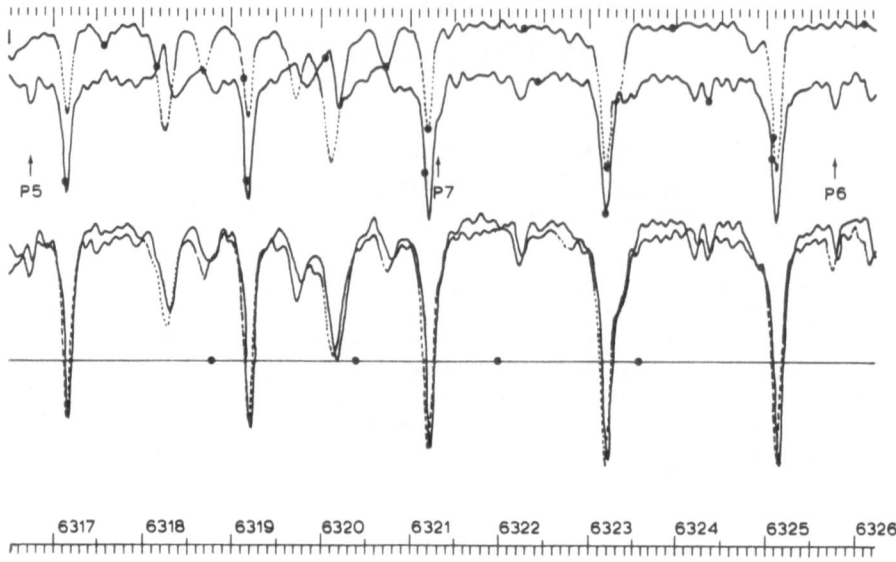

Fig. 2.   Part of the 3-0 band of CO with computed positions of the P_6, P_7 and P_8 lines marked with arrows. Lower curves (and lower baseline) are two independent averages of Mars spectra (with mean secant $Z = 2.4$). Upper curve (and upper baseline) is solar spectrum. Middle curve (and upper baseline) is ratio of Mars to solar spectrum. All scales are in units of cm^{-1}. The differences between the two spectra of Mars is partly due to noise and partly due to differences in the Doppler shift and telluric absorption.

References

[1] Kaplan, L. D., Connes, J., and Connes, P.: 1969, 'Carbon Monoxide in the Martian Atmosphere', *Astrophys. J.* **157**, L187–192.
[2] Young, L. D. G.: 1969, 'Interpretation of High Resolution Spectra of Mars: I. CO_2 Abundance and Surface Pressure Derived from the Curve of Growth', *Icarus*.

OBSERVATIONS OF THE MARTIAN 1.2 μ CO₂ BANDS

EDWIN S. BARKER

McDonald Observatory, The University of Texas, Fort Davis, Tex., U.S.A.

Abstract. A method to determine independently the Martian surface pressure from measurements of individual lines in the 1.2030 and 1.2177 μ CO₂ bands is presented. Observations obtained during the 1967 apparition and some preliminary results of the 1969 apparition observations yield CO₂ abundances near 100 m-atm and surface pressures of 4–8 mb.

The Martian surface pressure and CO₂ abundance can be determined from observations of the 1.2030 and 1.2177 μ CO₂ bands if the resolution and dispersion used are great enough to resolve the individual Martian CO₂ lines.

Since April 1967 we have been using an RCA Carnegie infrared image tube at the A camera coudé focus of the 82-inch Struve reflector at McDonald Observatory. The resultant dispersion of 3.1 Å/mm allows us a resolution of approximately 0.2 to 0.3 Å, which is quite sufficient to resolve the Martian CO₂ lines.

Figure 1 not only shows the wavelength of the major CO₂ bands between 0.87 μ and 1.6 μ but also indicates their change in strength with increasing pressure. Observations of the weaker bands below one micron, being essentially independent of pressure, give the CO₂ abundance; the stronger bands above 1 μ give a pressure-abundance product, in which the effect of the pressure becomes more dominant as the number of

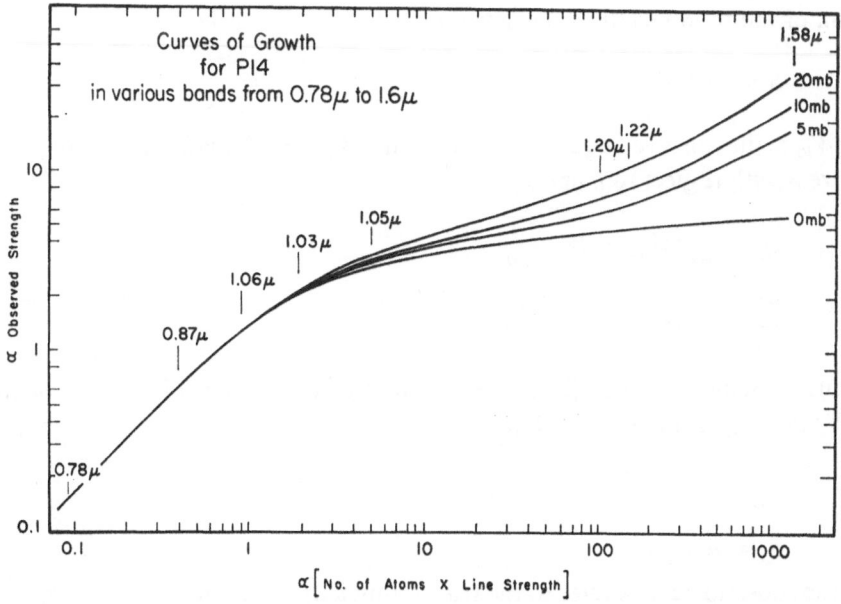

Fig. 1. Curves of growth for P14 in various CO₂ bands from 0.78 μ to 1.6 μ.

14—P.A.

Sagan et al. (eds), *Planetary Atmospheres*, 191–195.

CO_2 molecules and line strength increases. The 1.2 μ bands lie in a transition region, since there are lines in the band that are near both the linear and square root regions.

1. Method

A non-linear least squares program has been developed which allows one to independently determine the 'best values' for the surface pressure and CO_2 abundance that fit the observed equivalent widths in a 1.2 μ band (Barker, 1969). This method linearizes the standard formula for the theoretical equivalent width of a line which is given in Equation (1).

$$W = \int_{-\infty}^{\infty} [1 - \exp{(-P^1\eta\omega H(a, \xi))}] \, d\xi \tag{1}$$

The iterative solution is on the two curve of growth parameters; the total CO_2 abundance, $x = \eta\omega$, and the value of the surface pressure or a. a is the ratio of the Lorentz half-width at half-power, r_L, to the Doppler width r_D at which the absorption coefficient has dropped to $1/e$ of the absorption at the line center.

The equivalent width of line (i) as given by Equation (2) is

$$W_i = 2r_D \int_0^y [1 - \exp{(-kS_iH(a, \xi))}] \, d\xi \tag{2}$$

where S_i is the line strength, k is a constant, and

$$H(a, \xi) = \frac{a}{\pi} \int_{-\infty}^{\infty} \frac{e^{-y2}}{a^2 + (\xi - y)^2} \, dy \tag{3}$$

The sum of the residuals $(0-C)$ is given by

$$S = \sum_i (W_{i0} - W_i)^2 \tag{4}$$

where W_{i0} is the observed equivalent width and W_i the calculated value. We want to minimize S with respect to a and x.

$$\frac{\partial S}{\partial a} = \sum_i (W_{i0} - W_i) \frac{\partial W_i}{\partial a} \tag{5}$$

$$\frac{\partial S}{\partial x} = \sum_i (W_{i0} - W_i) \frac{\partial W_i}{\partial x} \tag{6}$$

Because Equations (2) and (3) are such complicated functions of a and x, we linearize W by making estimates for a and x

$$x_t = \dot{x}_e + X_c \tag{7}$$

and

$$a_t = \dot{a}_e + A_c \tag{8}$$

where the subscripts t, e, c refer to the true, estimated, and corrections values, respectively.

We can express W_i in a Taylor series about $\dot{a}$ and $\dot{x}$ (neglecting higher order terms),

$$W_i = W_i(\dot{x}, \dot{a}, S_i, k) + \left.\frac{\partial W_i}{\partial x}\right|_{\dot{x},\dot{a}} X + \left.\frac{\partial W_i}{\partial a}\right|_{\dot{x},\dot{a}} A \qquad (9)$$

As a first approximation we can express the partial derivatives as

$$W_{i,\dot{x}} = \left.\frac{\partial W_i}{\partial x}\right|_{\dot{x},\dot{a}} \simeq \frac{W_i(\dot{x} + hx, \dot{a}) - W_i(\dot{x}, \dot{a})}{hx} \qquad (10)$$

$$W_{i,\dot{a}} = \left.\frac{\partial W_i}{\partial a}\right|_{\dot{x},\dot{a}} \simeq \frac{W_i(\dot{x}, \dot{a} + ha) - W_i(\dot{x}, \dot{a})}{ha} \qquad (11)$$

Substituting

$$W_i = \dot{W}_i + W_{i,\dot{x}}X + W_{i,\dot{a}}A \qquad (12)$$

Now we can rewrite Equations (5) and (6), setting them equal to zero and introducing weights for each observed equivalent width, t_i.

$$\sum_i (W_{i0} - \dot{W}_i - W_{i,\dot{x}}X - W_{i,\dot{a}}A)(-W_{i,\dot{x}})t_i = 0 \qquad (13)$$

$$\sum_i (W_{i0} - \dot{W}_i - W_{i,\dot{x}}X - W_{i,\dot{a}}A)(-W_{i,\dot{a}})t_i = 0 \qquad (14)$$

Equations (13) and (14) are in the form of regular weighted normal equations and can be solved for the correction terms X and A. To improve the convergence properties of the solution, a weight matrix was introduced to damp the corrections and these weights were decreased, then set equal to zero before the final solution was obtained. This weight matrix was added to the normal equation matrix with the weights being the proper order of magnitude (starting out at 10% of value of input estimate).

Then the corrections were added to the initial estimates according to Equations (7) and (8). This type of iteration process was carried out until convergence in both a and x was obtained.

2. Observations

Only one good plate was obtained during the 1967 opposition, primarily due to the eight hours of exposure time required. This plate was taken when the Doppler shift was sufficient to separate the Martian and telluric CO$_2$ absorptions. Figure 2 shows the results obtained from this plate of the 1.2030 μ band.

This year in June we obtained six high quality plates of the 1.2 μ bands. Since reductions and observations are still in progress, one plate has been reduced as a sample for presentation at this meeting. Figure 3 shows the observed equivalent widths for the 1.2030 μ band, with only a preliminary correction having been made for the telluric CO$_2$ absorption which was super-imposed on the Martian absorption. The telluric absorption only affects the wings of the observed CO$_2$ lines since the Martian lines are black in the center. The equivalent widths for the telluric lines were calculated and

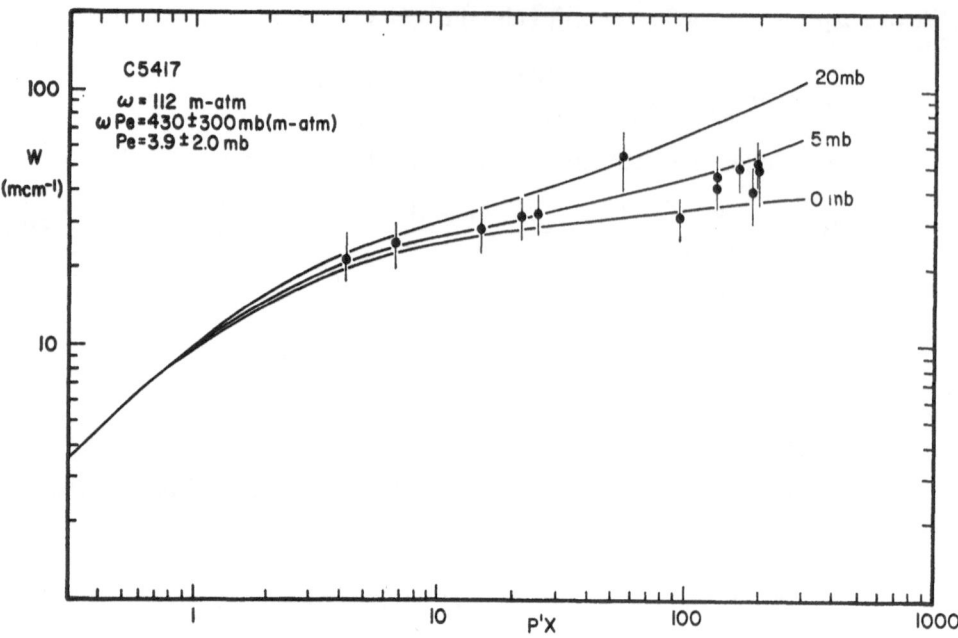

Fig. 2.   Curve of growth for plate no. 5431 (1.2030 μ band).

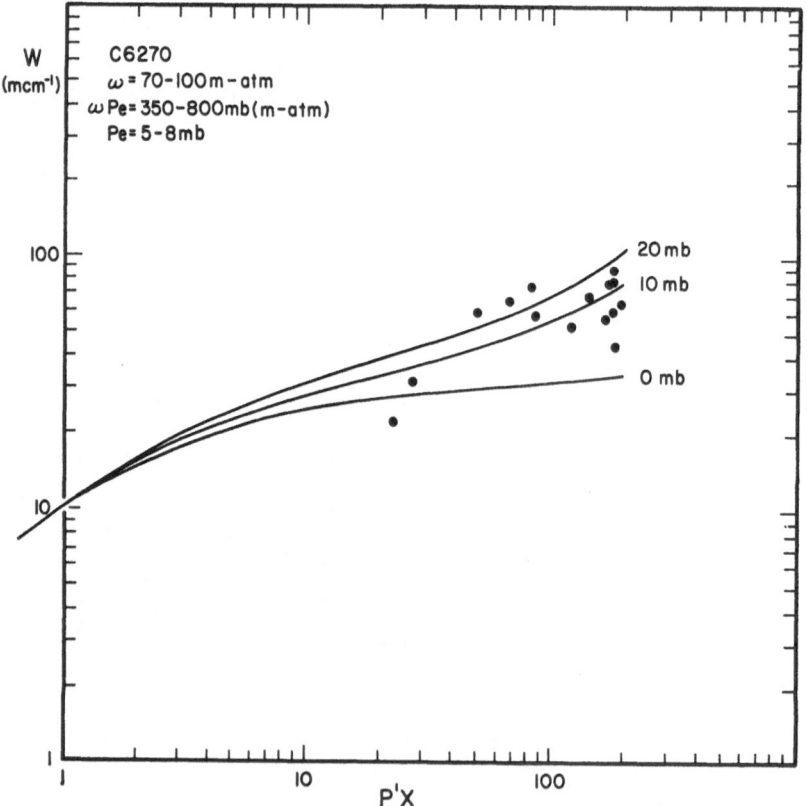

Fig. 3.   Curve of growth for plate no. 6270 (1.2030 μ band).

the absorption in the wings of the telluric lines beyond the half-width of the Martian lines was subtracted from the observed equivalent widths. Obviously a more detailed analysis is required and will be used to completely remove the telluric absorption from the observed equivalent widths.

In late September when the Doppler shift was sufficient to separate the telluric and Martian absorptions, we obtained two excellent plates of the 1.2 μ bands which have not yet been reduced.

3. Summary

High-dispersion observations of the 1.2 μ bands give CO$_2$ abundances near 100 m-atm and surface pressures of 4–8 mb. But their observation is limited to periods of time when the Doppler shift is sufficient to resolve the telluric and Martian lines and when the lines are directly superimposed at opposition. Further reductions of the available spectrograms taken during the 1969 apparition will lead to improved values of the Martian CO$_2$ abundance and surface pressure at the corresponding periods during the Martian season.

Acknowledgment

The support of the National Aeronautics and Space Administration Office of Lunar and Planetary Programs is gratefully acknowledged.

Reference

Barker, E. S.: 1969, unpublished Ph.D. Dissertation, The University of Texas.

VARIATIONS OF THE MARTIAN CO₂ ABUNDANCE WITH MARTIAN SEASON

EDWIN S. BARKER

McDonald Observatory, The University of Texas, Fort Davis, Tex. U.S.A.

Abstract. All previous published equivalent widths of the Martian CO_2 bands including the 1967 apparition coverage at McDonald Observatory are reduced to CO_2 abundances using the same curve of growth parameters. The corresponding CO_2 partial pressures (assuming a pure CO_2 atmosphere), along with the regression curves of the north polar cap, are presented as functions of Martian heliocentric longitudes, L_s. A correlation is noted between the maximum CO_2 abundance and minimum north cap diameter and the decrease in CO_2 abundance when the northern polar haze begins to reform. Preliminary results of the 1969 apparition coverage at McDonald Observatory are also presented.

The CO_2 abundance is most easily determined from the 0.87 μ and 1.05 μ bands and practically all ground-based measurements have been made with these bands. The first such observations, dating from 1963 and 1965, happened to be limited to brief periods; hence, it was simply not possible to detect any seasonal variation in CO_2 abundance. During 1967 the seasonal coverage was extended an additional four months, primarily by our observations at McDonald Observatory.

Our present program at McDonald Observatory is to obtain as long a baseline as possible in Martian season to measure any seasonal variation in the atmospheric constituents of Mars, primarily the CO_2 abundance and the water vapor content. Numerically, Martian season can be represented as heliocentric longitude, L_s. Whenever telescope time on the 107-inch reflector or 82-inch Struve reflector is available (between other observations or during twilight or daytime hours), spectrographic observations of the CO_2 bands between 0.87 μ and 1.22 μ are made.

Many different abundance observations have been made since 1963. Table I contains all published abundances and corrected abundances which have been obtained by reducing the respective equivalent widths under the same conditions (broadening factors, temperature, band strengths, and atmospheric composition). These abundances were determined using the same curve of growth analysis which provides the 'best fit' CO_2 abundance and surface pressure. (For a more complete and detailed description of the procedures and computer programs used, see Barker (1969).) For simplicity, the results for a pure CO_2 atmosphere will be presented, although the trends are similar in nature when a fixed pressure-abundance product is assumed.

It is straightforward to convert from CO_2 abundances to CO_2 partial pressures. The partial pressures corresponding to spectra taken at McDonald Observatory during the 1967 apparition are shown in Figure 1 as a function of Martian season, or L_s. (L_s is defined as 0° at the northern spring equinox.) The polar cap measurements were by Capen (1966, 1969) of the Jet Propulsion Laboratory with several telescopes, including the 82-inch Struve reflector. Figure 2 combines all previous measurements

TABLE I

CO$_2$ abundances determined during 1963, 1965, and 1967 apparitions

Author	Plate number	L_s (°)	Published CO$_2$ abundance (m-atm)	Corrected CO$_2$ abundance (m-atm)	CO$_2$ partial pressure (mb)
Kaplan et al. (1964)	—	76	50 ± 20	43 ± 10	3.1 ± 0.7
Kliore et al. (1966)	—	139	90 ± 20	90 ± 20	6.7 ± 1.4
Belton and Hunten (1966)	—	—	68 ± 26	72 ± 15	5.3 ± 1.2
Owen (1966)	C4365	82	56	56 ± 11	4.1 ± 0.8
	C4367	82	65	65 ± 13	4.8 ± 1.0
	C4368	82	62	62 ± 12	4.7 ± 0.9
	C4376	84	65	65 ± 13	4.8 ± 1.0
	C4381	85	52	52 ± 10	3.8 ± 0.7
Spinrad et al. (1966)	C4290	47	93	93 ± 33	6.9 ± 2.4
	C4298	49	82	80 ± 46	5.9 ± 3.6
	C4304	49	88	90 ± 14	6.6 ± 1.0
	C4347	59	118	129 ± 43	9.5 ± 3.2
	C4387	110	111	121 ± 46	8.9 ± 3.4
	C4392	110	110	115 ± 41	8.5 ± 3.0
	C4396	112	125	140 ± 15	10.2 ± 1.1
	C4406	114	111	120 ± 42	8.8 ± 3.1
	EC4037a	72	112	134 ± 63	9.9 ± 4.7
	EC4037b		64	73 ± 23	5.4 ± 1.7
	EC4090	81	70	74 ± 64	5.4 ± 4.8
	EC4120a	84	98	95 ± 61	7.0 ± 4.5
	EC4120b		61	67 ± 37	4.9 ± 2.7
	EC4272a	112	51	74 ± 64	5.5 ± 4.7
	EC4272b		45	42 ± 23	3.1 ± 1.7
	EC4287a	122	111	114 ± 46	8.4 ± 3.4
	EC4287b		51	75 ± 68	5.5 ± 5.0
Belton et al. (1968)		120–122	75–77	71 ± 19	5.3 ± 1.5
Giver et al. (1968)		131–135	61	58 ± 10	4.3 ± 0.8
Carleton et al. (1969)		110–122	83 ± 11	83 ± 11	6.1 ± 0.8
Munch (1969)		120	76 ± 9	81 ± 18	6.0 ± 1.3
Owen (1968)	—	120	75 ± 15	—	—

of the CO$_2$ partial pressure given in Table I with the 1967 McDonald data. The following conclusions can be tentatively drawn from the data.

The maximum CO$_2$ abundance occurs at L_s values, when the north polar cap has finished the major portion of its annual shrinking; and as the northern polar cap begins to reform as a polar haze in the Martian autumn, the CO$_2$ abundance drops significantly to a minimum near $L_s = 220°$. Our 1969 observations of the CO$_2$ abundance are filling in the region from $L_s = 140°$ to 300°. (A mean correction for elevation differences based on the 1967 radar data of Pettengill et al. (1969) has been applied to the 1967 McDonald data and the result is to slightly increase the scatter in the partial pressures, but the conclusions remain the same.)

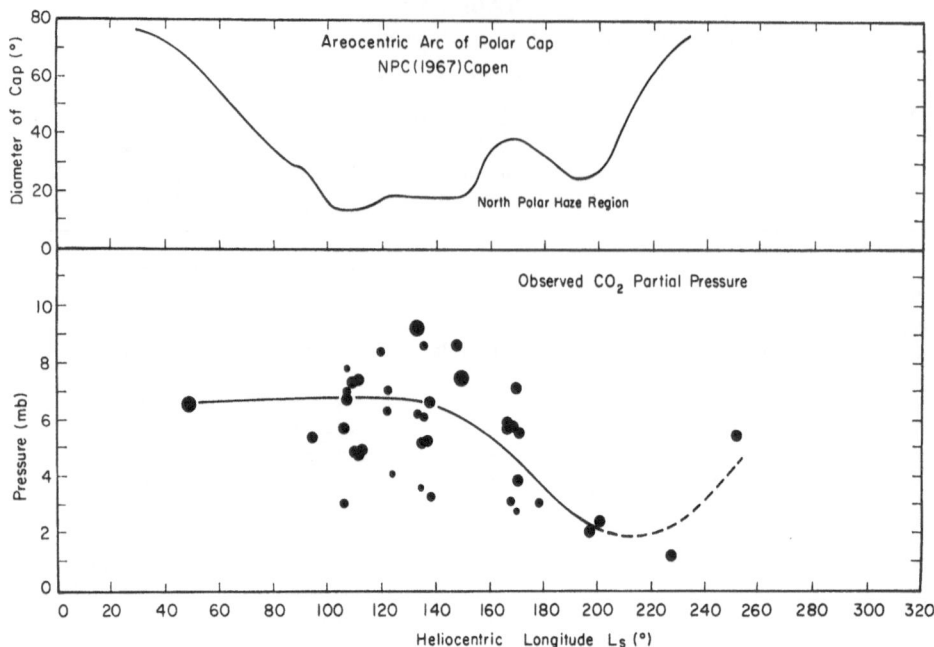

Fig. 1.   Regression curve for north polar cap in 1966–67. CO_2 partial pressures obtained from
1967 McDonald spectra as a function of L_s. Barker (1969).

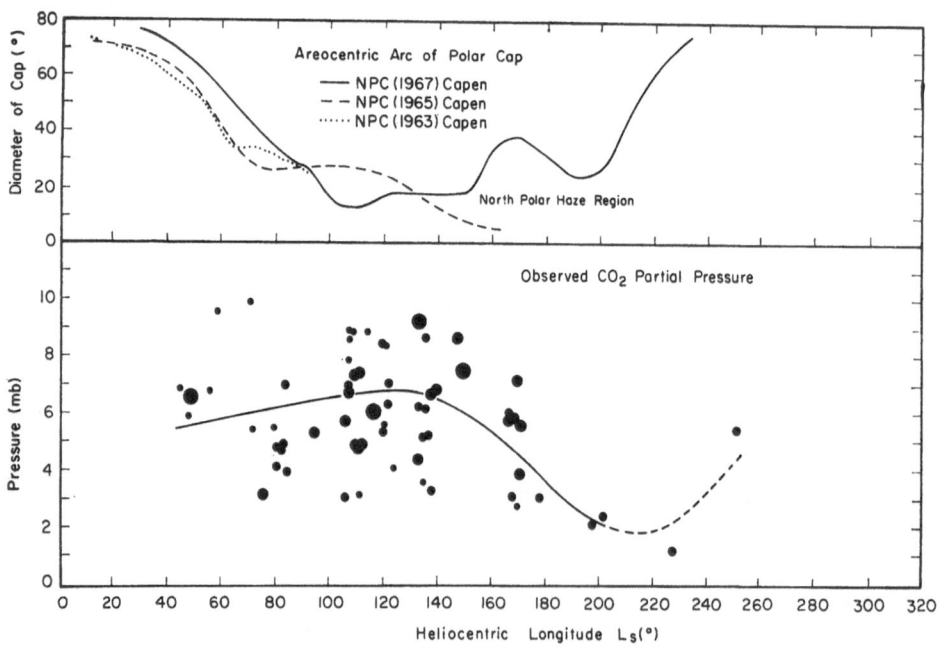

Fig. 2.   Regression curves for north polar cap and all CO_2 partial pressures obtained during the
1963, 1965, and 1967 Martian apparitions.

Our Mars observations at McDonald Observatory during the 1969 apparition began at an L_s of 56° (October 1968) and will continue in 1970. The 1969 McDonald spectral coverage of Mars in the 0.87, 1.03, 1.05, and 1.2 μ CO₂ bands as a function of L_s is shown schematically, with each dot referring to a usable spectrum in Figure 3. The 8700 Å bands were recorded on hypersensitized IV-N emulsions at 2.0 to 4.0 Å/mm. The 1 μ bands were observed with the RCA Carnegie infrared image tube at a dispersion of 3.6 Å/mm and a resolution of 0.15 Å.

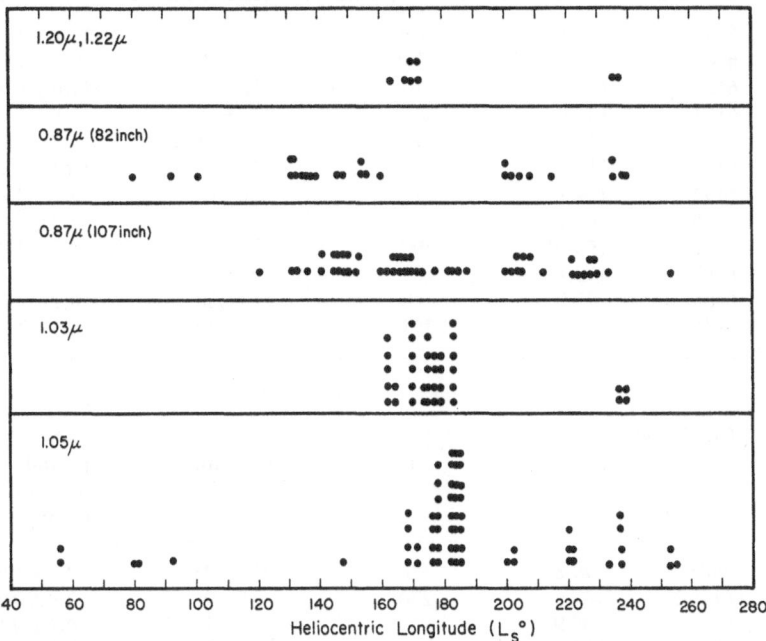

Fig. 3.   1969 McDonald coverage of the 0.87, 1.03, 1.05, and 1.2 μ CO₂ bands. Each dot represents a usable spectrum.

Because of the press of observing time and very recent reduction of most of the data, only a few representative spectra have been reduced by this time for presentation. The CO₂ abundances and corresponding partial pressures (assuming a pure CO₂ atmosphere), appear in Table II. The CO₂ partial pressures are shown in Figure 4, along with Capen's (1969) measurements of the north polar cap that have been reduced at the present time. These partial pressures have been obtained from spectra of the full planet and represent the mean value over the meridional strip accepted by the spectrograph slit. The filled square refers to a mean of ten 1.05 μ spectra taken within 2° of L_s near opposition.

The trend is similar to that observed in 1967, and indicates a possible decline toward a minima beyond an L_s of 200°. Only further observations and a complete reduction of the spectra will show the actual minima. The presented data were obtained prior to the middle of August. Many spectra have been taken since then, and we plan to follow Mars into early 1970, which corresponds to L_s values of 300 to 320°.

One must take into account the effect of the Martian topography which transits

TABLE II

CO$_2$ abundances determined during 1969 apparition from full planet spectra

Coudé No.	L_s (°)	L_{cm} (°)	ω (m-atm)	CO$_2$ partial pressure (mb)
			0.87 μ band	
5902	79.8	283	151 ± 32	11.1 ± 2.3
6147	132.6	224	132 ± 40	9.8 ± 3.0
6152	134.5	170	141 ± 26	11.6 ± 1.9
6163	145.9	306	106 ± 16	7.8 ± 1.2
6166	146.7	290	134 ± 33	9.9 ± 2.4
6185	153.6	181	142 ± 39	10.4 ± 2.9
6194	154.1	174	131 ± 43	9.7 ± 3.2
76	167.9	264	99 ± 34	7.3 ± 2.5
77	168.0	300	104 ± 26	7.7 ± 2.0
6393	199.1	82	96 ± 24	7.4 ± 1.7
6406	203.9	8	110 ± 30	8.1 ± 2.2
146	211.8	254	115 ± 30	8.5 ± 2.2

Coudé No.	L_s (°)	L_{cm} (°)	ω (m-atm)	CO$_2$ partial pressure (mb)
			1.05 μ band	
5903	79.8	336	103 ± 23	7.6 ± 1.9
6175B	147.9	270	35 ± 9	2.8 ± 0.7
6251	168.9	261	56 ± 11	4.1 ± 0.8
6269A	170.5	162	97 ± 15	7.2 ± 1.3
6358	181.7	157	40 ± 9	2.9 ± 0.7
6359A	181.7	10	93 ± 24	6.9 ± 1.9
6359C	181.7	40	134 ± 35	10.0 ± 2.8
6363A	182.2	334	93 ± 28	6.9 ± 2.2
6363B	182.2	348	57 ± 14	4.2 ± 1.1
6363C	182.2	3	99 ± 19	7.3 ± 1.6
6364A	182.2	18	95 ± 23	7.0 ± 1.8
6364C	182.3	48	81 ± 26	6.0 ± 1.9
6371A	182.8	321	127 ± 27	9.4 ± 2.2
6371B	182.8	335	72 ± 17	5.3 ± 1.4
6403	201.5	32	90 ± 18	6.6 ± 1.5
6404	201.5	65	34 ± 5	2.4 ± 0.5

under the spectrographic slit during the exposure. No correction for mean topographic differences has been applied at this time to the CO$_2$ partial pressures shown in Figure 4. In fact the central meridian values, L_{cm}, for the three low partial pressures shown in Figure 4, correspond to central meridians where highlands were indicated in the central regions of the Martian disk by Woszczyk and Belton (see papers presented during this symposium). An attempt will be made to remove the effect of elevation differences when the results of Woszczyk and Belton appear in final form. The

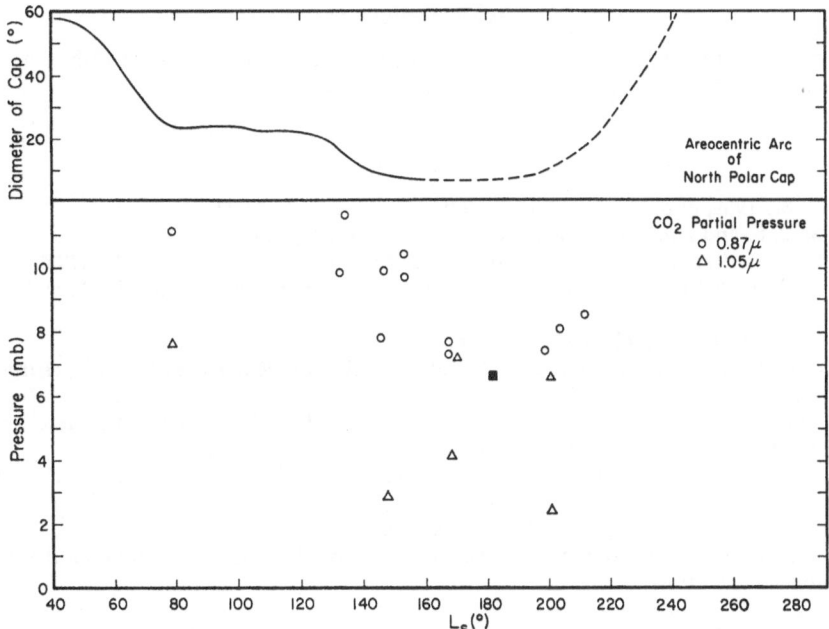

Fig. 4. Regression curve for north polar cap and some CO_2 partial pressures obtained during the 1969 Martian apparition.

radar topographic profiles of Pettengill *et al.* (1969) and Rogers *et al.* (1969) will also be used in this correction procedure. The effect of topographic differences may be so large that it may be impossible to assign a total CO_2 abundance to the Martian atmosphere at a given seasonal date without a thorough knowledge of the topography for the entire planetary surface.

Such a phenomenon as the CO_2 partial pressure variation was predicted by Leighton and Murray (1966) based on their CO_2 polar cap model. They predicted a double-sine wave variation over the Martian season based on the assumption that the polar caps were primarily composed of CO_2 and the total atmospheric CO_2 abundance is determined by the season and extent of the polar caps. Their variation is similar in shape, but the magnitude of the variation is smaller and the minima in the CO_2 partial pressure occurs at an L_s of 160°. The aim of this long-range program is to develop a complete set of data, which can be used to test models of the polar cap composition such as Leighton and Murray's.

Summary

Since CO_2 is the major constituent of Martian atmosphere (probably 70% or greater), any variation in the CO_2 partial pressure must constitute a proportional variation in the total atmospheric pressure. Based on the spectra reduced up to the present, we believe that the CO_2 abundance varies by a factor of two; thus, presumably the Martian surface pressure varies by almost a factor of two, and this variation is correlated with the waxing and waning of the polar caps or Martian season.

Acknowledgment

The support of the National Aeronautics and Space Administration Office of Lunar and Planetary Programs is gratefully acknowledged.

References

Barker, E. S.: 1969, unpublished Ph.D Dissertation, University of Texas.

Belton, M. J. S. and Hunten, D. M.: 1966, *Astrophys. J.* **145**, 454.

Belton, M. J. S., Broadfoot, A. L., and Hunten, D. M.: 1968, *J. Geophys. Res.* **73**, 4795.

Capen, C. F.: 1966, *The Mars 1964–1965 Apparition*, Jet Propulsion Laboratories Tech. Rept. No. 32-990.

Capen, C. F.: 1969, personal communication.

Carleton, N. P., Sharma, A., Goody, R. M., Liller, W. L., and Roesler, F. L.: 1969, *Astrophys. J.* **155**, 323.

Giver, L. P., Inn, E. C. Y., Miller, J. H., and Boese, R. W.: 1968, *Astrophys. J.* **153**, 285.

Kaplan, L. D., Munch, G., and Spinrad, H.: 1964, *Astrophys. J.* **139**, 1.

Kliore, A., Cain, D. L., and Levy, G. S.: 1966, *Moon and Planets* (ed. by A. Dollfus), North-Holland Publishing Co., Amsterdam, p. 226.

Leighton, R. B. and Murray, B. C.: 1966, *Science* **153**, 136.

Munch, G.: 1969, 'Interferometric Measurement of Weak CO_2 Absorption in the Martian Atmosphere', preprint.

Owen, T. C.: 1966, *Astrophys. J.* **146**, 257.

Owen, T. C. and Mason, H. P.: 1968, 'Spectroscopic Studies of the Atmospheres of Mars and Venus', Paper presented at Symposium on the Physics of the Moon and Planets, Kiev, U.S.S.R., October 15–22.

Pettengill, G. H., Counselman, C. C., Rainville, L. P., and Shapiro, I. I.: 1969, *Astron. J.* **74**, 461.

Spinrad, H., Schorn, R. A., Moore, R., Giver, L. P., and Smith, H. J.: 1966, *Astrophys. J.* **146**, 331.

Rogers, A. E. E., Ash, M. E., Counselman, C. C., Shapiro, I. I., and Pettengill, G. H.: 1969, 'Radar Measurements of Martian Surface Topography and Roughness', Paper presented at the Joint IAU/URSI Symposium on Planetary Atmospheres and Surfaces, Woods Hole, Mass., 11–15 August.

RELATIVE ELEVATION DIFFERENCES REVEALED
BY NEAR INFRARED CO$_2$ BANDS ON MARS

ANDRZEJ WOSZCZYK*

McDonald Observatory, The University of Texas, Fort Davis, Tex., U.S.A.

Abstract. High dispersion spectroscopic observations of Mars, carried out with the new 107-inch and 82-inch Struve telescopes at McDonald Observatory, were used for detection of relative elevation differences on the surface of Mars. The dispersion of the spectra was 4.4 Å/mm for the 0.87 CO$_2$ band and 3.5 Å/mm for the CO$_2$ bands in the 1 μ region. Spatial resolution during the 1969 opposition was about 20° by 10° for 0.87 μ and 15° by 15° on the planet for the 1.03 μ and 1.05 μ bands. The results are shown in Tables I, II, III and Figures 2 and 3. The observed differences in elevation are on the order of 10 km, which is in good agreement with the latest radar measurements of the topographic features on the Martian surface.

1. Introduction

The growing possibility for study of the planetary system by direct means of space probes provides us with new, very accurate, and frequently surprising information about physical conditions on planets and in interplanetary space. On the other hand, these direct means require better and better initial data to insure the success of planetary orbiters, fly-bys, probes, and especially soft-landers. The physical properties of planetary atmospheres still remain rather poorly known. For example, successful realization of the projected unmanned soft landing on Mars in 1973 demands a special effort in acquisition of the data.

The 1969 opposition of Mars provided a good occasion for this. During more than a month the geocentric distance of the planet was less than 0.5 AU, its diameter greater than 19 arcsec, and declination not too low for observation at McDonald Observatory. Furthermore, the new 107-inch telescope and its excellent coudé spectrograph went into operation three months before opposition. Large blocks of observing time being available on both the 107-inch and 82-inch telescopes at McDonald Observatory for the planetary work, a large number of spectroscopic observations of Mars, chiefly in the region 8000–12 200 Å, were obtained. This material can be used not only for total-disc abundance and pressures determination of Martian CO$_2$ and its seasonal variations, but also, due to relatively large disc size and good seeing, for differential measures of CO$_2$ over different aerographic regions of the planet.

2. Recent Determination of the Relative Elevation Differences

Relative altitudes of the dark and bright areas on Mars have been subjects of great interest during the last few years. Traditionally, the prevailing opinion of most observers of Mars has been that the dark areas are lowlands and the bright areas highlands. This opinion, in lack of direct supporting evidence, was based mainly on

---

* On leave from Toruń Observatory, Poland.

Sagan et al. (eds.), Planetary Atmospheres, 203–211.

analogies with terrestrial weather patterns: lowlands would be warmer and more likely to support the life forms which could be responsible for the regenerative properties of the dark areas. The observed 8° temperature difference in favor of the dark areas gave some support for this opinion. However, there is strong evidence that the differences in temperature are connected rather more closely with the local surface albedo than with altitude difference (Sagan and Pollack, 1968). Also, the regenerative properties can be explained as purely mechanical, wind-driven phenomena (Rea, 1963, 1964; Sagan *et al.*, 1967), or as meteorological effects (Wells, 1965, 1966).

To interpret early radar-reflectivity observations, Sagan *et al.* (1966a, 1966b) developed models of the Martian surface in which elevations up to 15 km existed. The dark areas were highlands or continental blocks, the bright areas were low basins filled with sand, and the canals were mountain ranges. In disagreement with these last hypotheses are the results of the direct time-delay radar measurement of Mars by Haystack Microwave Facility operated by M.I.T. Lincoln Laboratory (Counselman *et al.*, 1968; Pettengill *et al.*, 1969). These observations, made along the subterrestrial latitude of 21°N, also showed elevation differences of up to 12 km, but indicated a definite lack of correlation either with radar-reflectivity or with optical reflectivity. Similar results were obtained by this laboratory during the 1969 opposition (Rogers *et al.*, 1969) for Martian latitude between 3°N and 12°N.

The spectroscopic method provides a completely independent method for checking the different elevation hypotheses. The surface pressure is simply the summation of the partial pressures produced by the various particles in a vertical column from the surface, and a change in the number of particles in a unit area column implies a change in the total surface pressure. This method was used by Belton *et al.* (1968) and Belton and Hunten (1969) to measure the relative altitude differences between different dark and bright regions of Mars, and by Barker (1969). Their results agree roughly with Pettengill's (Pettengill *et al.*, 1969) radar observation in the conclusion that the highlands are not restricted to the dark areas and that the vertical scale of height variation is on the order of 10 km.

In this paper we will present some preliminary results obtained at McDonald Observatory in order to check the relative elevation differences by means of the spectroscopic method.

3. Spectroscopic Method of the Detection of Relative Altitude

The preliminary result of the spectroscopic method is the equivalent width of lines, which gives the abundance of CO_2 over the area accepted by the entrance slit of the spectrograph.

The equivalent width of a line which has been corrected for saturation is given by

$$W = \sigma(S_1 \eta \omega)^b \qquad (1)$$

where b is the slope of the curve of growth, σ the saturation factor, η the total airmass, and S_1 the line strength given by

$$S_1 = \frac{|m| S_b}{Q_{roT}} \left[\exp - m(m-1) \frac{hcB}{KT} \right], \qquad (2)$$

ω is the number $N(h)$, of absorbing molecules in a given rotational state in a unit vertical column through the atmosphere at a constant gravitational potential.

Observed equivalent width is

$$W_0 = W/\sigma = [S_1 \eta N(h)]^b. \tag{3}$$

The differential of this equation gives us an expression for the difference in the observed equivalent widths:

$$\frac{\Delta W}{W} = b \left[\frac{\Delta \eta}{\eta} + \frac{\Delta N}{N} + \frac{\Delta S_1}{S_1} \right] \tag{4}$$

where ΔW and $\Delta \eta$ can be obtained from the observational data.

By assuming the atmosphere is exponentially distributed with a scale height H, $\Delta N/N$ is obtained from a combination of the perfect gas law $(P = NKT)$, the change in pressure with height $(P = \rho g h)$, and the definition of scale height $(H = kT/\mu m_H g)$:

$$\frac{\Delta N}{N} = - \left[\frac{\Delta T}{T} + \frac{\Delta h}{H} \right]. \tag{5}$$

By taking the differential of Equation (2) with respect to temperature we obtain

$$\frac{\Delta S_1}{S_1} = \frac{hcB}{KT} m(m-1) \frac{\Delta T}{T}. \tag{6}$$

Substituting the last two equations into (4):

$$\frac{\Delta W}{W} = b \left[\frac{\Delta \eta}{\eta} - (1 - \frac{hcB}{KT} m(m-1)) \frac{\Delta T}{T} - \frac{\Delta h}{H} \right]. \tag{7}$$

This expression gives the fractional change in the observed equivalent width between two regions in terms of the fractional changes in airmass, temperature, and height. Solving for Δh, we obtain an expression for the elevation difference between two areas:

$$\Delta h = H \left\{ \frac{\Delta \eta}{\eta} - \left[1 - \frac{hcB}{KT} m(m-1) \right] \frac{\Delta T}{T} - \frac{1}{b} \frac{\Delta W}{W} \right\}. \tag{8}$$

This equation was applied to our data.

4. McDonald Spectroscopic Observations

The long term systematic spectroscopic observations of Mars at McDonald Observatory during the 1969 opposition provide a good coverage of the planet in the 0.87, 1.03, and 1.05 μ regions with reasonably good resolution in both latitude and longitude.

This actual coverage is schematically shown in Figure 1. Each line represents one spectrum, and its length, proportional to the time of exposure, gives an idea of the resolution in longitude. Effective resolution is of course affected by seeing conditions and guiding errors. These plates were obtained in the period between April 25 and October 25, 1969.

The Martian 8700 Å CO_2 band could be recorded in a reasonably short exposure

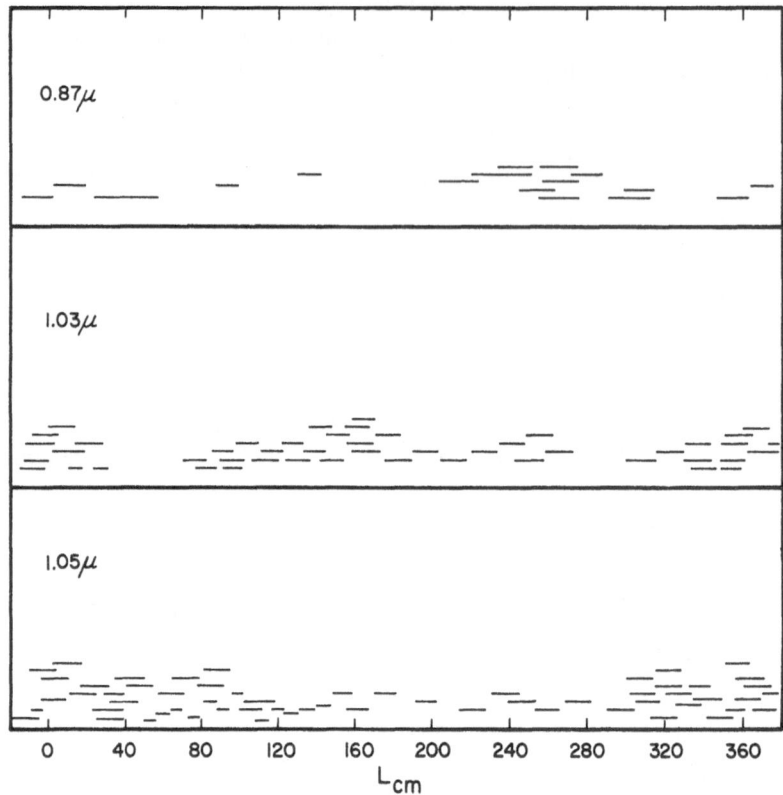

Fig. 1.   Actual coverage of Martian longitude by McDonald Observatory plates in the regions of 0.87, 1.03, and 1.05 μ CO_2 bands. Each line represents one spectrum and its length the resolution in longitude.

(resolution in longitude) only with the 107-inch telescope. The six-foot camera and grating 'B' (1200 grooves/mm, blazed at 6000 Å in first order) were used. The plate scale is about 9 arcsec/mm and the dispersion 4.4 Å/mm. Exposure time was sixty to ninety minutes and resulting resolution in longitude about 15°–20°.

Observations in the 1 μ region were carried out on the 82-inch Struve telescope. A Carnegie RCA infrared image tube with S1 photocathode was placed in offset focus of the 'A' coudé camera. Grating I (600 grooves/mm, blazed at 2.5 μ in the second order), giving a dispersion of about 3.5 Å/mm, was used. The scale on the plate is about 8 arcsec/mm. Exposure time was thirty to, in a few cases, ninety minutes.

On both telescopes the planet was aligned visually with the direction pole to pole on the slit of the spectrograph and guided as carefully as possible with the slit over the central meridian. Tracings of a small part of the width of such spectra give us information about the CO_2 behavior on different latitudes on Mars. Many of the plates are so good that it is possible to have a good tracing of one-fifth of the image of the planetary disc.

Eight spectrograms of Barker's (1969) observations of Mars in the 8700 Å and one of 1.03 μ regions during the 1967 opposition also had sufficient areographic resolution for detection of topographic features. They were taken in an interval of about five months on hypersensitized IV–N emulsion (1.03 μ band with image tube) with the 'A' camera of the 82-inch telescope's coudé spectrograph. The required exposure time was two to more than five hours, which gives a resolution in longitude of about 30° to 70° in extreme cases. Central microphotometric scans of one-third of the spectrum, combined with seeing and guiding errors, provided a resolution in latitude of about 30° on the planet, centered roughly on Martian latitude 20°N. By analysis of the curve of growth, the abundance and pressure over the corresponding areographic areas were determined and interpreted in terms of relative elevation differences on the planet. Because of the long period of observations, a correction for seasonal variation (see Barker's paper presented at this symposium) of the Martian abundance of CO_2 was applied. A small correction to the calculated central meridian for the middle of the exposure, due to the effect of guiding on the red image of the planet (RG8 filter blocking overlapping spectral orders), the increased atmospheric extinction during the later part of the exposure, and the effect of phase defect on guiding, was also applied. The final results are given in Table I. Figure 2 shows a

TABLE I

Relative elevation differences from 1967 observations

Coudé Number	L_s (°)	L'_{cm} (°)	Longitude resolution (°)	$\Delta h(L_s)$ ($H=9$ km) (km)
5023b	108	90	35	+0.2 ± 1.1
5167b	134	212	75	+0.7 ± 2.0
5187b	138	130	55	+1.6 ± 1.5
5187c	138	130	55	+2.3 ± 1.2
5212b	150	300	75	+0.8 ± 1.9
5243b	167	347	45	0.0 ± 1.1
5246b	167	305	55	+1.4 ± 1.4
5258b	170	294	40	−3.0 ± 1.7
5262b	171	246	25	−1.3 ± 1.5

comparison of these results with Pettengill's *et al.* (1969) radar profile smoothed to 45° resolution. Agreement of this spectroscopic determination of relative differences in altitude with the radar data is quite good. The correlation coefficient is 0.66.

Only a small part of our observations during 1969 opposition has been reduced. These reduced spectra cover roughly a period of one month of observation (about fifteen days of a Martian season) in both 8700 Å and 1.05 μ regions.

To avoid uncertainty in the determination of the pathlength in high Martian latitude, only the central scans of the spectrum were used, and Martian airmass was assumed to be 2.2. Because of the short period of observations, no corrections for seasonal variation of the amount of Martian CO_2 seemed to be necessary. The central thirds of the 8700 Å band spectra (resolution in latitude 20°) were traced in density, and both R and P branches (calibrated on solar lines) were used in analysis of the curve of growth.

15—P.A.

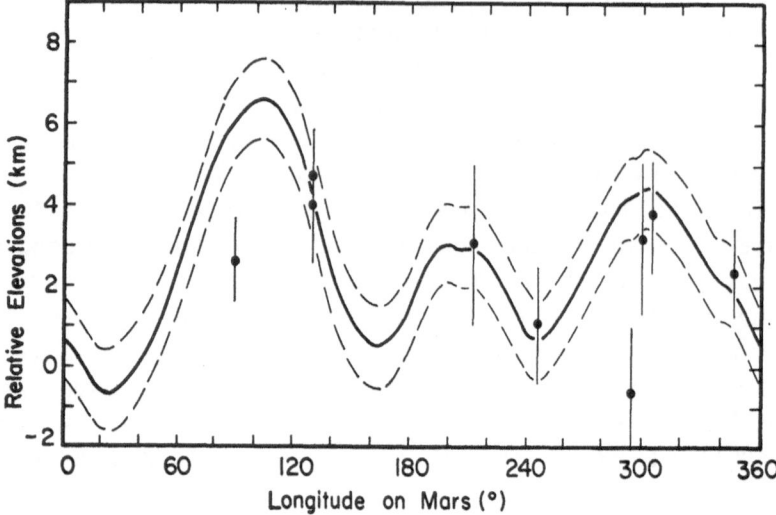

Fig. 2.   Relative elevation differences as a function of the longitude from 1967 opposition observations using the CO_2 abundances corrected for seasonal variation. Solid line represents Pettengill's radar profile smoothed to 45° resolution; dashed line, the error limits of ±1 km on the radar profile.

Because of its weakness, this band is unsuitable for temperature determination, and a rotational temperature of 200 K was assumed for all of these plates. The results obtained are shown in Table II and Figure 3 (open circles).

The 1.05 μ band was traced in intensity mode over the central quarter of the Martian disc. The resolution in latitude is about 15° on the planet. Only the P branch was used for abundance and temperature determination. In calculating elevation differences,

TABLE II

Relative elevation differences from 1969 observations of 0.87 μ band

Coudé No.	L_s (°)	L_{cm} (°)	ω (m-atm)	CO_2 partial pressure (mb)	Relative elevation (km)
104	184.0	10 ± 7	132 ± 43	9.8	0.0 ± 2.7
70	162.0	40 ± 18	165 ± 68	12.2	−1.8 ± 3.5
74	165.7	353 ± 9	102 ± 55	7.6	+2.4 ± 2.5
80	168.4	252 ± 11	125 ± 39	9.2	+1.4 ± 2.2
82	168.4	281 ± 11	118 ± 32	8.7	+1.2 ± 2.0
84	169.0	266 ± 11	98 ± 37	7.2	+2.6 ± 1.8
87	169.5	213 ± 11	150 ± 57	11.1	−1.2 ± 2.5
88	169.5	240 ± 11	139 ± 61	9.9	+0.5 ± 2.7
89	169.6	264 ± 11	81 ± 32	6.0	+4.0 ± 1.5
91	170.1	243 ± 11	67 ± 23	4.9	+5.1 ± 1.4
96	176.7	135 ± 7	102 ± 33	7.5	+2.4 ± 2.2
98	181.2	92 ± 6	108 ± 66	8.0	+1.8 ± 2.6

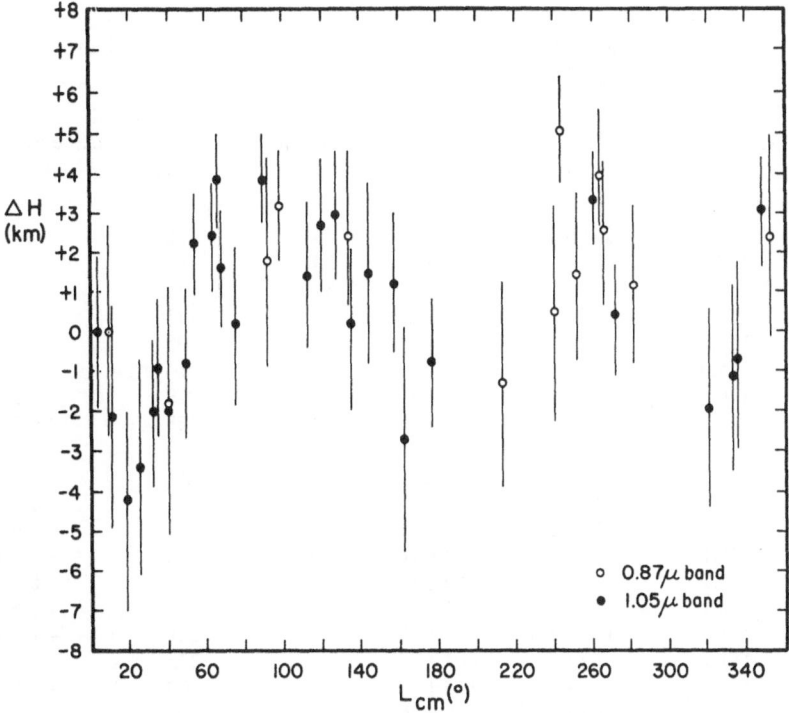

Fig. 3. Relative elevation differences at about 10°N Martian latitude versus Martian longitude from observations of the 0.87 μ (open circles) and 1.05 μ (filled circles) CO$_2$ bands during the 1969 opposition.

only lines from *P6* to *P26* were used in order to be on approximately the same portion of the curve of growth.

Table III and Figure 3 (filled circles) show the results. The subearth point of Mars was between 8° and 12°N during these observations.

One can see easily that there are no systematic differences between the results obtained from the 8700 Å and the 0.05 μ bands. The observed altitude differences are on the order of 10 km. A comparison with the latest radar data (Rogers *et al.*, 1969) shows that there is a good agreement of both results in the region from 0° to about 200° in Martian longitude. There is a lowland region at about 20°–30° of longitude, and there is a big highland area at 100°–120°. Regarding the longitudes of about 200°–300°, the correlation is not as good. However, this region is chiefly covered by our 8700 Å plates, with relatively large internal errors. We have good coverage of this region of Martian longitudes by the observations in the 1.03 μ band; these have not yet been reduced.

5. Summary

By improvements in the Struve reflector's coudé spectrograph and construction of the new 107-inch telescope and its large spectrograph, observations of Mars with greater spatial resolution were possible. This allowed the study of the Martian atmosphere

ANDREJ WOSZCZYK

TABLE III
Relative elevation differences from 1969 observations of 1.05 μ band

Coudé No.	L_s (°)	L_{cm} (°)	ω (m-atm)	CO_2 partial pressure (mb)	Rotational temperature (°K)	Relative elevation (km)
6251	168.9	261 ± 7	54 ± 18	4.0	220	+3.9 ± 1.3
6252	168.9	276 ± 7	79 ± 10	5.8	185	+0.4 ± 1.7
6269A	170.5	162 ± 7	113 ± 19	8.3	216	−2.7 ± 2.8
6269B	170.5	177 ± 7	69 ± 10	5.1	173	−0.9 ± 1.8
6301A	176.0	53 ± 3.5	64 ± 11	4.7	205	+2.2 ± 1.4
6301B	176.0	63 ± 3.5	64 ± 11	4.7	214	+2.5 ± 1.4
6301C	176.0	675 ± 3.5	69 ± 13	5.1	228	+1.5 ± 1.5
6302A	176.0	75 ± 3.5	79 ± 15	5.9	225	+0.2 ± 1.9
6302C	176.1	90 ± 3.5	50 ± 15	3.7	230	+3.9 ± 1.3
6302D	176.1	97 + 3.5	53 ± 16	3.9	211	+3.2 ± 1.4
6303A	176.1	112 ± 3.5	83 ± 14	6.2	251	+1.4 ± 1.8
6303B	176.1	119 ± 3.5	60 ± 14	4.3	250	+2.6 ± 1.7
6303C	176.1	127 ± 3.5	58 ± 16	4.3	212	+2.9 ± 1.7
6303D	176.1	135 ± 4	85 ± 16	6.3	196	+0.2 ± 2.1
6303E	176.1	144 ± 5	71 ± 32	5.3	281	+1.5 ± 2.2
6358	181.7	157 ± 5.5	64 ± 18	4.7	221	+1.7 ± 1.5
6359A	181.7	10 ± 7	101 ± 27	7.5	225	−2.0 ± 2.7
6359B	181.7	25 ± 7	113 ± 17	8.4	200	−3.5 ± 2.5
6359C	181.7	40 ± 7.5	98 ± 39	7.3	200	−2.0 ± 3.1
6363A	182.2	334 ± 7	91 ± 27	6.7	220	−0.6 ± 2.1
6363B	182.2	348 ± 7	56 ± 17	4.2	237	+3.1 ± 1.4
6363C	182.2	3 ± 7	85 ± 18	6.3	195	+0.0 ± 1.8
6364A	182.2	18 ± 7	118 ± 32	8.7	217	−4.2 ± 2.6
6364C	182.3	32 ± 7	83 ± 23	6.3	275	−0.8 ± 1.8
6371A	182.8	321 ± 7	102 ± 24	7.5	177	−1.9 ± 2.5
6371B	182.8	335 ± 7	95 ± 20	7.0	159	−1.1 ± 2.2
6403	201.5	31 ± 7	87 ± 19	6.4	229	−2.0 ± 1.9
6404	201.5	65 ± 7	52 ± 7	3.8	202	+3.9 ± 1.2

over particular regions of the planet and the detection of large scale irregularities in the surface by means of spectrographic methods. The results presented here, though in agreement with radar data, have a relatively large internal error. The average error of the elevation determinations is about ±1.8 km for the 1.05 μ band and ±2.3 km for the 0.87 μ band. One could probably decrease the error for the 1.05 band by including in the solution the lines of the R branch. Furthermore, the spectroscopic elevation profile for the planet will be certainly better defined after reduction of our remaining observations. More precise observations in the future will require better spatial as well as spectral resolution.

Acknowledgments

The author's stay at the McDonald Observatory and this study were possible as a result of a University of Texas at Austin McDonald Fellowship. I wish to express my sincere gratitude to the director of the McDonald Observatory, Dr. Harlan J. Smith, and his staff for their kind hospitality. Special thanks are due to Dr. Ed Barker and Jay Bergstralh for their assistance in acquisition and reduction of the data.

The support of the National Aeronautics and Space Administration Office of Lunar and Planetary Programs for the McDonald Observatory planetary observation is also gratefully acknowledged.

References

Barker, E. S.: 1969, unpublished Ph.D. Dissertation, University of Texas.
Belton, M. J. S., Broadfoot, A. L., and Hunten, D. M.: 1968, *J. Geophys. Res.* **73**, 4795.
Belton, M. J. S., and Hunten, D. M.: 1969, *Science* **166**, 225.
Counselman, C. C., Pettengill, G. H., and Shapiro, I. I.: 1968, *Trans. Amer. Geophys. Union* **49**, 217.
Pettengill, G. H., Counselman, C. C., Rainville, L. P., and Shapiro, I. I.: 1969, *Astron. J.* **74**, 461.
Rea, D. G.: 1963, *Nature* **200**, 114.
Rea, D. G.: 1964, *Nature* **201**, 1014.
Rogers, A. E. E., Ash, M. E., Counselman, C. C., Shapiro, I. I., and Pettengill, G. H.: 1969, 'Radar Measurements of Martian Surface Topography and Roughness', Paper presented at the Joint IAU/URSI Symposium on Planetary Atmosphere and Surfaces, Woods Hole, Mass., 11–15 August.
Sagan, C. and Pollack, J. B.: 1966a, *Nature* **212**, 117.
Sagan, C. and Pollack, J. B.: 1966b, *Astron. J.* **71**, 178.
Sagan, C., Pollack, J. B., and Goldstein, R. M.: 1967, *Astron. J.* **72**, 20.
Sagan, C. and Pollack, J. B.: 1968, *J. Geophys. Res.* **73**, 1373.
Wells, R. D.: 1965, *Nature* **207**, 735.
Wells, R. D.: 1966, *Nature* **209**, 1338.

SPECTROSCOPIC DETERMINATION OF SURFACE PRESSURE AND ELEVATION DIFFERENCES ON MARS

D. P. CRUIKSHANK*

Lunar and Planetary Laboratory, University of Arizona, Tucson, Ariz., U.S.A.

Abstract. Observations of a carbon dioxide band at 1.2206 μ (8192.6 cm^{-1}) made at the opposition of Mars in June, 1969, are used with laboratory data to derive a value for the pressure at the surface of Mars. This band is especially suitable because it is sensitive to pressure, lying in the transition region of the curve of growth, but is relatively free of telluric contamination. The pressure derived is $5.3^{+2.2}_{-2.6}$ mb, corresponding to the Martian desert region Amazonis.

Using a five-channel spectrometer, the strengths of two carbon dioxide bands at 1.5753 and 1.6057 μ (6347.8 and 6227.9 cm^{-1}) were compared over a Martian dark area (Mare Acidalium) and a nearby bright area in Amazonis. The bands were 1.32 times stronger in the dark area than in the bright area. Interpreting this as evidence for elevation differences on the Martian surface, it is found that the dark Mare Acidalium is 2.5 km lower in elevation than the bright area with which it was compared.

1. Introduction

The observations described in this paper were made jointly with V. I. Moroz using his equipment on the 125-cm reflector of the Southern Station of the Sternberg State Astronomical Institute (Moscow, U.S.S.R.) in the Crimea, while the author was a guest of the Soviet Academy of Sciences. The laboratory measurements and the interpretations in this paper were made by the present author, and the appearance of the data here does not preclude their elaboration and further interpretation elsewhere, either by Moroz alone or jointly with other investigators. The author wishes to record his gratitude to V. I. Moroz for his generosity in making the observational material freely available.

2. Determination of the Surface Pressure

Previous spectroscopic investigations of the Martian surface pressure have been made in two steps. First, the CO_2 abundance is determined from observations of weak lines which are essentially independent of the pressure. Then, strong bands, frequently those near 1.6 and 2.0 μ, which are sensitive to pressure broadening are used for deriving the surface pressure. Reviews of earlier studies of this problem are given by Cann *et al.* (1965) and Chamberlain and Hunten (1965).

Contradictory and uncertain values for the Martian surface pressure resulting from analyses of various CO_2 bands made it seem desirable to study previously neglected bands in order to establish an independent pressure value. The study of faint bands requires more spectral resolution than can easily be achieved with scanning spectrometers in the infrared, and for this reason the bands we are considering here were not

* Present address: Institute for Astronomy, University of Hawaii, Honolulu, Hawaii, 96822, U.S.A.

Sagan et al. (eds.), Planetary Atmospheres, 212–219.
All Rights Reserved. Copyright © 1971 by the I.A.U.

treated in the lengthy investigations of Owen and Kuiper (1964), Chamberlain and Hunten (1965), and others. However, using a scanning spectrometer built by Moroz, and a Cu:Ge photoconductive detector cooled with liquid nitrogen, it was possible to obtain good spectra of Mars in the 1.2 μ region with resolution 1700, or about 7 Å. The 125-cm reflector of the Southern Station of the Sternberg State Astronomical Institute (Moscow) was used in its location in the Crimea at latitude 45°N. The northern latitude of the observatory made it difficult to observe Mars because of the planet's low declination.

Figure 1 shows representative spectra of Mars and the moon (at similar airmass) taken on June 6–7 and June 5–6, 1969, respectively. While the band designated 023

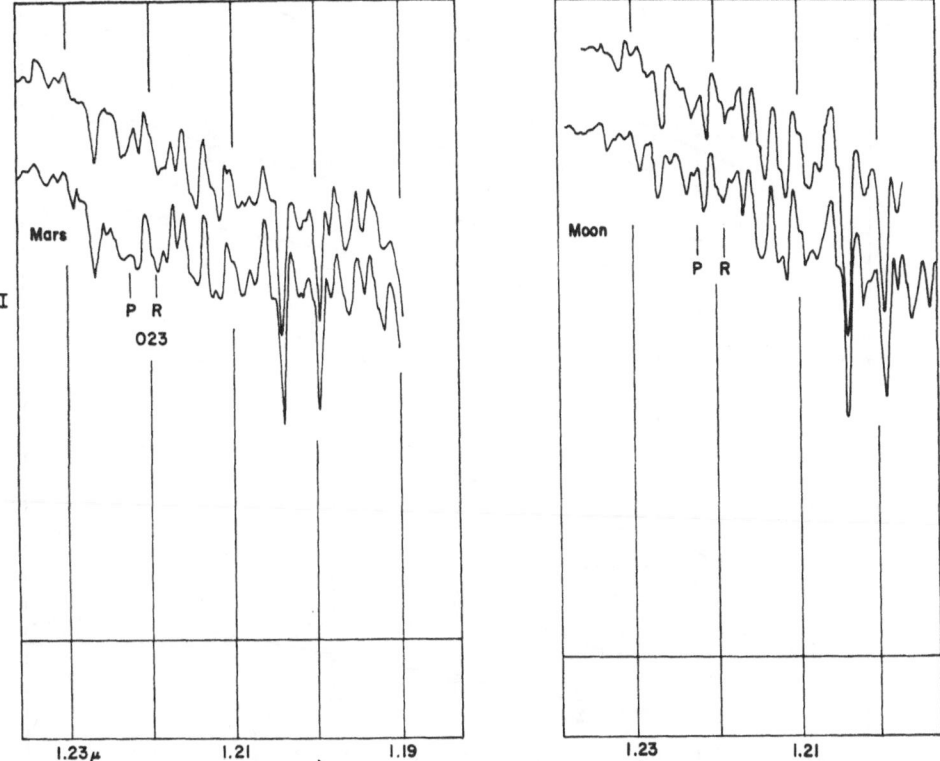

Fig. 1.   Spectra of Mars and the moon in the region of the bands of CO_2 at 1.2206 and 1.2055 μ. Made with scanning spectrometer having resolution 7 Å.

(old style) shows distinctly in the Martian spectrum, the companion band (103) at 1.2055 μ (8294.0 cm^{-1}) is too strongly blended with telluric water vapor lines to be disentangled for measurement. The 023 band was measured by Moroz on the original tracings, and he found 2.6 Å $\pm$ 0.5 Å for the equivalent width. It is upon this value that the following analysis is based. The spectra were taken with the slit accepting an equatorial strip across the planet.

In this analysis, the empirical approach used by Owen and Kuiper (1964) was adopted. Laboratory spectra of pure CO_2 were made at the Lunar and Planetary

Laboratory using a scanning spectrometer adjusted to give the same spectral resolution as that with which the Mars spectra were obtained. The gas was admitted to the 40 m multiple-pass White cell and for a given pressure, spectra were taken at three or four different total path lengths up to 2.47 km (64 traversals through the cell). Because the bands at 1.2206 and 1.2055 μ are weak, they did not show well on the spectra with

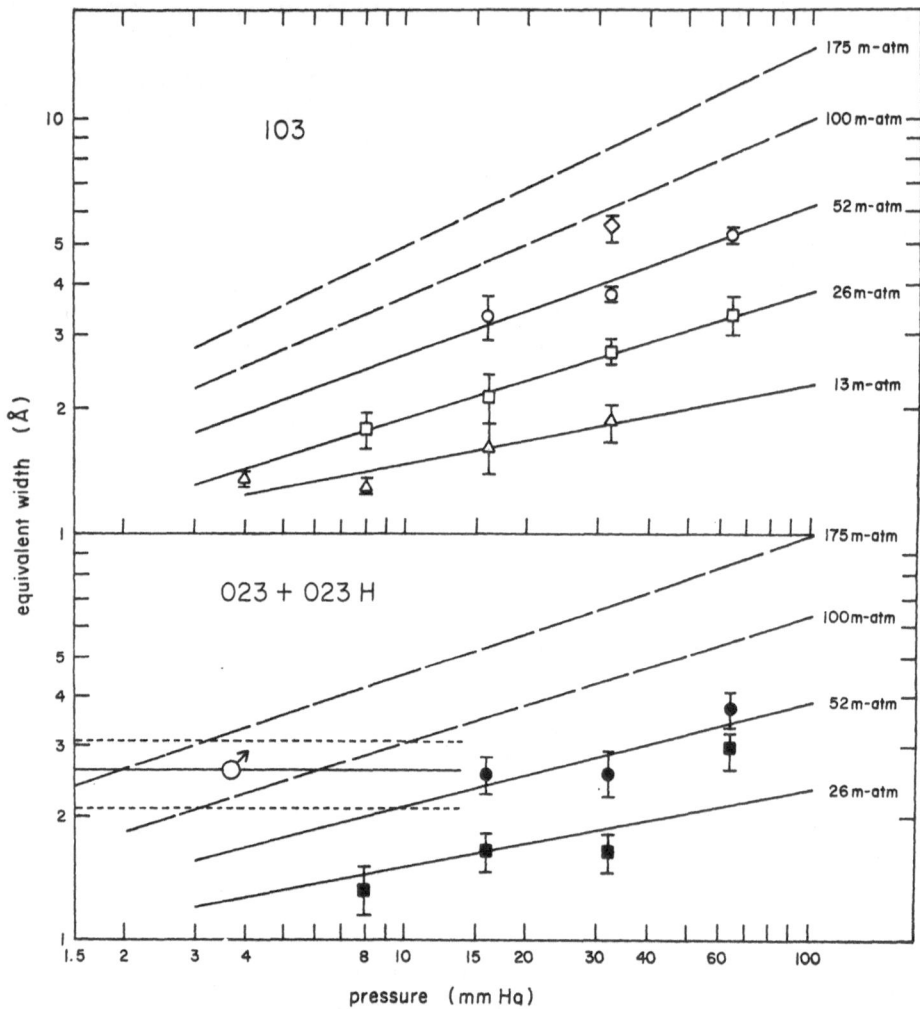

Fig. 2.   Curves of growth of the CO_2 bands at 1.2055 (103) and 1.2206 (023) microns from laboratory observations. The curves of growth for the 023 band include a small contribution from the hot band 023H. This contribution does not exceed about 10%.

only 4-mm Hg pressure in the cell, but at 8 mm pressure and higher they were easily seen and measured.

   The laboratory data are plotted in Figure 2 for both bands observed in the laboratory. While only the 1.2206 μ band is considered in this report, the data for the 1.2055 μ band are also included for possible future use. The data at first show that these two bands are in fact pressure sensitive – the same quantity of gas (in meter-atmospheres) gives smaller equivalent widths at lower pressures. The slope of the lines, shows, however, that the bands are neither on the 'straight-line' or 'square-root' portions of the curve of growth, but instead in the transition region.

As Owen and Kuiper (1964) pointed out, the amount of gas required in the laboratory simulation of the Martian atmosphere is given by

$$w \frac{T_{lab}}{273} \eta$$

where w is the CO_2 abundance (in meter-atmospheres) in the Martian atmosphere as determined by other spectroscopic investigations, and η is the effective airmass in the Martian atmosphere. Using the table of Owen and Kuiper (1964, p. 132) we find $\eta = 2.21$. If we adopt 75 m-atm as the Martian CO_2 abundance (Kaplan and Gray-Young, 1969), then the amount of gas required in the laboratory becomes 179 m-atm. In the range of pressures appropriate for the Martian surface (say 4 mm Hg), 179 m-atm CO_2 requires a path length of some 34 km. This was not possible in the LPL White cell, so observations were made at three different quantities of gas (four, in the case of the 103 band) and several different pressures. Then, as Owen and Kuiper found necessary, the curves were extrapolated to larger quantities of gas, as shown by the dashed lines in Figure 2. The 023 band was also observed at 13 m-atm but the data are not included in Figure 2 because of their inferior quality. They do not bear directly on the extrapolation to large quantities of gas.

Because of the gentle slope of the 175 m-atm line in Figure 2, it can be seen that the greatest source of error is the uncertainty in measurement of the equivalent width of the CO_2 band in the Mars spectrum. The error in extrapolation of the laboratory data to 175 m-atm is less significant. The horizontal line representing the equivalent width of the band in the Mars spectrum intersects the 175 m-atm line at a pressure value of 2 mm Hg (2.66 mb). Because the mean pressure along the absorbing path in the atmosphere of Mars is one-half the surface pressure (Curtis-Godson approximation), we double this value to give the *Martian surface pressure of* $5.3^{+2.2}_{-2.6}$ mb.

3. Elevation Differences

In 1965, when it became known from analysis of Mariner IV occultation data that the scale height in the lower Martian atmosphere was 9 km ± 1 km (Kliore *et al.*, 1965), speculation about elevation differences on Mars took on a new dimension. It was evident that differences in elevation of the order of a few kilometers should be detectable by the difference in atmospheric CO_2 absorption that would occur above regions of relatively different altitude.

No reliable spectroscopic results were reported from the 1967 apparition of Mars, but radar ranging data with high spatial resolution on the planet were obtained (Pettengill *et al.*, 1969) and interpreted (Binder, 1969) to show that the dark areas are uniformly lower in elevation that the bright areas. Further analysis of the radar shows that the correlation of topography with dark and bright areas is not direct. The first spectroscopic results were obtained at the 1969 apparition of Mars and reported in a preliminary form by Hunten and Belton (1969). They verified the lack of direct correlation between elevation and surface albedo, and reported a total range of altitude of more than 20 km.

In order to compare Martian atmospheric CO_2 absorption over the bright and dark areas of the planet, a 5-channel, simultaneously recording spectrometer was used with a multi-element PbS detector on the 125-cm reflector in the Crimea. The dispersion of the spectrometer and the spacing interval of the PbS detector elements were such that the two strong bands of CO_2 at 1.5753 and 1.6057 μ (6347.8 and 6227.9 cm^{-1}) fell on elements 2 and 4 of the five-element detector array. Elements, 1, 3, and 5 covered the regions outside and between the two strong bands, thus providing a measure of the intensity of the continuum. The normal entrance slit of the spectrometer was replaced by a plastic plate in which several tiny holes had been drilled and the front surface of which had been aluminized for high reflectivity. This made it possible for the observer to continuously monitor the image of Mars with the spectrometer guide microscope during the integrations. A diaphragm of diameter 3 sec of arc

Fig. 3.   Photograph of Mars showing the regions observed for comparison of CO_2 band strengths and the derivation of elevation difference. Photograph by John Fountain, June 18, 1969, 01 05 UT, Cerro Tololo 60-inch reflector, Kodak High Speed Infrared film with filter transmitted 6600–8700 Å. Composite of four images. Central meridian longitude 61.24°.

gave suitable signal levels and provided the highest practical resolution on the disk of Mars. Figure 3 shows the appearance of Mars during the measurements with the appropriate size of the diaphragm superimposed. The innermost circle is the actual projected diameter of the entrance diaphragm, and the outer circle represents the degree of smearing by seeing. The guiding of the telescope was exceedingly accurate.

The observed points on Mars were centered over a dark area (approximate co-ordinates 35°W longitude, +45°N latitude) and a light region (100°W longitude, +10°N latitude). Many integrations were made over the dark area (Mare Acidalium) and over the bright region which was equidistant from the limb of the planet (Amazonis desert), each with a sky background integration for comparison. An attempt at absolute calibration of the measurements was made by observing a bright star, but only the relative measurements are of interest here. From observations of the same bright and dark regions over two nights (June 8–9 and 9–10, 1969), there was clear indication that the intensity of the two CO_2 bands was greater over the dark Mare Acidalium than over the comparison desert region. Averaging the data of both nights, we find that the absorption in the dark area is stronger by a factor 1.32 than in the bright region.

In the simplest interpretation of these data, we can find the elevation difference knowing only this ratio of band intensities and the atmosphere scale height. If N_0 is the number of molecules in the atmosphere at some datum h_0, then the number of molecules at height h_1 is $N_0 \exp(-h_1/H)$ and at height h_2 is $N_0 \exp(-h_2/H)$. Then, the ratio of the absorption band strength at the two levels h_1 and h_2 is given by $\exp[-(h_1-h_2)/H]$. H is the scale height, taken here as 9 km. The difference in elevation, $h_2 - h_1 = H \ln 1.32 = 2.5$ km.

The uncertainties in this last quantity depend on the probable errors in the scale height (about 10%) and in the observed ratio of band strengths (10–12%). Thus, the observed elevation difference with its probable error is 2.5 ± 1.2 km.

Two further sources of error exist, both considered minor. First, the low altitude of Mars during the observation results in atmospheric dispersion of the image seen in the telescope. This means that the infrared image is displaced slightly from the visual light image, and while the observer positions the visual image correctly with respect to the entrance diaphragm, the infrared image (detected by the spectrometer) lies in another position. During the observations reported here, the effect of image displacement was tested by moving the image around with respect to the diaphragm and noting the signal level received by the detector. For the Mare Acidalium point, the signal level was the lowest that could be seen for any region of the planet, and from this it is concluded that atmospheric dispersion was a minor effect, displacing the infrared image of the dark region less than one diaphragm diameter from the visual image.

The second source of error is in the assumption of an isothermal lower atmosphere on Mars. This assumption considerably simplifies the analysis of the present observations and any reasonable departure from isothermal conditions will not significantly affect the results given here, especially since the observed elevation difference is relatively small.

It is instructive to compare the results of the present measurements with those reported by other investigators. The radar ranging data of Pettengill *et al.* (1969) indicate that the relative elevation difference at the longitudes of the regions observed here is 9 ± 2 km. The radar data refer to a rather narrow strip centered at latitude $+21°$N, and extrapolation to the latitudes of the spots observed spectroscopically is uncertain. In an independent analysis of an early version of Pettengill's data, Binder (1969) made an extrapolation in latitude, and found that the elevation difference near the two regions observed here was 7–8 km. The important thing is that analyses of both Pettengill *et al.* and Binder agree with the present study in the *sense* of the difference in elevation of these two regions, i.e. the dark Mare Acidalium is lower than nearby desert Amazonis.

Spectrographic detection of elevation differences near the regions discussed in the present paper were reported by Belton and Hunter (1969). On a solid relief map made from their observations of approximately 200 points on Mars, Mare Acidalium is seen as a low area compared with the region in the Amazonis desert at $100°$ longitude. The elevation difference estimated from the map is 5 km. The map was made from preliminary reductions of the observations and is subject to refinement.

Acknowledgments

Gratitude is expressed to V. I. Moroz for his generosity in sharing the observational data obtained in June, 1969. Miss L. V. Gromova assisted in making some of the Mars observations, and A. B. Thomson assisted in obtaining the laboratory data for CO_2. The Mars observations were made while the author was a National Academy of Sciences exchangee in the Soviet Union. The laboratory studies and interpretation were completed upon return to the Lunar and Planetary Laboratory, and were supported by NASA grant NsG 161-61.

Note Added in Proof. Two detailed studies of the observational data reported here with additions have been submitted for publication in the *Astronomicheskii Zhurnal* (U.S.S.R.). In the first, 'An Attempt to Determine Differences in Height on Mars from the Intensity of the CO_2 Bands at 1.6 Microns', by V. I. Moroz, N. A. Parfentev, D. P. Cruikshank, and L. V. Gromova, it is shown that dark regions on the planet can be higher, lower, or the same in elevation as the bright regions. The maria Acidalium and Cimmerium were compared with bright regions Amazonis, Isidis, and Cebrenia. The maximum elevation difference found was in the comparison of Mare Cimmerium with Cebrenia, the mare being 7.2 km higher in elevation.

In the second paper, 'A Spectroscopic Determination of the Pressure in the Atmosphere of Mars from the CO_2 Band at 1.22 Microns', by V. I. Moroz and D. P. Cruikshank, new laboratory spectra of carbon dioxide bands at 1.22, 1.6, and 2.1 microns were used to determine a composite curve of growth. This technique reduces the error of extrapolating the laboratory curve of growth to longer path lengths (corresponding to the Martian absorption) than are physically possible in the laboratory. The result of that analysis gives the Martian surface pressure for a pure CO_2 atmosphere not substantially different from the result of the more simple analysis in the present paper.

References

Belton, M. J. S. and Hunten, D. M.: 1969, 'Spectrographic Detection of Topographic Features on Mars', *Science* **166**, 225–227.

Binder, A. B.: 1969, 'Topography and Surface Features of Mars', *Icarus* **11**, 24–35.

Cann, M. W. P., Davies, W. O., Greenspan, J. A., and Owen, T. C.: 1965, 'A Review of Recent Determinations of the Composition and Surface Pressure of the Atmosphere of Mars', NASA Contractor Report CR-298, 176 pp.

Chamberlain, J. W. and Hunten, D. M.: 1965, 'Pressure and CO_2 Content of the Martian Atmosphere: A Critical Discussion', *Rev. Geophys.* **3**, 299–317.

Gray-Young, Louise D.: 1969, 'Interpretation of High-Resolution Spectra of Mars, I. *Icarus* **11**, 386–389.

Kliore, A., Cain, D. L., Levy, G. S., Eshleman, V. R., Fjeldbo, G., and Drake, F. D.: 1965, 'Occultation Experiment: Results of the First Direct Measurement of Mars' Atmosphere and Ionosphere', *Science* **149**, 1243–1248.

Kaplan, L. D. and Gray Young, L. D.: 1970, this volume, p. 189.

Owen, T. C. and Kuiper, G. P.: 1964, 'A Determination of the Composition and Surface Pressure of the Martian Atmosphere', *Commun. Lunar Planetary Lab.* **2**, 113–132.

Pettengill, G. H., Counselman, C. C., Rainville, L. P., and Shapiro, I. I.: 1969, 'Radar Measurements of Martian Topography', *Astron. J.* **74**, 461–482.

C. WATER VAPOR ABSORPTION

THE SPECTROSCOPIC SEARCH FOR WATER ON MARS:
A HISTORY

RONALD A. SCHORN

Jet Propulsion Laboratory, California Institute of Technology, Pasadena, Calif., U.S.A.

The discovery of water on Mars is a lot like the discovery of America. Both were done many times by many different people. In both cases many of the 'discoveries' were either fictitious or difficult to verify. Regarding the discovery of America, a grammar school teacher of mine once pointed out that Columbus was the *effective* discoverer of America. After he did it, people believed it was so. More importantly, people began to *act* on the basis of his discovery.

Curiously, the effective discovery of water on Mars did not take place until 1963. I say curiously, because for centuries, it seemed that definitive proof was 'just around the corner'. The history of this search for evidence is most interesting in itself. As a good deal of the observational work and speculation on the subject was wrong, it has the additional quality of being a tale with a moral (for astronomers and many others). Finally, it has a happy ending.

One hundred years ago, it seemed obvious that there was a great deal of water on Mars. There was plenty of water on Earth, and telescopic observations showed that Mars resembled the Earth. Mars had polar caps that varied in size with the seasons, clouds that came and went, continents (the bright or 'red' areas), and great bodies of water (the dark areas). While the Beer Sea and the Dawes Ocean are long gone, Mare Sirenum, Mare Tyrrhenum, and the like are more permanent reminders of this state of our knowledge.

Figure 1 reproduces a plate from *Other Worlds Than Ours* by the well known 19th-century popularizer of astronomy, Richard A. Proctor (1873). The chart was made from the observations of Dawes, and many permanent features of Mars are recognizable. For example, the Kaiser Sea is Syrtis Major, J. Herschel Strait is Sinus Sabaeus, etc. The interpretation of these features, however, has changed completely since the 1870's. Even in those days, however, there were those who, remembering the fate of the lunar 'seas', weren't too sure that the Martian seas were full of water. Of course, the existence of a Martian atmosphere and polar caps made the existence of water on Mars more probable than in the case of the Moon, but more direct evidence was needed. So, beginning in the 1860's, "The wonderful powers of the spectroscope have been applied to this question," to quote Proctor.

The first observations were made, of necessity, with the visual spectroscope. In 1867 Huggins, by comparing the spectra of Mars and the Moon in the orange region, became the first 'discoverer' of water vapor in the atmosphere of Mars. Thus began a dreary series of observations which were all wrong, misleading, indecisive, or negative. The important point about all of the late 19th- and early 20th-century observations was that the upper limit of the abundance of Martian atmospheric water vapor was

16—P.A.

Sagan et al. (eds.), Planetary Atmospheres, 223–236.

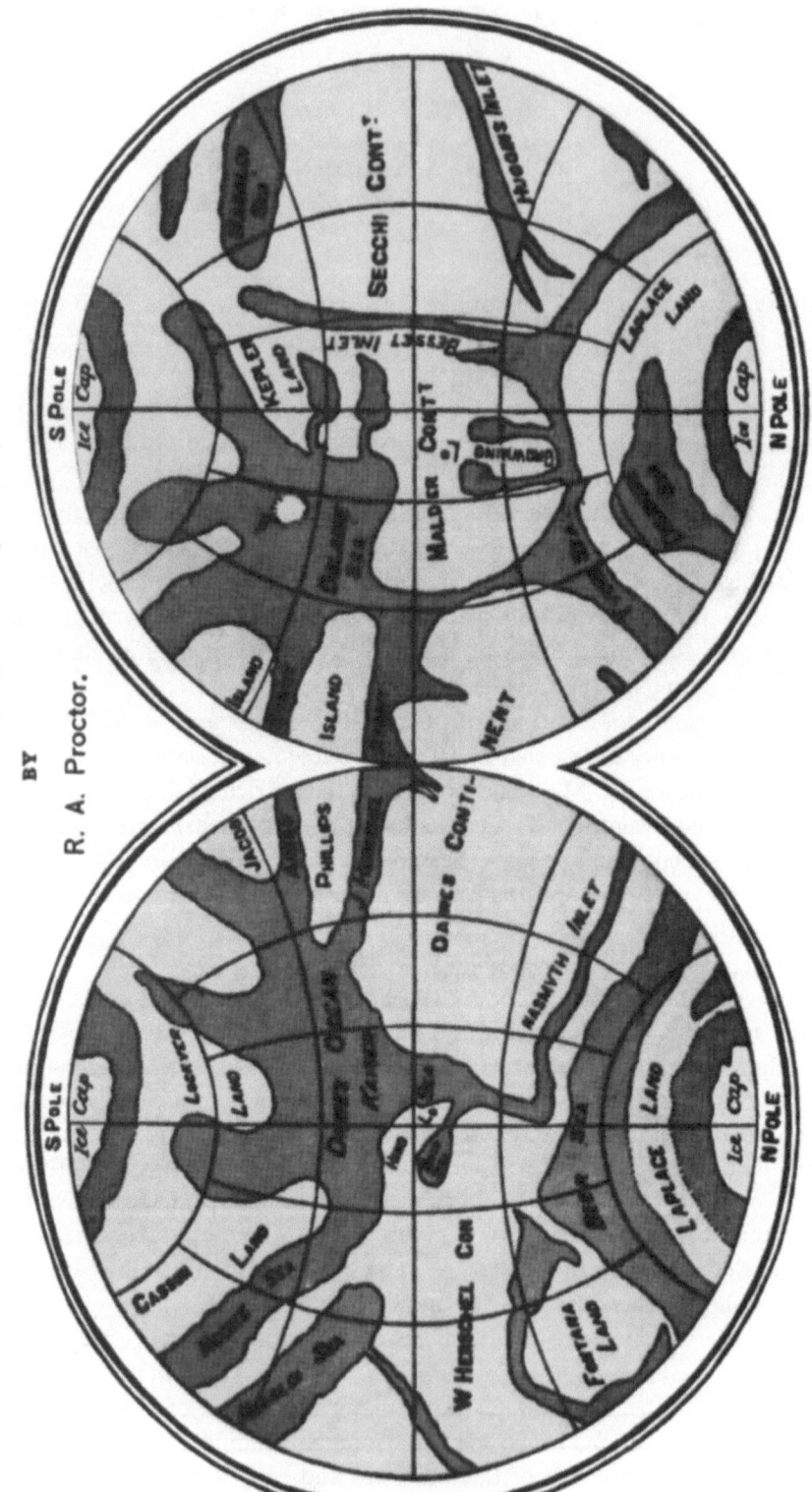

Fig. 1. A map of Mars, circa 1873. Most of the features can be identified with modern ones, but the interpretation of the dark areas has changed a great deal.

pushed steadily downward. The course of this story through the early years is given by Campbell (1909) and carried up to the middle of the twentieth century clearly and concisely by de Vaucouleurs (1954) in his landmark book on Mars, so will not give all the details here.

The photographic plate replaced the human eye as the recorder of spectra during this period, and observations were made at higher and dryer sites as time went on. Also, the comparison of 'band strengths' on low dispersion spectra gave way to high-dispersion spectra. On the latter, the Doppler shift (due to the relative radial velocity of Mars and the Earth) could be used to separate any Martian water lines from their terrestrial counterparts or, at least, to cause an asymmetry in the shape of the terrestrial water-vapor lines.

The Doppler-shift technique was first attempted by Lowell and Slipher (1905), and later by Campbell and Albrecht (1910), in the orange and red regions of the spectrum. Whether or not Martian water vapor was detected, exactly how much was, and what the upper limits were for these early studies, has never been settled. However, it is probably fair to say that all of these observations agreed that the amount of Martian water vapor per unit area (if it existed at all) was small in comparison to the amount in the Earth's atmosphere.

Meanwhile, the idea of Martian seas had been dying out as observational evidence against this concept accumulated. Russell et al. (1945) summarize this evidence. First, the brilliant reflection of the Sun from an open-water surface has never been seen. Second, the dark regions exhibit a great deal of internal structure. Finally, the dark areas show variation in shade and depth of color, and size and shape, during the march of the Martian seasons, and from year to year. This left the pole caps and various transient white 'clouds', 'frosts', and 'mists' as indirect evidence for water on Mars.

Considering the rapid shrinking in estimates of the amount of Martian water it is not surprising that, at the turn of the century, Stoney (1898) suggested frozen carbon dioxide for the composition of the polar caps and various white patches on Mars. At the time, and for long afterwards, there was no direct evidence that carbon dioxide (in large amounts) existed on Mars. But, during the next seventy years, the suggestion had varying degrees of popularity, and was never disproved.

There was a lack of definite spectroscopic evidence for the existence of *any* gas in the atmosphere of Mars (and in particular, water vapor) during the first half of the twentieth century. In this interval, of necessity, a great deal of work on possible and probable abundances was carried out based on 'less-direct' techniques. Some examples are photometric and polarimetric observations, calculations based on the existence and some assumed composition of the polar caps and clouds, and the like. Only a few examples of this large amount of work (large compared to the amount of spectrographic evidence) will be covered in the course of this review. First, there is not the time to do so here. Second, they are discussed in detail in de Vaucouleurs' book. Finally, I must say, most of these papers were not of much help in solving the problem of obtaining definite evidence for the existence of water on Mars. Usually they predicted amounts just below whatever the current spectroscopic limit was at the time.

The 'modern era' in spectroscopic observations of Mars began in 1925 when

Adams and St. John (1926) used the first coudé spectrograph – a six-prism instrument on the Mt. Wilson 60-inch reflector. Their resultant dispersion was 7.3 Å/mm near 6300 Å. They attempted to measure the displacement in wavelength of strong terrestrial H_2O lines which would be caused by any (presumably unresolved) weaker Martian H_2O components. Such a displacement was detected and, in present day terms, corresponded to about 450 μ of precipitable H_2O over unit area on Mars. This result was almost certainly wrong. For one thing, their measures of the O_2 band near 6300 Å showed Martian components one-third the strength of the terrestrial O_2 lines, which was certainly an error. Secondly, the H_2O lines available in the region of the spectrum they used are quite weak compared to those further out in the infrared (e.g. 8200 Å). The amount of water on Mars in the 1960's would have produced Martian components far below the detectability limit of the 1925 spectra. While it is possible that the amount of Martian atmospheric water vapor has decreased an order of magnitude in 40 yr, it is most unlikely.

The availability of speculum-metal diffraction gratings made possible a more ambitious test in 1933 (Dunham, 1952), which gave negative results for lines "just beyond the tail of the B band of oxygen." Finally, the 'modern' coudé spectrograph emerged in 1936, when Wood made available gratings ruled on aluminized Pyrex disks. Using one of these gratings at the coudé focus of the 100-inch reflector, Adams and Dunham obtained a number of spectra of Mars in the region of the 8200 Å H_2O band. The results were uniformly negative, and have been published in a number of places (Adams and Dunham, 1937; Adams, 1941; Dunham, 1952). This work was neatly summarized by Dunham in his 1952 paper. There he set a probable upper limit of about 10 μ of precipitable water vapor in the atmosphere of Mars. This upper limit was probably too low. Hess (1951, 1961) points out that the true limit from these old Mt. Wilson plates is probably much higher, somewhere between 10–100 μ. First of all, the measurement of line asymmetries was not made. Secondly, the curve of growth for H_2O was not employed (that for O_2 was used) and thirdly, the dependence of H_2O absorptions on pressure was not considered. Finally, Schorn et al. (1969) have shown that the old Mt. Wilson plates used by Adams and Dunham in the near-infrared are comparable to modern I–N emulsions in grain size and resolution. In the 8200 H_2O bands the strongest Martian H_2O lines so far observed have equivalent widths of about 5 mÅ. At the dispersion used by the Mt. Wilson workers (6 Å/mm @ 7200 Å) these lines would be invisible on I–N emulsions, a rough detection limit being 10 mÅ. Thus the old Mt. Wilson upper limits were doubtlessly too strict, 50 to 100 μ being more realistic. (In 1954, de Vaucouleurs suggested ≪350 μ, possibly 100 μ.) With this modest effort, high-dispersion spectroscopic work in this area ended for 20 yr.

In 1947–48 Kuiper (1947, 1952) was the first to record the spectrum of Mars between 1.0 and 2.5 μ, with a low-resolution (numerical resolution of 80) infrared spectrometer. Two CO_2 absorption bands near 1.6 μ, and three CO_2 bands near 2 μ, were considerably enhanced in the Martian spectrum when compared to Lunar spectra taken under similar circumstances. This indicated that the Martian CO_2 abundance was comparable to that of Earth, firmly establishing CO_2 as the first proven component of the Martian atmosphere. These are 'strong' CO_2 bands and their intensities (line profiles were not resolved in this pioneering work) depend on the abundance of CO_2 *and* the effective pressure. At the time, a CO_2 abundance about twice that of the Earth's atmosphere (twice 220 cm-atm or about 440 cm-atm), and a Martian surface pressure of a little less than 1/10 atm were reasonably consistent with Kuiper's data and our knowledge of Mars. This estimated CO_2 abundance led to a partial pressure of CO_2 at the surface of Mars of about $\frac{1}{4}$ mm Hg. This in turn led to a CO_2 equilibrium temperature of 129 K, much lower than any estimates of the cap temperature. At this time then, it seemed that dry-ice polar caps were ruled out, and water-ice caps strongly indicated. Of course, we now know that the CO_2 abundance is much higher than Kuiper's early estimate, and the surface pressure on Mars much lower than de Vaucouleurs 85 ± 4 mb estimate. The new determinations are consistent with Kuiper's original data and our present knowledge of Mars, and actually suggest CO_2 caps now (Leighton and Murray, 1968) – but this is a subject for a different review paper. Nevertheless, we have here a good example of how shaky 'solid' evidence can be – especially when it comes to interpretation.

As the Martian H_2O abundance was known to be much less than that of Earth, low-resolution spectroscopy from the Earth's surface was not expected to show any evidence of Martian H_2O absorption. However, Kuiper also obtained reflection spectra (at a low resolution of 9) of the polar caps themselves – a most difficult set of observations. The Martian cap spectra resembled the transmission spectrum of a water cell of 1 mm thickness. Kuiper noted that (thin) H_2O frost deposits on dry ice (-78 °C, 1 atm) also showed a water cell equivalent of 1 mm.* He concluded that the polar caps were frozen H_2O at temperatures much below 0 °C. Later, Moroz (1964) confirmed these observations. But, as Leighton and Murray (1968) pointed out, we do not know enough about the reflection spectrum of *both* solid H_2O and CO_2, under Martian conditions, to draw a definite conclusion.

At the same time Hess (1948, 1951, 1961) tackled the problem of whether the low spectroscopic H_2O abundance estimates would allow the production of the clouds seen on Mars. Considering low level clouds (morning limb haze, etc.), and assuming the water to be concentrated in a layer $\frac{1}{2}$ km deep near the ground (not at all impossible, in the light of 1969 observations) he derived an upper limit of 400 μ of precipitable H_2O. A treatment of possible high-level convective (afternoon) clouds led to an upper limit of 600 μ. He pointed out that the mean H_2O content would be considerably less, as such clouds are indeed rare. De Vaucouleurs, after discussing Hess' work, makes an order-of-magnitude estimate of the Martian atmospheric H_2O abundance based on the assumption that the polar caps are thin layers of water ice. His result was $<100 \mu$, quite consistent with the best spectroscopic results at the time. De Vaucouleurs quite candidly stresses the large uncertainties of his and other estimates, and cautions against taking agreement too literally.

In 1956 Kiess *et al.* (1957), in a most ingenious manner, obtained high-dispersion spectra of Mars *without* using a large telescope. They utilized a large grating spectrograph fed by a 12-inch uranostat. The lack of a large telescope enabled them to set up their equipment near the summit of Mauna Loa, Hawaii – the highest, driest site yet

* In the light of our present knowledge (1970) this statement is remarkably prophetic.

used for such work. Unfortunately, the resultant image size was quite small – 0.3 mm. This, together with the use of relatively grainy 103aF, (ammoniated) 103-U, and (ammoniated) I–N plates resulted in narrow spectra with a low signal to noise ratio. Thus, even though the terrestrial water lines were quite weak, the (as we now know) very weak Martian lines would still be invisible. The 'a' band of H_2O was observed at 2 Å/mm from Mauna Loa, but the stronger 8200 band was observed only at 5 Å/mm from Georgetown, Maryland, a much wetter location. By comparing their observations with laboratory spectra, Kiess et al. set an upper limit of 80 μ for the Martian H_2O abundance.

A number of less direct studies of water on Mars should now be mentioned. Observations by Lyot (1929) and Dollfus (1948, 1951, 1961) have shown that the polarimetric properties of a polar cap are exactly similar to that of finely granulated H_2O frost deposited in a cold (liquid air), low pressure (6 cm Hg) environment. Further, Dollfus found that the polarization of bright patches at the edge of a polar cap is closely matched by terrestrial ice clouds. However, Leighton and Murray later noted that, as Dollfus had *not* studied the polarization properties of CO_2 frost, his conclusions were far from definite.

Urey (1959), from the existence of (assumed H_2O) early-morning haze, suggested a lower limit to the H_2O atmospheric abundance of about 40 μ. From the sublimation rate of a (supposed water ice) polar cap, and the Martian greenhouse effect Sagan (1961) suggested values of 20–200 μ for the Martian water-vapor abundance. Lebedinskii and Salova (1962), considering the evaporation rate of the pole caps and the degree of atmospheric transparency on Mars, arrived at a mean H_2O abundance of 14 μ.

In 1962 Adamcik (1963) presented calculations "of the expected water vapor content of the Martian atmosphere on the assumption that it is determined by the equilibrium dissociation pressure of ferric oxide hydrate." Lower and upper limits of about 0.3–60 μ were obtained. These last few results are startlingly close to modern observational results, but the coincidence may be just that. When we know more about Mars we will be able to judge.

In this connection we must mention the interesting suggestion of Kiess et al. (1960, 1962, 1963). Based on their inability to detect any Martian H_2O, and their belief that absorption lines due to the oxides of nitrogen *were* visible on their plates (Kiess et al., 1962) they proposed that the latter were responsible for a wide variety of phenomena on Mars – including many of those used to 'prove' the presence of water (polar caps, clouds, etc.). Unfortunately, further spectroscopic tests (e.g. Sinton, 1961; Spinrad, 1963; Marshall, 1964) failed to detect any trace of the oxides of nitrogen. In particular Spinrad, working in the photographic region of the spectrum, as did Kiess et al., but using wider spectra and finer-grain plates, set an upper limit of 1 mm atm. Finally, Sagan et al. derived an upper limit of 1 mm atm for the NO_2 abundance by computing the photochemical equilibria of nitrogen oxides on Mars under the action of solar UV and visible light. They point out that the presence of water (which is now proven) would lower this limit further. It seems probable that Kiess and his collaborators were misled by their narrow, grainy spectra and did not, in fact, see any absorption features due to the oxides of nitrogen.

In 1962 Spinrad and Richardson (1963) analyzed an excellent coudé spectrum of Mars obtained with the Victoria 48-inch reflector. The terrestrial H_2O abundance was very much lower than usual at this usually sodden site. Unfortunately, the relatively weak 7200 Å H_2O band (weak compared to the 8200 Å H_2O band) and a low resolution (low compared to IV–N) I–N emulsion were used, and the Doppler shift was only -0.29 Å. As a result, all the authors could state was that there were no Martian H_2O components with strengths >9 mÅ. We now know that lines >9 mÅ in this band would not be expected.

At this point, in 1963, the full weight of modern, high-dispersion spectroscopy was, for the first time, turned on this problem. Between World War I and 1963 planets had been the stepchildren of astronomy – they definitely were not in the 'mainstream'. There were many reasons. First of all, extra-solar-system astronomy 'exploded': the structure of the Milky Way Galaxy was roughed out, the existence of external galaxies was demonstrated, the red-shift of galaxies was discovered, and stellar evolution studies began to show substantial results. After millennia of dealing almost entirely with the solar system, astronomers' attention was directed elsewhere. Secondly, planetary astronomy had hit an apparent dead end. New results were hard to come by (this review illustrates a particular case) and telescopes were few – too few to 'waste' much time on such a small portion of the universe. Finally, there was an element of irrational prejudice against planetary astronomy – no doubt generated by Lowell's Martians, Flammarion's poisonous comet tales, and Orson Wells' radio coup. Astronomers used to have a shy and retiring image, and 'life on Mars' may have been too much for them.

Regardless of the reasons, very little 'big-telescope' time was devoted to the planets during the (say) 1915–1965 period. Adams and Dunham, for example, note that most of their planetary spectra were taken when the seeing was *too poor* for stellar spectra! The catalyst which changed this situation was provided by Spinrad. His work on the old Mt. Wilson and Victoria plates convinced him that very fine-grained IV–N emulsions, used with spectrographs of the quality of the Mt. Wilson 100-inch coudé, would provide substantial new results. In addition, he had found that the IV–N emulsion, usually considered much too slow for useful astronomical work, could be ammonia-sensitized to a speed equal to that of the I–N emulsion.

While Spinrad was trying to get some coudé time Dollfus (1963) used conventional astronomical photometry of the strong 1.4 μ H_2O band at a high and dry mountain site. The location compensated for low spectral resolution of the photoelectric spectrophotometer and birefringent filter system. This was essentially a return to turn of the century techniques, with the important differences of a much dryer site and the use of a much stronger H_2O band. Dollfus' observations were made in January of 1963 when L_s* was about $40°$. The initial result (Dollfus, 1963) gave an average, over the visible disk of Mars, of 200 μ of precipitable water vapor. This result seemed much too large at the time, as it was *above* the Adams and Dunham abundance limit and

---

* L_s is the planetocentric longitude of the Sun. $L_s = 0$ at the vernal equinox, or beginning of spring in the northern hemisphere of Mars; $L_s = 90°$ at the beginning of summer in the northern hemisphere of Mars, etc.

also above that of Spinrad and Richardson set just the month before. It is important, in this regard, to note that detection and abundance estimates are two different things. Later, Dollfus (1964, 1965), using better laboratory data and newer, more accurate values of the Martian surface pressure, obtained values of 150 μ and, ultimately, 45 μ. This last result, as we shall see, still seems too high by about a factor of two, especially for the value of L_s involved, and this might illustrate an inherent limit to the accuracy of this method. And yet, in the light of our current knowledge of Mars, and the absence of any conflicting data from January, 1963, the result may be correct. It is however unfortunate that the first detection announcement was widely disbelieved because the abundance estimate was so high.

On April 12/13, 1963 ($L_s = 76°$) Spinrad et al. (1963), Kaplan et al. (1964) obtained a good IV–N plate of the 8200 Å H_2O band at 5.6 Å/mm using the 100-inch coudé spectrograph. The Doppler shift was $+0.414$ Å (at 8200 Å) at that time, and a number of 3–5 mÅ Martian H_2O lines were seen. A first estimate of 5–10 μ of H_2O on Mars was made by guessing at the line strengths of the H_2O lines involved and also by comparing the Martian and terrestrial lines, using the known amount of (terrestrial) water above Mt. Wilson that night, and assuming a curve of growth. Shortly thereafter Rank et al. (1964) measured the intrinsic strengths of a number of the lines in the 8200 Å band. Using these line strengths, the Mt. Wilson plate gave $14 \pm 7 \mu$ as a final result.

On this same plate the weak 8700 Å band of CO_2 was detected, showing at once that the Martian CO_2 abundance was much higher than previously thought. More importantly, the surface pressure was drastically reduced. In the Space Age this result had great practical importance, and the furor it generated led to a great deal of importance being laid on the *single* KMS plate. The KMS H_2O result shared in this attention, and can be considered the *effective* discovery of water on Mars. We shall see, however, that not everyone was convinced. Heyden et al. (1966) called particular attention to the possibility of weak telluric or solar lines confusing the issue. Clearly, more observations were needed.

During the 1964–65 apparition of Mars Schorn et al. (1967) carried out a program on the 8200 Å H_2O band using the 120-inch Lick and (newly improved) 82-inch McDonald Observatory coudé spectrographs. Using IV–N emulsions and dispersions of about 4 Å/mm, nineteen excellent spectra were obtained. The presence of Martian H_2O was confirmed, and variations with the season on Mars, and over the disk of the planet, were noted. Using Rank's values for line intensities the results were: (a) between $L_s = 5°$ and $L_s = 31°$ (i.e. early northern spring), $<15 \mu$ everywhere; (b) between $L_s = 47°$ and $L_s = 61°$, (i.e. late northern spring), $\sim 15 \mu$ in the northern hemisphere only (i.e. over the receding polar cap); between $L_s = 110°$ and $L_s = 122°$ (i.e. northern summer, southern winter), $\sim 10 \mu$ everywhere, with one plate at $L_s = 122°$ giving $\sim 25 \mu$ in the southern hemisphere. These results seemed to show that *some* of the pole cap, at least, was composed of frozen H_2O while the very presence of H_2O vapor implied an atmospheric temperature greater than 200 K in the absorbing layer.

We felt that the spectroscopic study of Martian H_2O was not finished by these results, but just begun. Higher spectral and areographic resolution, we felt, would give most interesting results (one superb spectrum obtained on December 24, 1964, showed H_2O *only* above the North polar cap and Mare Acidalium!). Increasing the range of L_s observed, and comparing the Martian weather from year to year were clearly desirable. Finally, I must admit that the evidence was far from overwhelming, and many remained unconvinced. For example, Goody (1969) referred to the 'debatable claim' of 15 μ of Martian H_2O. Further, variations over the planet might just be explained away as due to the varying albedo of Mars, the characteristic curve of the IV–N emulsion, etc., while supposed variations of the H_2O abundance estimates with L_s might be due to the varying terrestrial humidity, wishful thinking on the part of the estimators, etc. There was certainly room for improvement in the observations.

During the 1966–67 apparition Schorn and Moore and Owen used the McDonald 82-inch in an attempt to extend the 1964–65 work. Unfortunately, snow, rain, clouds, and high terrestrial humidity ruined the program. For the 1969 apparition, the outlook was considerably better. A larger collimator made a dispersion of 2 Å/mm attainable with the 82-inch coudé spectrograph.

At about this time Leighton and Murray (1966) and Leovy (1966) revived the CO_2-pole-cap hypothesis with some force, basing their work on new data for the CO_2 abundance and surface pressure on Mars. While this work threw no direct light on the validity of the recent H_2O detections, it certainly made them seem more doubtful in the eyes of many.

In 1967 Kuiper began using a series of jet-plane flights at high altitudes (around 40 000 ft), to perform astronomical observations in the infrared. This logical continuation of the high-altitude observatory idea is still continuing. Using interferometric techniques, he is observing a number of astronomical objects, among them Mars. While the observations have not yet been published, they show definite evidence for the presence of Martian water vapor in the 1.3 μ band. While Kuiper's observations could not give the areographic resolution attainable with large telescopes on the ground, they provided strong support for the feeling that continued study of Martian H_2O would indeed be useful.

Starting in late 1968 an intensive observational attack on the Martian 8200 Å H_2O band was carried out at McDonald Observatory. The 82-inch coudé spectrograph was used intensively and, when it became operational, the 107-inch coudé. Some of these results have been discussed in this symposium, other have been published, and the rest will shortly be in print.

In late November and early December of 1968 ($L_s \sim 80°$) I obtained the first usable spectra of this apparition at the 82-inch. Although the apparent size of Mars was small ($\sim$9 arc sec), the seeing poor, and the signal-to-noise ratio of the spectra low, Martian components were clearly seen. These early spectra were taken at 2 Å/mm *and* at 4 Å/mm (these latter duplicating the 1964–65 work), both dispersions showing the strongest Martian lines to be about 5 mÅ, stronger than in 1964–65.

During February, March, and April of 1969 ($L_s = 110 - 150°$) Little and Schorn, Owen, and Tull obtained a large number of spectra on the 82- and 107-inch telescopes. All were aided by superb, dry weather, a good Doppler shift (almost $+0.5$ Å at best), and excellent condition of the spectrographs. Owen and Mason (1969) obtained several spectra near $L_s = 120°$. The seeing was poor, so their results referred to a wide, 'pole-to-pole' band on Mars. Using Rank's data for line strengths, Owen obtained an abundance of $35 \pm 15\ \mu$. He attributed most of the apparent increase in Martian H_2O abundance over the 1964–65 results to an increase in the line equivalent widths caused by the greater spectral resolution of the later results. This is of course a very important point. However, the December, 1968 spectra (taken at 2 Å/mm *and* 4 Å/mm) showed no such systematic effect. As a check, several spectra of the weak lines in the 7820 Å CO_2 band of Venus were taken (as close together in time as possible!) later in 1969, at both dispersions. When reduced in the same way as the Martian 8200 Å spectra, there was no systematic difference in the derived equivalent widths between the two dispersions (Schorn and Gray, 1970). I feel that there probably *is* more water in the Martian atmosphere in 1969 than there was in 1964–65.

Other spectra made in this interval enjoyed better seeing, and the non-uniformity in the Martian humidity was striking. Schorn *et al.* (1970), in a preliminary step, analyzed spectra taken at $L_s = 111°$ and $122°$. As there were many more Martian H_2O lines visible in these plates than had been the case with earlier observations, it was imperative that additional lines (other than the four Rank's group had done) be observed in the laboratory. Farmer (1970) did this. Further, he corrected the laboratory strengths for temperature by using the lower state energies from Benedict's (1969) assignments for the atlas of Delbouille and Roland (1963). A surprisingly large correction was discovered. For example, the 8197.704 H_2O line increases in strength by a factor of 1.7 between room temperature and 200 K. The result is to *lower* all H_2O abundances based on Rank's laboratory data. The KMS result is reduced to about $10\ \mu$, and the 1964–65 values are also reduced to about two-thirds of their published values. Owen's result is pushed down toward $30\ \mu$. The two 1969 plates at $L_s = 111$ and $122°$ gave $26 \pm 5\ \mu$ in the northern (cap-receding) hemisphere, and less than one-third of that amount in the south. The visibility of so many lines enabled an estimate of 225 K to be made for the effective temperature. This is significantly higher than the 200 K temperature estimated by Young (1969) for the bulk of the Martian CO_2, and suggests that the water vapor exists in a layer close to the ground.

The results described by Little in this Symposium fill in the picture during the February–April period. The preponderance of water in the north was quite marked, and there was little apparent change in the situation over this interval. Also in this Symposium, and elsewhere (Tull, 1970), had discussed his 107-inch plates ($L_s = 132, 148°$). The greater areographic resolution, as some suspected, led to greater differences in observed H_2O abundances over the planet. It is important to note that all the observers involved feel that still greater abundance variations will be observed with better areographic resolution. All of these 1968-early 1969 studies, however, give a consistent picture of the situation on Mars at that time.

From May through July a hiatus in the H_2O observations was caused by the low Doppler shift of Mars. Starting with the Mariner VI and VII flybys, spectra were again obtained with the 82 and 107-inch coudé spectrographs by Barker, Woszczyk, Tull, and Schorn. In August and September ($L_s = 200 - 235°$) no Martian water was observed. The terrestrial humidity was high, and the Doppler shift low (about $\frac{1}{4}$ Å), so the abundance lower limit was relatively high. Nonetheless, it appeared that the mean Martian water vapor abundance was down by at least a factor of two, during this interval, from its preopposition value. In late December of 1969 ($L_s \sim 290°$) when the south polar cap was receding, 5 mÅ Martian lines once again appeared, indicating that the water vapor abundance was back to its early 1969 value. Although the Doppler shift had risen to about $\frac{1}{3}$ Å by mid-December, this slight increase does not appear to be the sole reason for the reappearance of the Martian H_2O lines. Probably the Martian water vapor did go away and then return. If so, it appeared later in the season than it did when the northern cap was receding in late 1968-early 1969. While this hemisphere-to-hemisphere difference in the seasonal behavior of H_2O on Mars needs much more observation to be firmly established, it would not be unexpected. The orbit of Mars is such that the properties of northern and southern hemispheric seasons are notably different – a difference which could be expected to manifest itself in the behavior of the Martian water-vapor abundance.

Observations with the 82-inch and 107-inch into March, 1970 (photographic and photoelectric scanner), as L_s passed 340°, show that the situation on Mars is similar to December 1969 (Barker et al., 1970). During this time the angular diameter of Mars was so small ($\sim$5 arc sec) that no effective areographic resolution is possible. The mean H_2O abundance is about 45 μ during this interval, a high value indeed!

In Figure 2, I have attempted to summarize the spectroscopic observations from 1963 through early 1970. No attempt has been made to indicate where on the planet various specific measures refer to. Rather the figure illustrates that a cyclical behavior of the Martian water vapor appears to be emerging from the observations.* The water vapor seems to appear when a pole cap has receded appreciably from its maximum extent, and disappears when a cap is at that maximum extent. If this correlation holds up under further observation** we will know that Martian water is cycled through the pole caps, in spite of the Mariner VI and VII evidence (Neugebauer et al., 1969) that those caps are predominantly CO_2. The Martian polar caps may indeed be vast deposits of frozen seltzer water!

The situation at present can be summed up as follows. Water exists on Mars. The water vapor varies with location on the planet, the season on Mars, and from year to year. The water appears to cycle through the polar caps, which are partly H_2O. The total amount of water in the atmosphere of Mars is at most a few cubic kilometers. Of course this does not include water which may be trapped as permafrost (as suggested by, among others, Leighton and Murray) or deposited in the permanent remnant of the north polar cap.

Does liquid H_2O exist on Mars? This is properly a topic for another review paper. *But*, though improbable (Owen, 1969), it is not impossible (Sagan et al., 1968; Schorn et al., 1970).

Sagan, et al. (1968) discussed the situation on the basis of modern water-vapor abundance estimates. In particular, they pointed out that, while open water surfaces

---

* The point due to Dollfus is troublesome. Perhaps, as was suggested earlier in this paper, it should be lowered further. At any rate, further observations should decide the matter.
** Further observations *are* needed.

234 RONALD A. SCHORN

(lakes, ponds, etc.) were improbable, the case for liquid water occurring in the interstices of loose soil (dirt, gravel, sand, etc.) is decidedly more favorable. Schorn *et al.* (1970) have suggested that, in low lying areas (surface pressure > 6.1 mb) when the terrain and temperature regime are favourable, there is a real possibility that liquid water exists – albeit only for short periods of time. This water might then be trapped for a substantial period of time – months perhaps – in the manner suggested by Sagan *et al.* In any case it should be perfectly clear that the possibility of liquid water on Mars cannot be dismissed out of hand on the basis of present evidence.

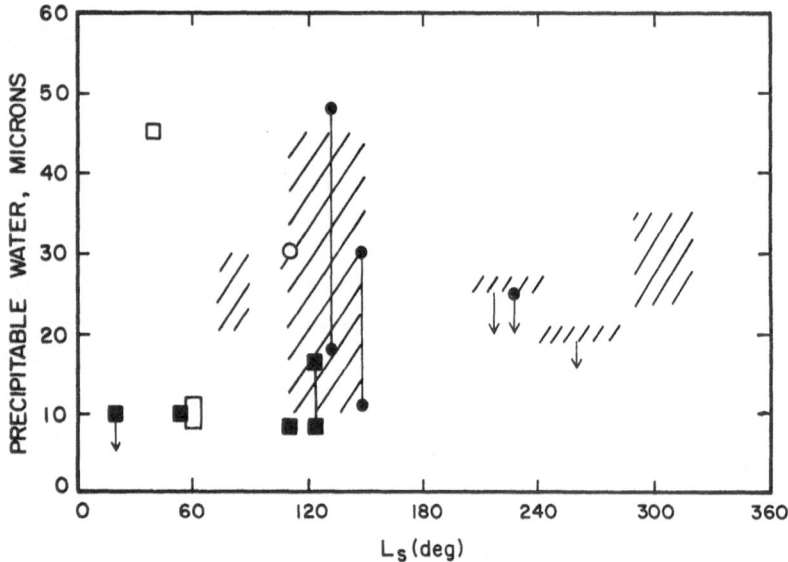

Fig. 2. Abundance determinations of Martian water vapor as a function of L_s. The open square represents Dollfus' 1963 observations; the open rectangle the 1963 results of Kaplan *et al.*; the filled squares the 1964–65 estimates of Schorn *et al.*; the open circle the 1969 observation of Owen and Mason; the filled circles the results of Tull in 1969. The individual results of a large number of McDonald spectra by Schorn, Farmer, Little, Barker, and Woszczyk are not plotted, but the approximate regions in which they fall are covered with diagonal lines. Vertical arrows point downward from observed upper limits and, in three cases, vertical lines connect the range of values found from a single plate. No attempt has been made to indicate probable or possible errors, which are substantial in all cases. All estimates have, however, been adjusted to utilize Farmer's improved line strengths for 225 K. This illustration shows the *general trend* of Martian H_2O abundances with L_s.

Finally, the obvious questions arise – where did this water come from, and why this small, but definite amount? It does seem improbable that we are living precisely at that time when the last few cubic miles of water are escaping from Mars. Perhaps the H_2O escape rate from Mars is much lower than we think. Perhaps water is added to the Martian atmosphere from the interior at just the right rate to make up for the escape rate. Perhaps a comet impact or igneous event has recently supplied the water. Perhaps. In my mind it does appear most likely to assume that in general, the origin of the water on Mars is similar to that of the origin of the water on Earth – it came from the body of the planet itself, whatever the time scales of production and escape may be. Such suggestions at present are, of course, just that. Mars, it appears, is still an engaging object for study and speculation.

References

Adamcik, J. A.: 1963, *Planetary Space Sci.* **11**, 355.
Adams, W. S.: 1941, *Astrophys.* **93**, 16.
Adams, W. S. and Dunham, T.: 1937, *Publ. Astron. Soc. Pacific* **49**, 209.
Adams, W. S. and St. John, C.: 1926, *Astrophys. J.* **63**, 133.
Barker, E. S., Schorn, R. A., Woszczyk, A., Tull, R. G., and Little, S. J.: 1970, *Science*, in press.
Benedict, W. S.: 1969, private communication to C. B. Farmer.
Campbell, W. W.: 1909, *Lick Obs. Bull.* **5**, No. 150.
Campbell, W. W. and Albrecht, S.: 1910, *Lick Obs. Bull.* **6**, No. 180, 11.
Delbouille, L. and Roland, G.: 1963, *Mem. Soc. Roy. Sci. Liege*, special Vol. 4, Liege.
de Vaucouleurs, Gerard: 1954, *Physics of the Planet Mars*, Faber and Faber, Ltd., London.
Dollfus, A.: 1948, *Compt. Rend. Sci.* **227**, 331.
Dollfus, A.: 1951, *Compt. Rend. Acad. Sci.* **233**, 467.
Dollfus, A.: 1961, in *The Solar System, 3: Planets and Satellites* (ed. by G. P. Kuiper and B. Middlehurst), University of Chicago Press, Chicago, p. 343.
Dollfus, A.: 1963, *Compt. Rend. Acad. Sci.* **256**, 3009.
Dollfus, A.: 1964, *Mem. Soc. Roy. Sci. Liege*, V^e ser., IX, 392.
Dollfus, A.: 1965, *Compt. Rend. Acad. Sci.* **261**, 1603.
Dunham, T.: 1952, in *The Atmospheres of the Earth and Planets*, 2nd ed., (ed. by G. P. Kuiper), University of Chicago Press, Chicago, p. 288.
Farmer, C. B.: *Icarus*, to be published.
Goody, R.: 1969, in *Ann. Rev. Astron. Astrophys.* **7** (ed. by L. Goldberg, D. Layzer, and J. G. Phillips), Annual Reviews, Inc., Palo Alto, Calif., 330.
Heyden, F. J., Kiess, C. C., and Willauer, W. R.: 1966, *Astrophys. J.* **143**, 595.
Hess, S. L.: 1948, *Publ. Astron. Soc. Pacific* **60**, 289.
Hess, S. L.: 1951, *Rep. No. 9*, Planetary Atmospheric Project, Lowell Obs., 172.
Hess, S. L.: 1961, in *Advances on Space Science and Technology*, Vol. 3, (ed. by F. I. Ordway), Academic Press, New York, p. 151.
Kaplan, L. D., Münch, G., and Spinrad, H.: 1964, *Astrophys. J.* **139**, 1.
Kiess, C. C., Corliss, C. H., Kiess, H. K., and Corliss, E. L. R.: 1957, *Astrophys. J.* **116**, 579.
Kiess, C. C., Corliss, C. H., and Kiess, H. K.: 1962, *Astron. J.* **67**, 579.
Kiess, C. C., Karrer, S., and Kiess, H. K.: 1960, *Publ. Astron. Soc. Pacific* **72**, 256.
Kiess, C. C., Karrer, S., and Kiess, H. K.: 1963, *Publ. Astron. Soc. Pacific* **75**, 50.
Kuiper, G. P.: 1952, in *The Atmospheres of the Earth and Planets* (ed. by G. P. Kuiper), University of Chicago Press, Chicago, p. 306.
Lebedenskii, A. I. and Salova, G. I.: 1962, *Soviet Astron.-AJ* **6**, 390.
Leighton, R. B. and Murray, B. C.: 1966, *Science* **153**, 136.
Leovy, C.: 1966, *Science* **154**, 1178.
Lowell, P. and Slipher, V. M.: 1905, *Lowell Obs. Bull.* No. **17**, 116.
Lyot, B.: 1929, *Ann. Obs. Meudon*, VIII, **1**, 51, 147.
Marshall, J. V.: 1964, *Comm. Lunar Planetary Lab.* **2**, 167.
Moroz, V. I.: 1964, *Soviet Astron.-AJ* **8**, 273.
Neugebauer, G., Münch, G., Chase, S. C., Hatzeubeler, H., Miner, E., and Schonfield, D.: 1969, *Science* **166**, 98.
Owen, T. and Mason, H. P.: 1969, *Science* **165**, 893.
Proctor, R. A.: 1873?, *Other Worlds Than Ours*, Hurst & Company, New York.
Rank, D. H., Fink, U., Folz, J. V., and Wiggins, T. A.: 1964, *Astrophys. J.* **140**, 366.
Russell, H. N., Dugan, R. S., and Stewart, J. Q.: 1945, *Astronomy: I. The Solar System*, rev. ed., Ginn and Company, Boston, p. 332.
Sagan, C.: 1961, *Astron. J.* **66**, 52.
Sagan, C., Hanst, P. L., and Young, A. T.: 1965, *Planetary Space Sci.* **13**, 73.

Sagan, C., Levinthal, E. C., and Lederberg, J.: 1968, *Science* **159**, 1191.

Schorn, R. A., Farmer, C. B., and Little, S. J.: 1970, *Icarus*, in press.

Schorn, R. A. J. and Young, L. D. G.: 1970, *Icarus*, to be published.

Schorn, R. A., Gray, L. D., and Barker, E. S.: 1969, *Icarus* **10**, 241.

Schorn, R. A., Spinrad, H., Moore, R. C., Smith, H. J., and Giver, L. P.: 1947, *Astrophys. J.* **147**, 743.

Sinton, W. M.: 1961, *Publ. Astron. Soc. Pacific* **73**, 125.

Spinrad, H.: 1963, *Publ. Astron. Soc. Pacific* **75**, 190.

Spinrad, H., Münch, G., and Kaplan, L.: 1963, *Astrophys. J.* **137**, 1319.

Spinrad, H. and Richardson, E. H.: 1963, *Icarus* **2**, 49.

Stoney, G. J.: 1898, *Astrophys. J.* **7**, 25.

Tull, R. G.: 1970, *Icarus*, in press.

Urey, H. C.: 1959, in *Handbuch der Physik* (ed. by S. Flügge), Springer-Verlag, Berlin, p. 395.

Very, F. W.: 1909, *Lowell Obs. Bull.*, Nos. 36 and 41.

Young, L. D. G.: 1969, *Icarus* **11**, 386.

THE LATITUDE VARIATION OF WATER VAPOR ON MARS

ROBERT G. TULL

The University of Texas at Austin, Austin, Tex. U.S.A.

Abstract. Spectra of Mars, centered on the 8200 Å band of H_2O, have been obtained using the coudé spectrograph of the 107-inch telescope at McDonald Observatory with reciprocal dispersion of 1.9 Å/mm. The plate scale (4.4 arc sec/mm) and angular resolution (3″–6″) were sufficient to measure the strength of the Doppler-shifted H_2O lines at 5 points across the disk.

The spectra were obtained in March and April 1969, when the apparent diameter of Mars was 10″ to 16″ during mid-summer of the northern hemisphere; the Doppler shift at 8200 Å varied from −0.42 Å to −0.28 Å. On the first plate, obtained on March 27, 1969 the abundance reached a maximum of about 48 μ precipitable H_2O at 30° to 40° north latitude and decreased to about 20 μ at 30° south latitude. The second plate, taken on April 28, 1969 showed the same north-south decrease in abundance but the total amounts were about two-thirds of the March abundances.

Observations

Three spectra of Mars have been obtained at 1.9 Å/mm dispersion and with a plate scale of 4.4 arc sec/mm, centered on the 8200 Å H_2O band, using the coudé spectrograph of the 107-inch (2.7 m) telescope at McDonald Observatory (Tull, 1969).

The observational parameters are summarized in Table I. Twenty-nine weak absorption lines have been detected on these plates, near the expected Doppler-shifted positions of unblended lines due to water vapor in the atmosphere of Mars. These are on the violet wings of strong telluric H_2O lines. Of these, 10 lines on plate 16, and 6 lines on plate 45, were of high enough quality for the measurement of equivalent widths, and hence of total atmospheric water abundance, as a function of latitude on Mars.

TABLE I

The observations

Plate	1969 date	L_s	Phase angle	Velocity	Slit	Emulsion
16	3/27	132°	35°	−15.6 km/sec	no rotator	Ammoniated IV–N
45	4/28	148°	24°	−11.2	N–S	Ammoniated IV–N
47	5/2	150°	22°	−10.4	E–W	Ammoniated IV–N

Reductions

The equivalent widths were measured with the aid of a Joyce-Loebl microdensitometer at 4 and 5 points across the spectrum and were calibrated against weak Fraunhofer lines of known equivalent widths (Moore *et al.*, 1966). These were corrected to total H_2O abundance along the double path through the atmosphere, using the tables of Janssen and Korb (1968) and the H_2O line strengths determined in the laboratory by Farmer (1971), and were then corrected to unit air mass, taking into account the seeing and guiding smear across the spectrum.

Sagan et al. (eds.), Planetary Atmospheres, 237–240.

Results

Figures 1 and 2 show the resultant water vapor distributions as a function of effective distance from the center of the disk, along the spectrograph slit. In Figure 2, this distance is expressed as latitude on Mars. For Figure 1, the image rotated on the slit over the area indicated in the diagram; the observed abundances, therefore, represent the integrations of the actual abundances over the rotation of the slit. The filled circles are reductions from microdensitometer tracings using a 60 μ slit width (projected on the plate), while the open circles result from a second reduction of the same plate, using a 30 μ slit width. Vertical error bars are probable error of the means; horizontal bars are effective microdensitometer slit length. The solid curve represents the resultant of an integration over the rotation of the slit, of an assumed H_2O distribution given in Table II.

The work presented here will be described in full in a forthcoming publication (Tull, 1970).

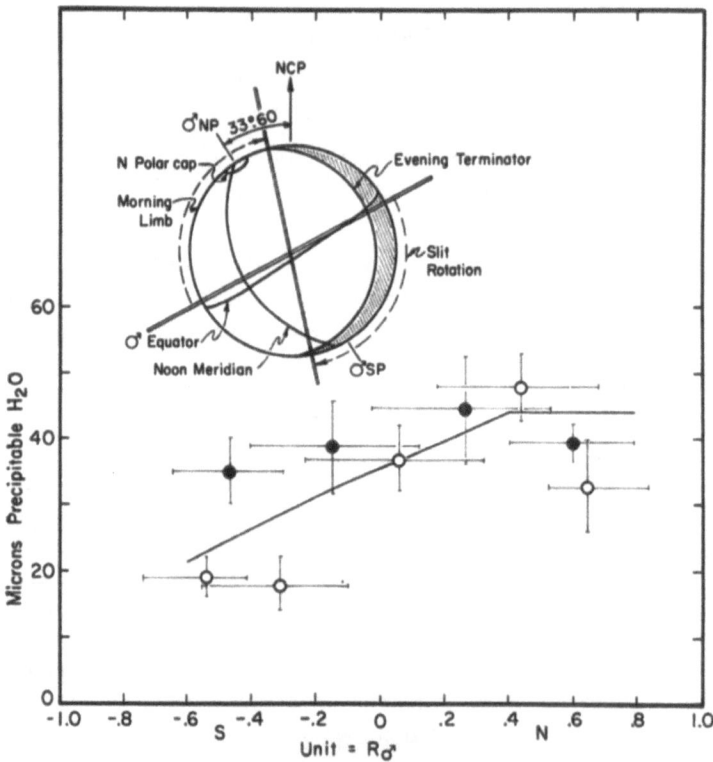

Fig. 1.   Geometrical configuration, and measured distribution of water vapor in the Martian atmosphere, from Plate 16 (27 March, 1969). Filled circles result from a microdensitometer tracing with a projected slit width of 60 μ; open circles were obtained using a 30 μ projected microdensitometer slit. The image rotated on the spectrograph slit through 105° during the 7-hour exposure, as indicated in the diagram. The solid curve represents an integration, over the rotation of the slit, of the assumed H_2O vapor distribution given in Table II. Vertical bars are probable error of the means; horizontal bars represent the effective location and length of the microdensitometer slit.

TABLE II

Assumed distribution of H_2O on March
27, 1969

Sine (Latitude)	Abundance
+0.8	45 μ precipitable H_2O
+0.6	46
+0.4	48
0.2	42
0	35
−0.2	28
−0.4	21
−0.6	13

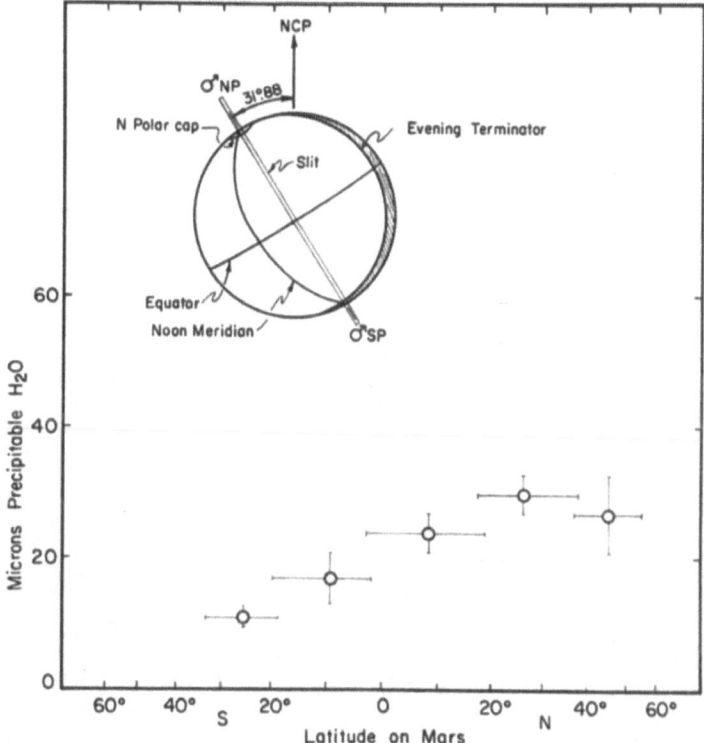

Fig. 2. Geometrical configuration and measured H_2O distribution from Plate 45 (28 April, 1969), as in Figure 1 except that the image was held fixed on the slit by an image rotator. Exposure 6 hours.

Acknowledgments

I wish to thank R. A. Schorn for helpful discussions and for suggesting this project, and C. B. Farmer for making available data on H_2O line strengths ahead of publication. The reductions were carried out during my summer appointment as a senior scientist at the Jet Propulsion Laboratory of California Institute of Technology.

17—P.A.

References

Farmer, C. B.: 1971, *Icarus*, to be published.
Jansson, P. A. and Korb, C. L.: 1968, *J. Quant. Spectry. Radiative Transfer* **8**, 1399.
Moore, C. E., Minnaert, M. G. J., and Houtgast, J.: 1966, NBS Monograph No. 61.
Tull, R. G.: 1969, *Sky Telesc.* **38**, 156.
Tull, R. G.: 1970, *Icarus* **13,** 1.

A REPORT ON MARTIAN ATMOSPHERIC WATER VAPOR NEAR OPPOSITION, 1969

S. J. LITTLE

Dept. of Astronomy, University of Texas at Austin, Austin, Tex. U.S.A.

Abstract. Little variation in Martian atmospheric H_2O abundance was observed during three months prior to the 1969 opposition.

This report covers the dates February 10 through April 25, 1969. Twenty coudé spectroscopic plates of Mars in the λ 8200 water vapor band at a dispersion of 2 Å/mm were taken by Dr. Ronald Schorn and the author with the 82-inch telescope of the McDonald Observatory. The best 10 of these plates have been reduced to give an abundance of water vapor in the Martian atmosphere. Several were of sufficient density to allow reduction of partial widths of the spectrum; and, since a pole-to-pole orientation of the slit was used for most plates, this enables one to determine a water vapor abundance in both the northern and southern hemispheres. Plates number 6132 and 6150 were not taken with an image rotator, and the slit trailed from parallel to the equator to about 45° from the equator. These plates therefore give an abundance of water vapor for northern and southern equatorial regions.

The abundances were determined using a curve of growth for Voigt profiles with small *a* values (implying a small Doppler half-width). The line strengths, taken from a recent experimental determination by Farmer (1971) vary as a function of temperature; the temperature of 225 K used in this reduction was adopted from an earlier study of plate number 6132 by Schorn, Farmer, and Little (1969). A presssure of 6 mb was adopted because the water vapor should be found near the Martian surface. The effective airmass was estimated for each tracing. The estimates of airmass could cause errors of as much as $\pm 20\%$ in the derived abundance, but the errors in airmass should cause little change in the relative abundance difference between the northern and southern hemispheres.

Density tracings of all the plates and partial widths of spectra on the plates were made with the Grant microdensitometer at the Astronomy Department of the University of California at Los Angeles. Equivalent widths were reduced from the tracings by two means: (1) the characteristic curve of the emulsion was plotted and points on the line profile were reduced to intensity, and (2) the square-counts of the Martian water vapor lines were calibrated to equivalent widths by referring them to a plot of equivalent width versus square-count for solar lines on the same plate. The two processes gave similar results, so the first method was used to reduce most of the plates.

The Doppler shift during February and March of 1969 was greater than 0.40 Å, and the terrestrial water vapor contribution was small enough to allow a clear separation of the Martian component (Figure 1). During April the separation became more difficult to detect because of the declining Doppler shift and because the telluric

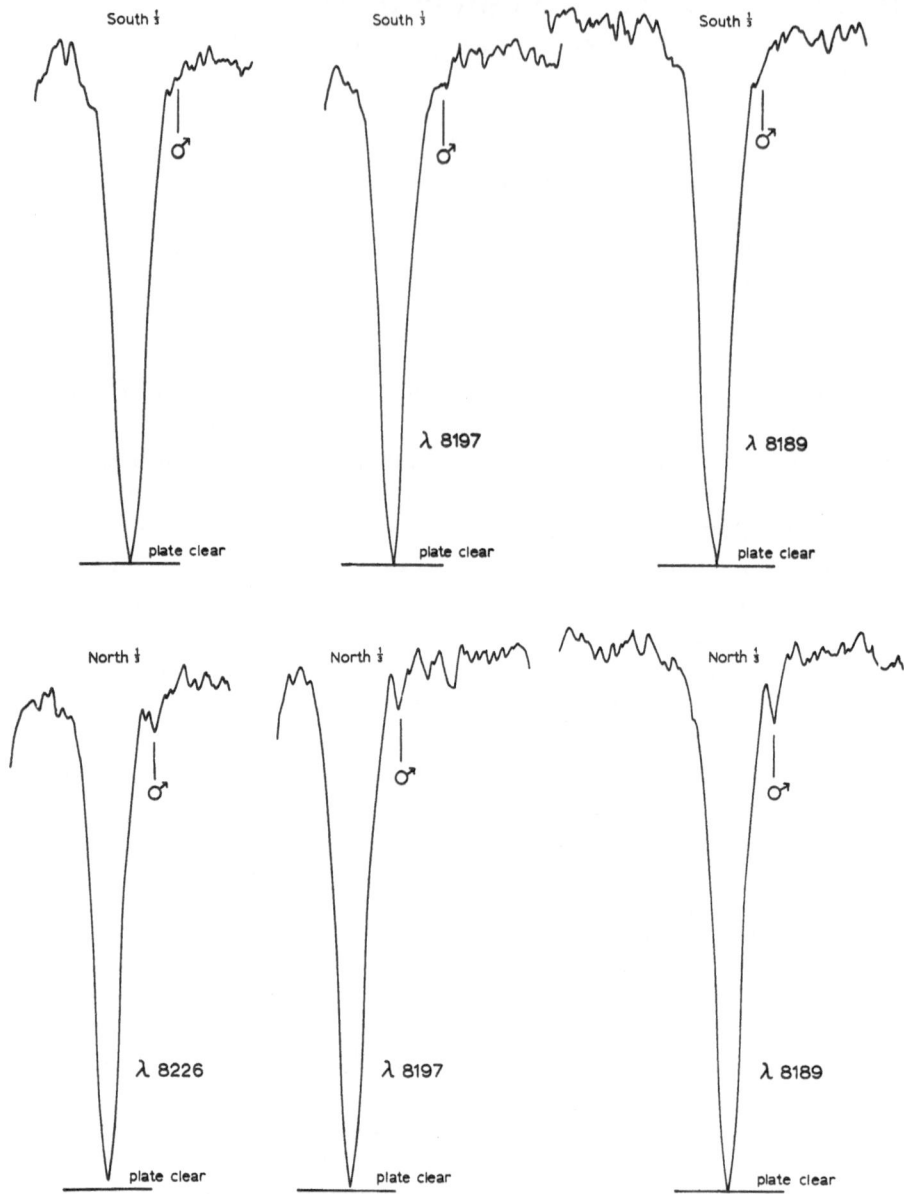

Fig. 1.   Density tracings from plate number 6147 of the northern (1b) and southern third of the planet Mars on March 27, 1969. The lines shown are λ 8189, λ 8197, and λ 8226 Å.

water vapor lines were broader. Plate number 6163 is the only good plate that was taken late in April. The error in equivalent width for an average individual line must be near ±50% on the better plates, and there is a possibility of spurious detection of lines on the poorer quality plates. The error of an abundance derived from 5 or 10 lines will be smaller than the error of any one line, so it is felt that the relative accuracy of the northern and southern hemisphere abundances is within ±20%. The uncertainty in the airmass estimates plus the equivalent width uncertainty means that the absolute abundance from any plate is probably uncertain by $\pm 50\%$.

The results are listed below in Table I. The derived abundances are subject to the

TABLE I

Water vapor abundance in microns

Plate no.	North polar	North equat.	Equat.	Whole plate	South equat.	South polar	Date
6094	43			32		24	2/10/69
6132		26			< 10		3/7/69
6146	40		33			30	3/27/69
6147	21		24			< 10	3/28/69
6149				46			3/29/69
6150		49			21		3/30/69
6152	39					< 10	4/1/69
6156				31			4/5/69
6159				33			4/7/69
6163	29		36			30	4/24/69

uncertainties described above. Figure 1 shows density tracings of the northern and southern one-third of Mars for plate number 6147 for lines $\lambda\,8189$, $\lambda\,8197$, and $\lambda\,8226$. The presence of a Martian component in the northern tracing is easily seen, and the absence of any strong component in the southern tracing contrasts sharply with the northern tracing. Figures 2, 3, and 4 show the results from Table I plotted for the northern, equatorial (and whole plate), and southern abundances respectively.

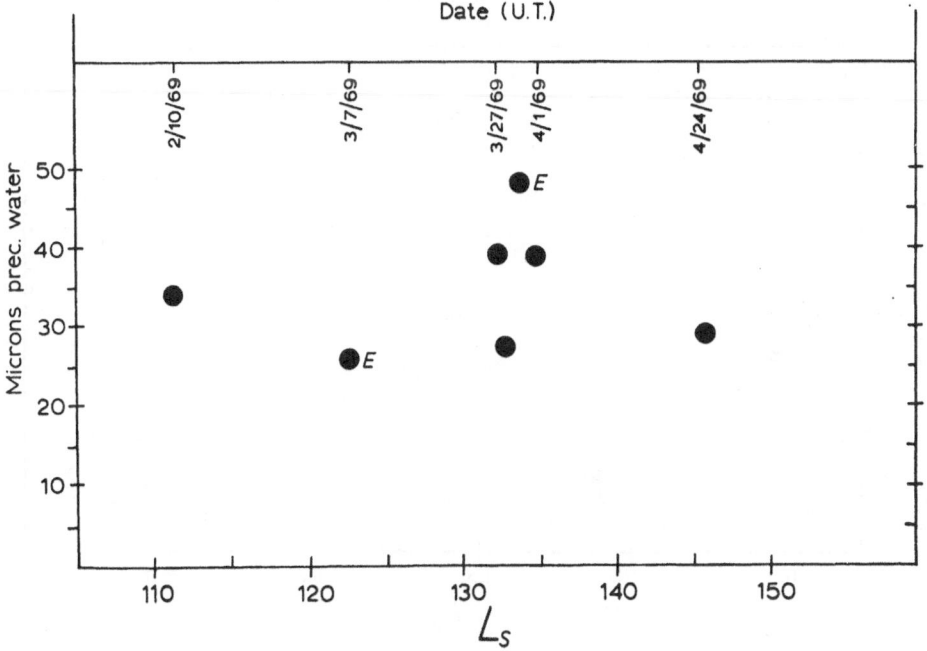

Fig. 2.   Abundance of water in north polar region of Mars. The north equatorial results are indicated by an E.

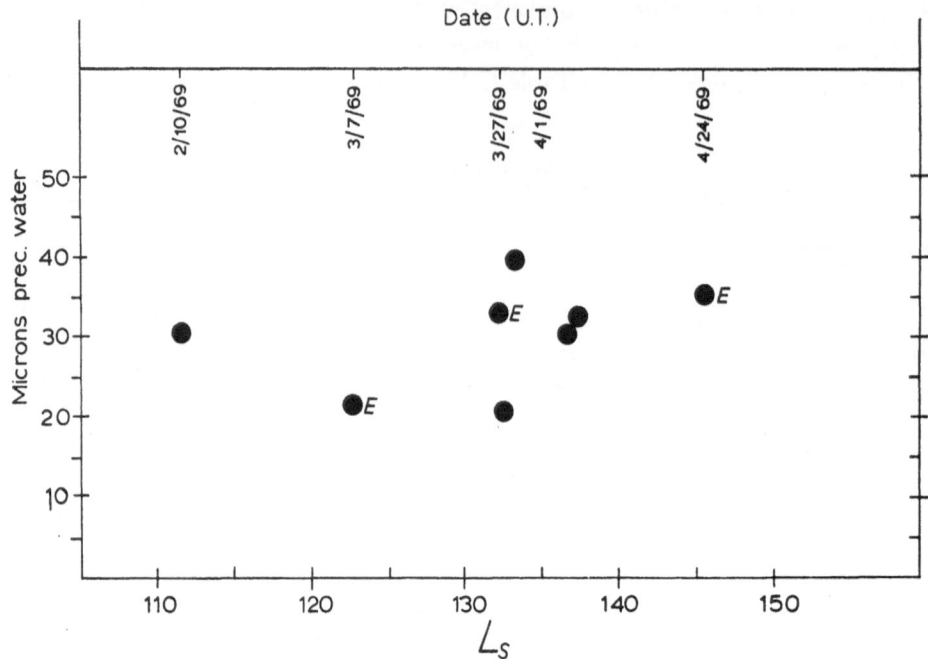

Fig. 3.   Abundance of water on the whole disc of Mars. If the abundance is only for the equatorial region, this is indicated by an *E*.

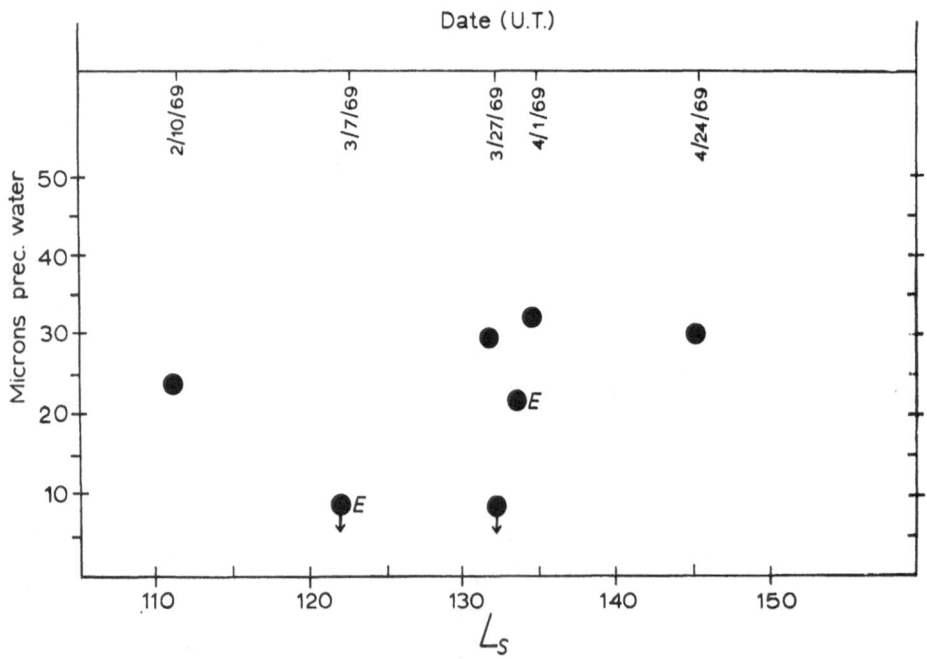

Fig. 4.   Abundance of water in south polar regions of Mars. The south equatorial regions are indicated by an *E*.

Considering the errors involved it appears that not much evidence exists for a variation of Martian water vapor abundance during the time span under study. The results of Owen (1969) in late February support this conclusion as do those of Tull given earlier in this conference. There is a definite predominance of water vapor in the northern hemisphere of Mars during this time in agreement with Tull's report.

Several good plates of the water vapor region have been recently obtained at McDonald Observatory with both the 82-inch and 107-inch telescopes. None of these plates show any trace of Martian water vapor upon visual inspection. The Doppler shift is not great, but it is felt that lines with an equivalent width corresponding to an abundance of 15 μ would have been seen had they been present. The Martian water vapor has apparently disappeared or decreased during opposition.

Acknowledgments

I thank Dr. Robert Tull for discussions about the abundance determination, and I appreciate Mr. J. Woodman's help in the reduction of equivalent widths.

References

Farmer, C. B.: 1971, *Icarus*, to be published.
Owen, T. and Mason, H. P.: 1969, *Science* **165**, 893.
Schorn, R. A., Farmer, C. B., and Little, S. J.: 1969, *Icarus* **11**, 283.

HIGH ALTITUDE INTERFEROMETER SPECTRA OF MARS

UWE FINK and G. P. KUIPER

Lunar and Planetary Laboratory, University of Arizona, Tucson, Ariz., U.S.A.

Abstract. Spectra of Mars were presented which were taken as part of the LPL high altitude program [1] with the NASA CV990 jet. The spectra were taken at about 40 000′ (12 km) altitude. Two sets of five flights each were made around May 1, 1969 and June 1, 1969. In each case the first flight was a Moon comparison flight and the other four were Mars flights, with some Moon if the Moon could be observed.

A 12″ telescope received light from a gyroscopically controlled heliostat and directed it into a Block Associates rapid scan interferometer. (For details see Reference [2].) The interferometer had a resolution of 5 cm^{-1}. The present spectra are, however, only reduced for a resolution of 8 cm^{-1} due to limitations in our computer operations.

The spectra taken around June 1 are of better quality since Mars was brighter at that time. The CO_2 bands at 1.4 μ, 1.6 μ, 2.0 μ, and 2.7 μ all show up very nicely. The isotopic C^{13} companion band of 003 as well as the ordinary forbidden isotopic band of $O^{16}C^{12}O^{18}$ (002) can be detected. No water vapor at 1.4 μ or 1.8 μ remains after taking the ratio with the Moon. Some possible trace remains at 2.7 μ. Quantitative analysis as well as reductions of the interferograms to higher resolution are in progress.

References

[1] *Comm. Lunar Planetary Lab.* Nos. 93–96.
[2] *Comm. Lunar Planetary Lab.* No. 100.

MARS: OCCURRENCE OF LIQUID WATER

ANDREW P. INGERSOLL

Division of Geological Sciences, California Institute of Technology, Pasadena, Calif., U.S.A.

Abstract. In the absence of juvenile liquid water, condensation and subsequent melting of ice are the only means of producing liquid water on the Martian surface. However, the evaporation rate is so high that the available heat sources cannot melt ice on Mars. Melting might occur only in concentrated solutions of strongly deliquescent salts.

The purpose of this study is to see whether sunlight or other heat sources can melt water ice on Mars; this is probably the most likely mechanism by which liquid water might naturally occur on the Martian surface. If all the water were to condense out of the atmosphere, it would cover the surface with a layer 10 to 20 μm thick (Kaplan *et al.*, 1964; Owen and Mason, 1969). On the other hand, if this amount were mixed uniformly with other atmospheric gases, condensation would occur at temperatures in the range 190 to 200 K (Leighton and Murray, 1966). This means that only ice will condense directly out of the atmosphere, and also that frosts at temperatures above 200 K will cool by evaporation unless there is an adequate heat source. Thus, the circumstances most favorable to melting are when the rate of evaporation of a frost at 0 °C is at its minimum.

In order to estimate this minimum rate, we shall assume there is no wind, and that the only atmospheric motions are those generated by the evaporation itself. However, water vapor is intrinsically lighter than carbon dioxide, the principal constituent of the Martian atmosphere, and so the saturated fluid layer near the ground is dynamically unstable. The situation is analogous to thermal convection above a heated horizontal surface, and therefore we shall use thermal convection data to estimate the evaporation rate.

The similarity between heat convection and mass exchange is described in many engineering textbooks (Jakob, 1949). It is based on the similarity of the equations for conservation of heat and diffusing substance, respectively, in a moving fluid, and on the fact that a smaller molecular weight has the same effect on buoyancy as a correspondingly higher temperature. The analogy is not exact, however, but the error leads to our underestimating the evaporative loss. In the first place, evaporation implies a velocity normal to the wall, for which there is no analogue in thermal convection. This normal velocity carries additional mass, so the evaporation rate should be greater than that estimated from the thermal convection analogy (Eckert and Drake, 1959). Second, when concentration gradients are large, the analogy fails because of thermal diffusion effects. Again this leads to underestimating the evaporation rate, since heat tends to diffuse toward fluid of lower molecular weight (Chapman and Cowling, 1939), thereby increasing the instability of the system.

Thus the estimate which follows will be a lower bound on the evaporation rate of water ice on Mars. The basic experimental data are measurements of heat flux above a heated horizontal plate in air as a function of the physical properties of air and the temperature difference between the plate and its surroundings. Using these data (Jakob, 1949) and the thermal convection analogy, we obtain

$$E = (0.17) \, \Delta\eta\rho D \, [(\Delta\rho/\rho) \, g/\nu^2]^{1/3}, \tag{1}$$

for the mass flux of water vapor E above an evaporating frost. Here $\Delta\eta$ is the difference between the mass concentration of water vapor at the evaporating surface and in the gas away from the surface. Since the fluid is saturated near the frost (at temperature T_0), and since the surroundings are almost completely dry, we have

$$\Delta\eta = \rho_w (T_0)/\rho, \tag{2}$$

where $\rho_w (T_0)$ is the saturation density of water vapor and ρ is the total density of the gas at the surface. The quantity D is the diffusion coefficient of water vapor in carbon dioxide; g is the acceleration of gravity on Mars; and ν is the kinematic viscosity of carbon dioxide. Finally, $(\Delta\rho/\rho)$ is the density difference between the ambient fluid and the fluid at the surface, divided by the density of the fluid at the surface. This may be expressed in terms of the Martian surface pressure P_0 and the saturation vapor pressure of water $e = e(T_0)$, provided both components of the mixture behave as ideal gases:

$$\Delta\rho/\rho = (m_c - m_w) \, e/[m_c P_0 - (m_c - m_w) \, e], \tag{3}$$

where m_w and m_c are the molecular weights of water and carbon dioxide, respectively. In this derivation the change in pressure across the saturated boundary layer is neglected; this is valid provided the layer is thin compared to the atmospheric scale height $H = RT/g$, where R is the gas constant of the atmosphere.

Combining these equations, and multiplying by the heat of vaporization of ice λ, we obtain an expression for λE, the rate of heat loss of a frost at temperature T_0 in a carbon dioxide atmosphere at pressure P_0. These results are presented in Table I, and

TABLE I

Heat fluxes (cal/cm²/min) necessary to maintain an evaporating frost deposit at constant temperature, for various temperatures and pressures

Temperature T_0 (°C)	Surface pressure P_0 (mb)			
	6	10	15	25
0	1.25	0.76	0.55	0.38
−5	0.62	0.41	0.30	0.21
−10	0.33	0.22	0.17	0.12

are based on published values of the gravitational acceleration on Mars, the vapor pressure and heat of vaporization of ice, the viscosity of carbon dioxide, and the mass diffusivity of water vapor in carbon dioxide (Boynton and Brattain, 1929). It is seen that the rate of evaporation, and hence the necessary heat flux, varies directly as the partial pressure of water vapor, and inversely as the Martian surface pressure.

From these data, the most favorable sites for the occurrence of liquid water are those at low elevations where the surface pressure is high. Slopes which face the sun directly during part of the day are also favored. However, the solar constant at the orbit of Mars is about 0.85 cal/cm^2/min, and the mean surface pressure is about 5–7 mb (Kliore *et al.*, 1965; Kliore, 1971). Even at the lowest points, the pressure is probably less than 10 mb (Belton and Hunten, 1969), and since the albedo of frost is high, it appears that water ice may never melt on the Martian surface. Under these circumstances, a frost exposed to sunlight simply evaporates at a temperature below the melting point when exposed to sunlight.

A separate issue concerns the lifetime of a frost of typical thickness compared to the time necessary to melt it. Even if all the atmospheric water vapor were to condense out during the Martian night, the morning frost layer would be only 10–20 μm thick. From Table I with $\lambda = 676$ cal/gm, the lifetime of such a frost at $-10\,°$C would be several minutes, and since the frost is likely to spend more time than this in warming from $-10\,°$C to $0\,°$C, it will probably disappear before the temperature reaches the melting point. The situation is more favorable at the polar caps, where larger amounts of frost might accumulate during the Martian winter (Leighton and Murray, 1966). However, the solar heating is also more gradual at the poles, following an annual rather than a daily cycle, so the frost lifetime is still short compared to the time necessary to heat the frost.

Another possibility concerns the melting of water in soil interstices (Sagan *et al.*, 1968), where evaporation is slowed by the close packing of soil above the melting level. Let us assume that this water collects by freezing, and must diffuse down from the atmosphere during the night. Then, since the partial pressure of water in the atmosphere is some 10^{-4} times the vapor pressure at $-10\,°$C, the lifetime of the accumulated deposit at $-10\,°$C, will be only 10^{-4} times the accumulation time. This lifetime is about $\frac{1}{2}$ h at the pole and 10 sec elsewhere, so this mechanism does not increase the likelihood of melting.

Note that this argument applies when the only source of water is the atmosphere. Liquid water from the interior of the planet might occasionally reach the surface to form hot springs (Sagan *et al.*, 1968), although there is no indication that such a process occurs on the earth. Most evidence suggests that terrestrial hot springs contain re-cycled rain water (Mason, 1966), although small amounts of juvenile water may also be present.

Let us now compare the effects of wind with the effects of compositional density differences already mentioned. Measured rates of evaporation under conditions of neutral stability on the earth give (Priestley, 1959):

$$E = (0.002)\rho_w U, \tag{4}$$

where ρ_w is the saturation density of water vapor and U is the wind velocity 1 m above the surface. However, even for $U = 100$ m/s, the additional evaporation due to wind is less than the rates implied in Table I. The opposite question is whether wind could supply the necessary heat to the surface, but here again the effect appears to be negligible. In fact, the turbulence necessary to mix heat downwards to the surface would also cause an increase in evaporation, leading to a net cooling. For, if ΔT is the temperature difference between the warm atmosphere and cool surface, then the flux of heat to the surface will be $\rho U_0 c_p \Delta T$, where c_p is the specific heat of the gas and U_0 is some velocity characteristic of the process. However, at the same time the evaporative cooling will be $\rho U_0 \lambda \Delta \eta$, which is larger than the heating for all reasonable choices of parameters, for frost temperatures greater than -10 °C.

All of the above remarks apply only to pure water ice. Dissolved salts have the effect of lowering both the melting temperature and the vapor pressure, and for certain substances these effects can be quite large. As an extreme example, $CaCl_2$ lowers the vapor pressure by a factor of 5, and the melting point by 50 °C (Hall and Sherrill, 1929). Under these conditions the evaporative cooling is negligible, and the lifetime is effectively infinite. Thus liquid water is possible on Mars, but it will probably occur only in concentrated solutions of strongly deliquescent salts. Pure liquid water may never occur on Mars, because the available heat sources cannot balance the evaporative cooling, and because any frost will evaporate completely before it reaches the melting point.

References

Belton, M. J. S., and Hunten, D. M.: 1969, *Science* **166**, 225.

Bircumshaw, L. L. and Stott, V. H.: 1929, in *International Critical Tables* (ed. by E. W. Washburn), McGraw-Hill, New York.

Boynton, W. P. and Brattain, W. H.: 1929, in *International Critical Tables* (ed. by E. W. Washburn), McGraw-Hill, New York.

Chapman, S. and Cowling, T. G.: 1939, *The Mathematical Theory of Non-Uniform Gases*, Cambridge University Press, p. 19.

Eckert, E. R. G. and Drake, R. M., Jr.: 1959, *Heat and Mass Transfer*, McGraw-Hill, New York.

Hall, R. E. and Sherrill, M. S.: 1929, in *International Critical Tables*, McGraw-Hill, New York.

Jakob, M.: 1949, *Heat Transfer*, Vol. 1, Wiley, New York.

Kaplan, L. D., Münch, G., and Spinrad, H.: 1964, *Astrophys. J.* **139**, 1.

Kliore, A.: 1971, in preparation.

Kliore, A., Cain, D. L., Levy, G. S., Eshelman, V. R., and Fjeldbo, C.: 1965, *Science* **149**, 1243.

Leighton, R. B. and Murray, B. C.: 1966, *Science* **153**, 136.

Mason, B.: 1966, *Principles of Geochemistry*, 3rd ed., Wiley, New York.

Owen, T. and Mason, H. P.: 1969, *Science* **165**, 893.

Priestley, C. H. B.: 1959, *Turbulent Transfer in the Lower Atmosphere*, University of Chicago Press.

Sagan, C., Levinthal, E. C., and Lederberg, J.: 1968, *Science* **159**, 1191.

Spencer, H. M.: 1929, in *International Critical Tables* (ed. by W. E. Washburn), McGraw-Hill, New York.

D. MARINER RESULTS

MARINER 6: ULTRAVIOLET SPECTRUM OF MARS UPPER ATMOSPHERE*

C. A. BARTH

Dept. of Astro-Geophysics and Laboratory for Atmospheric and Space Physics, University of Colorado, Boulder, Colo., U.S.A.

W. G. FASTIE

Dept. of Physics, Johns Hopkins University, Baltimore, Md., U.S.A.

C. W. HORD, J. B. PEARCE, K. K. KELLY, A. I. STEWART,
G. E. THOMAS, and G. P. ANDERSON

Dept. of Astro-Geophysics and Laboratory for Atmospheric and Space Physics, University of Colorado, Boulder, Colo., U.S.A.

and

O. F. RAPER

Space Sciences Division, Jet Propulsion Laboratory, Pasadena, Calif., U.S.A.

Abstract. Emission features from ionized carbon dioxide and carbon monoxide were measured in the 1900- to 4300-Å spectral region. The Lyman-α 1216-Å line of atomic hydrogen and the 1304-, 1356-, and 2972-Å lines of atomic oxygen were observed.

The flight of Mariner 6 past Mars on 31 July 1969 presented the first opportunity to measure the ultraviolet dayglow of that planet. The technique of using ultraviolet spectroscopy to study planetary atmospheres has been developed both theoretically and experimentally over the past 9 yr [1]. Rocket experiments have shown that the ultraviolet dayglow of the earth consists of the following emission features: the Lyman-α 1216-Å line of atomic hydrogen; the 1304-, 1356-, and 2972-Å lines of atomic oxygen; the 1200-, 1493-, 1744-, and 3466-Å lines of atomic nitrogen; the Lyman-Birge-Hopfield, Vegard-Kaplan, and second-positive bands of molecular nitrogen; the gamma bands of nitric oxide; and the first negative bands of ionized molecular nitrogen [2]. These emissions are produced in the earth's upper atmosphere by resonance and fluorescence scattering of ultraviolet solar radiation and by photo-electron impact excitation.

The Mariner ultraviolet spectrometer was specifically designed to measure emissions from the sunlit atmosphere above the limb of Mars. Extensive baffling in front of the telescope suppressed off-axis light from entering the 250-mm Ebert-Fastie spectrometer [1]. Two photomultiplier tubes simultaneously recorded the spectral

* The success of this experiment, which has been under preparation for 9 yr, is the result of the efforts of a large number of people at NASA headquarters, the Jet Propulsion Laboratory, the University of Colorado, Johns Hopkins University, and elsewhere in the scientific community. The large scientific return from the Mariner 1969 mission is due to the technical and managerial skills of H. M. Schurmeier and the Mariner project staff at JPL and NASA headquarters. Supported by NASA under JPL contract 951790 and NASA grant NGL 06-003-052. Reprinted from *Science* **165**, 1004–1005.

scans which occurred repetitively every 3 sec. The wavelength band at 1100 to 1900 Å was measured at a resolution 10 Å by a cesium iodide tube, and the 1900- to 4300-Å band was measured at a resolution of 20 Å with a bi-alkali tube.

The first observation of the sunlit atmosphere of Mars occurred when Mariner 6 was 8300 km from the planet's center, the slant range to the limb was 7600 km, and the solar zenith angle at the limb was 27°. The telescope baffling rejected the off-axis light from the disc sufficiently well so that a spectrum rich in emission features was obtained. In the spectral interval from 1900 to 4300 Å, the bi-alkali photomultiplier tube recorded the spectrum shown in Figure 1.

An initial spectroscopic analysis has been performed with the use of laboratory and theoretical data prepared before the Mars encounter. A comparison spectrum synthesized from three separate sources is shown in Figure 2 together with the individual spectra. Figure 2a was obtained from a spectrum of the Martian disc measured later in the flight by the Mariner spectrometer. This spectrum is the result of Rayleigh

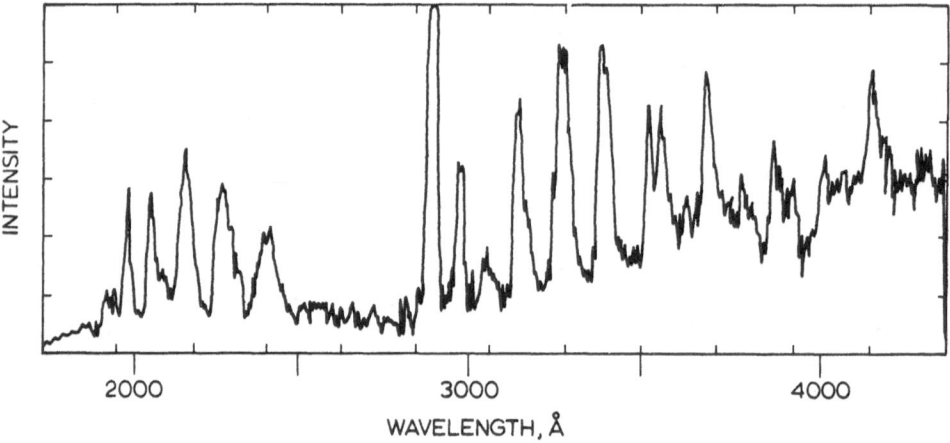

Fig. 1. Ultraviolet spectrum of Mars upper atmosphere. A nominal wavelength scale is shown on the abscissa. The ordinate is an arbitrary intensity scale uncorrected for the spectral response of the instrument. A number of spurious noise pulses have been edited out of this spectrogram.

scattering and ground reflection of solar radiation. The spectrum in Figure 2b was produced in the laboratory by the bombardment of carbon dioxide at 10^{-3} torr by electrons with an energy of 20 eV. This spectrum contains the prominent ionized carbon dioxide emission feature at 2890 Å and the Fox-Duffendack-Barker bands. The spectrum in Figure 2c was composed theoretically from the calculations of how the Cameron bands of carbon monoxide appear in the fluorescence scattering of sunlight [2], and it has been adjusted for the response of the instrument. The three spectra were normalized individually and summed to form the composite spectrum in Figure 2d. In view of the way in which we put together the synthetic spectrum in Figure 2d, the similarity between it and the Mars spectrum in Figure 1 is remarkable. However, the excitation mechanisms which occur in the upper atmosphere of Mars may or may not be the ones we used to produce the synthetic spectrum.

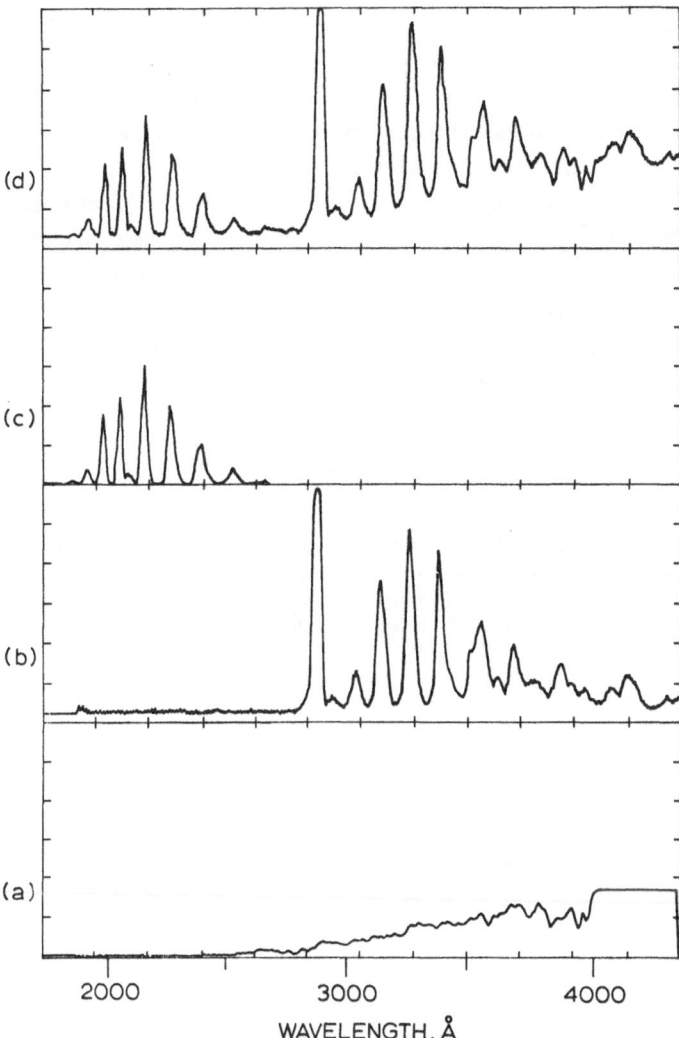

Fig. 2. Synthesis of comparison spectrum. (a) Off-axis light component from Mars disc. (b) Carbon dioxide spectrum from laboratory electron impact experiment. (c) Carbon monoxide spectrum from theoretical calculation of fluorescence scattering. (d) Composite spectrum synthesized by adding a, b, and c.

We have identified an additional feature which appears in the Mars spectrum as the 2972-Å line of atomic oxygen. There may be further unidentified features in this spectrum.

The limb spectrum in the 1100- to 1900-Å region was recorded by the cesium iodide photomultiplier tube. The principal emission features that were observed were: the Lyman-α 1216-Å line of atomic hydrogen, the 1304- and 1356-Å lines of atomic oxygen, and the fourth-positive bands of carbon monoxide.

One particularly important objective of the ultraviolet spectrometer experiment was to search for nitrogen in the atmosphere of Mars. This first analysis shows no evidence

18—P.A.

of nitrogen emissions in the ultraviolet spectrum of the upper atmosphere. The following emissions were searched for and found missing: second-positive and Lyman-Birge-Hopfield bands of molecular nitrogen, first-negative bands of ionized molecular nitrogen, gamma bands of nitric oxide, and 1200- and 1493-Å lines of atomic nitrogen. The final analysis of these data will allow an upper limit to be placed on the amount of nitrogen in the upper atmosphere of Mars.

Repetitive spectra were taken as the Mariner spectrometer crossed the limb of Mars. These data, which contain information about the scale height of individual spectral emissions will be used to construct a model of the Mars upper atmosphere. This first report is simply a record of our identifications in the ultraviolet spectrum of the upper atmosphere. The instrument also obtained spectra of the bright and dark parts of the disc, the terminator, and the atomic hydrogen corona.

References

[1] Barth, C. A.: 1969, *Appl. Opt.* **8**, 1295.
[2] Barth, C. A.: 1966, in *The Middle Ultraviolet – Its Science and Technology* (ed. by A. E. S. Green), Wiley, New York; Fastie, W. G., Crosswhite, H. M., and Heath, D. F.: 1964, *J. Geophys. Res.* **69**, 4129; Barth, C. A.: 1964, *J. Geophys. Res.* **69**, 3301; Barth, C. A. and Pearce, J. B.: 1966, *Space Res.* **6**, 381; unpublished results from a rocket flight, 13 June 1969.

MARINER MARS '69 CELESTIAL MECHANICS EXPERIMENT

JOHN D. ANDERSON

Jet Propulsion Laboratory, Pasadena, Calif., U.S.A.

Abstract. Spacecraft tracking data can be used to determine some parameters of importance to recent developments in celestial mechanics. Integrated Doppler data are available from both Mariner VI and VII. These data are phase coherent and are accurate to two one thousandths of a cycle at S-band over a 10 min integration time. In terms of a range rate error this is better than 0.2 mm/sec. Round trip range data are also available. They are accurate to 100 nsec in absolute accuracy and have a random error component of about 30 nsec, or about 4.5 m in one-way range.

Only data near the closest approach of the spacecraft to Mars are capable of yielding information on that planet, but other data are useful for investigations into the ratio of the masses of the Earth and Moon, the ephemeris of the Earth, and both orbital and propagation effects of general relativity. For purposes of this conference, however, the only results reported are those which pertain directly to Mars.

The problem in using the Mariner VI and VII encounter data is that significant non-gravitational forces were acting on the spacecraft.

On Mariner VI, the trajectory was essentially a gravitational one until about 35 minutes before closest approach when the system to cool the infrared spectrometer was turned on. This system expelled hydrogen and nitrogen gas through a Joule Thompson cryostat and as a result imparted a thrust on the spacecraft of something on the order of 100 dyne, or a total velocity change in the trajectory of about 0.2 m/sec. Unfortunately the system did not operate properly on Mariner VI; it is known that the thrusting from the infrared system occurred over a period of about six days after encounter. In normal operation, the thrust level would decrease to an insignificant level after a few hours.

On Mariner VII, the infrared cooling system worked properly but an unknown event which occurred some 5 days before encounter imparted a thrust to the spacecraft for at least several days prior to closest approach.

There is a possibility, in fact, that a thrust was applied to the spacecraft for several weeks after encounter. Combining this effect with the normal thrusting from the infrared system, makes the scientific analysis of the Mariner VII tracking data a near impossibility.

Despite these difficulties with Mariner VII, some interesting results have been obtained with Mariner VI. For example, the universal gravitational constant times the mass of Mars GM_δ has been determined from the acceleration on the spacecraft as it approached the planet. The best value of GM_δ is 42 828.3 ± 1.0. Also, by combining the range data from Mariner VI with data from the radar range measurements of Mars obtained by R. Goldstein this year at Goldstone, California, it has been possible to prepare a topographical map of Mars between 3° and 12° north latitudes. This map predicts that the immersion of Mariner VI into earth occultation should occur at a point on Mars which is 3393.6 km from the center of the planet. An independent calculation of this distance by A. Kliore, who used the Mariner VI occultation data, is in excellent agreement with the calibrated radar topography.

Sagan et al. (eds.), Planetary Atmospheres, 257.

MARINER 1969: RESULTS OF THE INFRARED RADIOMETER EXPERIMENT*

G. MÜNCH and G. NEUGEBAUER

California Institute of Technology, Pasadena, Calif., U.S.A.

and

S. C. CHASE

Santa Barbara Research Center, Santa Barbara, Calif., U.S.A.

Abstract. The energy radiated by Mars in the wavelength bands 8–12 and 18–25 μ has been measured during the flyby of Mariner 1969 with linear resolution of 50 km at closest approach. From the laboratory energy calibrations and assuming unit emissivity, the temperature of the surface materials has been derived. The main results are summarized as follows:

(a) The temperatures near the equator (Mariner 6), as function of local time, on the average agree with the values expected on the basis of the gross thermophysical properties of the soil determined from ground-based measurements.

(b) The temperature fluctuations around their mean seem to correspond to variations in albedo over the classical features of the planet, with the possible exception of one area.

(c) The lowest temperatures measured by channel 2 of Mariner 7, during the polar swath, were 148 K, while at the edge of the polar cap the temperature was 230 K.

A small amount of instrumental responsivity to off-axis radiation will change these nominal temperatures slightly, in the general sense that the low temperatures, when properly corrected, will decrease by a few degrees, while the high temperatures will increase by a similar amount.

* Reference: *Science* **166**, 98–99, October 3, 1969.

MARINER 6 AND 7 TELEVISION PICTURES:
PRELIMINARY ANALYSIS*

R. B. LEIGHTON, N. H. HOROWITZ, B. C. MURRAY, R. P. SHARP

California Institute of Technology, Pasadena, Calif., U.S.A.

A. H. HERRIMAN and A. T. YOUNG

Jet Propulsion Laboratory, Pasadena, Calif., U.S.A.

B. A. SMITH

New Mexico State University, Las Cruces, N.M., U.S.A.

M. E. DAVIES

RAND Corporation, Santa Monica, Calif., U.S.A.

and

C. B. LEOVY

University of Washington, Seattle, U.S.A.

Before the space era, Mars was thought to be like the earth; after Mariner 4, Mars seemed to be like the moon; Mariners 6 and 7 have shown Mars to have its own distinctive features, unknown elsewhere within the solar system.

The successful flyby of Mariner 4 past Mars in July 1965 opened a new era in the close-range study of planetary surfaces with imaging techniques. In spite of the limited return of data, Mariner 4 established the basic workability of one such technique, which involved use of a vidicon image tube, on-board digitization of the video signal, storage of the data on magnetic tape, transmission to the earth at reduced bit rate by way of a directional antenna, and reconstruction into a picture under computer control. Even though the Mariner 4 pictures covered only about 1% of Mars's area, they contributed significantly to our knowledge of that planet's surface and history [1, 2, 15, 17].

The objectives of the Mariner 6 and 7 television experiment were to apply the successful techniques of Mariner 4 to further explore the surface and atmosphere of Mars, both at long range and at close range, in order to determine the basic character of features familiar from ground-based telescopic studies; to discover possible further clues as to the internal state and past history of the planet; and to provide information germane to the search for extraterrestrial life.

The Mariner 6 and 7 spacecraft successfully flew past Mars on 31 July and 5 August 1969, respectively; first results of the television experiment, based upon qualitative study of the uncalibrated pictures, have been reported [3, 4]. The purpose of this article is to draw together the preliminary television results from the two spacecraft; to present tentative data concerning crater size distributions, wall slopes, and geographic distribution; to discuss evidences of haze or clouds; to describe new, distinctive types of topography seen in the pictures; and to discuss the implications of the results with respect to the present state, past history, and possible biological status of Mars.

* Reprinted from *Science* 166, No. 3901, 3 October 1969.

Sagan et al. (eds.), Planetary Atmospheres, 259–294.

The data presented here and in the two earlier reports were obtained from inspection and measurement of a partial sample of pictures in various stages of processing. As such, the results must be regarded as tentative, subject to considerable expansion and possible modification as more complete sets, and better-quality versions, of the pictures become available over a period of several months. They are offered at this time because of their unique nature, their wide interest, and their obvious relevance to the forthcoming Mariner 1971 (orbiter) and Viking 1973 (lander) missions.

1. Television System Design

The experience and results of Mariner 4 strongly influenced the basic design of the Mariner 6 and 7 television experiment. The earlier pictures showed Mars to be heavily cratered, but to have subdued surface relief and low photographic contrast, and possibly to have a hazy atmosphere. It was also found that a vidicon-type camera tube has a most important property: the 'target noise', analogous to photographic grain, is less than that of a photographic emulsion by perhaps a factor of 10 [2] and is the same from picture to picture. Thus the 64-level (6-bit) encoding scheme of Mariner 4 was able to cope with the extremely low contrast conditions because intensity calibration and contrast enhancement by computer techniques could be effectively applied to the data to produce pictures of useful quality.

Early design studies for Mariner 6 and 7 centered around 256-level (8-bit) encoding – at least a tenfold increase in data return over that from Mariner 4; overlapping two-color coverage along the picture track (similar to that of Mariner 4); use of two cameras of different focal lengths to provide higher-resolution views of areas nested within overlapping, wider-angle frames; and use of the camera of longer focal length to obtain a few full-disk photographs showing all sides of Mars as the spacecraft approached the planet. A third filter color, 'blue', was added to the 'red' and 'green' of Mariner 4 for the purpose of studying atmospheric effects.

Limitations of volume, money, and schedule prevented use of a suitable digital recorder system with the necessary data storage capacity, but, through a hybrid system which uses both digital and analog tape recorders, it appeared possible to achieve sufficient data storage capacity, albeit at the expense of complexity.

In its final form, the television experiment employed a two-camera system in which the picture formats and electronic circuits of the cameras were identical (for economy and for efficient use of the tape recorders); a digital tape recorder to store the six lowest-order bits of an 8-bit encoded word for every seventh picture element ('pixel') along each TV picture line (referred to as 1/7 digital data),* and a second, similar tape recorder to store analog data for all pixels [5].

In order to reduce the 'noise' introduced by analog recording, the average signal level was held nearly constant and the modulation index was increased by automatic gain control (AGC) and a cube-law 'contrast enhancement' circuit. This signal

* The 1/7 digital TV data for the central 20% of each line were replaced by encoded data from other on-board experiments. In this region, coarser, 1/28 digital data (6-bit-encoded for every 28th pixel), stored on the analog tape recorder, were available (see Figure 2).

processing increased the data return by a factor of about 5; it also made necessary an elaborate program of computer restoration of the pictorial data after receipt on the earth.

Some technical data relating to the camera system are given in Leighton *et al.* [3], and more complete data will be given elsewhere [5]. Briefly, one camera, called camera A, has a field of view $11° \times 14°$ and a rotary shutter which carries four colored filters in the sequence red, green, blue, green, and so on. Alternating exposures with camera A is camera B, which has a focal length 10 times as great and a field of view $1°.1 \times 1°.4$. Camera B carries only a minus-blue haze filter.

Certain further aspects of the camera-operation and on-board signal processing are shown in Figures 1 and 2. Figure 1 shows the cameras mounted on the spacecraft.

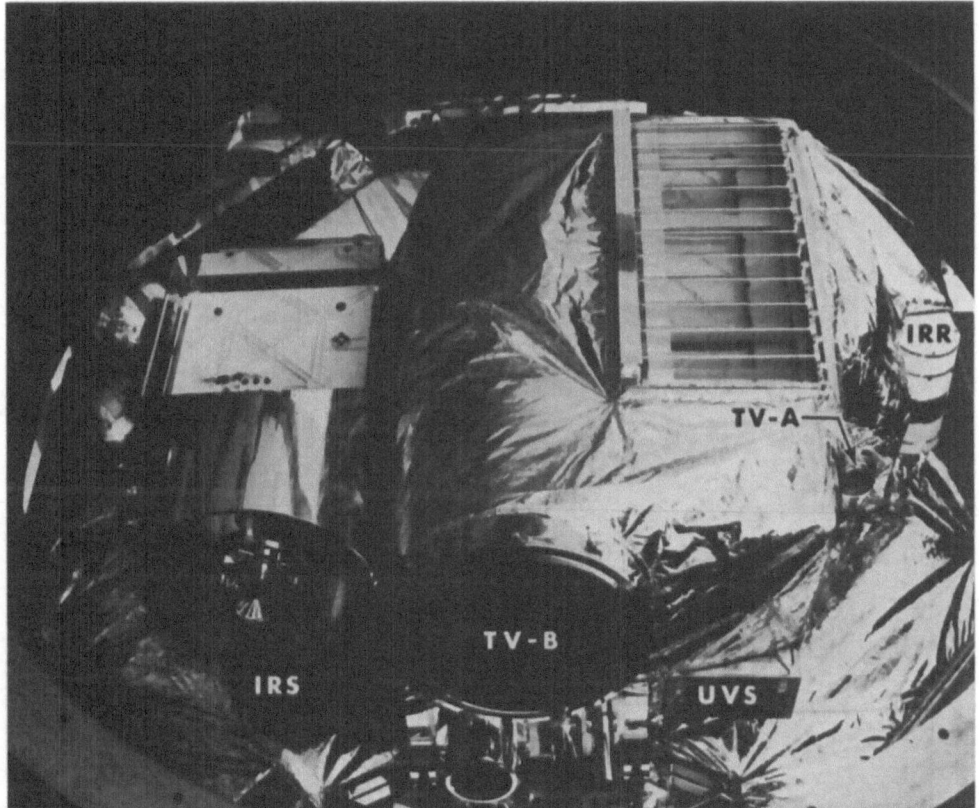

Fig. 1. Mariner 6 instruments and scan platform, mounted on the spacecraft. The five principal instruments are an infrared spectrometer (IRS), television camera B (TV-B), an ultraviolet spectrometer (UVS), television camera A (TV-A), and an infrared radiometer (IRR).

Figure 2a shows the effect of automatic gain control on the analog signal, an effect similar to that of a high-pass filter in that it diminishes the amplitude of long-spatial-wavelength (low-frequency) signals. The response time of the AGC, about 6 millisec, corresponds to about one-tenth of a picture line, and its characteristic effects are apparent in all near-encounter pictures, especially those of high contrast: the polar cap boundary, the planet limb, and the terminator.

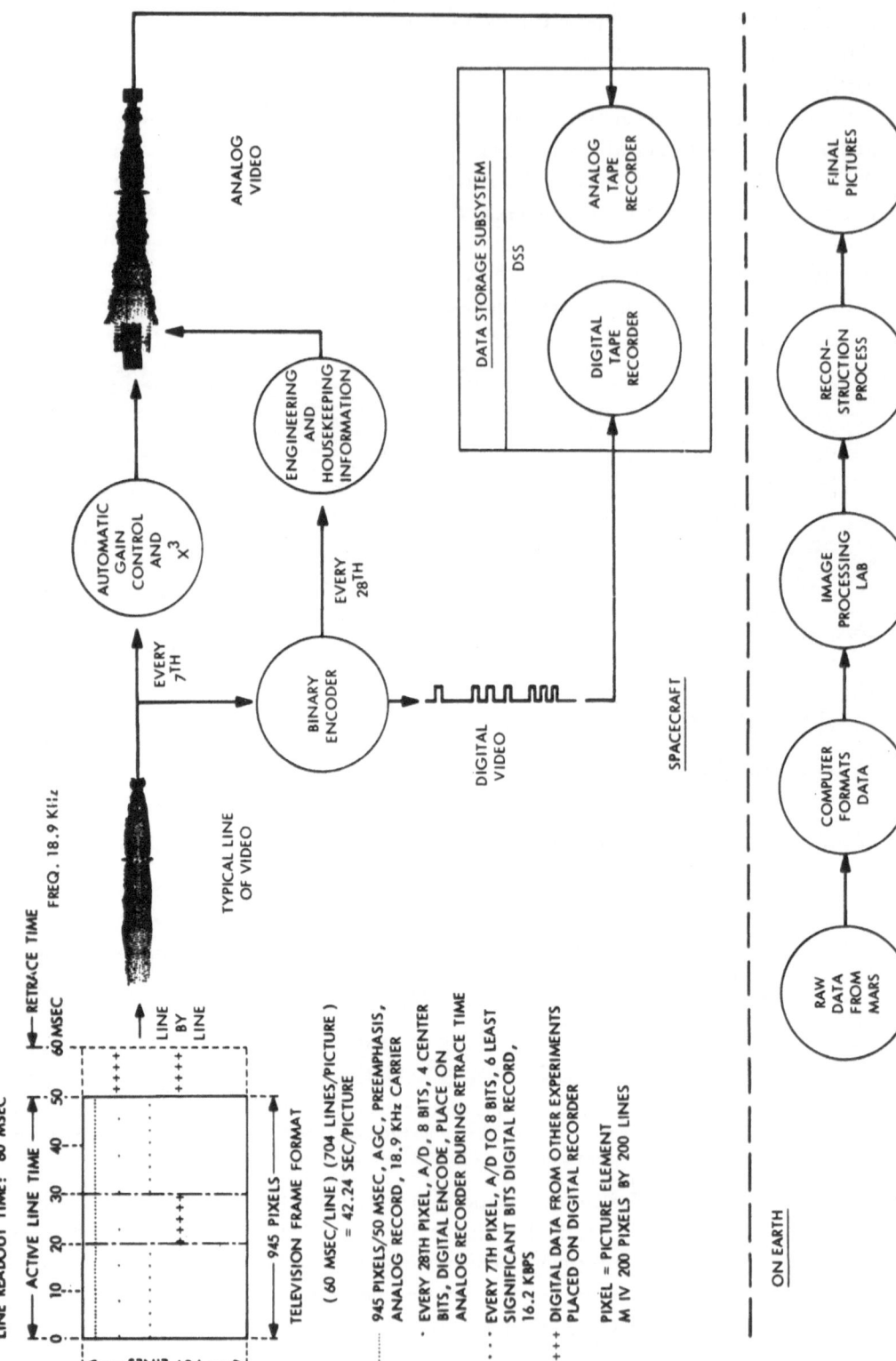

Fig. 2a.   Schematic diagram illustrating video signal processing.

SPECTRAL CURVES

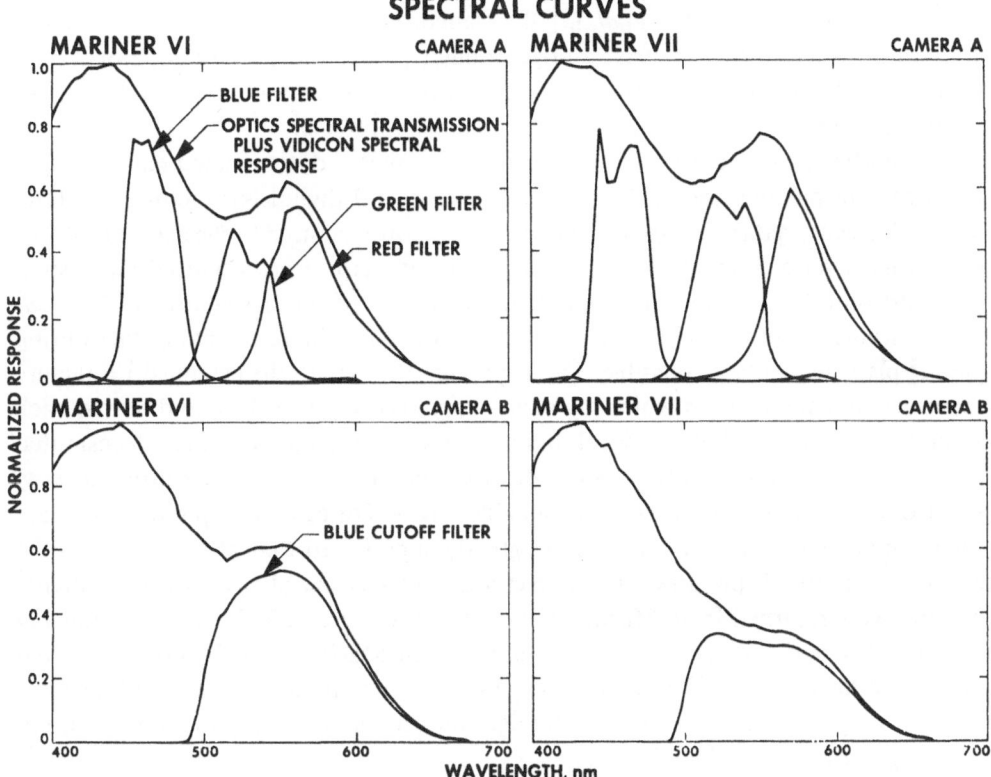

Fig. 2b. Spectral sensitivity curves of Mariner 6 and 7 cameras, with and without filters.

To illustrate the nature of the picture restoration process, we list some of the steps in the computer reduction: Restore the two highest-order bits to the digital data*; remove effects of AGC and 'cuber' in the analog data; combine digital and analog data; measure and remove electronic 'pickup' noise*; measure pixel locations of reseau marks on flight pictures and calibration pictures [6]; bring pictorial calibration and flight data, by interpolation, into agreement with the known reseau pattern; measure and correct for optical distortions; measure and remove effects of residual image from calibration and flight data; evaluate the sensitometric response of each pixel from calibration data and deduce the true photometric exposure for each flight pixel**; correct for the effects of shutter-speed variations and light leakage (camera B); and evaluate and correct for the modulation-transfer function of the camera system. Some of these steps have been applied to a few of the flight pictures on an experimental, ad hoc basis, pending receipt of complete, corrected telemetry data from the six playbacks of the recorded close-up pictures.

* This procedure is semiautomatic, subject to hand correction by the computer operator as necessary.
** For each spacecraft, this must be done for each filter of each camera and for all calibration temperatures, and the results must be corrected to the observed flight temperature.

2. Mission Design and Television Data Return

As was described in Leighton *et al.* [3], the planetary encounter period for each space-craft was divided into two parts: a far-encounter (FE) period beginning 2 or 3 days prior to, and extending to within a few hours of, closest approach, and a near-en-counter (NE) period bracketing the time of closest approach.

As the Mariner 6 and 7 mission was originally conceived, the analog data for eight FE B-camera pictures and 25 NE pictures, and the 1/7 digital NE picture data (and other NE science data) were to be recorded and later transmitted to the earth at 270 bits per second over a 5-day period for each spacecraft. However, a 60-fold increase in transmission rate was realized during the development of the spacecraft, so that *real-time* transmission of the 1/7 digital data stream, or of the digitized analog data during picture playback, became possible. This capability thus led to the extended FE picture sequences actually used, in which the analog tape recorder was filled and played back several times. In addition, several hundred 1/7 digital pictures were transmitted directly to the earth in real time. Some of these digital pictures, taken during the very late FE period of Mariner 7, contain valuable three-color camera-A photometric data for large areas of Mars at a resolution greatly superior to that attainable from the earth. In all, 50 FE pictures, 26 NE pictures, and 428 useful* real-time 1/7 digital pictures were returned from Mariner 6, and 93 FE pictures, 33 NE pictures, and 749 useful real-time digital pictures were returned from Mariner 7. This further ninefold increase in the number of FE pictures and 18% increase in the number of NE pictures over the original plan represents a total data return 200 times that of Mariner 4, not counting the real-time digital frames.

The pictures are designated by spacecraft, camera mode, and frame number. Thus '6N17' means Mariner 6 NE frame 17; '7F77' means Mariner 7 FE frame 77; and so on. The first NE picture from each spacecraft was a camera-A, blue-filter picture. Thus, in near-encounter, all odd-numbered frames are camera-A (wide-angle, low-resolution) frames.

During each day of the FE period, camera B was used to record a series of up to 33 full-disk analog pictures of Mars with the AGC clamped. These pictures were trans-mitted to the earth during each daily period when the 16.2-kilobits-per-second signal could be received by the 210-foot antenna of the Goldstone, California, station of the NASA deep-space net. These pictures showed all sides of the planet as it rotated each day, and, within the total 5-day series, each face of Mars was recorded at a number of different scales and under many different viewing conditions. The phase angle was nearly consant throughout the far-encounter for both spacecraft (25°, morning termi-nator visible).

The approximate near-encounter picture locations for the two spacecraft are shown in Figure 3, and the relevant data are given in Tables I and II. The picture tracks were chosen, in concert with investigators for other on-board experiments, on the basis of several considerations and constraints. First, the choice of possible arrival dates was

---

* Most of the real-time FE A-camera digital pictures were of no value because little or none of the image projected outside the central 20% blank area.

TABLE I

Mariner 6 near-encounter picture data at center of frame

Picture No.	Slant range (km)	View angle from vertical (deg)	Solar zenith angle (deg)	Latitude (deg)	East longitude (deg)	Area ('height × width') (km)
1	Center of picture does not intercept planet					Limb
2	7401	70.2	18.8	4.3	292.3	157.2 × 556.4
3	6614	57.3	6.8	− 2.0	303.2	Limb
4	6130	49.8	4.9	− 5.2	310.3	125.6 × 229.1
5	5682	41.8	11.1	− 8.6	317.3	Limb
6	5348	36.2	17.1	−10.6	323.0	109.1 × 158.7
7	5028	29.7	23.5	−12.9	329.1	1112 × 1597
8	4777	25.2	28.9	−14.1	334.4	97.4 × 125.7
9	4920	40.5	41.3	− 0.1	346.0	1679 × 1861
10	4737	39.4	45.8	− 1.2	350.7	119.1 × 122.1
11	4558	38.1	50.8	− 2.7	355.8	1309 × 1378
12	4439	38.8	55.6	− 3.0	0.6	113.9 × 108.4
13	4333	39.6	60.9	− 3.7	5.8	1191 × 1243
14	4832	59.5	20.5	−13.0	324.8	99.8 × 231.4
15	4382	49.3	30.6	−15.9	334.7	Limb
16	4103	42.1	38.2	−17.3	342.3	83.7 × 132.8
17	3868	34.6	45.7	−18.5	349.9	858 × 1289
18	3738	30.3	52.1	−16.6	356.6	77.1 × 102.5
19	3613	24.4	58.5	−16.9	3.1	773 × 1036
20	3543	20.4	64.2	−16.6	8.9	73.1 × 88.8
21	3498	16.3	70.3	−16.2	15.1	723 × 929
22	3497	14.9	75.8	−15.2	20.5	72.3 × 84.4
23	3522	15.0	81.7	−14.2	26.4	718 × 921
24	3584	17.6	87.2	−12.7	31.7	74.8 × 87.0
25	3622	19.1	90.0	−12.1	34.4	765 × 973
26	3680	21.3	93.0	−11.0	37.3	77.4 × 91.1

limited by engineering considerations to the interval 31 July to 15 August 1969. Second, on any given arrival date, the time of closest approach was limited to an interval of about 1 h by the requirement that the spacecraft be in radio view of Goldstone tracking station during a period of several hours which bracketed the time of closest approach. These two constraints and the approximate 24-h rotation period of Mars considerably limited the possible longitudes of Mars that could effectively be viewed; in particular, the most prominent dark area, Syrtis Major, could not be seen under optimum conditions. Fortunately, Meridiani Sinus, a prominent dark area almost as strong and permanent as Syrtis Major, and various other important features well known from earth observation, were easily accessible.

The cameras and other instruments were mounted on a two-axis 'scan platform' which could be programmed to point the instruments in as many as five successive directions during the near-encounter. The particular orbit and platform-pointing strategy adopted for each spacecraft was designed to achieve the best possible return of scientific data within a context of substantial commonality but with some divergence of needs of the various experiments. The television experimenters placed great weight upon viewing a wide variety of classical features, including the polar cap; continuity of picture coverage; substantial two-color overlap and some three-color overlap is possible; stereoscopic overlap; viewing the planet limb in blue light; viewing the same area at two different phase angles; and seeing the same area under different viewing conditions at nearly the same phase angle.

The Mariner 6 picture track was chosen to cover a broad longitude range at low latitudes in order to bring into view a number of well-studied transitional zones

TABLE II

Mariner 7 near-encounter picture data at center of frame

Picture No.	Slant range (km)	View angle from vertical (deg)	Solar zenith angle (deg)	Latitude (deg)	East longitude (deg)	Area ('height × width') (km)
1	Center of picture does not intercept planet					Limb
2	Center of picture does not intercept planet					Limb
3	9243	69.6	25.8	14.4	350.3	Limb
4	8533	59.7	15.6	5.3	354.9	175.3 × 412.5
5	7993	52.3	8.2	− 1.4	357.7	Limb
6	7533	46.1	2.5	− 7.0	1.3	152.9 × 262.0
7	7129	40.6	4.7	−12.0	3.8	Limb
8	6771	35.8	10.2	−16.4	7.3	137.7 × 200.4
9	6443	31.4	15.1	−20.5	10.0	1503 × 2233
10	6654	47.5	50.5	−53.1	328.3	197.0 × 165.7
11	6377	45.6	52.1	−57.1	332.4	2282 × 1672
12	6084	42.9	53.4	−60.6	338.9	166.6 × 147.5
13	5864	42.4	55.9	−64.2	344.1	2049 × 1912
14	5631	41.1	58.2	−67.4	352.3	148.8 × 139.4
15	5462	41.9	61.4	−70.7	0.1	2098 × 2058
16	5285	42.1	64.5	−73.2	12.5	137.4 × 138.9
17	5167	44.2	68.3	−75.9	26.6	1368 × 1561
18	5049	45.8	72.1	−76.8	46.3	131.69 × 147.2
19	4994	49.2	76.7	−77.3	68.6	1343 × 1862
20	4949	52.4	81.4	−75.4	89.4	133.5 × 171.7
21	5314	65.7	14.0	−20.6	6.1	Limb
22	4776	55.0	24.9	−28.5	14.2	97.9 × 201.3
23	4405	46.8	33.3	−34.3	21.2	1383 × 2488
24	4130	39.9	40.7	−38.3	28.8	84.5 × 128.9
25	3917	33.7	47.3	−41.7	36.2	874 × 1291
26	3759	28.4	53.6	−43.7	44.2	77.2 × 101.4
27	3638	23.6	59.5	−45.3	52.0	775 × 1035
28	3664	28.1	65.6	−40.4	62.1	80.7 × 92.2
29	3619	26.6	71.1	−40.4	69.3	823 × 995
30	3621	27.3	76.7	−39.1	76.3	81.5 × 87.7
31	3646	28.4	82.1	−37.7	83.1	831 × 987
32	3722	31.6	87.7	−35.0	89.5	87.1 × 90.8
33	3822	34.9	93.4	−32.2	95.5	996 × 1116

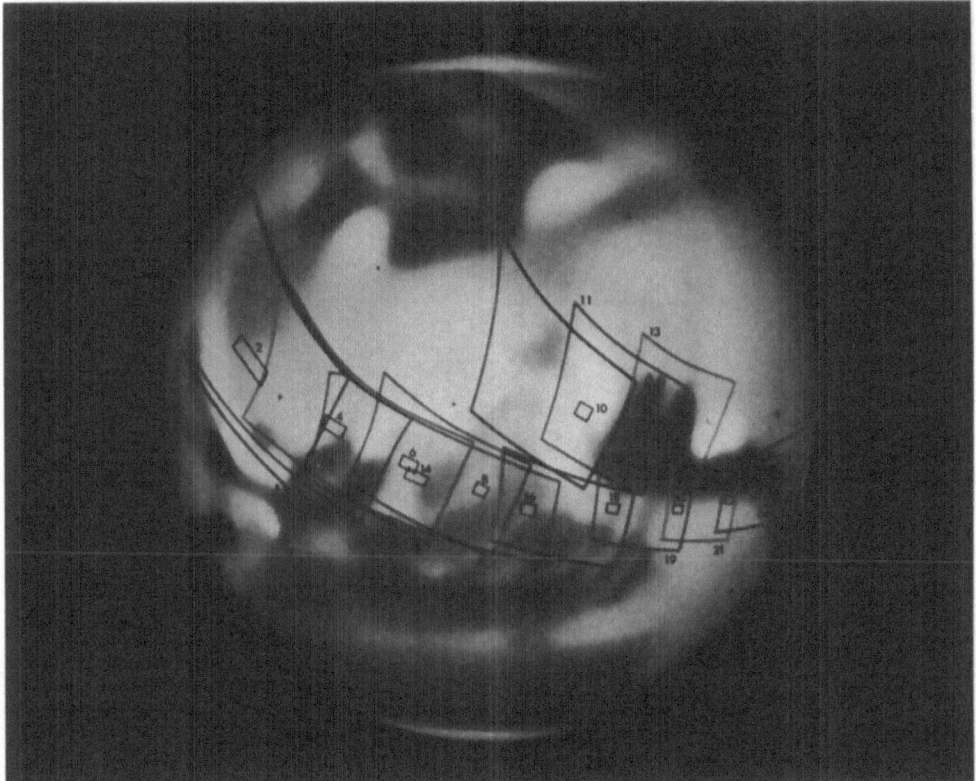

Fig. 3a. Mariner 6 NE picture locations, plotted on a painted globe of Mars. The first picture is taken with a blue filter. The camera-A filter sequence is blue, green, red, green, and so on. Wide-angle (camera A) frames and narrow-angle (camera B) frames alternate.

between light and dark areas, two 'oases' (Juventae Fons and Oxia Palus), and a variable light region (Deucalionis Regio). The picture track of Mariner 7 was selected so that it would cross that of Mariner 6 on the dark area Meridiani Sinus, thereby providing views of that important region under different lighting conditions. The track was also specifically arranged to include the south polar cap and cap edge, to intersect the 'wave-of-darkening' feature Hellespontus, and to cross the classical bright circular desert Hellas.

3. Camera Operation and Picture Appearance

With minor exceptions, both camera systems operated normally, well within expected ranges of the various ambient parameters. The principal exceptions were the following: the contrast of the FE pictures of Mariner 6 was lower than expected, because of an unaccountably low signal level; the electronic 'pickup' noise from the square-wave power system was somewhat greater than anticipated for both spacecraft; the first track of the Mariner 6 analog tape recorder showed a 50% drop in amplifier gain between FE and NE playback; and the fourth track of the analog tape recorder of Mariner 7 showed greater than normal 'dropout' noise. These deficiencies not only affected the subjective appearance of the pictures but also necessitate more elaborate

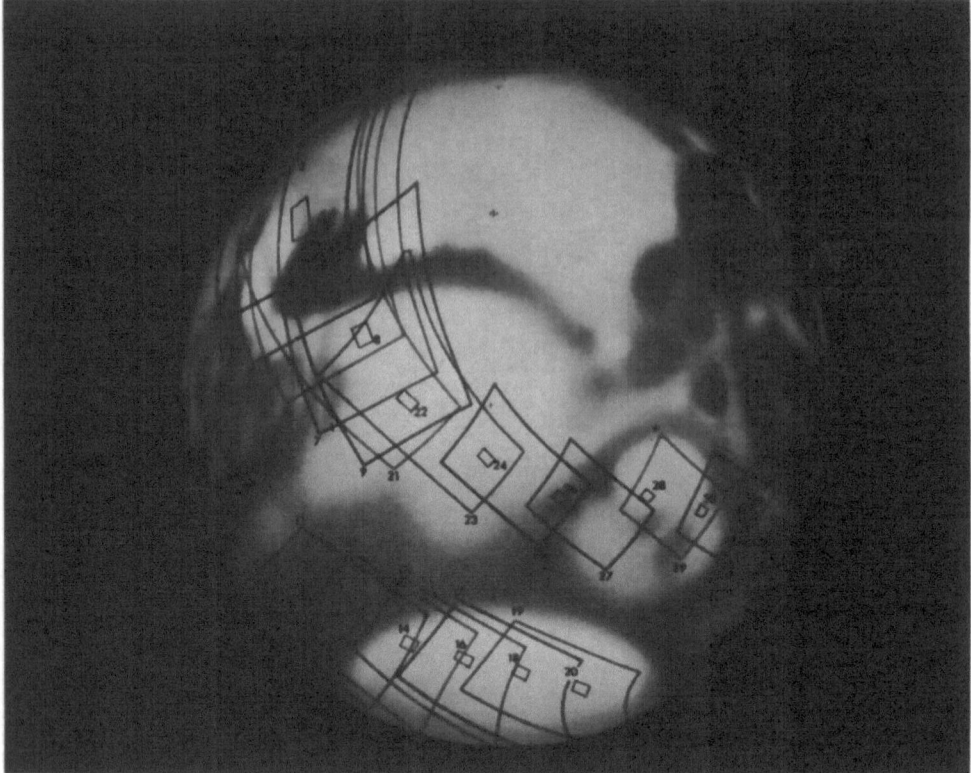

Fig. 3b.   Mariner 7 NE picture locations. The filter sequence is the same as for Mariner 6.

processing of the data before accurate, high-resolution photometric measurements can be made.

The first impression of Mars conveyed by the pictures is that the surface is generally visible and is not obscured by clouds or haze except perhaps in the polar regions and in a few areas marked by the appearance of afternoon 'clouds'. The classical martian features stand out clearly in the far-encounter pictures, and, as the image grows, these features transform into areas having recognizable relationships to the numerous craters which mark the surface. The near-encounter pictures seem to show a Moon-like terrain. However, one must bear in mind the fact that the camera system was designed to enhance the contrast of local brightness fluctuations by a factor of 3, and that the contrast of the pictures is often further enhanced in printing. Actually, although the surface is generally visible, its contrast is much less than that of the moon under similar lighting conditions. Fewer shadows are seen near the terminator.

The determination of true surface contrast depends critically upon the amount of haze or veiling glare in the picture field. Although the pictures appear to be free of such effects, more refined photometric measurements may well reveal the presence of veiling glare or a general atmospheric haze. Definite conclusions must await completion of the photometric reduction of the pictures, including corrections for vidicon dark current, residual images, shutter light leaks, and possible instrumental scattering.

4. Observed Atmospheric Features

A. AEROSOL SCATTERING

Clear-cut evidence for scattering layers in the atmosphere is provided by the pictures of the north-eastern limb of Mariner 7. The limb appears in frames 7N1, 2, 3, 5, and 7, and in a few real-time digital A-camera frames received immediately prior to frame 7N1. The limb appears again in frame 7N21 after the platform slew which began the track across Hellas. Thus the limb coverage includes each of the filters of the A camera and one B-camera frame.

Several characteristics of the scattering layer shown in Figure 4 are evident even at this early stage. (i) The scattering is distinctly stratified in horizontal layers, just as scattering from aerosol layers in the earth's atmosphere is. (ii) The intensity of the scattering varies substantially over distances of a few hundred kilometers and is more intense toward the west or toward earlier local times of day. (iii) The thickness of the scattering layer is about 10 km. (iv) The height of the layer is difficult to determine because of the difficulty of locating the true planetary limb, but it is estimated to be between 15 and 25 km in the region covered by frames 7N1 to 7N7 and up to 40 km in frame 7N21. (v) The layer is about 50% brighter in the blue-filter pictures than in the red or green. This is less difference in intensity than would be expected for Rayleigh scattering, but corresponds more closely to λ^{-2} wavelength dependence.

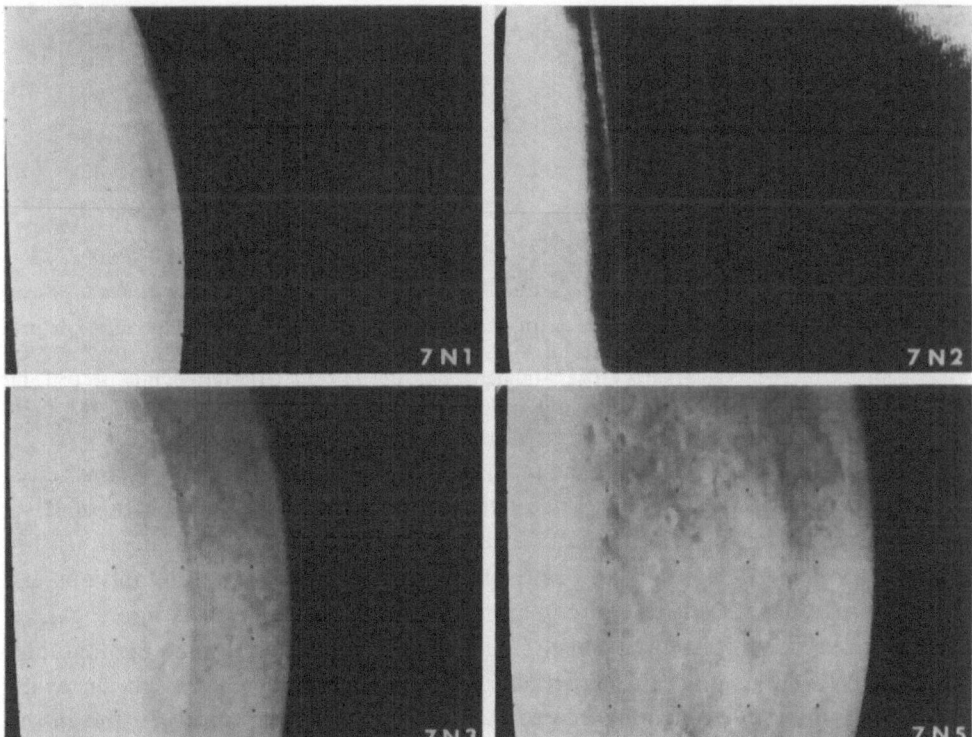

Fig. 4.   Mariner 7 limb frames 7N1, 2, 3, and 5. Note the sharp haze layer adjacent to the limb in frames 7N1, 3, and 5, and the magnified view (tenfold magnification) in 7N2. The prominent, cratered dark feature in frame 7N5 is Meridiani Sinus. North is approximately toward the right.

The relationship between this scattering layer and the martian tropopause is obviously of great interest and will be studied carefully as more refined data become available.

The normal-incidence optical depth, isotropic scattering being assumed, is estimated as 0.01 in the red and about 0.03 in the blue. A λ^{-2} dependence suggests that scattering should be predominantly forward, so these very small values should be under-estimates.

The real-time digital data reveal an apparent limb haze near the south polar cap, and over the regions of Mare Hadriaticum and Ausonia just east of Hellas. The haze over these regions is not as bright as the haze discussed above, so it is unlikely that it is sufficiently dense to obscure surface features seen at NE viewing angles. A faint limb haze may also be present in the Mariner 6 limb frames.

B. THE 'BLUE HAZE'

Despite these evidences of very thin aerosol hazes, visible tangentially on the limb, there is no obscuring 'blue haze' sufficient to account for the normally poor visibility of dark surface features seen or photographed in blue light and for their occasional better visibility – the so-called 'blue-clearing' phenomenon [7, 8].

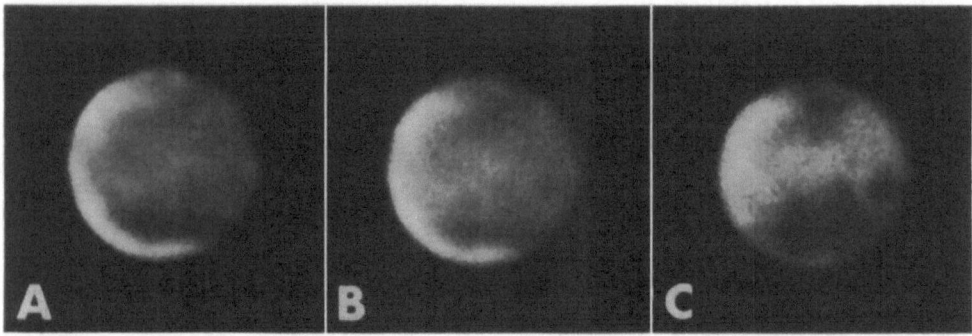

Fig. 5.   Photographs of Mars from the earth, taken to compare Mariner-type blue-filter pictures with 'standard' green and blue pictures of Mars. The pictures were taken 24 May 1969 at New Mexico State University Observatory. (A) 'Standard' blue (09 15 UT); (B) Mariner blue (09 05 UT); (C) standard green (08 44 UT). North is at the top.

The suitability of the Mariner blue pictures for 'blue haze' observations was tested by photographing Mars through one of the Mariner blue filters on Eastman III-G plates, whose response in this spectral region is similar to that of the vidicons used in the Mariner camera. Conventional blue photographs on unsensitized emulsions and green photographs were taken for comparison. A typical result is shown in Figure 5; the simulated TV blue picture is very similar to the conventional blue photographs. The effective wavelength of the actual blue TV pictures should be even shorter, owing to a lower ambient temperature and to the absence of reddening due to the earth's atmosphere.

The blue pictures taken by Mariners 6 and 7 clearly show craters and other surface features, even near the limb and terminator, where atmospheric effects are strong.

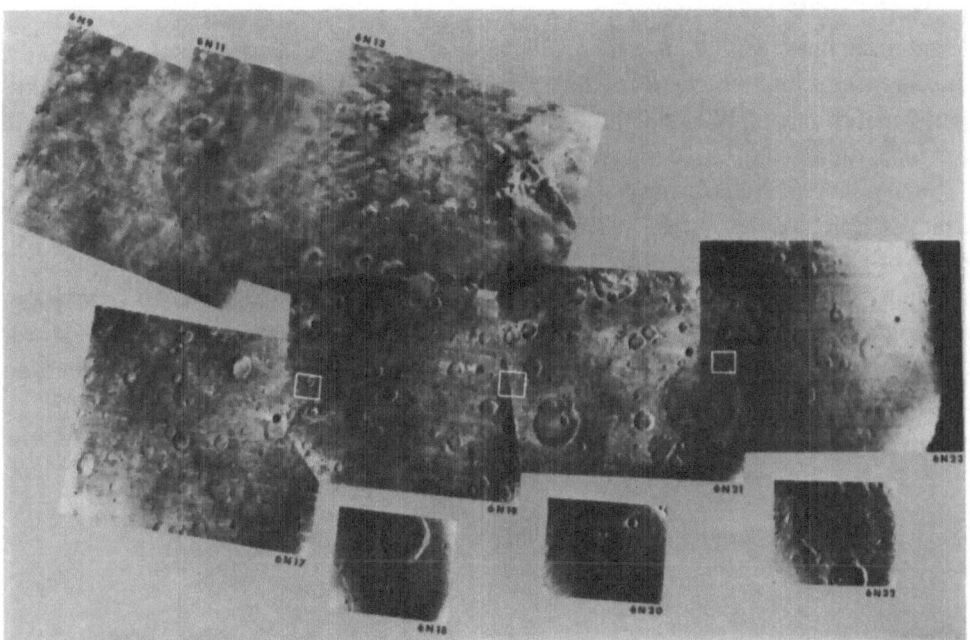

Fig. 6.   Composite of ten Mariner 6 pictures showing cratered terrain in the areas of Margaritifer Sinus (top left), Meridiani Sinus (top center), and Deucalionis Regio (lower strip). Large-scale contrasts are suppressed by AGC and small-scale contrast is enhanced (see text). Craters are clearly visible in blue frames 6N9 and 6N17, but albedo variations are subdued. Locations of three camera-B frames are marked by rectangles. North is approximately toward the top, and the sunset terminator lies near the right edge of 6N23.

Polar cap frame 7N17 shows sharp surface detail very near the terminator. The blue limb frame 6N1 shows surface detail corresponding to that seen in the subsequent overlapping green frame 6N3. Figure 6 includes blue, green, and red pictures in the region of Sinus Meridiani. Although craters show clearly in all three colors, albedo variations, associated both with craters and with larger-scale features, are much more pronounced in green and red than in blue. Blue photographs obtained from the earth during the Mariner encounters show the normal 'obscured' appearance of Mars.

C. SOUTH POLAR CAP SHADING

Another possible indication of atmospheric haze is the remarkable darkening of the south polar cap near both limb and terminator in the FE pictures (Figure 7). This darkening is plainly *not* due to cloud or thick haze since, during near-encounter, surface features are clearly visible everywhere over the polar cap. It may be related to darkening seen in NE Mariner 7 frames near the polar cap terminator, and to the decrease in contrast with increasing viewing angle between the cap and the adjacent mare seen in frame 7N11 (Figure 8b). The darkening may be due to optically thin aerosol scattering over the polar cap, or possibly to unusual photometric behavior of the cap itself. In either case, it may be complicated by systematic diurnal or latitudinal effects.

19—P.A.

D. NORTH POLAR PHENOMENA

Marked changes seem to have occurred, between the flybys of Mariners 6 and 7, in the appearance of high northern latitudes. Some of these changes are revealed by a comparison of frames 6F34 and 7F73, which correspond to approximately the same central meridian and distance from Mars (Figure 7). A large bright tongue (point 1 in frame 73) and a larger bright region near the limb (point 2) appear smaller and fainter in the Mariner 7 picture, despite the generally higher contrast of Mariner 7 FE frames. Much of the brightening near point 2 has disappeared entirely between the two flybys; in fact, it was not visible at all on pictures taken by Mariner 7 during the previous Mars rotation, although it was clearly visible in several Mariner 6 frames taken over the same range of distances. The bright tongue (point 1) increases in size and brightness during the martian day, as may be clearly seen from a comparison of frames 7F73 and 7F76 (Figure 7).

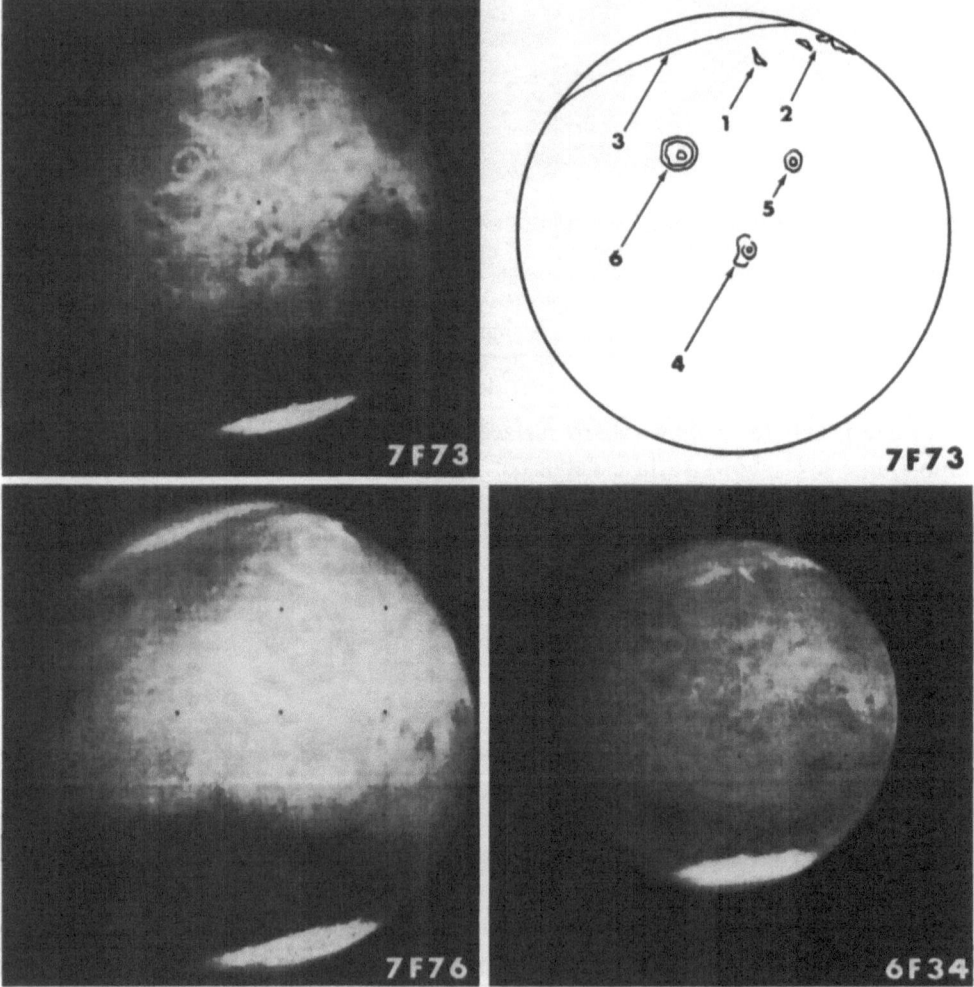

Fig. 7. Far-encounter pictures showing atmosphere and atmosphere-surface effects. Picture shutter times were as follows: 6F34, 30 July 07 23 UT; 7F73, 4 August 11 15 UT; 7F76, 4 August 13 36 UT.

The widespread, diffuse brightening covering much of the north polar cap region (point 3) apparently corresponds to the 'polar hood' which has been observed from the earth at this martian season (northern early autumn). The extent of this hood is smaller in Mariner 7 than in Mariner 6 pictures; the region between, and just north of, points 1 and 2 appears to be covered by the hood in the Mariner 6 frames, but shows no brightening in the Mariner 7 frames.

The different behaviors of the discrete bright regions and the hood suggest different origins for these features, although both apparently are either atmospheric phenomena or else result from the interaction of the atmosphere and the surface. The discrete bright regions have fixed locations suggesting either surface frost or orographically fixed clouds. The fluctuation in the areal extent of the diffuse hood suggests cloud or haze. An extensive cloud or haze composed of either CO_2 or CO_2 and H_2O ice would be consistent with the atmospheric temperature structure revealed by the Mariner 6 occultation experiment [9].

E. DIURNAL BRIGHTENING

Other variable bright features which may be indicative of atmospheric processes appear throughout the Tharsis, Candor, Tractus Albus, and Nix Olympica regions (Figure 7, frame 7F73; see also [3]). The brightness of these areas is observed to develop during the forenoon and increase during the martian afternoon, both in the Mariner pictures and in photographs taken from the earth. The structure and locations in the FE pictures do not change over the 6-day time span during which the region was observed. Particularly striking are several long light streaks near Nix Olympica (point 6 in frame 7F73) and numerous circular features resembling craters, which have bright centers and dark edges. Several of the circular features exhibit one or more concentric circles similar to, but less striking than, those near Nix Olympica. Two features of this type form two of the westernmost points of the classical 'W-cloud' (points 4 and 5). No morphology associated with clouds – such as waves, billows, or cirriform streaks – appear in this region, although at the 30-km resolution of these pictures such features would be visible in terrestrial clouds.

F. SEARCH FOR LOCAL CLOUDS AND FOG

All NE frames from both spacecraft were carefully examined for evidences of clouds or fog. Away from the south polar cap there are no evidences of such atmospheric phenomena. Over the polar cap and near its edge a number of bright features which may be atmospheric can be seen, although no detectable shadows are present and no local differences in height can be detected by stereoscopic viewing of overlapping regions whose stereo angles lie between 5° and 12°. Little or no illumination is evident near and beyond the polar cap terminator. On the other hand, frames 7N11, 12, and 13 (Figure 8) show several diffuse bright patches suggestive of clouds near the polar cap edge. Also, on the cap itself a few local diffuse bright patches are present in frames 7N15 (green) and 7N17 (blue). Unlike most polar cap craters, which appear sharp and clear, a few crater rims and other topographic forms appear diffuse (frames 7N17, 18, and 19). In frames 7N17 (blue) and 7N19 (green), remarkable curved, quasi-parallel

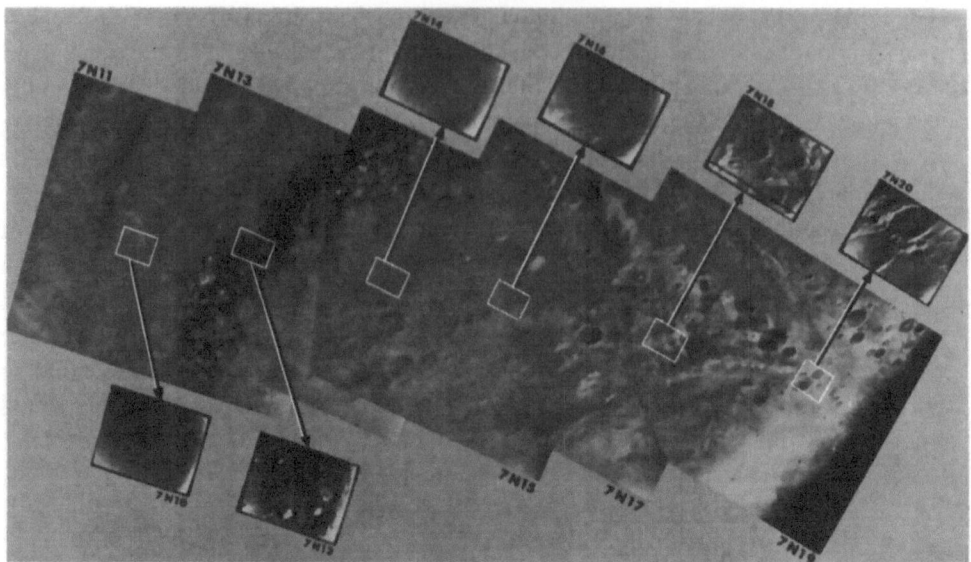

Fig. 8a.   Composite of polar cap frames 7N10 to 7N20. Effects of AGC are clearly evident near
the terminator (right) and at cap edge.

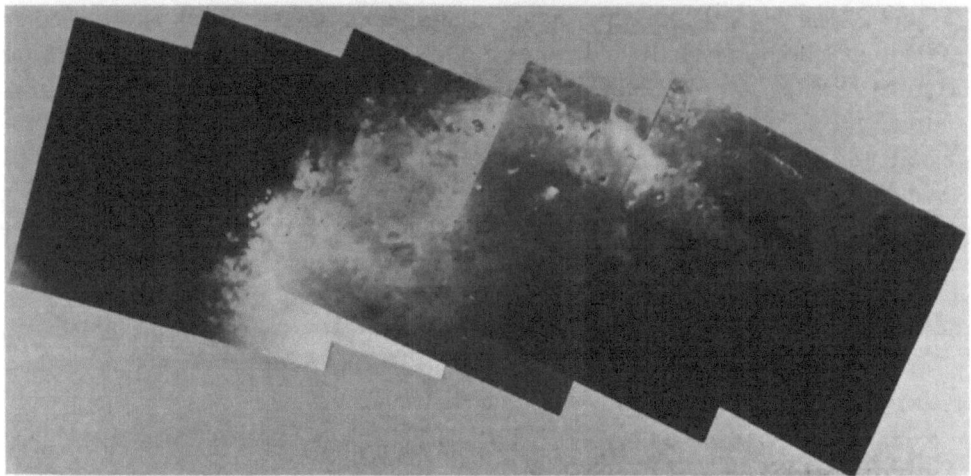

Fig. 8b.   Composite of polar cap camera-A frames 7N11 to 7N19. The effects of AGC have been
partially corrected, but contrast is enhanced. The south pole lies near the parallel streaks in the
lower right corner of frame 7N17.

bright streaks are visible near the south pole itself. While these show indications of
topographic form or control, including some crater-like shapes, their possible cloud-
like nature is suggested by lack of shading. Also frames 7N17 and 7N19 show faint
but definite streaks and mottles very near the terminator, superimposed on a sparsely
cratered surface.

5. Observed Surface Features

A primary objective of the Mariner 6 and 7 television experiment was to examine, at
close range, the principal types of martian surface features seen from the earth.

Mariners 6 and 7, while confirming the earlier evidence of a Moon-like cratered appearance for much of the martian surface, have also revealed significantly different terrains suggestive of more active, and more recent, surface processes than were previously evident. Preliminary analyses indicate that at least three distinctive terrains are represented in the pictures, as well as a mixture of permanent and transitory surface features displayed at the edge of, and within, the south polar cap; these terrains do not exhibit any simple correlation with the light and dark markings observed from the earth.

A. CRATERED TERRAINS

Cratered terrains are those parts of the martian surface upon which craters are the dominant topographic form (Figure 6). Pictures from Mariners 4, 6, and 7 all suggest that cratered terrains are widespread in the southern hemisphere.

Knowledge of cratered terrains in the northern hemisphere is less complete. Cratered areas appear in some Mariner frames as far north as latitude 20°. Nix Olympica, which in far-encounter photographs appears to be an unusually large crater, lies at 18°N. Numerous craters are visible in the closer-range FE frames. These are almost exclusively seen in the dark areas lying in the southern hemisphere, few being visible in the northern hemisphere. This difference may result from an enhancement of crater visibility by reflectivity variations in dark areas. However, poor photographic coverage, highly oblique views, and unfavorable sun angles combine to limit our knowledge of the northern portion of the planet.

Preliminary measurements of the diameter-frequency distribution of martian craters in the region Deucalionis Regio were made on frames 6N19 to 6N22 and are shown in Figure 9a. The curves are based upon 104 craters more than 0.7 km in diameter seen on frames 6N20 and 22, and upon 256 craters more than 7 km in diameter seen on frames 6N19 and 21. The most significant result is the existence of two different crater distributions, a dichotomy also apparent in morphology. The two morphological crater types are (i) large and flat-bottomed and (ii) small and bowl-shaped. Flat-bottomed craters and most evident on frames 6N19 and 6N21. The diameters range from a few kilometers to a few hundred kilometers, with estimated diameter-to-depth ratios on the order of 100 to 1. The smaller, bowl-shaped craters are best observed in frames 6N20 and 6N22 and resemble lunar primary-impact craters. Some of them appear to have interior slopes steeper than 20 degrees.

The diameter-frequency distribution of large flat-bottomed craters is compared in Figure 9b with the distribution of craters in the uplands on the far side of the moon near Tsiolkovsky. This particular lunar region was chosen for comparison because it is evidently a primordial surface, devoid of large post-upland features which might have modified the original crater distribution. The distribution of the small bowl-shaped craters is compared in Figure 9c with the distribution of craters on the lunar maria [10]. The distribution curve for small martian craters larger than about 1 km in diameter has a slope of about −2, which is similar to the curve for primary craters larger than 3 km in diameter on the lunar maria.

There are large variations in crater morphology among different cratered terrains.

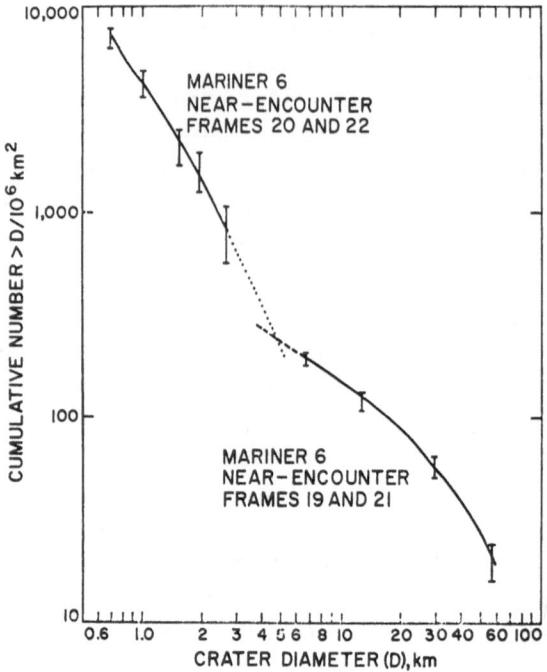

Fig. 9a.   Preliminary cumulative distribution of crater diameters. Solid curve at right is based upon 256 counted craters in frames 6N19 and 6N21 having diameters ≥ 7 km. The solid curve at left is based upon 104 counted craters in frames 6N20 and 6N22. The error bars are from counting statistics only ($N^{1/2}$).

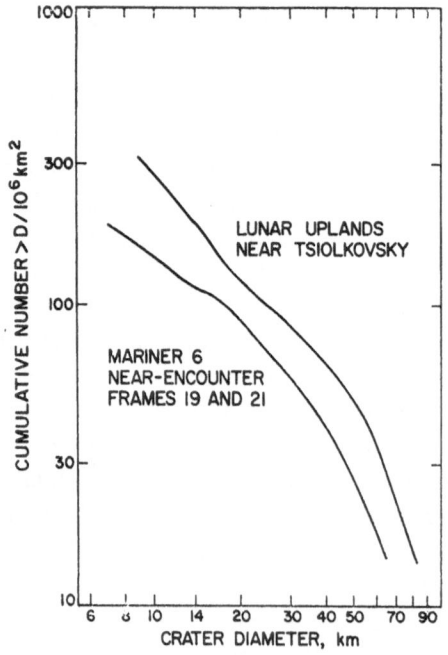

Fig. 9b.   Comparison of size distribution of large craters on Mars and on the lunar uplands.

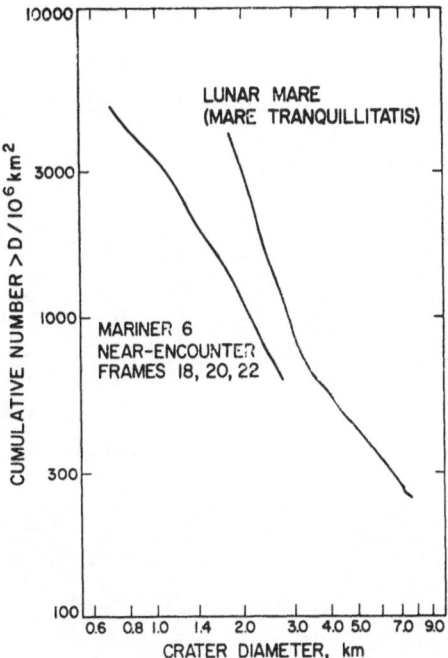

Fig. 9c.   Comparison of size distribution of small craters on Mars and on a lunar mare.

The craters of the dark area Meridiani Sinus (Figure 6) have more marked polygonal outlines and more central peaks than craters in some other martian areas; the especially distinctive lighter marking in the northwest portions of the crater floors is also, but less clearly, seen in some FE frames. Many of the primary and secondary features associated with large lunar impact craters can be found on the martian terrain (see, for example, frame 6N18); however, certain others, such as rays and secondary crater swarms, appear to be absent. Also, ejecta blankets appear to be much less well developed. The missing features are generally those most easily removed or hidden by erosion or blanketing – a pattern consistent with the observation that the martian craters are generally shallower and more smooth than lunar ones.

On frame 6N20 there are low irregular ridges similar to those seen on the lunar maria. However, no straight or sinuous rills have been identified with confidence. Similarly, no earth-like tectonic forms possibly associated with mountain building, island-arc formation, or compressional deformation have been recognized.

B. CHAOTIC TERRAINS

Mariner frames 6N6, 14, and 8 (Figure 10a) show two types of terrain – a relatively smooth cratered surface that gives way abruptly to irregularly shaped, apparently lower areas of chaotically jumbled ridges. This chaotic terrain seems characteristically to display higher albedo than its surroundings. On that basis, we infer that significant parts of the overlapping frames 6N5, 7, and 15 may contain similar terrain, although their resolution is not great enough to reveal the general morphological characteristics.

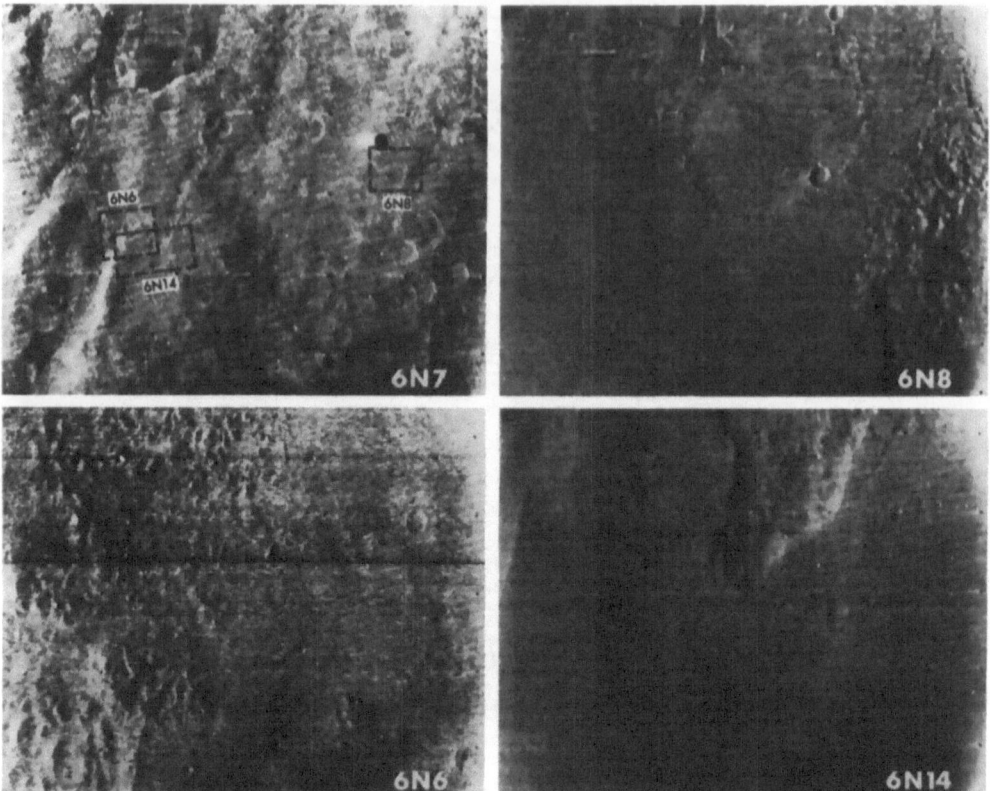

Fig. 10.

As shown in Figure 10a, frames 6N6, 14, and 8 all lie within frame 6N7, for which an interpretive map of possible chaotic terrain extent has been prepared (Figure 11).

About 10^6 km^2 of chaotic terrain may lie within the strip, 1000 km wide and 2000 km long, covered by these Mariner 6 wide-angle frames. Frames 6N9 and 10 contain faint suggestions of similar features. This belt lies at about 20°S, principally within the poorly defined, mixed light-and-dark area between the dark areas Aurorae Sinus and Margaritifer Sinus.

Chaotic terrain consists of a highly irregular plexus of short ridges and depressions, 1 to 3 km wide and 2 to 10 km long, best seen in frame 6N6 (Figure 10a). Although irregularly jumbled, this terrain is different in setting and pattern from crater ejecta sheets. Chaotic terrain is practically uncratered; only three faint possible craters are recognized in the 10^6-km^2 area. The patches of chaotic terrain are not all integrated, but they constitute an irregular pattern with an apparent N to N 30°E grain.

C. FEATURELESS TERRAINS

The floor of the bright circular 'desert', Hellas, centered at about 40°S, is the largest area of featureless terrain so far identified. Even under very low solar illumination the area appears devoid of craters down to the resolution limit of about 300 m. No area of comparable size and smoothness is known on the moon. It may be that all bright

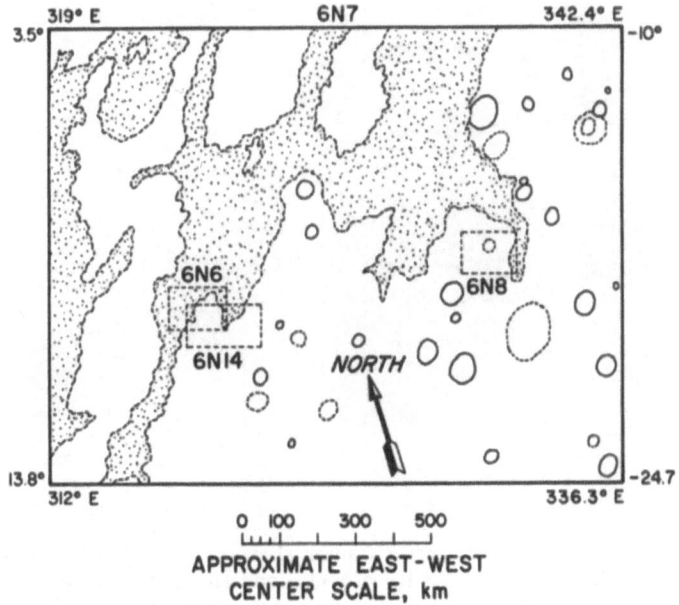

Fig. 11. Interpretive drawing showing the possible extent of chaotic terrain in frame 6N7.

circular 'deserts' of Mars have smooth featureless floors; however, in the present state of our knowledge it is not possible to define any significant geographic relationship for featureless terrains.

The Mariner 7 traverse shows that the dark area Hellespontus, lying west of Hellas, is heavily cratered. The 130- to 350-km-wide transitional zone is also well cratered and appears to slope gently downward to Hellas, interrupted by short, en echelon scarps and ridges (Figure 12). It gives way abruptly along an irregular foot to the flat floor of Hellas. Craters are observed within the transitional zone but abruptly become obscured within the first 200 km toward the center of Hellas.

The possibility that a low haze or fog may be obscuring the surface of Hellas, and that the featureless images are therefore not relevant to the true surface, has been considered. However, in frame 7N26 the ridges of the Hellas-Hellespontus boundary are clearly visible, proving that the surface is seen; yet there are virtually no craters within that frame. Thus the absence of well-defined craters appears to be a real effect.

D. SOUTH POLAR CAP FEATURES

The edge of the martian south polar cap was visible at close range over a 90° span of longitude, from 290°E to 20°E, and the cap itself was seen over a latitude range from its edge, at −60°, southward to, and perhaps beyond, the pole itself. Solar zenith angles ranged from 51° to 90° and more; the terminator is clearly visible in one picture. The phase angle for the picture centers was 35°. The superficial appearance is that of a clearly visible, moderately cratered surface covered with a varying thickness of 'snow'. The viewing angle and the unfamiliar surface conditions make quantitative comparison

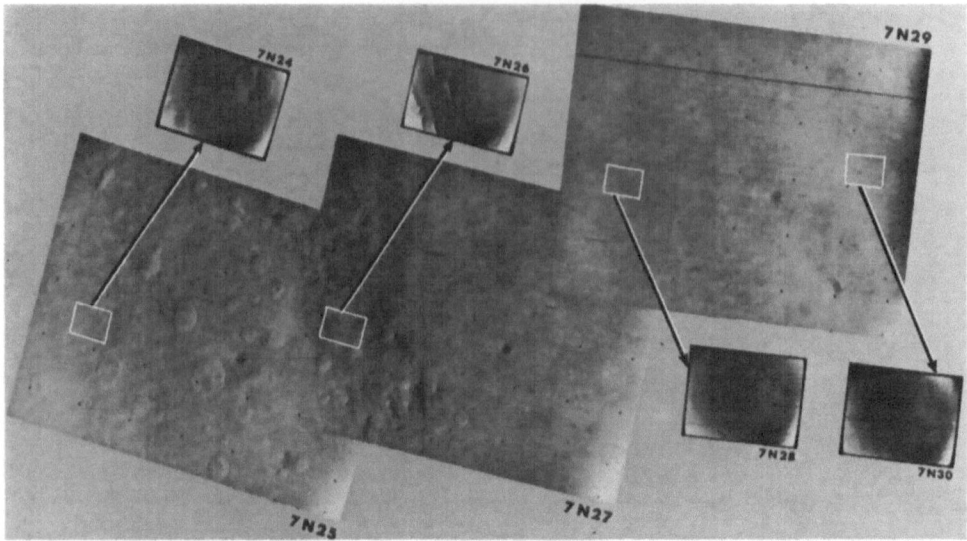

Fig. 12.   Composite of seven Mariner 7 frames, showing the cratered dark area Hellespontus, the ridged, broken boundary between Hellespontus and Hellas, and the featureless terrain of the bright, circular 'desert' Hellas. Large-scale variations in contrast are suppressed by AGC. Lighting conditions are similar to those of Figure 6, frames 7N17 to 7N21. North is approximately toward the top.

with other areas of Mars difficult with respect to the number and size distributions of craters. Discussion here is therefore confined to those qualitative aspects of the polar cap which seem distinctive to that region.

The edge of the cap was observed in the FE pictures to be very nearly at 60°S, as predicted from Lowell Observatory measurements [11]; this lends confidence to earth-based observations concerning the past behavior of the polar caps.

The principal effect seen at the cap edge is a spectacular enhancement of crater visibility and the subtle appearance of other topographic forms. In frames 7N11 to 7N13, where the local solar zenith angle was about 53°, craters are visible both on and off the cap. However, in the transition zone, about 2 degrees of latitude in width, the population density of visible craters is several times greater, and may equal any so far seen on Mars. This enhancement of crater visibility results mostly from the tendency, noted in Mariner 4 pictures 14 and 15, for snow to lie preferentially on poleward-facing slopes.

In frame 7N12 the cap edge is seen in finer detail. The tendency mentioned above is here so marked as to cause confusion concerning the direction of the illumination. There are several tiny craters as small as 0.7 km in diameter, and areas of fine mottling and sinuous lineations are seen near the larger craters. The largest crater shows interesting grooved structure, near its center and on its west inner wall, which appears similar to that in frame 6N18.

On the cap itself, the wide-angle views show many distinct reflectivity variations, mostly related to moderately large craters but not necessarily resulting from slope-illumination effects. Often a crater appears to have a darkened floor and a bright rim, and in some craters having central peaks the peaks seem unusually prominent. In frames 7N17 and 7N19 several large craters seem to have quite dark floors.

In contrast, the high-resolution polar cap frames 7N14 to 7N20 suggest a more uniformly coated surface whose brightness variations are mostly due to the effects of illumination upon local relief. Several craters are visible in each frame, some quite small; but, unlike most other areas of Mars, regions of positive relief are also visible, notably in frames 7N14 and 7N16. Also distinctive in these frames are areas of fine, irregular, quasi-parallel bright boundaries and irregular, shallow depressed regions. In frame 7N14, three such regions lie on the floor of a crater. Other irregular, shallow depressions apparently unrelated to craters appear in frames 7N15 and 7N17. Some of these are tens of kilometers in diameter and have no known counterparts elsewhere in the picture series of either spacecraft.

Finally, in frame 7N19 a curved, scalloped escarpment is seen, forming a boundary between a well-cratered area on its convex side and a relatively crater-free area on its concave side; this feature suggests the large circular structures associated with the mare basins on the moon.

E. RELATIONSHIP OF TERRAIN TO LIGHT AND DARK MARKINGS

The contrast of light and dark markings on Mars varies with wavelength, a fact long known from telescopic photography. In violet light, 'bright' and 'dark' areas are essentially indistinguishable, as they have approximately the same reflectivity. With increasing wavelength, contrast is enhanced as redder areas become relatively brighter. Frame 6N13 (Figure 6) shows craters in a dark area that have partially bright rims and floors, while craters in bright areas have rim and floor reflectivities similar to the reflectivity of the surroundings. These differences tend to increase the visibility of craters in dark areas, but only in photographs taken in red or green light. West of Meridiani Sinus (frames 6N9 and 6N11), and in some other parts of the planet, there are a number of dark-floored craters within bright areas, which, although otherwise relatively conspicuous, are difficult to see in blue light. The floors of these craters thus exhibit the same general dependence of reflectivity on wavelength that the larger dark areas do.

The distinction between bright and dark areas on the martian surface is generally more obvious in FE than in NE views. At higher resolution, the boundaries tend to become dispersed and indistinct. Exceptions are the sharp northeast and southern boundaries of Meridiani Sinus (frame 6N13) and the east edge of Hellespontus. The clearest structural relationship seen between a dark and a bright area is that of Hellespontus and Hellas (Figure 12).

Chaotic terrain appears to be lower and to have a somewhat higher reflectivity than adjacent cratered areas. Whether chaotic terrain is extensive enough to comprise any previously identified bright areas remains to be determined.

Some of the classical 'oases' observed from the earth have now been identified with single, large, dark-floored craters (such as Juventae Fons, see [4] and Figure 4) or groups of such craters (such as Oxia Palus, frame 7N5). At least two classical 'canals' (Cantabras and Gehon) have been found to coincide with quasi-linear

alignment of several dark-floored craters, shown also in frame 7N5 (Figure 13). As
reported elsewhere [4], other canals are composed of irregular dark patches. It is
probable that most canals will, upon closer inspection, prove to be associated with a
variety of physiographic features, and that eventually they will be considered less
distinctive as a class.

Some early drawings and 'maps' of Mars show a circular bright area within the dark
area south of Syrtis Major and east of Sabaeus Sinus, very nearly in the place occupied
by a large crater ([4] Figure 3; [12]). Further comparison of the Mariner pictures with
early maps and photographs may prove fruitful in revealing long-term aspects of
topographic associations of dark-area boundaries.

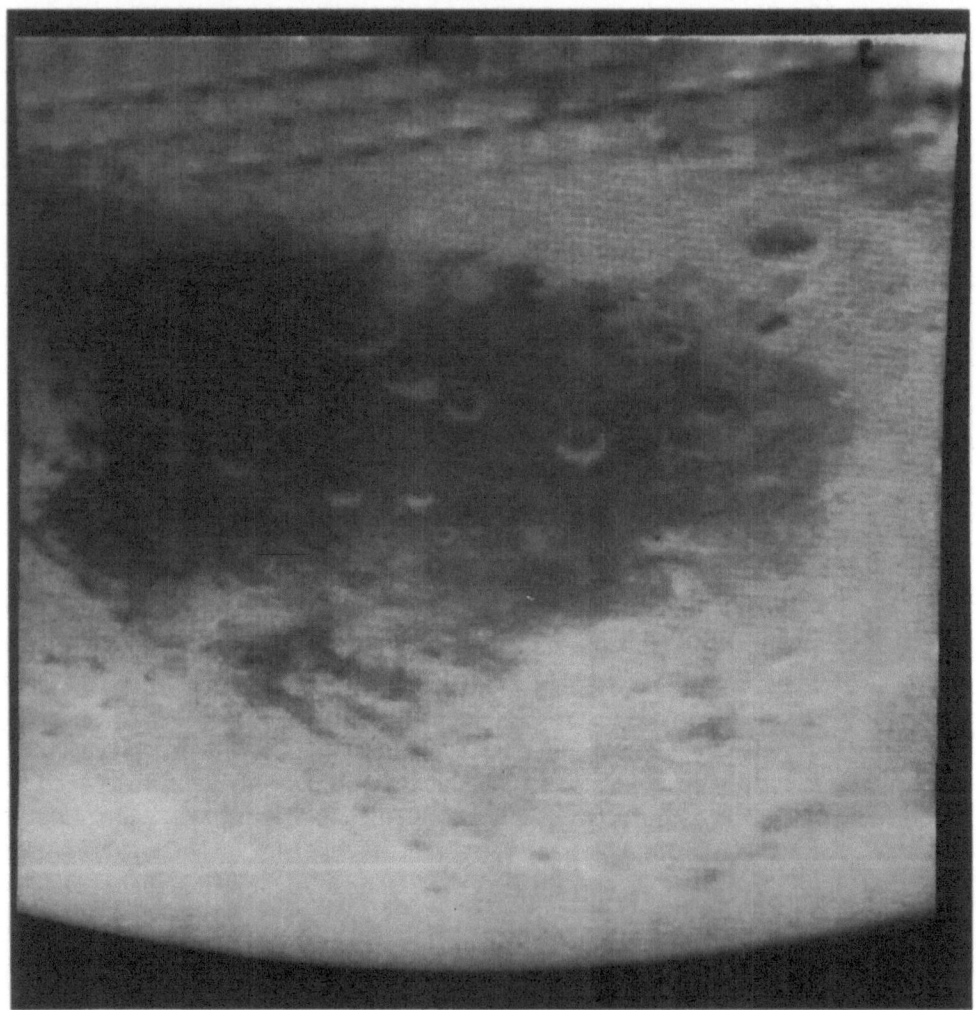

Fig. 13.   Partially reconstructed frame 7N5, showing Meridiani Sinus in the foreground. The
'oasis' Oxia Palus projects into the picture from the left edge, near the limb. Note the asymmetric
shading in several craters in Meridiani Sinus; the isolated, dark-floored craters in surrounding
bright areas; and the crater complex in Oxia Palus. The view is approximately toward N 20°W. The
banded appearance of the sky area is an artifact of the picture processing.

It is possible to make a few rough comparisons between the Mariner 6 and 7 pictures and estimates of Mars topographic elevations based on Mariner occultations [9], earth-based radar measurements [13], or CO_2 equivalent-width spectral measurements [14].

The long stretch of cratered terrain in Deucalionis Regio (Figure 6) is located on what earth-based radar and CO_2 measurements both suggest to be a very gradual slope rising westward. Here, at least, cratered terrain is not restricted to regionally high or to regionally low areas; like dark and light areas, it may not exhibit any particular correlation with planetary-scale relief.

The chaotic terrain viewed in Mariner 6 NE frames occurs in what appears to be the topographically lowest area surveyed by the spacecraft, if earth-based CO_2 measurements are correct. Thus, it is at least possible that chaotic terrain is related in some consistent way to planetary-scale relief.

Mariner 7's occultation point occurred near Hellespontica Depressio, near latitude 55°S, longitude 20°E. A very low surface pressure was measured there, suggesting highly elevated terrain. If Hellespontica Depressio is elevated, Hellespontus may be also. This relationship would be consistent with the impression gained from the Mariner pictures that the floor of Hellas, with its featureless terrain, is a local basin rimmed by higher areas.

6. Inferences concerning Processes and Surface History

The features observed in the Mariner 6 and 7 pictures are the result of both present and past processes; therefore, they provide the basis of at least limited conjecture about those processes and their variations through time. In this section we consider the implications of (i) the absence of earth-like tectonic features; (ii) the erosion, blanketing, and secondary modification evidenced in the three principal terrains; and (iii) the probable role of equilibrium between CO_2 solid and vapor in the formation of features of the south polar cap. We also consider the possible role of equilibrium between H_2O solid and vapor as an explanation of the diurnal brightenings observed in the FE photographs and biological implications.

A. SIGNIFICANCE OF THE ABSENCE OF EARTH-LIKE FORMS

The absence of earth-like tectonic features on Mars indicates that, for the time period represented by the present large martian topographic forms, the crust of Mars has not been subjected to the kinds of internal forces that have modified, and continue to modify, the surface of the earth.

Inasmuch as the larger craters probably have survived from a very early time in the planet's history, it is inferred that Mars's interior is, and probably has always been, much less active than the earth's [15]. Furthermore, a currently held view [16] is that the earth's dense, aqueous atmosphere may have formed early, in a singular event associated with planetary differentiation and the origin of the core. To the extent, therefore, that surface tectonic features may be related in origin to the formation of a dense atmosphere, their absence on Mars independently suggests that Mars never had an earth-like atmosphere.

B. AGE IMPLICATIONS OF CRATERED TERRAINS

At present, the ages of martian topographic forms can be discussed only by comparison with the moon. Both the moon and Mars exhibit heavily cratered and lightly cratered areas, which evidently reflect in each case regional differences in the history of, or the response to, meteoroidal bombardment over the total life-span of the surfaces. The existence of a thin atmosphere on Mars may have produced recognizable secondary effects in the form and size distribution of craters, by contrast with the moon, where a significant atmosphere has presumably never been present. To the extent that relative fluxes of large objects impinging upon the two bodies can be determined, or a common episodic history established, a valid age comparison may be hoped for, except in the extreme case of a saturated cratered surface, where only a lower limit to an age can be found.

It is a generally accepted view that the present crater density on the lunar uplands could not have been produced within the 4.5-billion-year age of the solar system had the bombardment rate been no greater than the estimated present rate; that is, the inferred minimum age is already much greater than is considered possible. Indeed, it is found that even the sparsely cratered lunar maria would have required about a billion years to attain their present crater density. Unless this discrepancy is somehow removed by direct measurements of the crystallization ages of returned samples of lunar upland and mare materials, the previously accepted implication of an early era of high bombardment followed by a long period of bombardment at a drastically reduced rate will presumably stand.

In the case of Mars, a bombardment rate per unit area as much as 25 times that on the moon has been estimated [17]. However, even this would still seem to require at least several billion years to produce the density of large craters that is seen on Mars in the more heavily cratered areas [15]. Thus these areas *could also be primordial.* Further, were these areas to have actually been bombarded at a constant rate for such a time, at least a few very recent, large craters should be visible, including secondary craters and other local effects. Instead, the most heavily cratered areas seem relatively uniform with respect to the degree of preservation of large craters, with no martian Tycho or Copernicus standing out from the rest. This again suggests an early episodic history rather than a continuous history for cratered martian terrain, and increases the likelihood that cratered terrain is primordial.

If areas of primordial terrain do exist on Mars, an important conclusion follows: these areas have never been subject to erosion by water. This in turn reduces the likelihood that a dense, earth-like atmosphere and large, open bodies of water were ever present on the planet, because these would almost surely have produced high rates of planet-wide erosion. On the earth, no topographic form survives as long as 10^8 yr unless it is renewed by uplift or other tectonic activity.

C. IMPLICATIONS OF MODIFICATION OF TERRAIN

Although erosional and blanketing processes on Mars have not been strong enough to obliterate large craters within the cratered terrains, their effects are easily seen. On frames 6N19 and 6N21 (Figure 6), even craters as larger as 20 to 50 km in diameter appear scarce by comparison with the lunar uplands (a feature originally noted by Hartmann [15] on the basis of the Mariner 4 data), and the scarcity of smaller craters is marked. The latter have a relatively fresh appearance, however, which suggests an episodic history of formation, modification, or both. Such a history seems particularly indicated by the apparently bimodal crater frequency distribution of Figure 9.

Marked erosion, blanketing, and other surface processes must have been operating almost up to the present in the areas of featureless and chaotic terrains; only this could account for the absence of even small craters there. These processes may not be the same as those at work on the cratered terrains, because large craters have also been erased. The cratered terrains obviously have *never been* affected by such processes; this indicates an enduring geographic dependence of these extraordinary surface processes.

The chaotic terrain gives a general impression of collapse structures, suggesting the possibility of large-scale withdrawal of substances from the underlying layers. The possibility of permafrost some kilometers thick, and of its localized withdrawal, may deserve further consideration. Magmatic withdrawal or other near-surface disturbance associated with regional volcanism might be another possibility, but the apparent absence of extensive volcanic terrains on the surface would seem to be a serious obstacle to such an interpretation. It may also be that chaotic terrain is the product either of some unknown intense and localized erosional process or of unsuspected local sensitivity to a widespread process.

D. CARBON DIOXIDE CONDENSATION EFFECTS

The Mariner 7 NE pictures of the polar cap give no direct information concerning the material or the thickness of the polar snow deposit, since the observed brightness could be produced by a very few milligrams per square centimeter of any white, powdery material. However, they do provide important indirect evidence as to the thickness of the deposit and, together with other known factors, may help to establish its composition.

The relatively normal appearance of craters on the polar cap in the high-resolution frames, and the existence on these same frames of topographic relief unlike that so far recognized elsewhere on the planet, suggest that some of the apparent relief may be due to variable thicknesses of snow, perhaps drifted by wind. If it is, local thicknesses of at least several meters are indicated.

The structure of the polar cap edge shows that evaporation of the snow is strongly influenced by local slopes – that is, by insolation effects rather than by wind. On the assumption that the evaporation is entirely determined by the midday radiation balance, when the absorbed solar power exceeds the radiation loss at the appropriate frost-point temperature, one may estimate the daily evaporation loss from the cap. We find the net daily loss to be about 0.8 g/cm^2 in the case of CO_2, although the loss is reduced by overnight recondensation. In the case of H_2O, the loss would be about 0.08 gram per square centimeter, and it would be essentially irreversible because H_2O is a minor constituent whose deposition is limited by diffusion.

Since the complete evaporation of the cap at a given latitude requires many days, we may multiply the above rates by a factor between 10 and 100, obtaining estimates for total cap thickness of tens of grams per square centimeter for CO_2 and several grams per square centimeter for H_2O, on the assumption that the cap is composed of one or the other of these materials. The estimate for CO_2 is quite acceptable, but that for H_2O is unacceptable because of the problem of transporting such quantities annually from one pole to the other at the observed vapor density [18]. For the remainder of this discussion we assume the polar cap to be composed of CO_2, with a few milligrams of H_2O per square centimeter deposited throughout the layer.

Several formations have been observed which suggest a tendency for snow to be preferentially removed from low areas and deposited on high areas, contrary to what might be expected under quiescent conditions [19]. These formations include craters with dark floors and bright rims, prominent central peaks in some craters, and irregular depressed areas (frames 7N14, 15, and 17). While such effects might result simply from wind transport of solid material, it is also possible that interchange of solid and vapor plays a role.

The adiabatic lapse rate of the polar atmosphere is about 6 K/km, as compared to a frost-point lapse rate of about 1 K/km. Therefore, adiabatic heating or cooling of martian air blowing across sloping terrain may result in evaporation of low-lying solid and precipitation over high areas. The rate is determined by atmospheric density, latent heat, difference in elevation, lateral scale of relief, wind speed, turbulent-layer thickness, interfacial heat transfer, and net adiabatic gradient. For example, a wind of 20 m/sec blowing across a crater 20 km in diameter and 0.5 km deep, with a turbulent layer only 100 m thick, could differentially deposit about 0.1 g/cm^2 per day. Since this process would be effective during the entire time that snow is on the ground, several tens of grams per square centimeter could become systematically redistributed in the course of a few hundred days. A speculative possibility is that such a process, or perhaps merely a sustained accumulation of snow in a particular topographic 'trap', occasionally results in sufficient accumulation during the winter to survive summer evaporation. The increased albedo might then lead to a permanent accumulation of CO_2 snow within that topographic feature – that is, to formation of a martian 'ice field'. The very prominent snow-covered central peaks of some of the small craters located at high latitude might conceivably be the sites of such permanent deposits of solid CO_2. The permanent part of the north polar cap is presumably such a structure [18].

E. WATER: PROCESSES SUGGESTED BY BRIGHTENING PHENOMENA

Several of the brightening and haze phenomena described above could be related either to formation of H_2O frost on the surface or to formation of H_2O ice clouds in the atmosphere. In most of these instances, however, the phenomena could equally well be explained by condensation of CO_2. This is true of the bright tongues and polar hood in the north polar region, of the cloud-like features observed over and near the south polar cap, and of the limb hazes observed in tropical latitudes and over the Mare Hadriaticum and Ausonia regions.

On the other hand, the brightenings in the Nix Olympica, Tharsis, Candor, and

Tractus Albus regions cannot be explained by CO_2 condensation because their complete topographic control requires that they be on or near the surfaces where temperatures are well above the CO_2 frost point. An explanation of these phenomena in terms of H_2O condensation processes also faces serious difficulties, however. Most of the region is observed to brighten during the forenoon, when the surface is hotter than either the material below or the atmosphere above, so that water vapor could not diffuse toward the surface and condense on it, either from above or below. Thus a surface ice-frost is very unlikely. A few features in the area, parts of the 'W-cloud', for example, are observed to brighten markedly during the late afternoon, where H_2O frost could form on the surface if the air were sufficiently saturated. These features are not observed, from the earth, to be bright in the early morning, but a thin layer of H_2O frost persisting through the night would evaporate almost immediately when illuminated by the early morning sun, provided the air were then sufficiently dry. Under these conditions, the behavior of the 'W-cloud' could be due to frost.

The diurnal behavior of the bright regions throughout this part of Mars is consistent with a theory of convective H_2O ice clouds, but the absence of any cloud-like morphology and the clear topographic detail observed at the highest resolution available (frame 7F76) render this explanation questionable. Even very light winds of 5 m/sec would produce easily observable displacements of the order of 100 km in the course of the more than one-fourth of the Mars day during which these regions were continuously observed by each spacecraft. Since condensation and evaporation processes are slow at Mars temperatures and pressures, some observable distortion and streakiness due to these displacements should be seen in clouds, even if they are orographically produced. No such distortions or streakiness are observed.

An additional difficulty with an explanation of these phenomena in terms of H_2O condensation lies in the relatively rapid removal of water from the local surface. Water vapor evolved from the surface during the daytime would quickly be transported upward through a deep atmospheric layer by thermal convection, and most of it would be removed from the source region. Local permafrost sources should be effectively exhausted by this mechanism within a few hundred years at most, unless somehow replenished. Since most of this region lies near the equator, where seasonal temperature variations are small, it is difficult to see how any significant seasonal replenishment from the atmosphere could take place. The possibility of replenishment from a sub-surface source of liquid water is not considered here.

In summary, in our examination of the data thus far, we see no strong indications of H_2O processes involving vapor and ice. The brightenings seen in the tropics and sub-tropics at far-encounter are not easily explained by a mechanism involving H_2O. On the other hand, we have no satisfactory alternative explanation for these phenomena. Perhaps detailed exploration of these regions by the Mariner '71 orbiters will provide the answer.

F. BIOLOGICAL INFERENCES

No direct evidence suggesting the presence of life on Mars has been found in the pictures. This is not surprising, since martian life, if any, would probably be microbial

20—P.A.

and undetectable at a resolution of 300 m. Although inconclusive on the question of martian life, the photographs are informative on at least three subjects of biological interest: the general nature of the martian maria, the present availability of water, and the availability of water in the past.

One of the most surprising results so far of the TV experiment is that nothing in the pictures suggests that the dark areas, the sites of the seasonal darkening wave, are more favorable for life than other parts of the planet. On the contrary, it would now appear that the large-scale surface processes implied by the chaotic and featureless terrains may be of greater biological interest than the wave of darkening. We reiterate that these are preliminary conclusions; it may be that subtle physiographic differences between dark and bright regions will become evident when photometrically corrected pictures are examined.

With regard to the availability of water, the pictures so far have not revealed any evidence of geothermal areas. We would expect such areas to be permanently covered with clouds and frost, and these ought to be visible on the morning terminator; no such areas have been seen. A classically described feature of the polar cap which has been interpreted as wet ground – the dark collar – has likewise not been found. Other locales which have been considered to be sites of higher-than-average moisture content are those which show diurnal brightening. A number of such places have been observed in the pictures, but on close inspection the brightening appears not to be readily interpretable in terms of water frosts or clouds. Pending their definite identification, however, the brightenings should be considered possible indications of water.

The results thus reinforce the conclusion, drawn from Mariner 4 and ground-based observations, that scarcity of water is the most serious limiting factor for life on Mars. No terrestrial species known to us could live in the dry martian environment. If there is a permafrost layer near the surface, or if the small amount of atmospheric water vapor condenses as frost in favorable sites, it is conceivable that, by evolutionary adaptation, life as we know it could use this water and survive on the planet. In any case, the continued search for regions of water condensation on Mars will be an important task for the 1971 orbiter.

The past history of water on Mars is a matter of much biological interest. According to current views, the chemical reactions which led to the origin of life on the earth were initiated in the reducing atmosphere of the primitive earth. These reactions produced simple organic compounds which were precipitated into the ocean, where they underwent further reactions that eventually yielded living matter. The pictorial evidence raises the question of whether Mars ever had enough water to sustain an origin of life. If the proportion of water outgassed relative to CO_2 is the same for Mars as for the earth, then, from the mass of CO_2 now in the martian atmosphere, it can be estimated that Mars has produced sufficient water to cover the planet to a depth of a few meters. The question is whether anything approaching this quantity of water was ever present on Mars in the liquid state.

The existence of cratered terrains and the absence of earth-like tectonic forms on Mars clearly implies that the planet has not had oceans of terrestrial magnitude for a very long time, possibly never. However, we have only very rough ideas of how much ocean is required for an origin of life, and of how long such an ocean must last. An upper limit on the required time, based on terrestrial experience, can be derived from the age of the oldest fossils, $> 3.2 \times 10^9$ years [20]. Since these fossils are the remains of what were apparently highly evolved microorganisms, the origin of life must have taken place at a much earlier time, probably during the first few hundred million years of the earth's history. While one cannot rule out, on the basis of the TV data, the possibility that a comparably brief, aqueous epoch occurred during the early history of the planet, it must be said that the effect of the TV results so far is to diminish the a priori likelihood of finding life on Mars. However, it should be noted that if Mars is to be a testing ground for our notions about the origin of life, we must avoid using these same notions to disprove in advance the possibility of life on that planet.

7. Potentialities of the Data

Careful computer restoration of the pictures, starting with data recovered from six sequential playbacks of the near-encounter analog tapes, will be carried out over the next several months. This further processing will greatly enhance the completeness, appearance, and quantitative usefulness of the pictures. While it is not yet certain whether the desired 8-bit relative photometric accuracy can be attained, there are reasonable grounds for thinking that much new information bearing on the physiography, meteorology, geography, and other aspects of Mars will ultimately be obtained from the pictures. Some of the planned uses of the processed data are as follows.

A. STEREOSCOPY

Most of the NE wide-angle pictures contain regions of two-picture overlap, and a few contain regions of three-picture overlap. These areas can be viewed in stereoscopic vision in the conventional manner of aerial photography. Preliminary tests on pictures of the south polar cap (frames 7N17 and 7N19) indicate that measurement of crater depth, central-peak height, and crater-rim height is possible. However, accuracy can be estimated for the elevation determinations at this time.

B. PLANETARY RADII

Geometric correction of the FE photographs should make it possible to determine the radius of Mars as a function of latitude, and possibly of longitude. The geometric figure of Mars has been historically troublesome because of inconsistencies between the optical and the dynamical oblateness, a discrepancy amounting to some 18 km in the value for the difference of the equatorial and polar radii. It is possible that the darkening of the polar limb observed by Mariners 6 and 7, if it is a persistent phenomenon, might have systematically affected the earlier telescopic measurements of the polar diameter more than irradiation has, giving too large a value for the optical flattening. However, this cannot explain the large flattening obtained from surface-feature geodesy [21]. Although a fairly reliable figure for the polar flattening may be obtained from the Mariner data, it is unlikely that the actual radii will be determined with an accuracy greater than several kilometers because of the relatively low picture-element resolution in these frames and the difficulty in locating the limb.

C. CARTOGRAPHY

The large number of craters found on the surface of Mars makes it feasible to establish a control net which uses topographic features as control points, instead of surface markings based on albedo differences. This net should provide the basic locations for compiling a new series of Mars charts. The NE pictures, which cover 10 to 20% of the area of the planet, will constitute the basic material for detailed maps of these areas.

D. SATELLITES

We hope to detect the larger of Mars's satellites, Phobos, in two of the Mariner 6FE pictures taken when Phobos was just beyond the limb of the planet. The satellite should have moved between the two frames by about ten picture elements, and should appear as a 'defect' that has moved by this amount between the two pictures. If Phobos itself is not visible, its shadow (again detectable by its motion) should be. The shadow will be some five picture elements across and will have a photometric depth of about 10%. If the photometric depth of the shadow can be measured accurately, we can determine the projected area (and hence the diameter) of the satellite. A similar method has been used to measure the diameter of Mercury during solar transits.

E. PHOTOMETRIC STUDIES

We expect to derive the photometric function for each color, combining data from the two spacecraft. Observations by the current Mariners were made near 25°, 35°, 45°, and 80° phase. Since data obtained from the earth can be used to establish the absolute calibration at the smaller phase angles, we will also be able to relate the 80°-phase data to earth-based observations, thus doubling the range over which the phase function is determined. This information should then make possible the determination of crater slopes. Agreement for areas of overlap between different filters and between A-camera and B-camera frames can be used to check the validity of the results and possibly to measure and correct for atmospheric scattering.

The reciprocity principle may be useful in testing quantitatively for diurnal changes in the FE pictures. Such changes might include dissipation of frost or haze near the morning terminator and formation of afternoon clouds near the limb.

Overlap areas in NE pictures can be used to obtain approximate colors, even though these areas are seen at different phase angles in each color. In addition, color-difference or color-ratio pictures may be useful in identifying local areas of anomalous photometric or colorimetric behavior. Camera-A digital pictures obtained by Mariner 7 in late far-encounter will be very useful for making color measurements.

F. COMPARISON OF PICTURES WITH RADAR-SCATTERING AND HEIGHT DATA

The reflection coefficient of the martian surface for radar waves of decimeter wavelength shows marked variations at a given latitude as a function of longitude. Even though few of the areas of Mars so far observed by radar are visible at close range, some correlation of topography with radar reflectivity may become apparent upon careful study. Clearly, the Mariner pictures will become steadily more valuable in this connection as more radar results and other height data become available.

8. Effects on Mariner '71

The distinctive new terrains revealed in the Mariner 6 and 7 pictures, the relatively small fraction (10 to 20%) of the surface so far viewed even at moderate (A-camera) resolution, and the tantalizing new evidence of afternoon-brightening phenomena all emphasize the importance of an exploratory, adaptive strategy in 1971 as opposed to a routine mapping of geographic features. The fact that each of three successive Mariner spacecraft has revealed a new and unexpected topography strongly suggests that more surprises (perhaps the most important ones) are still to appear.

A primary objective should be to view nearly all of the visible surface at A-camera resolution (1-km pixel spacing), and to inspect selected typical areas at higher resolution, very early in the 90-day orbiting period. The true extent and character of cratered, chaotic, and featureless terrains, and of any new kinds of terrain, can thus be determined and correlated with classical light and dark areas, with regional height data, and so on.

A second objective should be to search for and examine, in both spatial and temporal detail, those areas which suggest the local presence of water, through the afternoon-brightening phenomena, morning frosts or fogs, or other behavior not now recognized. Certainly the known 'W-cloud' areas, Nix Olympica, and other, similar areas known from earth observation take on a new interest by virtue of the Mariner 6 and 7 results.

The complex structure found in the south polar cap calls for further examination, particularly with respect to separation of its more permanent features from diurnally or seasonally varying ones. The sublimation of the cap should be carefully followed, so as to detect evidence of variations in thickness of the deposit and especially evidence of the possible existence of permanent deposits. Study of the north polar cap at close range should also be exceedingly interesting.

9. Effects on Viking '73

If the effects of the Mariner 6 and 7 results on Mariner '71 are substantial, they at least do not require a change of instrumentation, only one of mission strategy. This may not be true of the effects on Viking '73. The discovery of so many new, unexpected properties of the martian surface and atmosphere adds a new dimension to the problem of selecting the most suitable landing site and may make Viking even more dependent on the success of Mariner '71 than has been supposed. Furthermore, since so much new information is revealed through the tenfold step in resolution afforded by the B-camera frames, a further substantial increase in resolution, not available to Mariner '71, may have to be incorporated in Viking in order to examine even more closely the fine-scale characteristics of various terrain types before a landing site is chosen.

10. Summary and Conclusions

Even in relatively unprocessed form, the Mariner 6 and 7 pictures provide fundamental new insights concerning the surface and atmosphere of Mars. Several unexpected results emphasize the importance of versatility in instrument design, flexibility in mission design, and use of an adaptive strategy in exploring planetary surfaces at high resolution.

The surface is clearly visible in all wavelengths used, including the blue. No blue-absorbing haze is found.

Thin, patchy, aerosol-scattering layers are present in the atmosphere at heights of from 15 to 40 km, at several latitudes.

Diurnal brightening in the 'W-cloud' area is seen repeatedly and is associated with specific topographic features. No fully satisfactory explanation for the effect is found.

Darkening of the polar cap in a band near the limb is clearly seen in FE pictures and is less distinctly visible in one or two NE frames. Localized, diffuse bright patches are seen in several places on and near the polar cap; these may be small, low clouds.

Widespread cratered terrain is seen, especially in dark areas of the southern hemisphere. Details of light-dark transitions are often related to local crater forms. Asymmetric markings are characteristic of craters in many dark areas; locally, these asymmetries often appear related, as if defined by a prevailing wind direction.

Two distinct populations of primary craters are present, distinguished on the basis of size, morphology, and age. An episodic surface history is indicated.

In addition to the cratered terrain anticipated from Mariner 4 results, at least two new, distinctive topographic forms are seen: chaotic terrains and featureless terrains. The cratered terrain is indicative of extreme age; the two new terrains both seem to require the present-day operation of especially active modifying processes in these areas. When seen at closer range, the very bright, streaked complex found in the Tharsis-Candor region may reveal yet another distinctive topographic character. Because of the afternoon-brightening phenomena long known here, this area provides a fascinating prospect for further exploration in 1971.

No tectonic and topographic forms similar to terrestrial forms are observed.

Evidences of both atmosphere-surface effects and topographic effects are seen on the south polar cap. At the cap edge, where the 'snow' is thinnest, strong control by solar heating, as affected by local slopes, is indicated. Crater visibility is greatly enhanced in this area.

On the cap itself, intensity variations suggestive of variable 'snow' thickness are seen. These may be caused by wind-drifting of the snow or by differential exchange of solid and vapor, or by both.

Snow thicknesses here of several grams or several tens of grams per square centimeter are inferred if the snow material is H_2O or CO_2, respectively. The possibility that the material is H_2O seems strongly ruled out on several grounds.

Variable atmospheric, and atmosphere-surface, effects are seen at high northern latitudes; these effects include the polar 'hood' and bright, diurnally variable circumpolar patches.

Several classical features have been successfully identified with specific topographic forms, mostly craters or crater remnants.

The findings are inconclusive on the question of life on Mars, but they are relevant in several ways. They support earlier evidence that scarcity of water, past and present, is a serious limiting factor for life on the planet. Nothing so far seen in the pictures suggests that the dark regions are more favorable for life than other parts of Mars.

Acknowledgments

We gratefully acknowledge the support and encouragement of the National Aeronautics and Space Administration. An undertaking as complex as that of Mariners 6 and 7 rests upon a broad base of facilities, technical staff, experience, and management, and requires not only money but much individual and team effort to be brought to a successful conclusion. It is impossible to know, much less to acknowledge, the important roles played by hundreds of individuals. We are deeply appreciative of the support and efforts of H. M. Schurmeier and the entire Mariner 1969 project staff. With respect to the television system, responsibility for the design, assembly, testing, calibration, flight operation, and picture data processing lay with the Jet Propulsion Laboratory. We gratefully acknowledge the contributions of G. M. Smith, D. G. Montgomery, M. C. Clary, L. A. Adams, F. P. Landauer, C. C. LaBaw, T. C. Rindfleisch, and J. A. Dunne in these areas. L. Malling, J. D. Allen, and R. K. Sloan made important early contributions. We are indebted to V. C. Clarke, C. E. Kohlhase, R. Miles, and E. Greenberg for their help in exploiting the flexibility of the spacecraft to achieve maximum return of pictorial data. We are especially appreciative of the broad and creative efforts of G. E. Danielson as Experiment Representative. The able collaborative contributions of J. C. Robinson in comparing Mariner pictures with earth-based photographs and of L. A. Soderblom and J. A. Cutts in measuring craters are gratefully acknowledged.

References

[1] Leighton, R. B., Murray, B. C., Sharp, R. P., Allen, J. D., and Sloan, R. K.: 1965, *Science* **149**, 627.

[2] Leighton, R. B., Murray, B. C., Sharp, R. P., Allen, J. D., and Sloan, R. K.: 1967, 'Mariner IV Pictures of Mars', *Tech. Rept. Jet. Propul. Lab., Calif. Inst. Technol. No. 32-884*, pt. 1.

[3] Leighton, R. B., Horowitz, N. H., Murray, B. C., Sharp, R. P., Herriman, A. G., Young, A. T., Smith, B. A., Davis, M. E., and Leovy, C. B.: 1969, *Science* **165**, 684.

[4] Leighton, R. B., Horowitz, N. H., Murray, B. C., Sharp, R. P., Herriman, A. G., Young, A. T., Smith, B. A., Davis, M. E., and Leovy, C. B.: 1969, *Science* **165**, 787.

[5] Montgomery, D. G.: in preparation.

[6] Danielson, G. E.: in preparation.

[7] Slipher, E. C.: 1937, *Publ. Astron. Soc. Pacific* **49**, 137.

[8] Pollack, J. B. and Sagan, C.: 1969 *Space Sci. Rev.* **9**, 243.

[9] Kliore, A. J., Fjeldbo, G., and Seidel, B.: in preparation.

[10] Trasle, M. J.: 1966, *Tech. Rept. Jet. Propul. Lab. Calif. Inst. Technol. No. 32-800*, p. 252.

[11] Fischbucher, G. E., Martin, L. J., and Baum, W. A.: 1969, 'Martian Polar Cap Boundaries', final report under Jet Propulsion Laboratory contract 951547, Lowell Observatory, May.

[12] Burgess, E.: private communication.

[13] Goldstein, R. M.: private communication; Councilman, C. C.: private communication.

[14] Belton, M. J. S. and Hunten, D. M.: *Science*, in press.

[15] Hartmann, W. K.: 1966, *Icarus* **5**, 565.

[16] Anderson, D. L. and Phinney, R. A.: 1967, in *Mantles of the Earth and Terrestrial Planets* (ed. by S. K. Runcorn), Interscience, New York, pp. 113–126.

[17] Anders, E. and Arnold, J. R.: 1965, *Science* **49**, 1494.

[18] Leighton, R. B. and Murray, B. C.: 1966, *Science* **153**, 136.

[19] O'Leary, B. T. and Rea, D. G.: 1967, *Science* **155**, 317.

[20] Engel, A. E. J., Nagy, B., Nagy, L. A., Engel, C. G., Kremp, G. O. W., and Drew, C. M.: 1968, *Science* **161**, 1005; Schopf, J. W. and Barghoorn, E. S.: 1967, *Science* **156**, 508.

[21] Trumpler, R. J.: 1927, *Lick Obs. Bull.* **13**, 19.

E. CLOUD MOTIONS AND ATMOSPHERIC DYNAMICS

ESTIMATES OF BOUNDARY LAYER PARAMETERS IN THE ATMOSPHERES OF THE TERRESTRIAL PLANETS*

G. S. GOLITSYN

Institute of Atmospheric Physics, Soviet Academy of Sciences, Moscow, U.S.S.R.

Abstract. The similarity theory of atmospheric boundary layers is applied to an estimate of the form of vertical profiles of average wind velocity and potential temperature in the atmospheres of the terrestrial planets in day- and night-time conditions.

It is then considered, as in the case of the earth, that the magnitude of the turbulent heat flux q_T during the day is about 0.1 of $q(1-A)$, where q is the solar constant for the planet and A is its albedo; at night, q_T is several times smaller still. The friction velocity u_* is taken equal to 2–5% (depending upon the stratification) of the mean wind velocity in the free atmosphere, which was adopted from previous calculations (Golitsyn, 1968).

The boundary layers in the atmospheres of Mars and Venus and in the hypothetical atmosphere of Mercury are examined in detail. Sharp temperature drops are characteristic of Mars within a few tens of meters from the surface, attaining a magnitude of several tens of degrees, especially during the day. Large changes of the wind velocity also take place in this thin lower layer. This effect results from the low density of the Martian atmosphere.

For Venus, owing to the very high density of the atmosphere, the stratification is close to neutral, i.e., the temperature profile is close to the adiabatic one and the wind profile is of a logarithmic shape.

Owing to high winds, the stratification on Mercury must also be close to neutral with respect to the wind (the profile being close to the logarithmic), but because of the expected low density, the temperature changes near the ground may still be very great.

1. The theory of an atmosphere's boundary layer has been sufficiently well worked out (Obukhov, 1946; Monin and Obukhov, 1965; Monin and Yaglom, 1965; Zilitin-kevich *et al.*, 1967). A large amount of empirical results has been assembled in terrestrial conditions, particularly in the lower part of the boundary layer, that is, the atmospheric surface layer, or the layer of constant turbulent momentum and heat fluxes, corroborating the conclusions of theory. For the terrestrial atmosphere, the main direction of research is the obtaining of estimates of turbulent fluxes of the momentum τ and heat q_T according to measurements of mean velocity $u(z)$ and temperature $T(z)$ profiles for the atmospheric surface layer. It is also possible to use the data on the geostrophic wind velocity U_g and on the potential temperature change $\delta\theta$ for the boundary layer of the atmosphere, in which Coriolis forces already exert a substantial influence. It would be interesting to obtain at least an estimate of the mean temperature and velocity profiles in the boundary layers of other planets.

Do we presently have the data required for such approximate estimates of the structure of the boundary layers on other planets? It seems to us that such data are already available. The mean characteristic velocities of motions in the atmospheres of the planets have already been estimated (Golitsyn, 1968) [see also Golitsyn, 1969]. Consequently, we may evaluate the friction velocity $u^* = \sqrt{\tau/\rho}$, where ρ is the density.

---

* Translated and reprinted from: Izv. Akad. Nauk. S.S.S.R., *Fizika Atmosfery i Okeana*, Tom 5, No. 8, pp. 775–781, Izdatel'stvo 'NAUKA', 1969.

Depending upon the stratification (Zilitinkevich *et al.*, 1967), in the terrestrial atmosphere we have $u_*/U_g \approx 2\text{--}5\%$ (the first value being associated with strong stability, and the second with strong instability, i.e., convection).

A limitation exists on the magnitude of the second parameter determining the structure of the boundary layer, i.e., on the turbulent heat flux: it cannot exceed $q(1-A)=q_A$, where q is the solar constant for the planet and A is its albedo. For the earth, even in conditions of strongly developed convection, the ratio q_T/q_A is of the order of 0.1. At stable stratification, when the atmosphere is warmer than the ground, which is usually observed at night, $q_T < 0$, i.e., the heat flux is directed toward the soil and the modulus of the ratio q_T/q_A is generally considerably smaller than in the daytime. For other planets (Mars, Venus, and perhaps Mercury) there is no reason to expect very substantial departures from the regularities associated with the terrestrial atmosphere. Moreover, one may qualitatively estimate in which direction any such departures could be occurring in these planets.

Therefore, we are confronted by a problem which in a certain sense is the inverse of the problem in the terrestrial atmosphere: having some kind of idea about the magnitude of the momentum and heat fluxes, one must estimate the thickness of the boundary layer and determine the mean vertical profiles of velocity and temperature.

2. According to the general theory (Obukhov, 1946; Monin and Obukhov, 1954; Monin and Yaglom, 1965), the structure of turbulence in a temperature-stratified medium is determined by the following parameters: $q'=q_T/c_p\rho$ is the normalized turbulent heat flux, $u_* = \sqrt{\tau/\rho}$ is the friction velocity, and the buoyancy parameter $g\beta$, where g is the gravitational acceleration and β is the volumetric expansion coefficient, equal to $1/T_0$ for an ideal gas, where T_0 is the characteristic temperature of the medium. From these parameters one may construct the scale of length

$$L = - u_*^3/(\kappa g\beta q_T/c_p\rho), \tag{1}$$

usually called the 'Monin-Obukhov' scale, and the scale of temperature

$$T_* = q_T/c_q\rho\kappa u_*, \tag{2}$$

where κ is the von Karman constant.

The vertical profiles of the mean velocity and potential temperature $\theta=T+\gamma_a z$, where γ_a is the adiabatic temperature gradient, are universal functions of dimensionless height $\zeta=z/L$, whereupon

$$u(z) = \kappa^{-1}u_*[f_u(z/L) - f_u(z_0/L)], \tag{3}$$

$$\theta(z) = \theta_0 + T_*[f_0(z/L) - f_\theta(z_0/L)], \tag{4}$$

where z_0 is the roughness parameter. For the universal functions f_u and f_θ we have the following expressions (Monin and Obukhov, 1954; Monin and Yaglom, 1965):

$$f_u(\zeta) = f_\theta(\zeta) = \begin{cases} \ln \zeta + \beta\zeta, & 0 < \zeta, \\ \ln |\zeta| + \beta'\zeta_1, & \zeta_1 \leqslant \zeta \leqslant 0, \\ a + C\zeta^{-1/3}, & \zeta < \zeta_1. \end{cases} \tag{5}$$

According to careful statistical processing of a vast amount of empirical data (Zilintikevich and Chalikov, 1968), $\kappa = 0.43$; $\beta = 9.9$; $\beta' = 1.45$; $\zeta_1 = 0.16$; $a = 0.24$; $C = 1.25$.

These formulas are valid for the atmospheric surface layer, where one may neglect the variation of τ and q_T with height. An estimate is given in Monin and Obukhov (1954) of the thickness H of the atmospheric surface layer under that assumption:

$$H < \alpha u_*^2(0)/lU_g, \tag{6}$$

where $\alpha = [u_*^2(0) - u_*^2(H)]/u_*^2(0)$ is the relative variation of friction stress τ, l is the Coriolis parameter, and U_g is the geostrophic wind velocity. For the terrestrial atmosphere, we obtain at $\alpha = 20\%$ and $u_*/U_g \approx 5\%$, $H \approx 50$ m. For Mars, with the same value for α, we obtain $H \approx 100$–200 m, since the Coriolis parameter has the same value, while the mean wind velocities are two to four times higher (Golitsyn, 1968; Golitsyn, 1969). For the slowly-rotating Venus and Mercury we may take for the thickness of the atmospheric surface layer, or, to be more precise, of the boundary layer, the altitude at which the wind velocity is comparable with that in the free atmosphere. Usually, as will be seen below, this thickness is of the order of a few units of the Monin-Obukhov scale L.

For the earth and Mars, one may determine the planetary boundary layer inside which the wind velocity varies little in modulus compared with the surface layer, but where, owing to the action of Coriolis forces, a notable wind turn with altitude takes place. The thickness of this layer may be determined as (Zilitinkevich et al., 1967):

$$L_* = \kappa u_*/l. \tag{7}$$

For the earth, L_* is of the order of 1 km, for Mars, it is 2 to 4 times greater. The wind's rotation angle with height depends on the stratification parameter $\mu = L_*/L = \kappa^2 \beta T_*/lu_*$. In terrestrial conditions (Zilitinkevich et al., 1967), the total angle of wind rotation with height is of the order of several degrees under convective conditions, and attains a value of approximately 40° under conditions of strong stability (rise of potential temperature with altitude).

A parameter whose value is entirely unknown for other planets enters into formulas (3) and (4), viz., the height of the dynamic roughness z_0 of the planet's surface. Fortunately, it enters logarithmically, and thus for our purposes an approximate estimate of its magnitude is sufficient. In terrestrial conditions, we have on the average for dry land $z_0 \approx 1$ cm; for oceans, depending upon the sea state, it may be substantially lower; even for a forest $z_0 \lesssim 1$ m. Bearing in mind that for the other planets there are neither oceans nor forests, we shall assume $z_0 \approx 1$ cm.

Being aware of temperature and velocity profiles, we may determine the stability parameter, namely, the Richardson number

$$Ri = g\beta(d\theta/dz)/(du/dz)^2 = \zeta\phi(\zeta), \tag{8}$$

where the universal function $\theta(\zeta)$ is defined as

$$\phi(\zeta) = \kappa z u_*^{-1}\, du/dz = zT_*^{-1}\, d\theta/dz. \tag{9}$$

At the same time, it is assumed that the coefficients of turbulent exchange for the momentum K and heat K_T, introduced according to the equalities

$$\tau = \rho K(du/dz), q_T = - c_p \rho K_T(d\theta/dz),$$

are identical. Note that during conditions of strong stability this is specifically not so, and one must then introduce in the denominator of the right-hand part of formula (8) the multiplier α, which is the inverse Prandtl turbulence number ($\alpha = K_T/K$). The universal function for the velocity and temperature will also differ by that multiplier ($f_u = \alpha f_\theta$, see Monin and Yaglom (1965)). In view of the great uncertainty of a series of other factors and the estimative character of the present paper, we shall not take this effect into account here.

The turbulent exchange coefficient or turbulent eddy viscosity $K = Ku_*LRi$ is expressed by the following formulas:

$$K = \kappa u_* z, \qquad\qquad |L| \to \infty, \tag{10}$$

$$K = \kappa u_* z(1 + \beta z/L)^{-1}, \quad |L| < \infty, \tag{11}$$

$$K = 3C^{-1}u_* z(z/L)^{1/3}, \qquad \zeta = Z/L < \zeta_1. \tag{12}$$

3. Table I contains the values of solar energy flux q_A arriving at each planet's surface: these values are for Mars, Venus, and Mercury, and for the sake of comparison, the earth. The Table also contains the values of the characteristic scale of temperature T_* and velocity of atmospheric motions U, taken from Golitsyn (1968) or Golitsyn (1969), the normalized turbulent heat flow $q_T' = q_T/c_p\rho$, the buoyancy parameter g/T_0 and the Monin-Obukhov scale L. The value of q_T/q_A was taken equal to 0.1, which, as was earlier noted, is valid in the case of the terrestrial atmosphere for noontime in conditions of strong convection. During the night, q_T and T_* will be several times smaller, with another sign, and L will be considerably larger. In the morning and evening the stratification becomes close to neutral, and then $L \to \infty$, i.e., the boundary layer becomes logarithmic. For Mars we assumed the minimum atmosphere model with a surface pressure $p_0 = 5$ mb, and for Venus we adopted $p_0 = 100$ atm. For the hypothetical atmosphere of Mercury we considered that $p_0 = 1$ mb. The ratio u_*/U was taken equal to 3%.

TABLE I

Planet	q_A cal/cm² min	U m/sec	q_T' deg cm/sec	T_* deg	$g\beta$ cm/sec² deg	$-L$ m
Mars	0.6	40	600	10	2	50
Venus	0.9	0.7	0.03	0.04	1.2	150
Mercury	12	200	10	40	1	600
Earth	1.2	10	7	1	3.3	20

The data of Table I show that the basic parameters determining the structure of the atmospheric surface layer, the friction velocity u_*, and particularly the scale of temperature T_* for the planet under consideration differ strongly owing to sharp differences of the fundamental atmospheric parameters, especially the density, so that the atmospheric surface layer on each planet must have its own well expressed features.

Let us now pass to a more detailed examination of these singular features.

A. MARS

The dynamic and, more particularly, the thermal structure of the lower part of the atmosphere of Mars was considered at fairly great length in Gierasch and Goody (1968). Vertical profiles of temperature and the conditions for convection were computed in detail for various latitudes and seasons, and even for different times of day. One should note that the very same estimate of the mean wind velocity of 40 m/sec for a model atmosphere with $p_0 = 5$ mb was obtained there by another method than that used in Golitsyn (1968). However, the vertical profiles of the mean wind could not be found in the numerical model considered there, and the authors limited themselves to a very rough estimate of Richardson numbers for various conditions.

In conditions of convection, the 'logarithmic + the linear law' for wind and temperature profiles (5) is valid to values of $\zeta_1 = -0.16$, i.e., at $L = -50$ to 8 m altitude from planet's surface. At the same time ($z_0 = 1$ cm)

$$u(z) = 3[\ln(100z) - z/35],$$
$$\theta(z) \approx T(z) = T_0 + 10°[1(100z) - z/35],$$

where $u(z)$ is expressed in m/sec and z in m. At 8 m altitude, $u \approx 20$ m/sec, and $\Delta T = T(0) - T(8\ m) \approx 60°$. At the same time the number $Ri \approx -0.03$. Therefore, over an atmospheric layer of the order of 10 m in all, the velocity attains about one half of the value characteristic of the free atmosphere, while the temperature change reaches $60°$! (here the difference between the usual temperature T and the potential temperature θ, equal to $\theta = T + \gamma_a z$, where for Mars $\gamma_a \approx 5$ deg km^{-1} is entirely insignificant). Very sharp, though somewhat smaller temperature variations in daytime and in the lowermost atmospheric layer were also found in Gierasch and Goody (1968). Note that such sharp variations as those found here could not have been obtained in Gierasch and Goody (1968), for in the computation developed there the adopted vertical spacing was 100 m.

Above 8 m it was necessary to make use of the last formula (5), describing the condition of free convection. At the same time the mean velocity approaches asymptotically its limiting value: the wind velocity in the free atmosphere. According to Monin and Yaglom (1965), this takes place for $\zeta = 5$, i.e., $z \approx 250$ m. The calculation by the last formula (5) shows that for $\zeta \approx 3$ the velocity reaches about 90% of its limiting magnitude at infinity. The turbulent exchange coefficient increases rapidly with height. For $z \approx 250$ m we shall have, according to (12), $K \approx 10^7$ cm^2/sec. In Gierasch and Goody (1968), for the lower kilometer layer a value $K \approx 10^8$ cm^2/sec was obtained. It is apparent that these two estimates are to some degree in accord.

Under conditions of stable stratification (night) we shall assume $L = 250$ m and $T_* = 2°$. Then the velocity 40 m/sec will be attained at an altitude of about 200 m, and the temperature drop will be $40°$. The turbulent exchange coefficient will be of the order of 10^5 cm^2/sec, according to (11). As shown by Gierasch and Goody (1968), at such a value of K, the effects of radiative attenuation of temperature can already prove to be substantial. This must lead to a certain decrease of q_T, i.e., to the increase of L, and, by the same token, to a decrease of temperature drop for the given altitude. Note that in Gierasch and Goody (1968), the temperature profiles during the night were determined from purely radiational computations, completely ignoring the turbulence.

In both cases, the velocity above a level of about 200 m will be nearly constant, but the wind will turn with increasing altitude, approaching the direction of the geostrophic wind at an elevation of 2–4 km. In the daytime, the total rotation angle is small (a few degrees), while at night it may reach several tens of degrees.

B. VENUS

Owing to great values of $|L|$ one should expect that profiles of potential temperature and velocity should be rather close to logarithmic. Let us estimate at the outset at what altitude the velocity, computed by the formula $u(z) = \kappa^{-1} u_* \ln (z/z_0)$ is comparable with the mean velocity of 0.7 m/sec. This altitude is estimated by the formula $z = z_0 \exp (\kappa u / u_*)$. Hence, for $z_0 \approx 1$ cm, we have $z \approx 2$ km. At the same time, the variation of potential temperature at a distance of the order of 2 km due to the small value of T_* will constitute an entirely insignificant quantity: $\Delta \theta \approx 0.1°$, i.e., the temperature profile must be adiabatic to a high degree of precision. At a finite value of L, a certain departure from purely logarithmic profiles especially for the velocity, can be observed. It will be manifested mainly in the lowering of the altitude at which the velocity of 0.7 m/sec is attained. Thus, under stable conditions (at night), when q_T is several times smaller than in the daytime (say 4 times), $L \approx 600$ m and the velocity profile will have the form $u(z) = 0.06 (\ln 100z + z/60)$, where z is expressed in m and $u(z)$ in m/sec. The velocity $u(z) \approx 0.7$ m/sec will be attained for $z \approx 100$ m. At the same time, $\Delta \theta \approx 0.1°$.

For unstable conditions (daytime) $L = -150$ m and the 'logarithmic + linear law' will be observed to $z = -0.16L \approx 20$ m. At this altitude the velocity will reach 0.5 m/sec. Above this level the velocity profile (and that of potential temperature) will be described by the law $z^{-1/3}$, and if one considers that the constant value of velocity is attained for $\zeta = 5$, we shall have for the thickness of the layer encompassed by convection $z \approx 750$ m. At that distance the variation of potential temperature will be on the order of 0.4°.

Therefore, owing to the great thickness of Venus' atmosphere and the relatively small flux of incident solar radiation, the state of the atmosphere should be close to neutral, i.e., the profile of temperature should be adiabatic and that of velocity logarithmic. The situation will change little especially for the temperature even if we take the limiting, and entirely unrealistic, case $q_T = q_A$. But, generally speaking, one must bear in mind that if part of the solar radiation is absorbed in the atmosphere of Venus, which is quite probable, then our assertion on the closeness of stratification to a neutral condition will be more correct. But if all the radiation is absorbed in the planet's atmosphere, the boundary layer will be purely logarithmic.

The coefficient of vertical turbulent exchange at an altitude on the order of 1 km will be $K \approx \kappa u_* z \approx 10^5$ cm²/sec. Let us note that for the scale $L = 1$ km we could obtain by the Richardson-Obukhov formula a quantity of the same order: $K \approx 0.1 \epsilon^{1/3} L^{4/3}$, if we take $\epsilon \sim 10^{-3}$ cm²/sec³, which is the quantity obtained in Golitsyn (1969) from global estimates of the effectiveness of Venus' atmosphere as a whole in the reprocessing of the solar energy arriving at the planet into mechanical power, i.e., into the generation of kinetic energy of atmospheric motions. Then $K \approx 5 \times 10^4$ cm²/sec.

C. MERCURY

If an atmosphere indeed exists on Mercury, the processes in it will prove to be the most exotic of all the planets of the solar system. The enormous wind velocities, of the order of 200 m/sec, the differences in temperature between the dark and illuminated sides of the planet of the order of 500°, the length of the day and night corresponding to 180 terrestrial days, all this is instrumental in rendering the boundary layer unique also. Owing to great friction, say $u_* \approx 12$ m/sec (!) the characteristic Monin-Obukhov scale L will be very great also. This is why the velocity profile is found to be close to logarithmic although certain small deflections from the purely logarithmic profile may take place. The velocity reaches a magnitude of the order of 200 m/sec over the extent of the lower 100–200 m. At the same time, the high value of T_* results in high temperature drops. Thus, in the daytime, the jump of potential temperature, determined by formulas (3)–(5), using the parameters of Table I, yields $\Delta T \approx 400°$ over the extent of the lower 100 m (at a surface temperature of the order of 650°); at night, with a surface temperature of approximately 150°, this drop is of the order of 50° for the lower 200 m. These estimates appear to be extremes, for owing to a very long duration of the day and of the night, the radiation may strongly diminish the temperature drops. In other words, on Mercury, the fraction q_T/q_A is probably notably less than 0.1. At the same time, the scale of L will be still greater, i.e., the velocity profile will be still closer to logarithmic, and the boundary layer will be correspondingly thicker.

References

Gierasch, P. and Goody, R. A.: 1968, *Planetary Space Sci.* **16**, No. 5.

Golitsyn, G. S.: 1968, *Izv. AN SSSR, Fizika atmosfery i Okeana* **4**, No. 11.

Golitsyn, G. S.: 1969, this volume, p. 304.

Monin, A. S. and Obukhov, A. M.: 1954, *Trudy Geofiz. In-ta AN SSSR*, No. 24.

Monin, A. S. and Yaglom, A. M.: 1965, *Statisticheskaya gidromekhanika*, part 1, ch. 4, Izd-vo 'Nauka'.

Obukhov, A. M.: 1946, *Trudy Inst. teoret. geofiziki AN SSSR*, **1**.

Zilitinkevich, S. S. and Chalikov, D. V.: 1968, *Izv. AN SSSR, Fizika atmosfery i Okeana* **4**, No. 3.

Zilitinkevich, S. S., Laykhtman, D. L., and Monin, A. S.: 1967, *Izv. AN SSSR, Fizika atmosfery i Okeana* **3**, No. 3.

THE THEORY OF SIMILARITY FOR LARGE-SCALE MOTIONS
IN PLANETARY ATMOSPHERES

G. S. GOLITSYN

Institute of Atmospheric Physics, Soviet Academy of Sciences, Moscow, U.S.S.R.

Abstract. A general similarity theory for dynamics of planetary atmospheres, which gives results consistent with those of more complex theories, is briefly outlined.

1. One of the most important problems of contemporary meteorology is the theory of the general circulation of the atmosphere. This subject is currently attracting interest because of its relevance to problems of planetary physics in connection with the attainments of modern space technology. Attempts at calculation of characteristics of atmospheric circulation (as yet only for Mars) have already been undertaken by using an appropriate numerical model. However this method requires very laborious and extensive calculations, as well as very definite assumptions about a series of atmospheric characteristics, such as the absorptivity of the optically active constituents, knowledge of their relative concentration, etc. At present we do not possess such detailed knowledge. Another method is to obtain a few mean quantitative estimates on the basis of energetic and thermodynamic considerations and by the methods of the theory of similarity.

One such attempt was undertaken in Golitsyn (1968) where a formula for the mean rate of generation (and dissipation) of kinetic energy per unit mass by large scale motions was introduced:

$$\varepsilon = \left(\frac{k\delta T}{T_1}\right)\left(\frac{q_A}{4M}\right). \tag{1}$$

Here k is a numerical coefficient of the order of 0.1 (according to actual data) for earth's atmosphere, δT is a characteristic temperature difference on the surface or in the atmosphere of the planet, T_1 is the temperature of the most strongly heated regions, $q_A = q(1-A)$ where q is the solar constant for the planet and A is the planet's albedo, and $M = P_s/g$, the mass of a unit column of the atmosphere. For the characteristic scale L of synoptic processes Obukhov's scale $L_0 = c/l \approx c/\Omega$ (Obukhov, 1949) was adopted for rapidly rotating planets in Golitsyn (1968) (here c is the speed of sound, l is the Coriolis parameter), or, for a slowly rotating planet, the planetary radius. For the terrestrial planets, characteristic velocities $U \approx (\varepsilon L)^{1/3}$ and times $\tau \approx L^{2/3}\varepsilon^{-1/3}$ were estimated. For the calculation the value of 0.1 was adopted for k for all planets, and the values of δT and T_1 were taken from observations (Moroz, 1967).

However, in any closed theory of general circulation the quantities δT and T_1 should be determined, and not left to be assigned. We will now attempt to construct such a theory using the methods of similarity theory and dimensional analysis (Sedov, 1967).

2. Our discussion is based on the following physical principles: that the intensity of the circulation and its driving temperature difference are interconnected and self-consistent and that they are determined by the energy input to the atmosphere and by the mass of the atmosphere and its thermal properties.

The heat balance is established by long-wave radiation to space, since the atmosphere is considered to be gray – that is, the temperature of thermal radiation is determined by the energy input q_A. For the sake of simplicity we will not take into account the absorption of direct radiation by the atmosphere itself. This procedure is permissible for an atmosphere that is not very thick.*

We will first investigate the case of a non-rotating planet. The determining parameters in the equation describing the dynamic behavior of the atmosphere, averaged over height, are found to be the surface-averaged inflow of energy $q_A/4$ [gm sec^{-3}], the heat capacity per unit mass of the atmosphere c_p [cm^2 sec^{-2} deg^{-1}], the mass of a unit column of the atmosphere M [gm cm^{-2}], the radius of the planet r [cm] and the Stefan-Boltzmann constant $\sigma = 5.67 \times 10^{-5}$ gm sec^{-3} deg^{-4}. From these five parameters, a single nondimensional quantity can be formed:

$$\mathcal{M} = \sigma^{3/8} c_p^{-3/2} \left(\frac{q_A}{4}\right)^{5/8} r M^{-1}. \tag{2}$$

Under the adopted assumptions and constraints, the motions in the atmospheres of planets with equal values of the number $\mathcal{M}$ should be similar. The values of $\mathcal{M}$ for the terrestrial planets are 1.1×10^{-3} (earth), 3.4×10^{-2} (Mars, $p_s = 5$ mb), and 1.3×10^{-5} (Venus, $p_s = 90$ atm). These numbers are all seen to be small. Consequently, the dependence on $\mathcal{M}$ should not be significant. The smallness of $\mathcal{M}$ is equivalent to the largeness of M; thus in a first approximation the mass M can be neglected. This means that the atmospheres of the planets, with respect to some global circulation characteristics, are self-similar by mass.**

From the remaining four parameters we can form combinations with the dimensions of velocity, time, and energy. The first has the form:

$$w^{1/2} = c_p^{1/2} \sigma^{-1/8} (q_A/4)^{1/8} = C_e (\chi - 1)^{-1/2}. \tag{3}$$

---

* In general it is necessary to consider the nondimensional parameter I/H, where I is a characteristic mean free path of a photon, and H is the effective thickness of the atmosphere. This parameter can affect, for example, the quantity k in (1). However, since thermodynamics imposes some general restrictions that do not depend on the nature of the substance ($k < 1$!), we can hope that the estimates obtained will be fairly universal. These effects will be indirectly surveyed in Section 5 when the intensity of the circulation of Venus is considered.

** The author was asked many times why g, the gravity acceleration, does not enter the consideration. Drs. F. P. Bretherton and N. A. Phillips, to whom the author is very grateful, made the point especially clear stating that if $g = 0$ then no motion will arise through the inhomogeneous heating of an atmosphere which is the main cause of the circulation. The answer is this: if we add g then a new independent nondimensional quantity can be formed:

$$c_p \left(\frac{q_A}{4\sigma}\right)^{1/4} / gr \approx RT_e / \mu g r = H/r$$

which is the ratio of an atmosphere's scale height to the planetary radius. This ratio is usually very small unless g is large enough. This means the neglect of g is a new kind of self-similarity of circulations, i.e. independence of the exact value of g. Of course if $g \to 0$ the theory becomes invalid. The same kind of argument shows why an exact value, say, of the molecular viscosity is not essential, etc.

Recalling that for an ideal gas $c_p = \chi(\chi-1)^{-1}R/\mu$ where $\chi = c_p/c_v$, R is the universal gas constant, μ is the molecular weight, and that $(q_A/4\sigma)^{1/8} = T_e^{1/2}$, the effective radiational temperature of the atmosphere, we see that $w = C_e^2(\chi-1)^{-1}$ is the enthalpy and $C_e = (\chi-1)^{1/2}w^{1/2}$ is the velocity of sound (to within a multiplicative factor which takes into account the difference between T_e and the mean temperature of the atmosphere).

The quantity with the dimension of time

$$\tau_e = r/C_e = r(\chi - 1)^{1/2}c_p{}^{-1/2}(4\sigma/q_A)^{1/8} \tag{4}$$

is the relaxation time for atmospheric pressure or density disturbances on a global scale.

The quantity with dimensions of energy has the form

$$E = B\sigma^{1/8}c_p{}^{-1/2}(q_A/4)^{7/8}r^3, \tag{5}$$

where B is some numerical coefficient. This coefficient can change a little from planet to planet, even though there is still a series of factors, not considered here, that may have an effect on the intensity of atmospheric processes. Taking (3) and (4) into account, we write (5) as

$$E = [B(\chi - 1)^{1/2}/4\pi]q_A\pi r^2 \cdot r/C_e = [B(\chi - 1)^{1/2}/4\pi]Q_A\tau_e. \tag{6}$$

Thus the energy E is, to within a multiplicative constant, equal to the full power of the solar radiation Q_A that falls on the planet multipled by the relaxation time τ_e.

The total enthalpy of the atmosphere should depend in an explicit manner on M. It is natural to suppose that (5) determines the total kinetic energy of the atmosphere. The value of E/B, according to (5), is for earth 1.12×10^{27} erg and for Mars 0.97×10^{26} erg. The magnitude of the total kinetic energy of the terrestrial atmosphere varies from season to season and is equal to 6–9×10^{27} erg (Borisenkov, 1963). According to calculations in Leovy and Mintz (1966), the total kinetic energy of Mars' atmosphere is 1.2–1.6×10^{26} erg. Thus, to within a multiplicative constant of the order of unity, (5) does in fact determine the total kinetic energy. A series of consequences follows from this:

(1) the kinetic energy of a unit volume $\rho(U^2/2)$ does not depend on M or p_s;

(2) the mean velocity of atmospheric motions is

$$U = (E/2\pi r^2 M)^{1/2} = (B/2\pi)^{1/2}\sigma^{1/16}c_p{}^{-1/4}(q_A/4)^{7/16}(r/M)^{1/2}; \tag{7}$$

(3) the nondimensional parameter $\mathcal{M}$ is, to within multiplicative factor of the order of unity, equal to the square of the Mach number Ma:

$$\mathcal{M} = [2\pi(\chi - 1)/B]U^2/C_e^2 \approx \text{Ma}^2; \tag{8}$$

(4) the time scale of atmospheric motions is of the order of

$$\tau_u \approx r/U \approx (2\pi/B)^{1/2}c_p{}^{1/4}\sigma^{-1/16}(q_A/4)^{-7/16}(rM)^{1/2}; \tag{9}$$

(5) the total rate of generation (dissipation) of kinetic energy in the entire atmosphere of a planet is on the order of

$$\epsilon \approx E/\tau_u \approx (B^3/2\pi)^{1/2}\sigma^{3/16}c_p{}^{-3/4}(q_A/4)^{21/16}r^{5/2}M^{-1/2}, \tag{10}$$

from which in the calculation for a unit mass we have

$$\varepsilon \approx \epsilon/4\pi r^2 M \approx \tfrac{1}{2}(B/2\pi)^{3/2}\sigma^{3/16}c_p{}^{-3/4}(q_A/4)^{21/16}r^{1/2}M^{-3/2}. \tag{11}$$

Comparing (11) and (1), we can write the efficiency of the atmosphere in the form

$$\eta = k\,\delta T/T_1 \approx \tfrac{1}{2}(B/2\pi)^{3/2}\sigma^{3/16}c_p{}^{-3/4}(q_A/4)^{5/16}(r/M)^{1/2} \approx \text{Ma}. \tag{12}$$

Considering $T_1 \approx T_e$, we can evaluate the characteristic temperature difference

$$\delta T \approx \eta T_e/k \approx (2k)^{-1}(B/2\pi)^{3/2}\sigma^{-1/16}c_p{}^{-3/4}(q_A/4)^{9/16}(r/M)^{1/2}. \tag{13}$$

Sometimes we should introduce $\alpha = T_e/T_1$, and then in (13) it is necessary to substitute k for $k\alpha$. We note that Equation (1) does not appear as a result of the theory of similarity developed here, but follows from other considerations. It is also related to Equation (13).

3. We now include in the number of determining parameters the angular velocity of actual rotation of the planet, Ω (sec^{-1}). The quantity $\mathcal{M} \approx$ Ma^2, we let be small, as before. Then from the parameters σ, c_p, $q_A/4$, r, and Ω we can compose only one dimensionless combination

$$\lambda = (\chi - 1)^{-1/2}c_p{}^{-1/2}(4\sigma/q_A)^{1/8}\Omega r = \Omega(r/C_e) = \Omega\tau_e = r/L_0. \tag{14}$$

Formula (5) must now be multiplied by a function $f(\lambda)$, where $f(0) = 1$. From general considerations it follows only that $f(\lambda)$ must increase with an increase in λ. We cannot successfully obtain explicit formulas of the type (7) through (13), however the dependences $U \sim M^{-1/2}$, $T_u \sim M^{1/2}$, $\varepsilon \sim M^{-3/2}$, $\eta \sim M^{-1/2} \sim \delta T$ are preserved here, as is the independence of $\rho U^2/2$ from M. In the general case when $\mathcal{M} \gtrsim 1$, equation (5) must be multiplied by the function from the two dimensionless parameters $f(\mathcal{M}, \lambda)$. From the point of view of similarity, this case is the most complex and it is difficult to obtain here even qualitative relationships.

4. We will now demonstrate another derivation of the basic relationships of Section 2. We write the equal of the balance of heat for a spherical atmosphere, averaged over height in the most simplified form as

$$Mc_p u_i \frac{\partial T}{\partial x_i} = \sigma T_e^4, \tag{15}$$

i.e. the advection of heat by large scale motions is balanced by the atmosphere's cooling into space. To an order of magnitude the advection is $u_i\,\partial T/\partial x_i \approx U\delta T/(\pi r/2)$, where $U \approx (\varepsilon r)^{1/3}$. Then upon taking into account Equation (1), we obtain from (15)

$$\delta T \approx (\pi/2)^{1/2}k^{-1/4}\sigma^{-1/16}c_p{}^{-3/4}(q_A/4)^{9/16}(r/M)^{1/2}. \tag{16}$$

Putting δT in Equation (1), we determine ε, then U, and then E. Furthermore the structure of all the formulas obtained turns out to be same as that of the formulas in Section 2. Comparing them, we have

$$B \approx \pi^2 k^{1/2} \tag{17}$$

by which the theory of similarity and the concepts developed in Golitsyn (1968) are naturally married. From Equation (16), taking δT as fixed, we can obtain

$$k \approx \frac{\pi^2}{4} \delta T^{-4} \frac{(q_A/4)^{9/4}}{\sigma^{1/4} c_p^3} \left(\frac{r}{M}\right)^2. \tag{18}$$

Therefore the whole theory is determined with precision except for one empirical quantity, for which we can select B, k, or δT, or U, etc.

Since from thermodynamic considerations $k < 1$ ($k = 1$ only for an ideal heat engine in a Carnot cycle), then from (16) there follows the inequality

$$\delta T > \frac{\pi^{1/2}}{2} \frac{(q_A/4)^{9/16}}{\sigma^{1/16} c_p^{3/4}} \left(\frac{r}{M}\right)^{1/2}. \tag{19}$$

Similar inequalities can be obtained for U, ε (from above) and for τ_u (from below).

For temperature there is still one obvious inequality: $\delta T < T_1 \approx T_e$. This can give some information in the case of a planet with a very thin atmosphere (for example, the hypothetical atmosphere of Mercury), when the formal use of expression (16) can lead to its violation, that is to $\delta T > T_e$. In this case in Equation (13) instead of B it is necessary to put $B f_1(\mathcal{M}) \approx \pi^2 k^{1/2} f_1(\mathcal{M})$ and then the inequality $\delta T < T_e$ gives the following estimate from above:

$$f_1(\mathcal{M}) < \mathcal{M}^{-1/3} \quad \text{for } \mathcal{M} \gtrsim 1. \tag{20}$$

5. We shall now see what it is possible to obtain using the dependences derived here for planets of the terrestrial type, if in the case of the earth and Mars we ignore the rotation of the planet. We take in all cases $k = 0.1$, as it was evaluated in Golitsyn (1968) for the terrestrial atmosphere.

Results of calculations according to the formulas in Section 2 are given in Table I, together with the quantities U and δT observed for the earth, or computed in Leovy

TABLE I

Planet	ρ_s atm	U m/sec	U, obs. or comp.	δT K	δT, obs. or comp.	τ_u days	ε cm^2/sec^3
Earth	1	12	17	20	40	5	4
Mars	5×10^{-3}	50	40	100	110	1	~ 100
Venus	90	0.7	?	2	< 15	100	3×10^{-4}

and Mintz (1966) for Mars. For these two planets the theoretical predictions are satisfactory. In the case of Venus, the most striking thing turns out to be the small magnitude of the calculated characteristic temperature difference. This agrees with the practical absence of a phase effect in the Venus radio emission, and also agrees with the results of radio interferometric measurements at 11 cm, about which Dr. David Morrison spoke at the present symposium. According to these measurements, there is no difference of temperature between the equator and poles on the surface of Venus within an error of less than 15 K.

Concerning the winds of Venus, we have only observations of the motions of ultra-violet clouds, the velocities of which reach 100 m/sec. This, obviously, is a mesopheric phenomenon, originating at a level where the pressure is of the order of 1 to 10 mb. Our value of the characteristic velocity relates to the entire thickness of the atmosphere. The motions of the deep atmosphere are so far unknown from experimental methods. However, the success of the theory in predicting the temperature difference gives faith in the correctness of the predicted order of magnitude of both the velocity and time scale τ_u.

If it turns out that all or most of the direct solar radiation is absorbed by the atmosphere of Venus, and does not reach the surface, then we are brought to the analogy with the oceans (Golitsyn, 1968; Goody and Robinson, 1966). We may think then, that this could lead to a reduction of k by roughly a factor of two or three orders of magnitude. In such a case the velocity is reduced by a factor of 3 to 5, and τ_u and δT are correspondingly increased.

A considerably more detailed discussion of the theory and its applications to the terrestrial planets as well as its possible extension to the case of giant planets can be found in Golitsyn (1970).

References

Borisenkov, E. P.: 1963, in *Works of the Arctic and Antarctic Institute*, No. 23.

Golitsyn, G. S.: 1968, *Izv. AN S.S.S.R., Atmospheric and Oceanic Physics* **4**, No. 11.

Golitsyn, G. S.: 1970, *Icarus* **13**, 1.

Goody, R. M. and Robinson, A. R.: 1966, *Astrophys. J.* **146**, No. 2.

Leovy, C. B. and Mintz, Y.: 1966, *A Numerical General Circulation Experiment for the Atmosphere of Mars*, RAND Corp., RM-5110-NASA.

Moroz, V. I.: 1967, *Physics of the Planets*, Nauka, Moscow (English translation in NASA TT F-515).

Obukhov, A. M.: 1949, *Izv. AN S.S.S.R., Series in Geography and Geophysics* **13**, No. 4.

Sedov, L. I.: 1967, *Methods of Similarity and Dimensions in Mechanics*, 6th ed., Nauka, Moscow.

CLOUD ACTIVITY ON MARS NEAR THE EQUINOX: COMPARISON OF THE 1937 AND 1969 OPPOSITIONS

G. de VAUCOULEURS

Department of Astronomy, University of Texas at Austin, Austin, Tex., U.S.A.

Abstract. Four classes of manifestations of cloud activity observed in 1937 and 1969 near the autumn equinox of the northern hemisphere ($165° < L_s < 192°$) are discussed: (1) formation of the north polar haze cap, (2) unusual cloud systems on the disk, (3) recurrent or persistent cloud activity at the sunset limb, (4) clouds seen in projection above the sunrise terminator.

This paper compares various manifestations of cloud activity on Mars near the equinox observed by the author near the corresponding oppositions of 1937 and 1969. Figure 1 shows the ranges of heliocentric longitudes L_s covered by the observations available at several oppositions, in particular 1937 and 1969; other relevant data including opposition dates (crosses), perigee dates and maximum apparent diameter, and the times of the Mariner 6 and 7 encounters (small circles) are also given. Figures 2a and 2b show

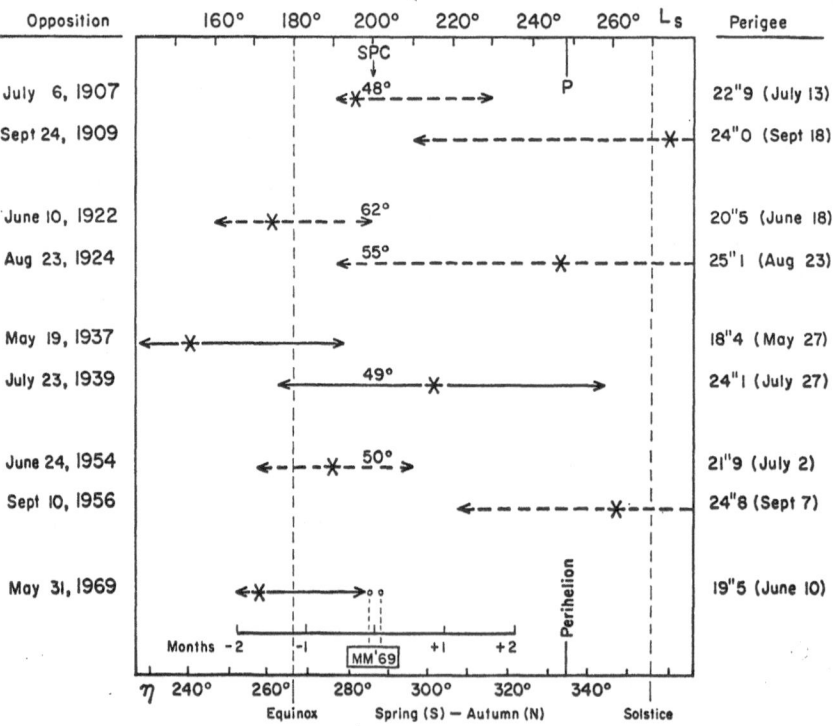

Fig. 1. Observations of Mars at several oppositions. Arrows indicate the range of heliocentric longitudes (η, L_s) covered by the author's observations (full line) and of others (dashed). Opposition dates (left and crosses), and perigee apparent diameter (right) are given. Mariner 6, 7 encounter dates and areocentric diameter of south polar surface cap are also shown.

Sagan et al. (eds.), *Planetary Atmospheres*, 310–319.

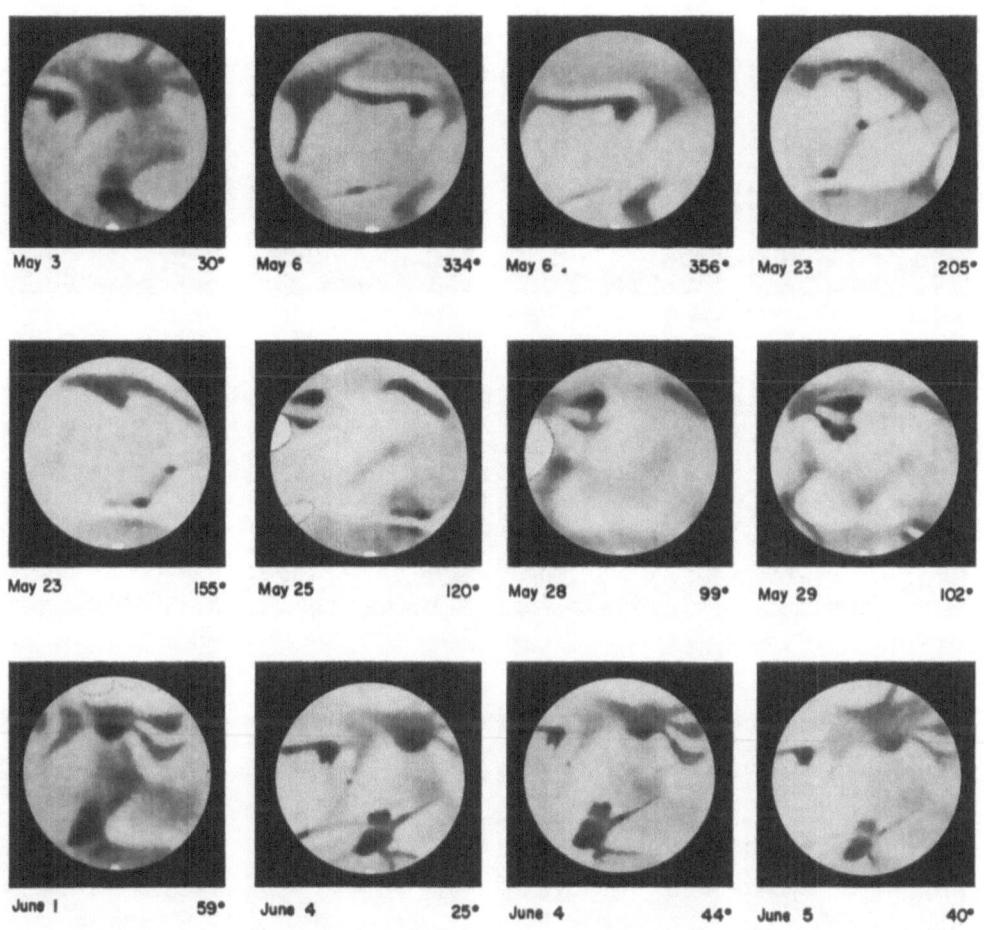

Opposition date 1937 May 19. Perigee diameter 18"4 (May 27)
Heliocentric longitudes L_S = 143° to 160°
Latitude of center of disk D_E = +9° to +15°

Fig. 2a.   Observations of Mars in 1937. (18-cm refractor, Observatoire de la Société Astronomique de France, Paris.)

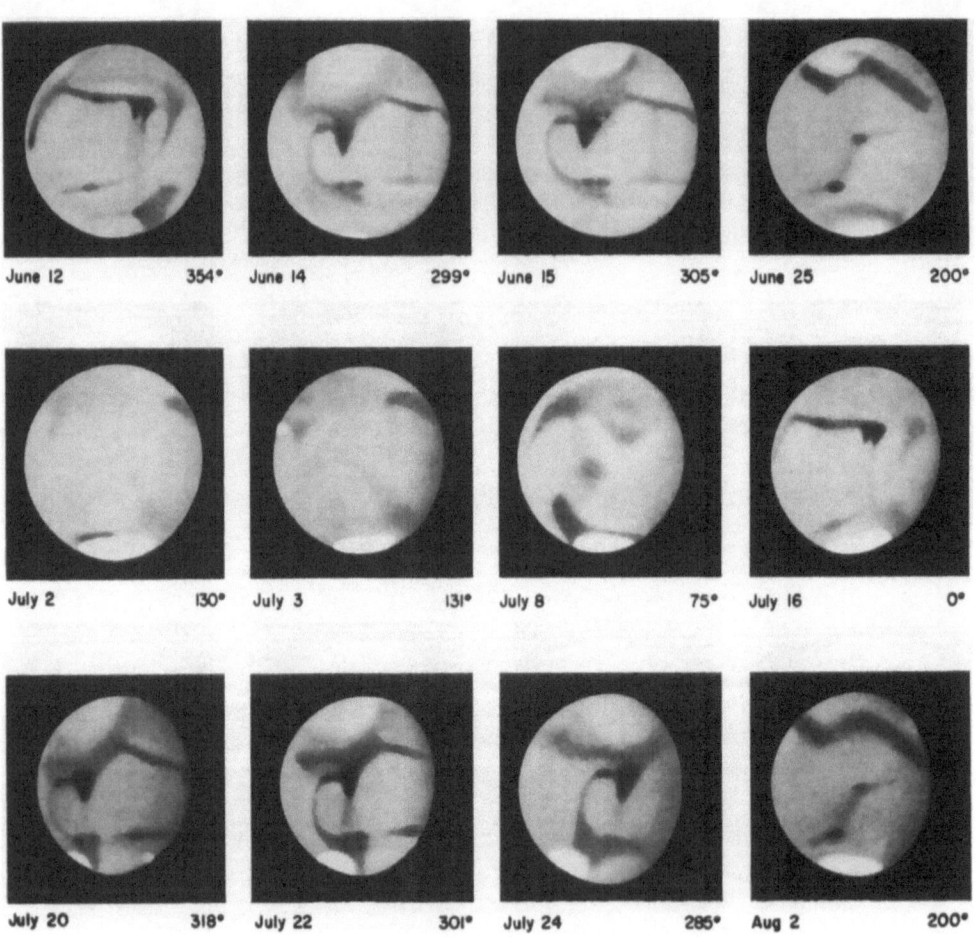

Opposition date 1937 May 19. Perigee diameter 18"4 (May 27)
Heliocentric longitudes L_s = 164° to 192°
Latitude of center of disk D_E = +14° to +17°

Fig. 2b.

24 drawings made in 1937 ($143° < L_s < 192°$) with an 18-cm refractor ($\times 180$ to $\times 365$) in Paris. Figures 3a, 3b, and 3c show 38 drawings made in 1969 ($165° < L_s < 191°$) with a 23-cm refractor ($\times 370$) in Austin. South is at the top. In 1937 the observations were made in 'white' light ($\lambda_e \simeq 0.56\ \mu$); in 1969 an orange filter (Wr21, $\lambda_e \simeq 0.59\ \mu$) was used, supplemented by inspections through green (Wr58, $\lambda_e \simeq 0.52\ \mu$) and red (Wr25, $\lambda_e \simeq 0.63\ \mu$) filters to help in the discrimination of atmospheric and surface phenomena.

Cloud phenomena observed visually fall into 4 classes: (1) formation of the haze cap over the north polar regions near the autumn equinox, (2) unusual cloud systems on the disk, (3) persistent or recurrent cloud activity at the sunset limb, (4) clouds seen in projection above the sunrise terminator.

1. Formation of the North Polar Haze Cap

This is a seasonal phenomenon which repeats with remarkable regularity from year to year. Near the end of summer the north polar snow cap is very small (diam. $<5° = 300$ km),* often difficult to see and occasionally invisible in poor seeing, but the atmosphere is quite clear and transparent over the polar regions. This situation prevailed in 1937 until June 15 ($L_s = 166°$) and in 1969 until June 14 ($L_s = 173°$). The atmospheric haze over the north polar regions was first observed on June 25 ($L_s = 170°$) in 1937 and on June 13 ($L_s = 172°$) in 1969. Thus, visible condensation of the polar haze cap begins at $L_s \simeq 171°$ or about 2 weeks before the equinox. For a period of several weeks the polar haze is evidently still tenuous, and varies greatly in extent and density. Note the day-to-day variations in Figures 3b and 3c. Frequently the brightest part is not centered above the pole but forms over preferred areas at the Western limb (i.e. in the afternoon); note especially the striking repetition of a bright whitish cloud formation over Utopia (lower left) on July 22–24, 1937 ($L_s = 186°–188°$) and July 6–10, 1969 ($L_s = 185°–187°$). During this period immediately following the equinox a great deal of variable cloudiness was observed in 1969 over the northern sub-polar areas extending from Mare Acidalium (June 28–30) through Dioscuria-Cydonia (July 1–5) to Umbra-Utopia (July 5–9) and Panchaia-Lemuria (July 10–16). Note in particular the broken-up cloudy area covering part of Mare Acidalium on June 30 (compare with clear views of June 28, 29), another concealing Arethusa Lacus and the Callirrhoe streak on July 2 and 3 (compare with July 4), the small isolated cloud mass north of Coloe Palus on July 5 (compare with July 4), the interruption of Casius by a complex cloud system extending also to the East over Aetheria and West over the north tip of Isidis Regio on July 8 (compare with July 7 and 10, a remnant of the western mass was still present on the 9th), finally notice the bright pointed prominence of the main north haze cap over Herculis Pons on July 15 (compare with July 16).

It is important to note that these cloud masses and haze-covered areas are whitish and become always more conspicuous through a green filter, they should not be confused (as some observers have done in recent years) with the normal brighter areas of

---

* Corrected for diffraction and seeing (8° apparent).

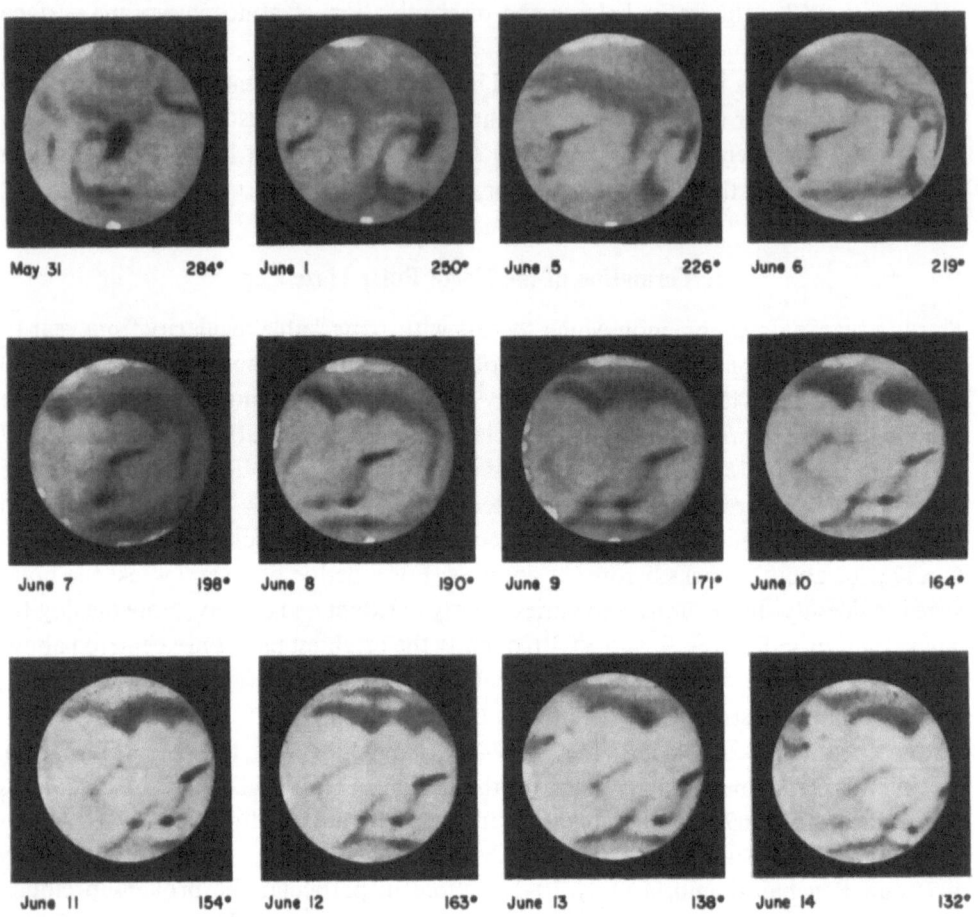

Opposition date 1969 May 31. Perigee diameter 19"5 (June 9)
Heliocentric longitudes L_s = 165° to 173°
Latitude of center of disk D_g = + 7° to 10°

Fig. 3a.   Observations of Mars in 1969. (23-cm refractor, University of Texas Students'
Observatory, Austin.)

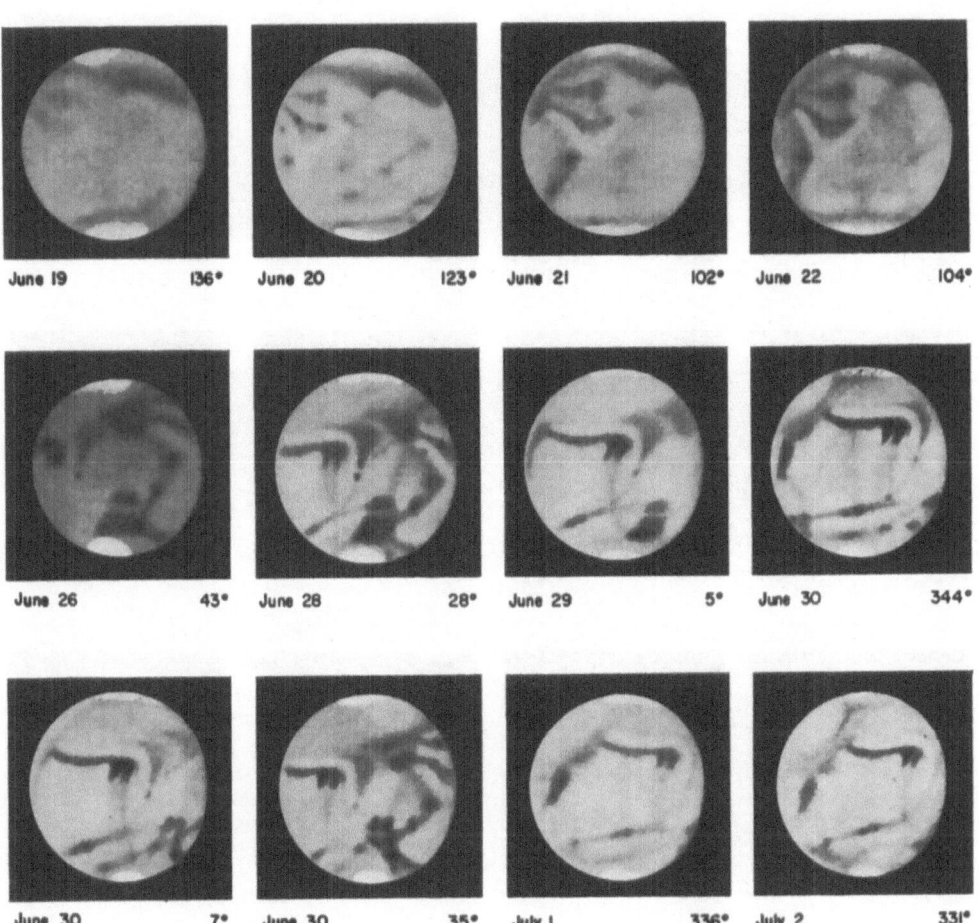

Opposition date 1969 May 31. Perigee diameter 19″5 (June 9)
Heliocentric longitudes $L_S = 175°$ to $182°$
Latitude of center of disk $D_E = +10°$ to $+12°$

Fig. 3b.

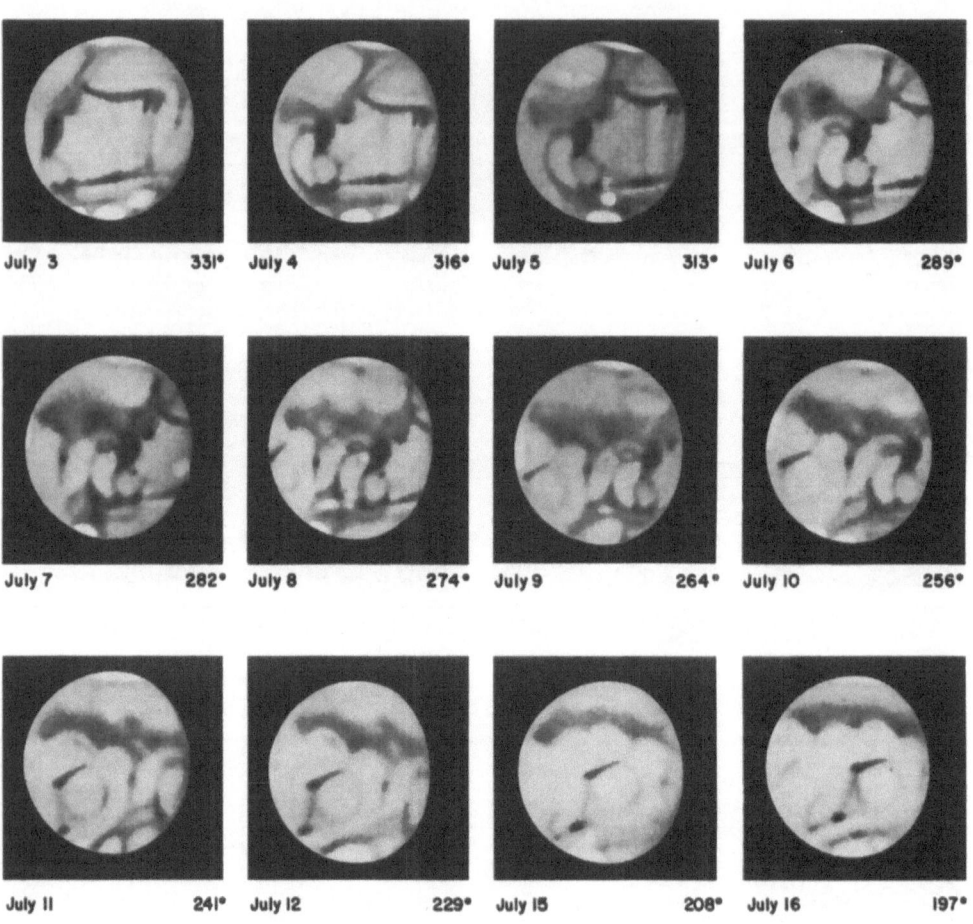

July 3   331°   July 4   316°   July 5   313°   July 6   289°

July 7   282°   July 8   274°   July 9   264°   July 10   256°

July 11   241°   July 12   229°   July 15   208°   July 16   197°

Opposition date 1969 May 31. Perigee diameter 19"5 (June 9)
Heliocentric longitudes L_S = 183° to 191°
Latitude of center of disk D_E = +12°

Fig. 3c.

the martian topography which have the characteristic pink or reddish tint of the permanent surface dust. An example is the very stable brighter band on the southern half
of Dioscuria-Cydonia running parallel to the Protonilus-Ismenius Lacus-Deuteronilus
axis; it was well seen in 1937 (June 4 to 15, July 22) and 1969 (July 1 to 8) and at
other favorable oppositions; it is always pinkish, not white, and is well seen through
the red filter.

After a month or so of unsteady growth ($170° < L_s < 190°$) the north polar haze cap
assumes a more stable, denser form, and covers the totality of the north polar regions
which it will cover solidly throughout the cold season and until the vernal equinox.
While the solid frost deposit condenses at the surface under cover of this atmospheric
hood, occasional disturbances, perhaps similar to our polar winter storms, are
observed at times in the form of wisps or 'tongues' of haze projecting out of the main
body into the temperate zone and down to fairly low latitudes; good examples were
observed in 1941 (Sept. 19–21, $L_s = 276°$) and 1958 (Nov. 4, $L_s = 320°$) stretching out
to $+30°$ or $+35°$ latitude over the Niliacus Lacus region (Reference [1], Figure 29).

The transition from the residual surface deposit to the large atmospheric haze cap
in the range $150° < L_s < 200°$ is summarized in Figure 4 where the intermingling of the
1937 and 1969 data points emphasizes the close repeatability of the seasonal pattern.

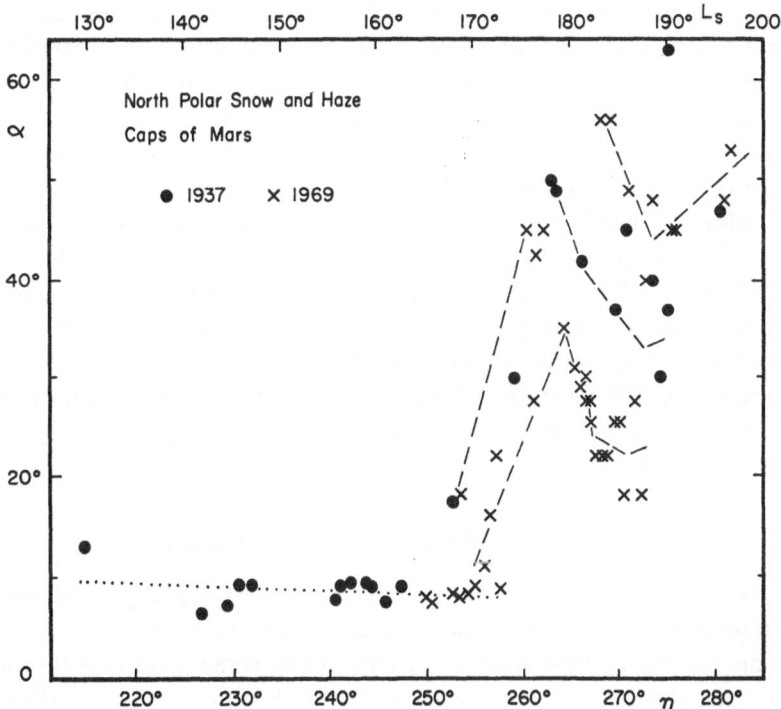

Fig. 4.   Regression curve of surface cap and growth curve of haze cap near the autumn equinox
of the northern hemisphere of Mars in 1937 and 1939. Lower points after $L_s = 170°$ refer mainly to
temporary cloud formations not centered at North Pole.

2. Cloud Systems and Widespread Haze on the Disk

Large persistent cloud systems or widespread haze obscuring the surface details are relatively rare, outside the haze cap over the polar regions and the recurrent condensations at the morning or evening limbs (see Sections 3 and 4). In 1937 a large whitish cloud system was observed on June 25 ($L_s = 171°$) over Phaethontis-Electris-Atlantis and the east end of Mare Cimmerium; gaps in the observations prevented following its motion and evolution, but on July 2 ($L_s = 177°$) the region of Phaethontis-Aonius Sinus was still covered by clouds. On the same date the disk was exceptionally pale and devoid of details east of Sirenum Mare and over the region of Solis Lacus and Tithonius Lacus which were again barely detectable on July 3 and Solis Lacus remained abnormally faint at least until July 8 (compare with May 25–29, Figures 2a and 2b). In 1969, on June 10 ($L_s = 171°$) the east end of Mare Cimmerium was covered by a large cloud mass spreading southward from the bright desert area of Zephyria where none was present the day before (Figure 3a); scattered remnants of this probable dust storm were observed on the following days over Electris-Phaethontis and perhaps parts of Mare Sirenum (June 11–14). The Solis Lacus region had low contrast on June 19–20 ($L_s = 175°$), but poor seeing may have been a contributing factor.

3. Cloud Activity at the Sunset Limb

This is a very common and well-documented phenomenon which appears to involve two distinct classes of clouds: (a) bright, white cloud systems in mass motion, which may be high-level convective ice clouds; (b) whitish, recurrent but stationary haze activity which may be low level churning of surface dust. Some areas (Isidis R., Tharsis) seem especially prone to this type of late-afternoon cloud formation. Good examples of type (a) were observed in 1937 (Figure 2a) between May 2 and 6 ($L_s = 142°$–144°) over Syrtis Major and Libya-Isidis Regio, slowly drifting east at 10 km/h (12° in 3 days), and between May 25 and 29 ($L_s = 153°$) over Candor-Ophir, drifting north-east at 13 km/h (20° in 4 days) while spreading and fading over Xanthe-Chryse.

No type (a) cloud was seen in 1969, but bright condensations were frequently seen at the sunset limb (Figures 3a–c), in particular over the Tharsis-Amazonis area, a favorite location for this phenomenon (June 8–14, July 9–16). Similar effects have been observed at several other oppositions, in particular 1939 (July 18–25) and 1954 (Reference [3], p. 116).

4. Cloud Projecting at the Sunrise Terminator

This rare, but well-documented phenomenon (Reference [2], plate IX) proves that all early-morning white patches observed on Mars are not frost deposits on the surface; at least some are due to high-altitude icy haze (or in some cases dust layers) whose elevation can be estimated when the cloud is completely or, more often, partly detached from the terminator.

Occasionally, a bright haze or frost patch at sunrise will appear to protrude slightly beyond the terminator (for example see 1937 July 20 and Aug. 2), but most or all of the effect is caused by diffraction or seeing spread. However, when the prominence is not significantly brighter than surrounding areas at the terminator the projection must be caused by its elevation above the surface. A remarkable example was observed on July 7, 1969 (Figure 3c) between 3 05 and 3 34 UT; this partly-detached cloud was fully developed and immediately noticed, at the beginning of the observation at 3 05 UT; it extended from a faint, diffuse haze patch on the disk, some 15° south-south-west of Ismenius Lacus. The prominence was still perceptible at 3 25 UT and the last trace was seen at 3 34 UT. The haze patch on the disk remained visible at least until 3 45 or 3 50 UT. The geometric elevation of the top of the formation is estimated at some 15 ± 3 (p.e.)km; similar observations in the past have given elevations ranging from 5 to 30 km (Reference [1], p. 48; Reference [2], plate IX).

Acknowledgments

The 1969 Mars observations and their analysis have been supported by the U.S.A.F. Office of Aerospace Research.

References

[1] de Vaucouleurs, G.: 1959, *Sky Telesc.* **18**, 484.
[2] de Vaucouleurs, G.: 1954, *Physics of the Planet Mars*, Faber, London.
[3] Slipher, E. C.: 1962, *The Photographic Story of Mars*, Sky Publ. Corp., Cambridge.
[4] Antoniadi, E. M.: 1930, *La planète Mars*, Hermann, Paris.

CLOUD MOTIONS ON MARS

W. A. BAUM and L. J. MARTIN

Planetary Research Center, Lowell Observatory, Flagstaff, Ariz., U.S.A.

Abstract. A search of several thousand plates in the Lowell Observatory collection yielded 28 groups of plates on which the positions of well-defined transient bright spots (often assumed to be clouds) could be followed on a nearly daily basis. These groups of plates were from 15 different oppositions of Mars, starting from 1907 and ending with 1958. All but two of these spanned four nights or more, and the maximum interval covered was thirty nights. Whether they appeared to show motion or not, the successive positions and shapes of all apparently associated bright spots or clouds were plotted on Mercator projections with the use of a projection plate reader especially designed at the Planetary Research Center for planet image studies of this kind. Clouds near the limb were avoided.

The 28 groups of plates yielded 95 cloud histories. More than half appeared to be relatively stationary. Others showed definite motion well in excess of observational error but sometimes followed paths that partly doubled back upon themselves. The mean velocity for non-stationary clouds was found to be 5.6 km per hour, and the most commonly occurring direction of motion was eastward, particularly at high latitudes. The range of velocities found by this mapping procedure is nearly an order of magnitude smaller than values that have been estimated earlier by others from visual observations. These earlier observations are evidently in error, unless there exist clouds at high elevation, visible only on the limb, that can move much faster than those that were mapped from this photographic survey. More clouds were found in the northern hemisphere than in the southern, and there seemed to be an avoidance of the relatively darker areas of the Martian surface. Certain regions seem to be more favored than others. A few recurrences at identical positions suggest the existence of related topographic features.

1. Introduction

The title of this paper may turn out to be a misnomer, because the objects we have studied may not necessarily be clouds of suspended particles in the Martian atmosphere. It is not the purpose of this paper to attempt making any distinction between true clouds and surface phenomena, and it is entirely possible that our various cases include both. We have applied the term 'cloud' to any distinct transient local brightening that is not a normal tone of the surface feature in the area where it occurs. Many of these transient bright spots show no detectable movement at all, particularly in such well known areas as Hellas, Elysium, Tharsis, and Amazonis. On the other hand, the relatively stationary brightenings do not always occur in exactly the same positions and configurations.

The Lowell Observatory plate collection includes about 5000 plates of Mars covering a 60-yr period through 1965. More recent observations are being made on cine film and will be studied separately. A search was made through the plate collection to find cases in which the local positions of well-defined clouds could be followed on a nearly daily basis for significant lengths of time. This search yielded 28 groups of plates from fifteen different oppositions of Mars starting with 1907 and ending with 1958. Further work remains to be done on 1956, 1961, 1963, and 1965. All but two of the 28 selected cases spanned four nights or more, and the maximum interval covered was 30 nights.

Sagan et al. (eds.), Planetary Atmospheres, 320–328.

This study of the Martian clouds was limited to the smaller and more discrete types, because those covering broader areas are more difficult to map to the accuracy needed for comparing the positions and configurations seen on various plates. The study was also limited to clouds well within the visible disk so as to maintain a high order of accuracy. Limb clouds were avoided, because their positions cannot be determined with sufficient accuracy. Indeed, our inferred velocities differ by an order of magnitude from those that have been estimated earlier by others on the basis of limb cloud observations.

2. Procedure

The first step was a plate-by-plate search for those Mars plates which had clouds that could be mapped. Small sketch maps showing the approximate configurations and locations of all clouds were made for each plate. All readable plates were included, regardless of the filter color. The sketch maps were then reviewed to determine which plates showed clouds that might be related to clouds seen on other plates within a reasonable time span. All possible relationships were included, regardless of the motion or non-motion that appeared to be indicated.

A specially designed optical projector was used for measuring cloud positions on the images. Each projected Mars image was adjusted in size, orientation, and position to fit a transparent coordinate graticule superimposed on the projection screen. This graticule was an orthographic projection with latitude and longitude lines at ten-degree intervals. Different orthographic graticules were used to accommodate different tilt angles of the Martian axis in two-degree steps with respect to the line of sight.

The clouds were first outlined with a grease-pencil on a transparent plastic sheet covering the projection screen. This outline was then transferred to a Mercator projection by drawing it on a transparent plastic sheet covering a sliding Mercator graticule on top of a Mercator map of the Martian surface. The sliding Mercator graticule, with ten-degree intervals in both coordinates, was placed so that the location of its central meridian on the map corresponded with the central meridian in the photograph. Thus the Mercator projection lines corresponded with the orthographic projection lines.

Final maps were compiled from groupings of these plastic overlays by plotting together the successive positions of apparently related clouds. All together, 29 such maps were prepared, one for each of 28 selected groupings of overlays, except for one case that required two maps to represent the situation completely. When several plates taken during the same day showed the same cloud in about the same position, these cloud outlines were averaged in position and size to show a single symbol for that date. Clouds that could not be related to the others were usually omitted.

3. Cloud Maps

Sample maps are shown in Figures 1 and 2. Each covers only a part of Mars. The ten-degree intervals of the Mercator projection are equivalent to about 590 km at the equator. As observed from the earth, such an interval subtends between 1 and 2 sec

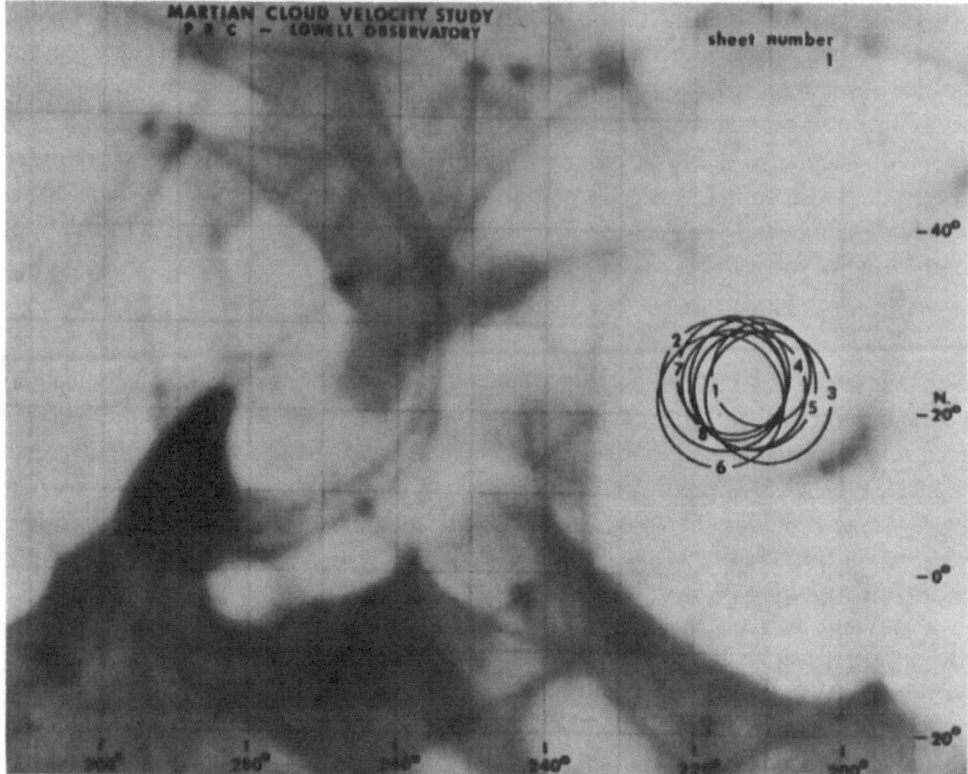

Fig. 1. A typical stationary bright spot in Elysium as it appeared on 33 photographs taken in 1907 (11 July through 18 July). The scatter in the positions of the circles (one for each night) is believed to represent our mapping accuracy.

of arc during the months close to a Martian opposition. We would not be able to say that any of our clouds are smaller than 300 km, although they may in fact be. Like the examples shown in Figures 1 and 2, the maps have all been oriented with north at the top and areographic east at the right. Longitude is numbered from 0 to 360 degrees in a westerly direction. Since Mercator projections were used, azimuths may be drawn as straight lines. The base map of the Martian surface is taken from the NASA-Air Force chart 'Mariner 69 Mars Chart (MEC-2)' drawn under an ACIC contract at the Lowell Observatory. The projection lines at ten-degree intervals have been included from this map, but feature names have been omitted.

Each different number on a symbol represents a different Martian day in chronological order. These days are often consecutive but not always. Gaps between dates arise, either because plates were lacking or because the available plates were not of readable quality. There was no clear-cut case of a cloud totally disappearing one day and reappearing the next. The lower-case letters are an aid in referencing groups or clusters of symbols. The grouping of symbols and the assigning of letters to these groups have necessarily been somewhat arbitrary.

Although a few cases may be ambiguous, most of the transient bright spots or clouds on our maps can be divided into two groups: moving and non-moving. Figure 1

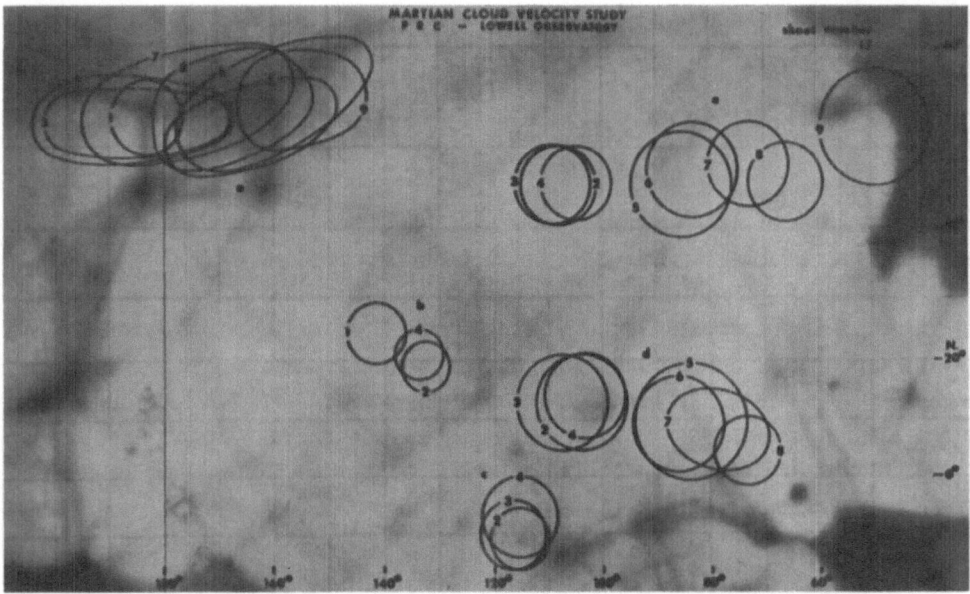

Fig. 2. An array of bright spots or clouds showing various degrees of sequential displacement, as they appeared on 19 photographs taken in 1937 (1 June through 13 June). Rates of motion inferred from the two cases shown top-left and top-right are about 3 km per hour and 9 km per hour, respectively.

shows an apparently non-moving local brightening in Elysium as it appeared in 1907. Insofar as we can tell, the scatter in the positions of the circles in Figure 1 is merely a representation of our mapping accuracy. Any progressive motion during the eight-day period covered by these observations was less than 1 km per hour, although brief random motions could conceivably have exceeded that and remained undetected.

Figure 2 shows an array of transient brightenings having various degrees of motion and non-motion. In the lower half of the figure is the well known 'W-cloud' region where transient brightenings of this kind frequently occur. It is very interesting to compare the appearance of this region in various months and years. For example, the bright spot at about 10°S latitude and 120°W longitude (scantily documented in this particular case) tends to be stationary and to reappear frequently at the same place. Other parts of the configuration in the lower half of Figure 2 tend to vary much more in position and apparent motion. Stationary clouds and moving clouds are in fact found to occur at exactly the same places at different times.

Figure 2 is of interest primarily because it includes two good illustrations of progressive eastward displacement at about 50°N latitude. The overlaid oval shapes at the left side of Figure 2, in the Propontis region, are especially unambiguous. However, this cloud and the one at the right of it both illustrate the jerky manner in which such clouds appear to move. To some degree, they act as though they jump from one favored spot to another, suggestive of terrain relationships. There are also cases in which such motion seems to double back upon itself, or to split into two or more spots in nearby positions. Observations now being obtained on a much more continuous schedule under a cooperative Planetary Patrol Program involving six observatories should be of considerable help in reducing future ambiguities of the kind just described.

4. Summary of Results

From an examination of all 29 maps, we find that there appear to be 95 usable cloud histories. Although this selection involves some degree of interpretation, we tried to apply consistent criteria of association in identifying the 95 cases.

Of these, 52 appear to be relatively stationary, while the remaining 43 show evidence of sequential displacement. A cloud was regarded as stationary if its mapped locations from a sequence of dates remained unchanged within the accuacy of measurement, like the example shown in Figure 1. In particular, motion was not regarded as real unless the first and last positions differed by at least 400 km. Since many of the moving clouds are found on the same sets of plates as stationary ones, there can be no question about the reality of the motion of one relative to the other.

For the 43 clouds regarded as showing evidence of sequential displacement, velocity vectors were measured on the basis of the first and last positions. As already mentioned, such clouds tend to move jerkily. In some instances they also double partly

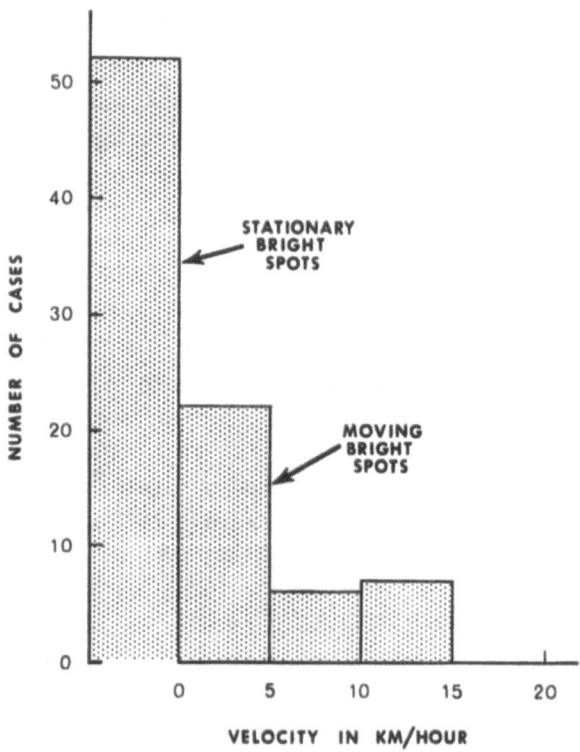

Fig. 3.   Distribution of Martian 'cloud' velocities inferred from mapped displacements based on sequences of photographs in the Lowell Observatory plate collection, 1907–1958. This sample was limited to clouds observed over a span of at least four nights.

back upon their tracks, so that the net displacement between the first and last positions can be less than the total excursion.

Since velocities based on the first and last positions of a cloud are vulnerable to mapping errors unless the time span of observation was long enough, we decided to confine velocity analysis to 35 cases (among the total 43) for which observations spanned at least four nights. These 35, taken together with the 52 clouds regarded as stationary, provide a high-accuracy sample of 87 clouds that yield the histogram in Figure 3. The mean velocity of all 87 clouds is 2.25 km per hour. The mean velocity for the 35 non-stationary cases alone is 5.60 km per hour.

The histogram in Figure 4, which includes all 95 original cases, shows that the transient bright spots or clouds did not occur randomly over the whole planet. Only about

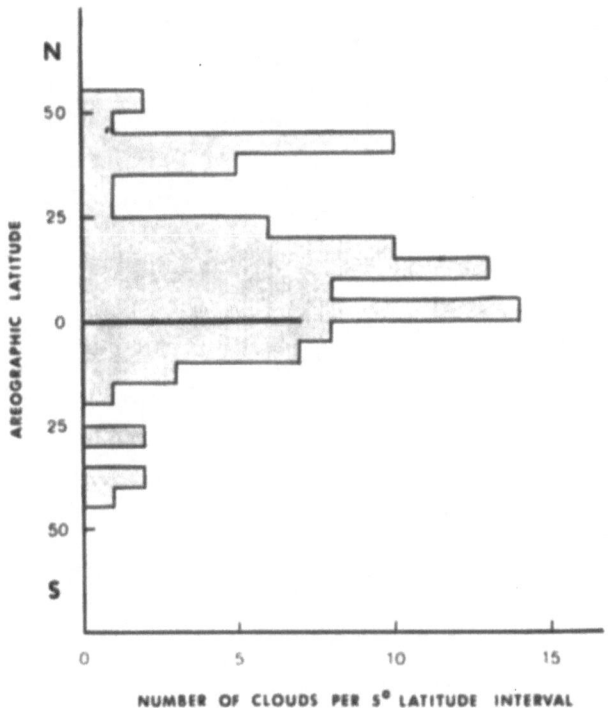

Fig. 4. Latitude distribution of the same Martian 'clouds' that yielded Figure 3, except that cases observed only two nights were included.

one-fourth of them occurred in the southern hemisphere. Almost half of the clouds are found in a relatively narrow belt between the equator and 20°N latitude. This result differs from Wells' (1967) summary of earlier cloud sightings that probably contained a stronger seasonal selection effect. In Figure 5 the locations of our clouds are mapped, the dots representing cloud positions on individual nights. It is interesting to note that there seems to be a partial avoidance of the darker areas on Mars. The latitude effects in Figure 4 could, of course, be merely a manifestation of this dark-area avoidance,

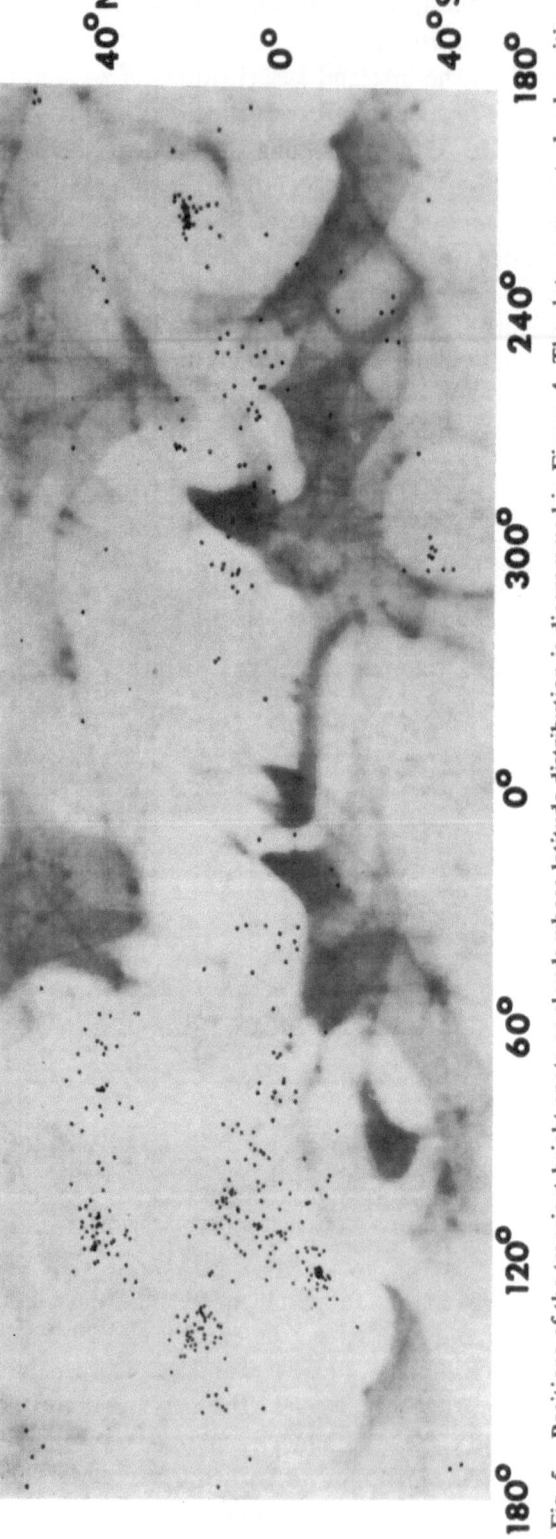

Fig. 5.   Positions of the transient bright spots or clouds whose latitude distribution is diagrammed in Figure 4. The dots represent cloud positions on individual nights.

because a large percentage of the dark area of Mars lies in the southern hemisphere. This apparent avoidance of dark areas cannot be an observational selection effect, because the visibility of bright clouds will be greater – not less – over dark areas than over light areas.

Figure 6 shows that somewhat more clouds move eastward and westward than northward and southward. This diagram represents the number of cases falling within each 45-degree sector of the compass. The sample was limited to the 35 non-stationary clouds observed over a time span of four nights or more. On average, those moving eastward tend to have the higher velocities and to occur farther from the equator than those moving in other directions. The sample is not yet large enough to relate these motions clearly to an atmospheric circulation pattern.

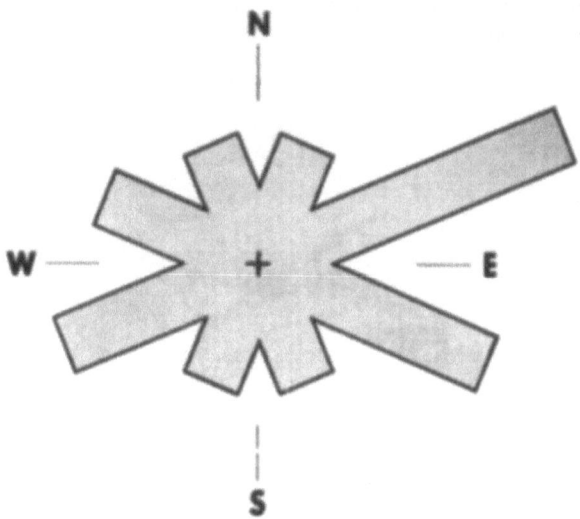

Fig. 6.   Directional distribution of Martian 'cloud' motions. This sample includes only the 35 cases for which photographic observations spanned at least four nights and which moved more than 400 km.

Previous observations of Martian cloud movement were reviewed by Gifford (1964). Owing to the comprehensive coverage of Gifford's fine review, we do not present a bibliography of earlier work here. Gifford tabulated 36 velocities and directions of motion that had been reported by various observers since 1873. Many were derived from visual observations, and 20 of the 36 were based on the appearance of clouds at the limb or terminator. The mean of Gifford's tabulated velocities is 36 km per hour, which is many times larger than the mean of our values derived from the actual mapping of progressive positions in the present study. The highest single value in Gifford's list is 124 km per hour, credited to the observatoires Jarry-Desloges, and it is more than eight times larger than the highest velocity found in any of our 95 mapped cases. It now seems likely that these earlier observations are largely in error, unless there exists a faster moving class of Martian clouds, possibly at high elevation, that can be seen only when on the limb or terminator. It was because of the large uncertainty associated with limb or terminator observations that those regions were intentionally avoided in the present study.

Acknowledgments

Our work was carried out with the joint support of the Jet Propulsion Laboratory and of NASA Headquarters, and we gratefully acknowledge both. All 29 of our cloud-history maps are contained in a contract report submitted to the Jet Propulsion Laboratory (Martin and Baum, 1969).

References

Gifford, F. A., Jr.: 1964, *Monthly Weather Rev.* **92**, 435.
Martin, L. J. and Baum, W. A.: 1969, *Final Report B*, under JPL Contract 951547, August.
Wells, R. A.: 1967, *Astrophys. J.* **147**, 1181.

F. UPPER ATMOSPHERES

THE EFFECT OF ATMOSPHERIC DYNAMICS
ON THE UPPER ATMOSPHERE PHENOMENA OF MARS AND
VENUS

MIKIO SHIMIZU

Dept. of Physics, Ochanomizu University, Tokyo, Japan and Institute of Space and Aeronautical Science, University of Tokyo, Komaba, Meguro-ku, Tokyo, Japan

Abstract. By the observations of Mariners and Veneras, oxygen atom deficiencies and very low exospheric temperatures were found in the upper atmospheres of Mars and Venus. There are two types of interpretations for these phenomena: photochemical ones (CO_3 hypothesis), and dynamical ones. In this paper, the latter point of view is summarized. It was emphasized that the effect of atmospheric mixing due to general circulation and/or eddy diffusion is important for the understanding of the Cytherean and Martian atmospheric phenomena.

A series of 'Mariner' space probes measured the electron densities in the Martian and Cytherean ionospheres by using the S-band radio occultation technique and found that the altitudes of the peaks of the density profiles were very low and that the widths of the profiles were very narrow. The characteristics may be correlated with the fact that these atmospheres are mainly composed of CO_2. After the success of Mariner 4 observation, E, F_1, and F_2 models have been presented to interpret these characteristics. Recently, however, the students in this field have reached a general agreement that the ionospheres are of the F_1-layer type (the observed temperature profiles in the lower atmospheres are too high for the F_2-layer model, and the E-layer model has difficulty in explaining the non-existence of the F_1 layers).

If these ionospheres are of F_1-layer type, the amounts of oxygen atoms in the upper atmospheres of these planets should be so small as to depress their F_2 layers, yet it is certain that CO_2 is dissociated to CO and O by Schumann-Runge radiation. Therefore the whereabouts of O atoms on these planets raises a new question. The deficiency of O atoms may be approached from two standpoints: the photochemical explanation and the dynamical one. Recently McElroy and Hunten (1970) speculated that $O(^1D)$ from the dissociation of CO_2 reacts with CO_2 to make CO_3^* and that this activated complex might easily be deactivated to its ground state by emitting photons. Then CO_3 may react with CO to make $2CO_2$ and the overall process result in a very rapid recombination of CO and O. CO_3^* may, however, be much more rapidly decomposed to CO_2 and $O(^3P)$. In other words, the reaction between $O(^1D)$ and CO_2 may be of the pre-dissociation type, as in the case of $O(^1D)$ and N_2; therefore the CO_3 explanation may not work in planetary upper atmospheres.

This paper presents an alternate, dynamical, interpretation for the O-atom deficiency: the dilution of the dissociation products in the upper atmosphere by atmospheric mixing. In order to discuss a problem of this type, we must solve a set of equations of continuity and motion simultaneously. The coupling between these equations is in particular strong in the case of the dissociation of the major atmospheric component. (It should be noted that the flux equation used by McElroy and Hunten – Equation

(3) of §3 in their paper – can apply only to the case of dissociation of a minor component.) We have already solved this problem under proper boundary conditions (Shimazaki and Shimizu, 1970; Shimizu, 1969a). Here the physical implication of the results will be explained schematically.

At first we shall discuss the upper atmospheric phenomena by taking into account the photochemical processes and the effect of molecular diffusion, and by neglecting the effect of turbulence. The dissociation cross-section of CO_2 is much smaller than that of O_2. Thus the Schumann-Runge radiation penetrates deeper into the atmosphere in the case of Mars and Venus than it does on earth. The upper part of Figure 1 describes this situation. It is noteworthy that there is a subtle difference between Mars and Venus, because of the difference of their distances from the sun. The effect of molecular diffusion decreases the densities of the dissociation products on Mars, while it has no influence on Venus.

Next we shall take the effect of turbulence into account. The planetary atmospheres

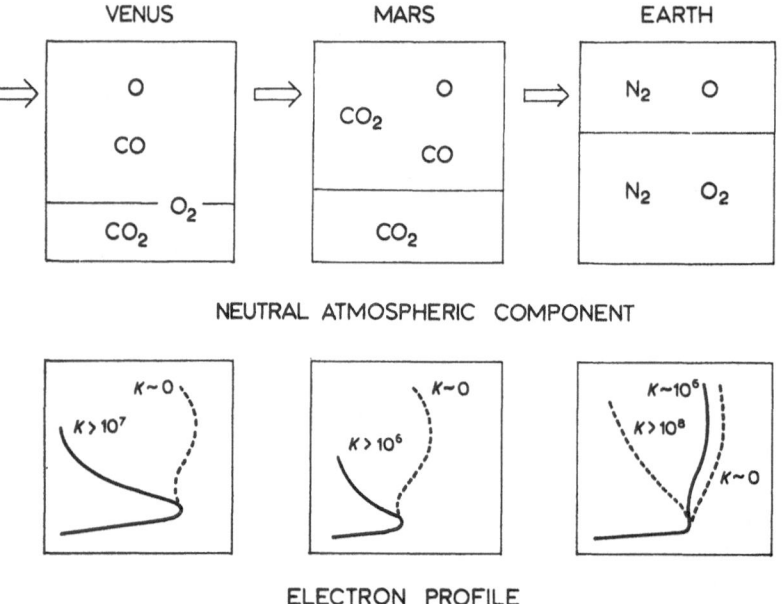

Fig. 1. A comparison of upper atmospheres and ionospheres of Venus, Mars, and Earth. In the upper part, the distributions of atmospheric constituents in the upper atmospheres of the planets are schematically shown (see text). In the lower part, the calculated electron density profiles for various values of κ, eddy diffusivity, are shown. The bold lines correspond to the observational curves.

are usually well mixed under their turbopauses, the positions of which are marked by the horizontal arrows in Figure 1. It may be intuitively clear that the effect of turbulence works more efficiently on Mars and Venus than on Earth. In the lower part of Figure 1, we plotted the calculated electron density profiles for various eddy diffusion coefficients κ on each planet. In the case of Earth, an F_2 layer exists even in the case of $\kappa = 10^6$ cm²/sec. In order to depress the F_2 layer, a large value of κ, 10^8 cm²/sec, is required. On the other hand, κ of the order of 10^6

cm²/sec is enough to depress the F_2 layer on Mars. In the case of Venus, 10^7 cm²/sec is necessary for κ, because the dissociation of CO_2 is more complete on this planet.

In the case of Mars, $O(^3P)$ and CO may be transported downwards to the level where the time constants of mixing and recombination become nearly equal (Figure 2).

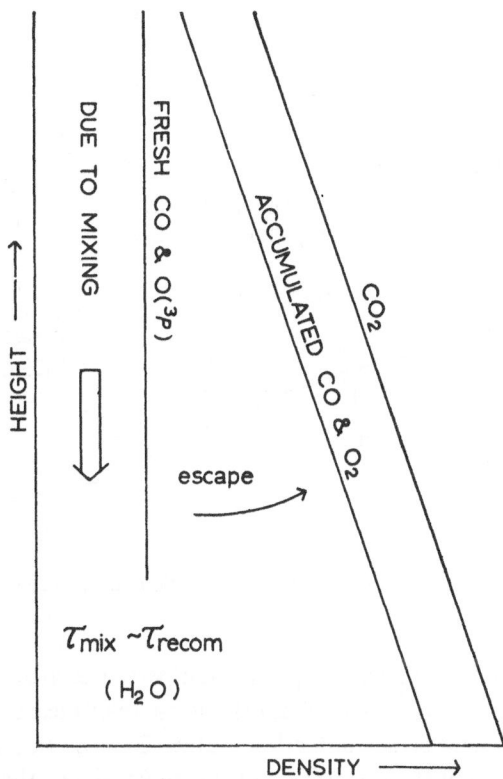

Fig. 2.   Recombination and accumulation mechanism of CO in the lower atmosphere of Mars.

At this level they will be almost completely recombined back to CO_2 by the catalytic reaction with H_2O (Reeves *et al.*, 1966). This reaction works only in the lower atmosphere where the atmospheric density is high. A very small production of O_2 and CO may be expected, and they will accumulate during a long time. As for Venus, the high temperature of its lower atmosphere may play a role to keep the concentration of O_2 and CO at the thermochemical equilibrium values.

Now, we shall consider another evidence for a strong dynamical effect in the upper atmosphere of Mars. From the electron density profile obtained by Mariner 4, a very low plasma temperature in the Martian ionosphere, 300 K, was obtained. The exospheric temperature, T_{exo}, can be a little lower than this value. McElroy (1969) calculated T_{exo} by taking into account the cooling of the atmosphere due to CO_2 infrared emission near the mesopause; the cooling causes a steep temperature gradient between the exosphere and mesopause. The heat deposited at the exosphere may be transported to the lower region by molecular heat conduction. Then T_{exo} may be decreased (Figure 3). McElroy obtained a value however, of about 500 K, much higher than the observed one. As the mesopause temperature was already low enough, it was in principle difficult to decrease T_{exo} less than this value by this mechanism. We think that one way to solve this puzzle is to increase the conductivity between these regions by taking into account the heat transport due to turbulence.

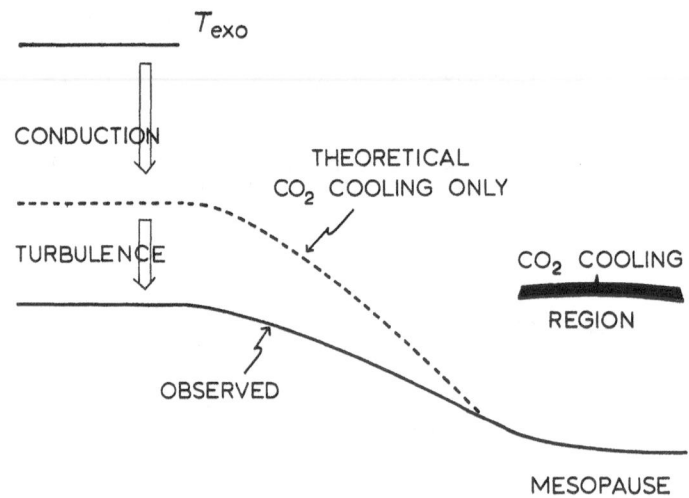

Fig. 3. Explanation for the low exospheric temperature of Mars.

At this symposium Kliore suggested plasma temperatures around 450 K at the time of Mariner 6 and 7. The T_{exo} derived from the ultraviolet measurement (probably of O atoms) appears to suggest a similar low value. But for this time the calculated value of McElroy was reported as 700 K, again much higher than observed. Such a situation was already predicted in our paper (Shimizu, 1969b) before the success of Mariner 6 and 7. By taking into account the heat transport by the eddy diffusion, we have solved the heat balance equation. By comparing the results with the new data presented in this symposium, we estimate the value of the eddy diffusivity κ in the Martian upper atmosphere as $7-10 \times 10^7$ cm²/sec. Horizontal atmospheric mixing due to a wind in the upper atmosphere of the velocity of several hundred kilometers/sec can equivalently explain the low exospheric temperature (see Shimizu, 1969a).

Gierasch and Goody (1968) derived theoretically so high an eddy diffusivity in the lower atmosphere of Mars as 10^8 cm²/sec. The internal gravity waves from this region may go up to the upper atmosphere and will dissipate there by interaction with the wind shear. It has already been suggested by Goody and Belton (1967) that the Martian upper atmosphere can easily be turbulent due to radiative instability. Thus strong turbulence with $\kappa \sim 10^8$ cm²/sec may be expected in the upper atmosphere. It may be argued that the temperature near mesopause may be a little increased by this forced mixing effect. The CO_2 cooling mechanism works, however, more effectively as the temperature increases. Therefore, it can be a good thermostat for the mesopause temperature.

In the Cytherean case, two different T_{exo} values have been presented. From the deuterium model (Barth, 1968), it might be 650 K. If we take, however, the hot-atom model (Barth, this symposium), it should be 325 K. At first we shall assume that T_{exo} is 650 K. This value is in good agreement with the calculated value by McElroy (1969) for a pure CO_2 atmosphere. (This does not mean that the CO_3 scheme works; a strong mixing effect can also explain the apparent depression of CO_2 dissociation.) This suggests that κ may be less than 10^7 cm²/sec on Venus, because in this case the time constant of the mixing effect is longer than that of the CO_2 cooling effect and we can expect no effect of turbulence or general circulation on T_{exo}. Therefore, together with the result of the discussion on the Cytherean ionosphere, we may conclude that the value for κ on the Venusian upper atmosphere is about 1×10^7 cm²/sec. (Of course, wind with a velocity of 100 m/sec has a similar effect.) If the lower value reported by Barth in this symposium is right, however, the situation may be changed and we must assume a stronger wind or a larger eddy diffusivity in the upper atmosphere of Venus.

In summary, we have explained many observational facts in the upper atmospheres of Mars and Venus, by using only one parameter κ. Without the recognition of the strong dynamical effect, we might adopt many artificial assumptions. For instance, McElroy and Hunten assumed a speculative CO_3 scheme. Then the formation of O_2 and CO will be forbidden in the upper atmosphere. Therefore they could not help assuming very complex chemical processes to explain the existence of CO in the Martian atmosphere. Furthermore, in order to resolve the discrepancy in the Martian exospheric temperature, Cloutier et al. (1969) speculated that the solar wind might change the electron density profile of Mars. The reason why the same solar wind fails to affect the Cytherean ionosphere is not yet clearly explained. Furthermore, they neglected an important effect, the rotation of Mars, in their discussion.

Finally we hope to remark here that we have already obtained the result that a similar type of strong dynamical effect may exist in the Jovian upper atmosphere as well as in the Martian and Cytherean ones (Shimizu, 1970).

Note added in proof: Recently, Slanger and Black confirmed directly by resonance fluorescent technique that the quantum yield of O (P^3) in CO_2 photolysis is unity due to very rapid deactivation. Shimizu showed that the wind due to solar EUV differential heating decreased T_{exo} of Venus and Mars effectively (ISAS report, University of Tokyo, Nos. 455 and 456.

References

Barth, C. A.: 1968, 'Interpretation of the Mariner 5 Lyman Alpha Measurements', *J. Atmospheric Sci.* **25**, 564.

Cloutier, P. et al.: 1969, 'Modification of the Martian Ionosphere by the Solar Wind', *J. Geophys. Res.*

Gierasch, P. and Goody, R.: 1968, 'A Study of the Thermal and Dynamical Structure of Martian Lower Atmosphere', *Planetary Space Sci.* **16**, 615.

23—P.A.

Goody, R. and Belton, M. J. S.: 1967, 'Radiative Relaxation Times on Mars', *Planetary Space Sci.* **15**, 247.

McElroy, M. B.: 1969, 'Structure of Venus and Mars Atmospheres', *J. Geophys. Res.* **74**, 29.

McElroy, M. B. and Hunten, D. M.: 1970, 'The Photochemistry of CO_2 in the Atmosphere of Mars', to appear.

Reeves, R. R. *et al.*: 1966, 'Photochemical Equilibrium Studies of Carbon Dioxide and Their Significance for the Venus Atmosphere', *J. Phys. Chem.* **70**, 1637.

Shimazaki, T. and Shimizu, M.: 1970, 'Model Calculation of the Martian Upper Atmosphere and Ionosphere with Transport Effects', *Rept. Ionos. Space Res. Japan* **24**, 80.

Shimizu, M.: 1969a, 'A Model Calculation of the Cytherean Upper Atmosphere', *Icarus* **10**, 11.

Shimizu, M.: 1969b, 'Atmospheric Structures of Planets', *Ōyō-butsuri* **38**, 284 (in Japanese).

Shimizu, M.: 1970, 'The Upper Atmosphere of Jupiter: Dissociation and Ionization of Hydrogen' *Icarus*.

CO_2^+ DAYGLOW ON MARS AND VENUS

A. DALGARNO

Harvard College Observatory and Smithsonian Astrophysical Observatory, Cambridge, Mass., U.S.A.

and

T. C. DEGGES*

Technology Division, GCA Corporation, Bedford, Mass., U.S.A.

Abstract. Quantitative calculations are presented of the intensities of the CO_2^+ bands present in the dayglows of Mars and Venus and it is argued that fluorescent scattering by CO_2^+ ions and photoionization of CO_2 are the main sources of excitation. An estimate is made of the intensity of 3914 Å emission that would arise if the atmosphere contained N_2.

1. Introduction

The ground-based observations of Kozyrev (1954) indicate the occurrence of the $(X^2\Pi_g - A^2\Pi_u)$ emission bands of CO_2^+ in the dayglow of Venus (Polyakova *et al.*, 1963) and the Mariner 6 observations of Barth *et al.* (1969) have established that the $(X^2\Pi_g - A^2\Pi_u)$ and $(X^2\Pi_g - B^2\Sigma_u^+)$ emission bands are important components of the dayglow of Mars. The excited $A^2\Pi_u$ and $B^2\Sigma_u^+$ levels of CO_2^+ can be populated by direct photoionization of CO_2 by solar radiation

$$CO_2 + h\nu \rightarrow CO_2^+(A^2\Pi_u, B^2\Sigma_u^+) + e \tag{1}$$

by fluorescent scattering of solar radiation by pre-existing CO_2^+ ions

$$CO_2^+ (X^2\Pi_g) + h\nu \rightarrow CO_2^+ (A^2\Pi_u, B^2\Sigma_u^+) \tag{2}$$

and by simultaneous excitation and ionization of CO_2 by photoelectron impact

$$e + CO_2 \rightarrow e + CO_2^+ (A^2\Pi_u, B^2\Sigma_u^+) + e. \tag{3}$$

In this paper we shall investigate each of the three mechanisms and calculate their contributions to the CO_2^+ dayglow on the planets Mars and Venus.

2. Photoionization of CO_2

Cross sections for the absorption of radiation by CO_2 have been measured by Tanaka *et al.* (1960), by Tanaka and Ogawa (1962), by Nakata *et al.* (1965), by Cook *et al.* (1966) and by Dibeler and Walker (1967).

Photons of wavelengths λ shorter than 902 Å can ionize CO_2 leaving CO_2^+ in its ground electronic state

$$CO_2 + h\nu(\lambda < 902 \text{ Å}) \rightarrow CO_2^+(X^2\Pi_g) + e. \tag{4}$$

For $\lambda < 716$ Å, it is energetically possible to populate also the first excited state of CO_2^+ in a photoionizing transition

$$CO_2 + h\nu(\lambda < 716 \text{ Å}) \rightarrow CO_2^+(A^2\Pi_u) + e \tag{5}$$

* Now at Visidyne Inc., Woburn, Mass., U.S.A.

Sagan et al. (eds.), Planetary Atmospheres, 337–345.

and for $\lambda < 687$ Å the second excited state

$$CO_2 + h\nu(\lambda < 687 \text{ Å}) \rightarrow CO_2^+(B^2\Sigma_u^+) + e. \tag{6}$$

For wavelengths shorter than 640 Å, it is possible to produce CO_2^+ ions in the third excited state

$$CO_2 + h\nu(\lambda < 640 \text{ Å}) \rightarrow CO_2^+ (C^2\Sigma_g^+) + e. \tag{7}$$

The total photoionization cross sections for CO_2 have been measured by Nakata *et al.* (1965), by Cairns and Samson (1965), by Cook *et al.* (1966), and by Dibeler and Walker (1967), and the branching ratios appropriate to the $X^2\Pi_g$, $A^2\Pi_u$, $B^2\Sigma_u^+$, and $C^2\Sigma_g^+$ states of CO_2^+ have been measured by Bahr *et al.* (1969).

If we assume tentatively that the neutral components of the atmospheres of Mars and Venus are effectively pure CO_2, the total rates of population of the electronic levels of CO_2^+ by photoionization can be calculated directly from the cross section data and the incident solar flux. Our adopted solar flux is based on the 1967 measurements reported by Hinteregger (1970) for which the solar 10.7 cm flux was $F_{10.7} = 144$. We scaled these data appropriately for Mars and Venus and for the 10.7 cm flux $F_{10.7} = 162$, adjusted for burst, that was measured at the time of the Mariner 6 flight past Mars.

The results for Mars and for Venus are presented in Table IA and IB respectively for various solar zenith angles Z. The table contains the rates of population in a vertical column. If no deactivation or cascading occurred, the rates would be also the emission intensities in photons cm^{-2} sec^{-1}. Deactivation of the $A^2\Pi_u$ and $B^2\Sigma_u^+$ states is

TABLE IA

Total production rates of CO_2^+ states by photoionization on Mars in units of 10^9 cm^{-2} sec^{-1} at various solar zenith angles $Z°$

Z	$X^2\Pi_g$	$A^2\Pi_u$	$B^2\Sigma_u^+$	$C^2\Sigma_g^+$
0	6.3	2.8	4.9	1.5
30	5.5	2.4	4.2	1.3
50	4.1	1.8	3.2	0.9
60	3.2	1.4	2.4	0.7
70	2.2	1.0	1.7	0.5
80	1.1	0.5	0.9	0.3

TABLE IB

Total production rates of CO_2^+ by photoionization on Venus in units of 10^9 cm^{-2} sec^{-1} at various solar zenith angles $Z°$

Z	$X^2\Pi_g$	$A^2\Pi_u$	$B^2\Sigma_u^+$	$C^2\Sigma_g^+$
0	25.0	11.2	19.4	6.5
30	21.7	9.7	16.8	5.7
50	16.1	7.2	12.4	4.2
60	12.5	5.6	9.7	3.3
70	8.6	3.8	6.6	2.2
80	4.4	1.9	3.4	1.1

negligible since the radiative lifetimes are about 10^{-7} sec (Schwenker, 1965; Anton, 1966; Hesser, 1968) but the $\mathbf{C}^2\Sigma_g^+$ state decays to both the $\mathbf{A}^2\Pi_u$ and $\mathbf{B}^2\Sigma_u^+$ states giving rise to emission in the red and infrared region of the spectrum.

We assume arbitrarily that the branching ratio for cascade into the $\mathbf{A}^2\Pi_u$ and $\mathbf{B}^2\Sigma_u^+$ states is 4:1. The predicted intensities in kilorayleighs of the $\mathbf{X}^2\Pi_g - \mathbf{A}^2\Pi_u$ and $\mathbf{X}^2\Pi_g - \mathbf{B}^2\Sigma_u^+$ emission systems are given in Table II.

TABLE II

Predicted intensities in kilorayleighs of the $\mathbf{X}^2\Pi_g - \mathbf{A}^2\Pi_u$ and $\mathbf{X}^2\Pi_g - \mathbf{B}^2\Sigma_u^+$ band systems resulting from photoionization on Mars and Venus at various solar zenith angles $Z°$

Mars			Venus	
Z	$\mathbf{A}^2\Pi_u$	$\mathbf{B}^2\Sigma_u^+$	$\mathbf{A}^2\Pi_u$	$\mathbf{B}^2\Sigma_u^+$
0	4.0	5.2	15.8	20.5
30	3.5	4.5	13.7	17.8
50	2.6	3.4	10.2	13.2
60	2.0	2.6	7.9	10.3
70	1.4	1.8	5.4	7.0
80	0.7	0.9	2.7	3.6

3. Fluorescent Scattering by CO$_2^+$

The rate of population for each molecular ion of CO$_2^+$ by fluorescent scattering is given by

$$g = 8.85 \times 10^{-21} \, If\lambda^2 \, \text{sec}^{-1}, \tag{8}$$

where I is the solar continuum intensity at the absorption wavelength λ Å measured in photons sec^{-1} Å^{-1} and f is the band oscillator strength. For the lifetime of the $\mathbf{A}^2\Pi_u$ state, Schwenker (1965) has measured a value of 1.39×10^{-7} sec and Hesser and Dressler (1966) and Hesser (1968) a value of 1.1×10^{-7} sec. The values are in harmony with that derived by Anton (1966) from pressure quenching measurements. The system oscillator strength corresponding to a lifetime of 1.1×10^{-7} sec is $f = 1.5 \times 10^{-2}$. Taken in conjunction with the relative probabilities of Poulizac and Dufay (1967) and the solar fluxes at the band heads, this oscillator strength gives a g value for the $\mathbf{A}^2\Pi_u$ population of 1.1×10^{-2} at Mars and 4.9×10^{-2} at Venus.

For the $\mathbf{B}^2\Sigma_u^+$ state of CO$_2^+$, Hesser and Dressler (1966) and Hesser (1968) have measured a lifetime of 1.19×10^{-7} sec consistent with an absorption oscillator strength of 5.3×10^{-3}. The corresponding g value at Mars is 1.2×10^{-3} and the g value at Venus is 5.2×10^{-3}. In making these estimates of scattering efficiencies, Fraunhofer absorption has been ignored.

If we assume tentatively that the ionic components of the atmospheres are pure CO$_2^+$, the total rates of emission of the two CO$_2^+$ band systems can be calculated directly from the measured electron densities. For Mars, the total electron content derived from Mariner 4 occultation data was about 4.2×10^{11} cm^{-2} for a solar zenith angle of 67° (Fjeldbo et al., 1966, Fjeldbo and Eshleman, 1968) and from Mariner 6 data it was about 1.2×10^{12} cm^{-2} for a solar zenith angle of 56° (Fjeldbo *et al.*, 1969). For Venus the total electron content derived from Mariner 5 occultation data was about 1.6×10^{12} cm^{-2} for a solar zenith angle of 33° (Kliore *et al.*, 1967).

The measured Venus ionization profile is consistent with a pure CO_2 atmosphere and dissociative recombination of CO_2^+ (McElroy, 1969). We have extended the equilibrium calculations of McElroy (1969) to other zenith angles and we obtain the total ionization contents of Table III, scaled linearly to that observed at 33°.

TABLE III

Total ionization content in 10^{11} cm^{-2} on Mars and Venus
at various solar zenith angles $Z°$

Mars			Venus
Z	Mariner 4	Mariner 6	Mariner 5
0	6.4	15.1	17.7
30	6.0	14.1	16.8
50	5.2	12.4	14.7
60	4.8	11.2	13.2
70	3.9	9.4	11.2
80	3.0	7.3	8.4

We adopt a similar model for Mars, but scaled linearly to the observed total electron contents at 56° or 67°. The corresponding ionization contents on Mars at various solar zenith angles are included in Table III. The linear scaling procedure is an arbitrary one.

The contributions to the $X^2\Pi_g - A^2\Pi_u$ and $X^2\Pi_g - B^2\Sigma_u^+$ emission intensities from fluorescent scattering by CO_2^+ are given in kilorayleighs in Table IV.

TABLE IV

Predicted intensities in kilorayleighs of the $X^2\Pi_g - A^2\Pi_u$ and
$X^2\Pi_g - B^2\Sigma_u^+$ band systems resulting from fluorescent scattering on Mars and Venus at various solar zenith angles $Z°$. The CO_2^+
contents are based upon the Mariner 5 and Mariner 6 data

Mars			Venus	
Z	$A^2\Pi_u$	$B^2\Sigma_u^+$	$A^2\Pi_u$	$B^2\Sigma_u^+$
0	15.9	1.8	86.7	8.8
30	15.3	1.6	82.3	8.4
50	13.9	1.5	72.0	7.3
60	12.3	1.3	64.6	6.6
70	10.2	1.0	55.2	5.7
80	7.6	0.9	41.1	4.2

4. Electron Impact Excitation

The photoelectrons produced by photoionization lose energy through excitation and ionization of CO_2 and by elastic collisions with the ambient electrons. Some fraction of the ionizations leaves CO_2^+ in excited states. Precise predictions would involve detailed studies of the energy degradation of the photoelectrons of the kind carried out for the earth by Green and Barth (1967) and by Dalgarno *et al.* (1969). We can easily demonstrate however that photoelectron impact is a comparatively small source of population of the $A^2\Pi_u$ and $B^2\Sigma_u^+$ states of CO_2^+ on Mars and Venus.

The thresholds for ionization to the $X^2\Pi_g$, $A^2\Pi_u$, and $B^2\Sigma_u^+$ states occur at respectively 13.8 eV, 17.3 eV, and 18.1 eV. The energy flux of photoelectrons with energies greater than 17.3 eV is about 3.8×10^{11} eV cm^{-2} sec^{-1} on Mars and about 1.5×10^{12} eV cm^{-2} sec^{-1} on Venus for the sun at zenith. For fast electrons absorbed by CO_2 the mean energy per ion pair is about 35 eV, a value which must increase ultimately with decreasing initial electron energy. Thus the number of ionizations produced by electrons with energies in excess of 17.3 eV does not exceed 1.1×10^{10} cm^{-2} sec^{-1} on Mars and does not exceed 4.3×10^{10} cm^{-2} sec^{-1} on Venus. According to McConkey *et al.* (1968), the excitation functions of the $A^2\Pi_u$ and $B^2\Sigma_u^+$ states are similar in shape to the total ionization function (Rapp and Englander-Golden, 1965), the ratios being approximately 1/5 and 1/15 respectively. The corresponding upper limits to the $X^2\Pi_g - A^2\Pi_u$ and $X^2\Pi_g - B^2\Sigma_u^+$ emission intensities are respectively 2 kR and 700 kR on Mars and 9 kR and 3 kR on Venus, all values referring to $Z = 0°$.

The actual intensities may be of the order of half the upper limits. The estimated intensities from photoelectron impact are given in Table V for various solar zenith angles.

TABLE V

Estimated[a] intensities in kilorayleighs of the $X^2\Pi_g - A^2\Pi_u$ and $X^2\Pi_g - B^2\Sigma_u^+$ band systems resulting from photoelectron impacts on Mars and Venus at various solar zenith angles $Z°$

Mars			Venus	
Z	$A^2\Pi_u$	$B^2\Sigma_u^+$	$A^2\Pi_u$	$B^2\Sigma_u^+$
0	1.1	0.4	4.3	1.4
30	0.9	0.3	3.7	1.2
50	0.7	0.2	2.8	0.9
60	0.5	0.2	2.1	0.7
70	0.4	0.1	1.5	0.5
80	0.2	0.1	0.7	0.2

[a] Upper limits are obtained by doubling the entries. The ratio of *any* pair of entries is much more accurate than the individual entries.

5. Comparison with Mars Observations

Table VI summarizes the contributions from the three sources for Mars at a solar zenith angle of 30°. Photoionization and fluorescence scattering are important sources for both transitions. Photoionization is more important for one system and fluorescence scattering for the other. Fluorescence scattering assumes a larger role for both planets with increasing solar zenith angle and it is relatively more important on Venus than on Mars.

Photoelectron impact contributes not more than 5% to either transition on either planet at any solar zenith angle. Barth *et al.* (1969) have remarked that their Mars

TABLE VI

Theoretical intensities in kilorayleighs of the $X^2\Pi_g - A^2\Pi_u$ and $X^2\Pi_g - B^2\Sigma_u^+$ band systems resulting from photoionization, fluorescence scattering and photoelectron impact on Mars at 30° solar zenith angle

System	$X^2\Pi_g - A^2\Pi_u$	$X^2\Pi_g - B^2\Sigma_u^+$
Photoionization	3.5	4.5
Fluorescence scattering	15.3	1.6
Photoelectron impact	0.9	0.3
Total	19.7	6.4

spectrum is similar to that produced in the laboratory by the bombardment of CO_2 with electrons of 20 eV energy. According to the cross section data of McConkey et al. (1968), the ratio of system intensities produced by 20 eV electrons is somewhat greater than 3 consistent with our predicted ratio on Mars at 30° of 3.1, produced by a combination of photoionization and fluorescence scattering.

Our ratio refers to the total intensities, whereas the measurements of Barth et al. (1969) give the intensities above particular altitudes. If we have correctly identified the excitation mechanisms, the ratio of the intensity of the $X^2\Pi_g - A^2\Pi_u$ system to that of the $X^2\Pi_g - B^2\Sigma_u^+$ system should increase with increasing altitude because photoionization decreases with the scale height of the neutral atmosphere and fluorescent scattering decreases with the scale height of the ionized component (if the major ion remains CO_2^+). A detailed study of the Mars data of Barth et al. (1969) may provide a test of the solar wind model of Cloutier et al. (1969).

The increasing importance of fluorescent scattering with increasing altitude should be reflected also in a changing vibrational distribution within the $X^2\Pi_g - A^2\Pi_u$ electronic transition.

6. Vibrational Distributions

Measurements of photoionization cross sections of CO_2 in which the individual vibrational levels of the product electronic state are resolved have been carried out by Turner and May (1967) and by Spohr and Puttkamer (1967) for a wavelength of 584 Å. The values for the $A^2\Pi_u$ state decrease more slowly with increasing v' than do the theoretical Franck-Condon factors of Sharp and Rosenstock (1964).

Contributions to photoionization from autoionizing levels of CO_2 are significant in the region between 830 Å and 600 Å (Dibeler and Walker, 1967) and the vibrational populations in the atmosphere are modified by cascading. Despite the uncertainties, we proceed on the assumption that the 584 Å ratios of Table VII give the relative rates of population of the different vibrational levels by photoionization in the atmospheres of Mars and Venus.

The relative efficiencies with which fluorescent scattering populates the individual vibrational levels of the $A^2\Pi_u$ state can be computed from the solar flux intensities and the Franck-Condon factors for the $X^2\Pi_g$ $(v''=0) - A^2\Pi_u(v')$ transitions. The

TABLE VII

Relative probabilities for populating vibrational levels v' of the $A^2\Pi_u$ state of CO_2^+ by photoionization, by fluorescent scattering and by electron impact

v'	Photoionization	Fluorescent scattering	Electron impact
0	0.08	0.30	0.09
1	0.18	0.35	0.21
2	0.20	0.25	0.20
3	0.22	0.07	0.23
4	0.16	0.015	0.21
5	0.12	0.005	0.06

results are included in Table VII as are the approximate relative efficiencies for electron impact that follow from the data of Nishimura (1966) and McConkey *et al.* (1968).

Individual band intensities resulting from the three sources of excitation can be calculated approximately by combining Table VII with the relative transition probabilities of Poulizac and Dufay (1967). The relative band intensities are listed in Table VIII together with the approximate wavelengths of the mean band heads for

TABLE VIII

Approximate relative intensities and wavelengths in Å of the bands of the $X^2\Pi_g - A^2\Pi_u$ transition of CO_2^+ produced by (1) photoionization of CO_2, (2) fluorescent scattering by CO_2^+ and (3) electron impact of CO_2

v''/v'		0	1	2	3	4	5
0	(1)	.030	.092	.092	.038	.008	.002
	(2)	.077	.125	.080	.009	.000	.000
	(3)	.028	.094	.082	.036	.007	.001
	λ (Å)	3508	3374	3250	3136	3031	2935
1	(1)	.043	.022	.035	.052	.043	.021
	(2)	.107	.031	.030	.011	.003	.000
	(3)	.041	.024	.031	.047	.051	.010
	λ (Å)	3669	3523	3392	3267	3154	3046
2	(1)	.024	.036	.062	.022	.043	.025
	(2)	.060	.051	.055	.006	.003	.001
	(3)	.023	.039	.055	.021	.051	.011
	λ (Å)	3845	3686	3540	3399	3280	3168
3	(1)	.013	.054	.013	.046	.017	.003
	(2)	.034	.076	.011	.010	.001	.000
	(3)	.013	.056	.011	.042	.021	.001
	λ (Å)	4038	3863	3704	3553	3420	3300
4	(1)	.006	.036	.032	–	.021	.010
	(2)	.017	.051	.045		.001	.000
	(3)	.006	.039	.028		.021	.004
	λ (Å)	4264	4059	3883		3573	3440
5	(1)	–	.011	.025	.024	–	–
	(2)		.016	.036	.006		
	(3)		.013	.022	.023		
	λ (Å)		4301	4102	3925		

$2\Pi_{1/2}$ and $^2\Pi_{3/2}$ final states. Table VIII shows that in the altitude region where CO_2^+ is the major positive ion bands originating in $v'=0$ and 1 become relatively more intense with increasing altitude than bands originating in $v'=2$.

7. Other Constituents

Molecular nitrogen may be a minor constituent of the atmospheres of Mars and Venus. Of the many emission band systems, the appearance of which would establish the presence of N_2, the 3914 Å band of the first negative system of N_2^+ may be the most sensitive. The $B^2\Sigma_u^+$ level of N_2^+ can be excited by the same three mechanisms that we have studied for the CO_2^+ emissions.

The N_2^+ ions produced by photoionization can be removed by dissociative recombination

$$N_2^+ + e \rightarrow N + N \tag{9}$$

with a rate coefficient of about 3×10^{-7} cm^3 sec^{-1} (cf. Biondi, 1969) and by conversion into CO_2^+ through the reaction

$$N_2^+ + CO_2 \rightarrow N_2 + CO_2^+ \tag{10}$$

which has a rate coefficient of 9×10^{-10} cm^4 sec^{-1} (Fite, 1969). For illustrative purposes we have adopted model atmospheres for Mars containing 9% N_2 and 91% CO_2 with an exospheric temperature of 487° (McElroy, 1969). The atmosphere is assumed to be either completely mixed at all altitudes or such that diffusive separation begins at 120 km. The total abundance of N_2^+ ions lies in the range from 10^9 to 10^8 cm^{-2} and the fluorescent scattering intensity on Mars lies between 20 rayleighs and 2 rayleighs.

The contribution from simultaneous excitation and ionization in a direct photoionization process can be computed straightforwardly. It is 120 rayleighs for the diffusive model and 70 rayleighs for the mixed model.

The contribution from photoelectron impact can be estimated using arguments similar to those of Section 4. It is unlikely to exceed 50 rayleighs for the diffusively separated atmosphere or 20 rayleighs for the mixed atmosphere.

The intensity of 3914 Å emission that we predict for a CO_2–N_2 composition ratio of ten is accordingly two or three hundred rayleighs for the entire atmosphere. The predicted intensity could be substantially reduced by the presence of a lighter constituent such as atomic oxygen.

References

Anton, H.: 1966, *Ann. Physik* **18**, 178.
Bahr, J. L., Blake, A. J., Carver, J. H., and Kumar, V.: 1969, *J. Quant. Spectr. Rad. Trans.* **9**, 1359.
Barth, C. A., Fastie, W. G., Hord, C. W., Pearce, J. B., Kelly, K. K., Stewart, A. I., Thomas, G. E., Anderson, G. P., and Raper, O. F.: 1969, *Science* **165**, 1004.
Biondi, M. A.: 1969, *Can. J. Chem.* **47**, 1711.
Cairns, R. B. and Samson, J. A. R.: 1965, *J. Geophys. Res.* **70**, 99.
Cloutier, P. A., McElroy, M. B., and Michel, F. C.: 1969, *J. Geophys. Res.* **74**, 6215.
Dalgarno, A., McElroy, M. B., and Stewart, A. I.: 1969, *J. Atmospheric Sci.* **26**, 753.

Dibeler, V. H. and Walker, J. A.: 1967, *J. Opt. Soc. Amer.* **57**, 1007.

Fite, W. L.: 1969, *Can. J. Chem.* **47**, 1797.

Fjeldbo, G. and Eshleman, V. R.: 1968, *The Atmospheres of Mars and Venus* (ed. by J. C. Brandt and M. B. McElroy), Gordon and Breach, New York.

Fjeldbo, G., Fjeldbo, W. C., and Eshleman, V. R.: 1966, *J. Geophys. Res.* **71**, 2307.

Fjeldbo, G., Kliore, A., and Seider, B.: 1970, *Radio Sci.* **5**, 381.

Green, A. E. S. and Barth, C. A.: 1967, *J. Geophys. Res.* **72**, 3975.

Hesser, J. E.: 1968, *J. Chem. Phys.* **48**, 2518.

Hesser, J. E. and Dressler, K.: 1966, *J. Chem. Phys.* **45**, 3149.

Hinteregger, H. E.: 1970, *Ann. Geophys.* **26**.

Kliore, A., Levy, G. S., Cain, D. L., Fjeldbo, G., and Rasool, S. I.: 1967, *Science* **158**, 1683.

Kosyrev, N. A.: 1954, *Astrophys. Obs. Crimea* **12**, 169.

McConkey, J. W., Burns, D. J., and Woolsey, J. M.: 1968, *J. Phys. B. (Atom. Mol. Phys.)* **1**, 71.

McElroy, M. B.: 1969, *J. Geophys. Res.* **74**, 29.

Nakata, R. S., Watanabe, K., and Matsunaga, F. M.: 1965, *Sci. Light* **14**, 54.

Nishimura, H.: 1966, *J. Phys. Soc. Japan* **21**, 564.

Polyakova, G. N., Fogel, Ya. M., and Yu Mei, Ch'iu: 1963, *Soviet Astron. AJ.* **7**, 267.

Poulizac, M. C. and Dufay, M.: 1967, *Astrophys. Letters* **1**, 17.

Schwenker, R. P.: 1965, *J. Chem. Phys.* **42**, 2618.

Sharp, T. E. and Rosenstock, H. M.: 1964, *J. Chem. Phys.* **41**, 3453.

Spohr, R. and von Puttkamer, E.: 1967, *Z. Naturforsch.* **22a**, 705.

Turner, D. W. and May, D. P.: 1967, *J. Chem. Phys.* **46**, 1156.

PART III

OUTER PLANETS

VARIATIONS IN THE COLOR OF JUPITER

NEIL B. HOPKINS and WILLIAM M. IRVINE

Dept. of Physics and Astronomy, University of Massachusetts, Amherst, Mass. U.S.A.

Abstract. Observations of Jupiter by multicolor photoelectric photometry in 10 narrow bands between 3150 Å and 1.06 μ and in UBV showed a brightening for shorter wavelengths in 1965 relative to 1963. An opposite effect occurred for the band at 7300 Å. These results are consistent with observed activity in the Jovian atmosphere. No obvious correlation could be found between brightness fluctuations and longitude of the central meridian, indicating that the activity was uniform in longitude or occurred on time scales short compared to a month.

1. Introduction

Results of a program of multicolor photoelectric photometry of the planet Jupiter between the dates June 3, 1963, and December 19, 1965, have been reported by Irvine *et al.* (1968a; Paper II) and Irvine *et al.* (1968b; Paper III). Observations were made at both the Le Houga Observatory in France and the Boyden Observatory in South Africa, using 10 narrow bands isolated by interference filters between 3150 Å and 1.06 μ, and also in UBV. The narrow bands were given the designations *v-u-s-p-m-l-k-h-g-e* as indicated in Table I. The present paper examines these data for time variations over the three year observing period.

TABLE I

Effective wavelengths of passbands[a]

Band	v	u	s	p	m	l	k	h	g	e
λ_{eff} (Å)	3147	3590	3926	4155	4573	5012	6264	7297	8595	10 635
Halfwidth	145	120	45	90	85	90	160	200	90	770

[a] For details, see Young and Irvine (1967).

2. Analysis

Using a linear least squares fit to the data of Papers II and III we found generally good agreement for the magnitudes at unit distance and zero phase $m(1, 0)$ for each filter between the two sites. In order to investigate Jovian brightness variations over yearly periods we determined $m(1, 0)$ for the three years separately. For each year, there was again fairly good agreement between sites when the dates of observation covered about the same period. The data from the two sites were thus combined. Corresponding values of $m(1, 0)$ and geometric albedo p for each year are given in Table II. The geometric albedo is computed from

$$\log_{10} p = 0.4[m_\odot - m(1, 0)] - 2 \log_{10} \sin \sigma'_1, \tag{1}$$

TABLE II
Spectral reflectivity of Jupiter for 1963–65

Filter	1963		1964		1965	
	$m(1, 0)$	p	$m(1, 0)$	p	$m(1, 0)$	p
v	−8.61	0.248	−8.71	0.270	−8.66	0.258
	±0.02		±0.03		±0.03	
u	−8.81	0.297	−8.88	0.317	−8.93	0.330
	±0.01		±0.03		±0.02	
s	−9.00	0.355	−8.98	0.348	−9.11	0.390
	±0.01		±0.03		±0.02	
p	−9.08	0.382	−9.08	0.380	−9.20	0.423
	±0.01		±0.02		±0.02	
m	−9.27	0.455	−9.24	0.442	−9.36	0.495
	±0.01		±0.02		±0.02	
l	−9.35	0.487	−9.33	0.480	−9.43	0.526
	±0.01		±0.01		±0.02	
k	−9.49	0.555	−9.47	0.547	−9.48	0.553
	±0.02		±0.01		±0.01	
h	−9.31	0.470	−9.17	0.413	−9.13	0.398
	±0.01		±0.02		±0.02	
g	−8.75	0.282	−8.86	0.312	−8.80	0.295
	±0.01		±0.01		±0.02	
e	−8.83	0.303	−8.86	0.310	−8.78	0.289
	±0.02		±0.01		±0.02	
U	−8.11	0.323	−8.03	0.301	−8.14	0.332
	±0.01		±0.02		±0.02	
B	−8.55	0.427	−8.53	0.418	−8.63	0.458
	±0.01		±0.02		±0.02	
V	−9.30	0.502	−9.38	0.502	−9.46	0.541
	±0.01		±0.02		±0.02	

where $m_{\odot}$ and $\sigma_1' = 95.''07$ were taken from Paper II. Figure 1 shows the spectra for each of the three years separately. We observe a brightening at the shorter wavelengths in 1965 relative to 1963 except for filter v (λ 3150). At longer wavelengths ($\lambda \geqslant 6264$ Å) the brightness is more nearly constant from year to year except in filter h (7297 Å). At this wavelength Jupiter is systematically brighter in 1963 relative to 1964 by 0.14 m and relative to 1965 by 0.18 m. There therefore appear to be real color changes in Jupiter over the three year period.

Our observations are consistent with and give further information on activity patterns of the planet Jupiter. According to Focas and Banos (1964) and a well accepted atmospheric model by Wasiutinsky (1946), turbulence in the Jovian atmosphere results in the appearance of dark matter which usually takes the form of dark blotches, filamentary strips, and veils. It is obvious that the appearance and disappearance of dark matter, if not black, will result in a color change of the planet. Focas and Banos (1964) and Aksenov et al. (1967) photographed Jupiter over the periods 1957 through 1963 and 1964 to early 1965, respectively, in order to measure intensity variations of the planet. They used an activity factor defined by Focas and Banos which is a measure of the amount of dark matter observed on the disk. The

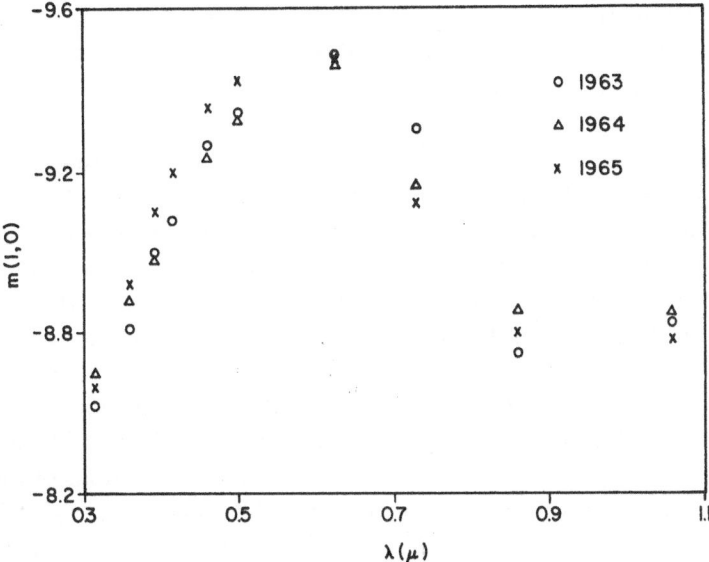

Fig. 1.   Magnitudes $m(1, 0)$ of Jupiter at unit distance and zero phase for the years 1963–65.

amount of dark matter increased from 1961 to 1963 when the activity factor reached a maximum and our filter h appeared anomalous. There was then a decrease in activity until early 1965. According to Aksenov *et al.* it seems that Jupiter had returned in late 1965 to a 'quasi-normal' state. Note that solar activity was increasing in 1965 after being at a minimum in 1963–64. Khodyachikh as quoted by Aksenov *et al.* (1967) found no correlation between the Jovian activity factor and solar activity over a period from 1932 to 1959. However, Shapiro (1953) reports such a correlation in blue and yellow light for the period 1926–50.

The fraction of solar energy incident in the observed wavelength region and absorbed by the planet may be estimated from

$$p^* = \frac{\int_{\lambda_1}^{\lambda_2} d\lambda p(\lambda) F(\lambda)}{\int_{\lambda_1}^{\lambda_2} d\lambda F(\lambda)}, \tag{2}$$

where $\lambda_1 = 0.315 \, \mu$, $\lambda_2 = 1.06 \, \mu$, and $F(\lambda)$ is the solar spectrum (taken from Allen (1963)). We find

$$\begin{aligned} p^* \, (1963) &= 0.411, \\ p^* \, (1965) &= 0.427, \end{aligned} \tag{3}$$

so that there was an estimated change of 4% in the reflected radiation and a corresponding change of 3% in the absorbed radiation $(1 - p^*)$ in this wavelength interval

24—P.A.

(an accurate determination of the spherical albedo would require a knowledge of possible changes in the phase integral $q(\lambda)$). The observed change is quite small, particularly when compared with Low's (1969) observations of a change in thermal emission by a factor of 2.5 within the last few years.

We investigated the possibility of a correlation between the detailed variation in the magnitudes at 3590 Å, 5012 Å, 7297 Å, and 8595 Å and the longitude of the central meridian ω. Residuals with respect to the mean Jovian phase curves were determined using a linear least squares fit for each year of observation. The central meridian was determined from time of observations with filter h and the *American Epherimeris and Nautical Almanac* for both systems of rotation I and II (most of the dark matter seems to appear in the System II area). Plots of the residuals versus ω for each year show no obvious correlations. Similar analyses for time periods of one month also showed no effect. This suggests that activity was rather uniform with respect to longitude, or that fluctuations of the order of our observed residuals ($\lesssim \pm 0.08$ m) occur on a time scale short compared to a month. Since variations over periods of the order of two or three months and also daily fluctuations (Aksenov *et al.*, 1967) have been observed, our results are not surprising. No correlation was found with the position of the Great Red Spot.

Acknowledgments

This research was supported by the National Aeronautics and Space Administration under grant NGR 22-010-023. The computations were performed at the University of Massachusetts' Research Computing Center.

References

Aksenov, A. N., Grigor'eva, A. N., Priboeva, N. V., Romanenco, A. G., and Teifel, V. G.: 1967, 'Atmospheric Activity of Jupiter in 1964–1965 According to Photometric Data', *Solar Systems Res. (U.S.S.R.)* **1**, 129–133.

Allen, C. W.: 1963, *Astrophysical Quantities*, Athlone Press, London, 2nd ed., Section 82.

Focas, J. H. and Banos, C. J.: 1964, 'Photometric Study of the Atmospheric Activity on the Planet Jupiter and Peculiar Activity in Its Equatorial Area', *Ann. Astrophys.* **27**, 36–45.

Irvine, W. M., Simon, T., Menzel, D. H., Charon, J., Lecomte, G., Griboval, P., and Young, A. T.: 1968a, 'Multicolor Photo-Electric Photometry of the Brighter Planets. II. Observations from Le Houga Observatory', *Astron. J.* **73**, 251–264 (Paper II).

Irvine, W. M., Simon, T., Menzel, D. H., Pikoos, C., and Young, A. T.: 1968b, 'Multicolor Photo-Electric Photometry of the Brighter Planets. III: Observations from Boyden Observatory', *Astron. J.* **73**, 807–828 (Paper III).

King, I.: 1952, 'Effective Extinction Values in Wide-Band Photometry', *Astron. J.* **67**, 253–258.

Low, F. L.: 1969, in *Third Arizona Conference on Planetary Atmospheres*.

Reese, E. V. and Solberg, H. G., Jr.: 1966, 'Recent Measures of the Latitude and Longitude of Jupiter's Red Spot', *Icarus* **5**, 266–273.

Shapiro, R.: 1953, 'A Planetary-Atmospheric Response to Solar Activity', *J. Meteorol.* **10**, 350–355.

Wasiutinsky, J.: 1946, 'Studies in Hydrodynamics and Structure of Stars and Planets', *Astrophys. Norvegica* **4**, 318–331.

Young, A. T. and Irvine, W. M.: 1967, 'Multicolor Photo-Electric Photometry of the Brighter Planets. I: Program and Procedure', *Astron. J.* **72**, 945–950 (Paper I).

THE EFFECTIVE TEMPERATURE OF JUPITER'S
EQUATORIAL BELT DURING THE 1965 APPARITION

L. M. TRAFTON

University of Texas at Austin, Austin, Tex., U.S.A.

and

R. WILDEY

U.S. Geological Survey, Center of Astrogeology, Flagstaff, Ariz., U.S.A.

Abstract. Preliminary results from our study of Jupiter's equatorial limb darkening at 8–14 μ indicate that the effective temperature of the equatorial belt lay between 140 K and 150 K. We find these temperatures to be insensitive to the likely range of the He/H$_2$ ratios. These results assume that Jupiter's only significant sources of extinction at these wavelengths are the pressure-induced absorption of H$_2$ and He and the absorption of saturated NH$_3$.

Limb-darkening observations of Jupiter's equatorial belt in the 8–14 μ spectral region were obtained at various times throughout the 1965 apparition. They consisted of diametric scans in right ascension and were collected from the 200-inch Hale telescope with the aid of a Hg-doped Ge photoconductor cooled with liquid H$_2$. The apparatus and technique have been described previously (Wildey, 1966 and 1968).

The diameter of the aperture was one eighth of Jupiter's apparent radius. The angle between a given scan and the Jovian equator averaged out to be only a few degrees.

We reduced the observations to obtain a mean limb-darkening profile corresponding to unit atmospheric thickness. The two points of each scan where the signal merged with the noise defined the end points of an abscissa having 40 equal intervals. We scaled the ordinate to provide unit area for the portion of each scan profile lying above the abscissa.

Let S_{ij} be the logarithm of the ordinate defined at the jth division of the abscissa and for the ith scan. For a given j, we evaluated the mean of S_{ij} over all the scans and fitted a least squares quadratic polynomial in sec Z to the difference between this mean and its corresponding set of ordinates S_{ij} $(i=1, \ldots, 41)$. We then computed the residual between each of these ordinates and the corresponding polynomial values and eliminated scan points whose residual was greater than 2.5 times the mean residual. Using the remaining scan points, we repeated this procedure, iterating until no more scan points were rejected. We did this for 38 values $(j=2, \ldots, 39)$ and used the resulting set of polynomials to generate a mean limb darkening profile at sec $Z=1$. Three cases retained all 41 scan points and one case had as few as 29 of the original 41 scan points surviving the iteration. This profile differed negligibly from the mean limb darkening profile. The statistical accuracy of each point of the final profile was better than one percent. This profile is illustrated in Figure 1.

In order to interpret this profile, we constructed non-gray model atmospheres for Jupiter in hydrostatic and thermal equilibrium (Trafton, 1967). The pressure-induced absorption of H$_2$ and He served as the thermal opacity. We added saturated 10 μ and 16 μ NH$_3$ bands to the models but neglected the opacity of the rotational band of

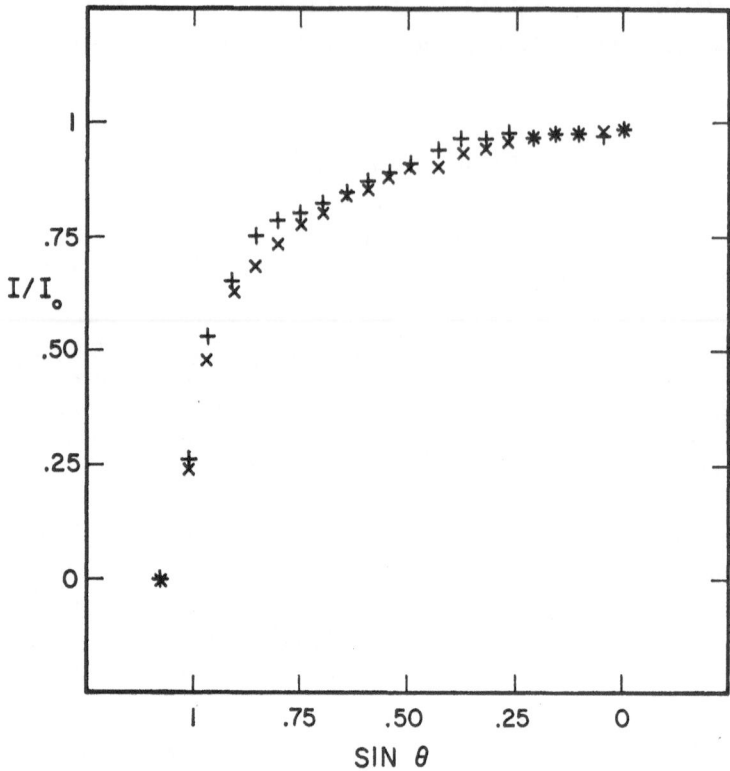

Fig. 1.   The reduced observational limb-darkening profile corresponding to unit atmospheric thickness. The pluses denote the sunrise limb and the crosses denote the sunset limb.

NH_3 and the extinction from any particle scattering in the upper Jovian atmosphere. Most of the 8–14 μ radiation comes predominantly from layers in Jupiter's atmosphere at a Planck mean optical depth near unity. The effective temperature (T_e) and He/H_2 ratio were left as free parameters to be determined by fitting a family of models to the observed limb darkening curves.

The shape of the limb darkening curve depends sensitively on the relative concentration of H_2 and NH_3 because of the different dependences on pressure and temperature. Because we assume NH_3 to be saturated in Jupiter's upper atmosphere, this relative concentration is sensitive to the boundary temperature and, therefore, to the effective temperature. Using limb darkening observations to deduce T_e requires a reliable band model for NH_3.

We constructed a model for the 10 μ and 16 μ NH_3 bands by the following procedure: We measured the equivalent widths and recorded the wavelengths of 550 lines from the published high resolution data of Garing *et al.* (1959) and Mould *et al.* (1959). Using the curve of growth theory of Ladenburg and Reiche (1913) for lines having Lorentz cores and assuming we could approximate all the lines to have the same half width, we derived line strengths for these bands. Most of the 10 μ laboratory lines were saturated.

Most of the radiation passes between the lines so that the mean transmission of the band is very sensitive to the shape of the lines in the wings except when the mean transmission is large. For the line shape in the wings we assumed a power law dependence on the frequency difference from the line center and determined this dependence by fitting the NH_3 band model to the low resolution data of France and Williams (1966). Their data consisted of NH_3 band absorptance measurements between 660 and 1300 cm^{-1} over many decades of path length and effective pressure, all obtained at room temperature. We found that the Lorentz line shape was invalid for the wings of NH_3. This shape would not allow the high resolution band model to fit the low resolution band data over the entire domain of laboratory pressures and path lengths. On the other hand, we did find a special power-law shape which caused better agreement. It falls off more slowly in the wing (as the 1.78 power of the frequency difference) than does the Lorentz shape. This indicates the sensitivity of line shape in the wings to the intermolecular potential.

After having established this dependence we were then forced by the absence of laboratory data to assume that essentially this same shape holds when NH_3 is broadened primarily by collisions with H_2 and He molecules rather than by other NH_3 molecules. This is the dominant mechanism for broadening NH_3 in Jupiter's upper atmosphere. Using laboratory data for the NH_3 line broadening coefficients for H_2 and He (Howard and Smith, 1950; Giver, 1968) helped to offset this uncertainty. The pressure dependence had to be derived approximately from consideration of the intermolecular potential (Townes and Schawlow, 1955).

We converted the line strengths to 130 K, the approximate temperature of the levels of formation of the observed 8–14 μ radiation, by altering the Boltzmann factors and partition function. We left the half widths expressed as a function of the local temperature. The effect of this was to reduce the 16 μ band strength by a factor of 200 and to narrow the 10 μ band by approximately 30%.

We computed a table of the mean transmission of these bands vs path length and pressure for homogeneous paths by integrating the monochromatic transmission over a frequency grid fine enough to delineate the band structure even at low pressures. These mean transmissions were evaluated at intervals of 10 wave numbers.

In order to determine the effective pressure and path length corresponding to the inhomogeneous path overlying a given atmospheric layer, we generalized the Curtis-Godson approximation to apply to the non-Lorentzian line shape. These parameters give the same mean transmission in both the weak band and strong band limits along a homogeneous path as along the actually inhomogeneous path. We evaluated these parameters for each layer of each model. Interpolation in the table of mean transmissions provided the values of the mean transmissions $T_v(NH_3)$ corresponding to the absorption of saturated NH_3 along the atmospheric path at direction cosine μ. Figure 2 illustrates the mean transmission of 10 cm-atm of NH_3 thinly diluted in 0.48 and 1.92 atm of H_2, respectively.

The expression for the limb darkening is

$$I_v(\mu) = \int_0^\infty B_v(\tau_v) \exp\left[-\frac{\tau_v(H_2 + He)}{\mu}\right] T_v(NH_3) \frac{d\tau_v}{\mu},$$

where the subscript v denotes an average value over a 10 wave-number interval evaluated at V. We calculated limb darkening profiles for families of T_e and He/H_2, doing the integrations to depths of $\tau_v > 12$. An exponential quadrature assured numerical accuracy. We weighted these curves by the spectral response of the instrument and by the mean atmospheric spectral transmission. After integrating over frequency, we convoluted the limb darkening signals with the Gaussian seeing function for values of the seeing in the neighborhood of 3″ arc. We also folded in the semi-elliptical function for the circular aperture of the instrument. We weighted Jupiter's apparent diameter by the number of scans obtained at that diameter to obtain a mean value for scaling the aperture. We then normalized the profiles to unit central intensity. The shape of the curve is sensitive to T_e but is relatively insensitive to fluctuations in the atmospheric transmission.

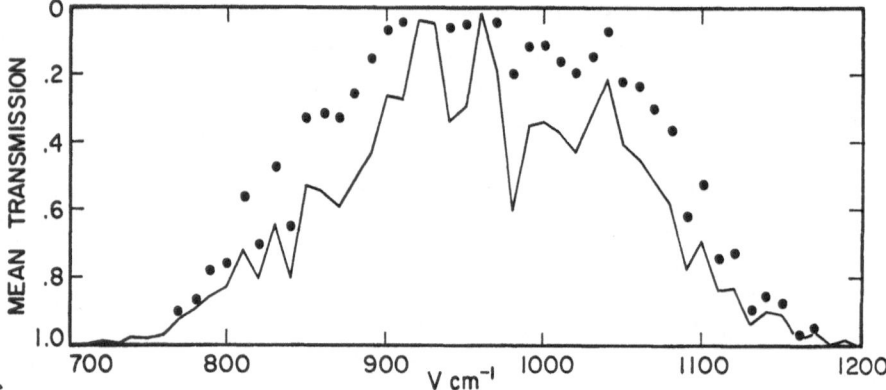

Fig. 2.   The mean transmission over intervals of 10 wave numbers for the model 10 μ and 16 μ NH_3 bands. The points correspond to 10 cm-atm of NH_3 thinly diluted in 0.48 and 1.92 atm of H_2, respectively, at 130 K.

We checked our assumption that the omission of Jupiter's rotational NH_3 band from the models would not significantly alter the shape of the limb darkening profile by inserting a very schematic model for this band and letting the model atmosphere relax to radiative equilibrium. The band model simply absorbed at the same depths and frequencies as the rotational NH_3 band. Calculating theoretical profiles by the above procedure, we found the difference to be small compared with that resulting from varying the effective temperature by 10 K. The effect of varying the He/H_2 ratio for values less than unity was also small.

To compare the observational and theoretical profiles, we normalized the former to unit central intensity and calibrated their abscissae in terms of $\sin \theta$ by equating their areas to the corresponding theoretical profiles.

Figure 3 illustrates the comparison for two temperatures. The models include the schematic NH_3 rotational band and assume $He/H_2 = 0$. The shaded region shows the $T_e = 150$ K model limb-darkening for extreme values of the seeing; namely, 1″ to 5″ arc. Note the rather small effect of the seeing. The crosses denote the corresponding observations for the sunset limb. The solid line shows the $T_e = 130$ K model for seeing

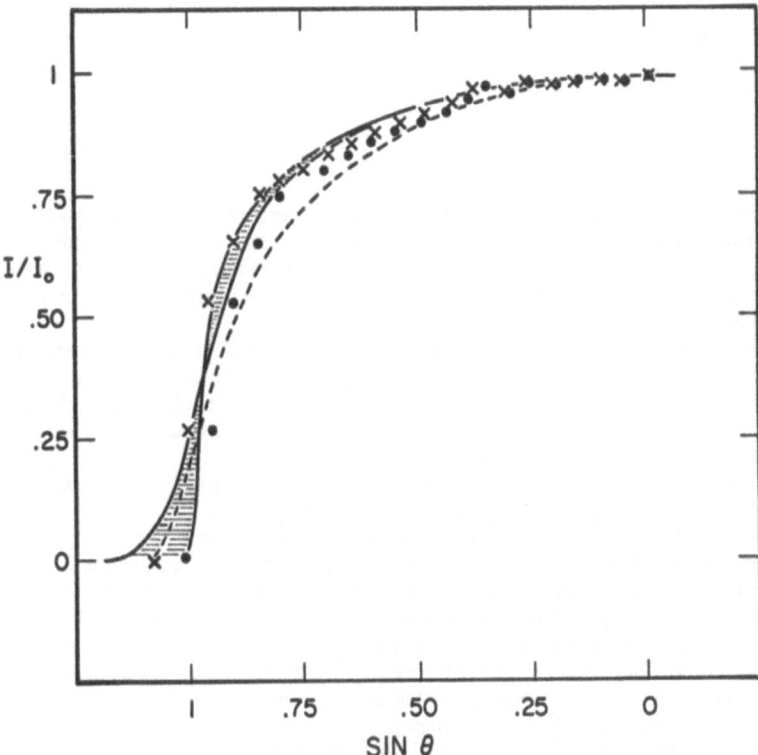

Fig. 3. Comparison of observational and theoretical profiles. The shaded region denotes the $T_e = 150$ K curves calculated for values of the seeing between 1″ and 5″ arc. The crosses are the correspondingly normalized and reduced observational data. The $T_e = 130$ K comparison is illustrated for seeing 3″ arc; the dots represent the observational data and the dashed line represents the model's profile.

equal to 3″ arc and the dots are the correspondingly normalized observations. Observe that the fit is distinctly better for the $T_e = 150$ K model. Varying He/H$_2$ causes only a minor effect. Our preliminary results indicate that the effective temperature of Jupiters' equatorial belt lies between 140 and 150 K.

In principle, we can extend this procedure to derive a He/H$_2$ ratio by examining the magnitude of the signal from the center of the disk. Whether the uncertainties in our knowledge of the earth's atmospheric transmission and the schematic rotational NH$_3$ band will permit us to derive improved limits on the He/H$_2$ ratio remains for our final analysis to reveal.

References

France, W. L. and Williams, D.: 1966, *J. Opt. Soc. Amer.* **56**, 70.
Garing, J. S., Nielson, H. H., and Rao, K. N.: 1959, *J. Molecular Spectrosc.* **3**, 496.
Giver, L.: 1968, private communication.
Howard, R. and Smith, W. V.: 1950, *Phys. Rev.* **79**, 132.
Landenburg, R. W. and Reiche, F.: 1913, *Ann. Phys.* **42**, 181.
Mould, H. M., Price, W. C., and Wilkinson, G. R.: 1959, *Spectrochim. Acta* **15**, 314.

358   L. M. TRAFTON AND R. WILDEY

Townes, C. H. and Schawlow, A. L.: 1955, in *Microwave Spectroscopy*, New York, Chapters 12–13.
Trafton, L. M.: 1967, *Astrophys. J.* **147**, 765.
Wildey, R. L.: 1966, *Z. Astrophys.* **64**, 32.
Wildey, R. L.: 1968, *Astrophys. J.* **154**, 761.

Discussion

Pollack: What did you say your estimated absolute temperature was and how do you interpret it?

Trafton: Between 140 and 150 K; this is the temperature which when raised to the fourth power and multiplied by σ, is equal to the total flux which departs from the planet.

Ingersoll: Did you investigate the dependence of this derived effective temperature on the various free parameters in your model?

Trafton: Yes, as illustrated in the second slide. The two main free parameters were the effective temperature and the helium-to-hydrogen ratio. The derived shape of the limb darkening curve depends very sensitively on the effective temperature but only insensitively on the helium/hydrogen ratio. We can bracket the latter from observations to lie between zero and unity; this is the main reason for the insensitivity arising from the latter.

Ingersoll: Did you have free parameters in preparing the line profile? And what about the ammonia itself; is that a free parameter, and how did you choose it?

Trafton: The only parameters describing the line profile were the effective temperature and the helium/hydrogen ratio; we assumed the ammonia was saturated in the atmosphere. If it turns out that ammonia is not saturated, then of course our results would not be valid. However, the close agreement between the first and final slide in the series would tend to indicate that the situation probably is not too different from what we expect it to be.

Pollack: If you allow for a slight deviation, then maybe the 130 K curve would give a better fit?

Trafton: No. The effective temperature is very sensitive to the relative absorption of the pressure-induced opacities and ammonia, because of their very different pressure and temperature dependence upon Jupiter's depth. The shape of the limb darkening curves is sensitive to the effective temperature.

Stephens: You say that temperatures correspond to unit optical depth. What sort of number densities would this correspond to?

Trafton: I would guess that this might be close to a third of an atmosphere. This absorption takes place relatively high in Jupiter's atmosphere, well above the cloud layers, and well into the radiative zone.

OBSERVATIONS OF JUPITER'S CLOUD STRUCTURE NEAR 8.5 μ

J. A. WESTPHAL

Mt. Wilson and Palomar Observatories, Carnegie Institution of Washington, California Institute of Technology and Division of Geological Sciences, California Institute of Technology, Pasadena, Calif., U.S.A.

Abstract. Measurements of the distribution of thermal flux along polar and equatorial scans across Jupiter suggest that the increase in brightness temperature near 8.5 μ observed by Gillett *et al.* may be caused by flux coming from near the cloud tops. Other possible sources are discussed and observational tests suggested which should clarify the thermal structure of the upper atmosphere.

In a recent paper, Gillett, Low, and Stein (1969) (referred to as GLS in this paper) have discussed a low-resolution spectrum of Jupiter. The observations indicate a large increase in brightness temperature from about 9.5 μ to 7.5 μ. In their discussion, they point out that in the region from 8.3–9.3 μ there is no obvious source of absorption above the cloud deck. The possibility that a window in the Jovian atmosphere might exist at these wavelengths led to an observational program at high spatial resolution.

On 4 August 1969, equatorial and polar scans of Jupiter were made with the 200-inch Hale telescope. Figure 1 shows data recorded at 8.2–13.5 μ (lower curve) and 8.2–9.2 μ (upper curve). The scans are displayed with a linear flux ordinate and a 8.1–9.2 μ brightness temperature scale. Also shown in Figure 1 is a photograph, on the same scale, of Jupiter taken in April 1969. These data were taken with a 2.5 arc sec circular diaphram.

1. Discussion

The correlation of the 8.2–9.2 μ scan with the gross cloud structure is remarkable and suggests that the flux structure is either closely related to the albedo variation in the visible region or that the effective radiating region is at a different height over the dark areas than over the light areas.

Several models of the temperature-opacity structure of Jupiter are compatible with this data. One model, following GLS, would separate the temperature-opacity structure above the clouds into three domains: (1) the region from 7.2–8.2 μ where the brightness temperature is ≈140 K, and where, since this is a region of high methane absorption, the opacity is probably also high; (2) the region from 8.2–9.5 μ where the temperature is ≈135 K, and the opacity is likely to be low. It is in this region that these observations have shown that the flux structure is strongly correlated with the visible albedo; and (3) a region from 9.5–12 μ where the temperature is ≈125 K, the opacity is probably due to ammonia and very high, and the flux structure is subdued but definitely correlated with the visible albedo.

Sagan et al. (eds.), *Planetary Atmospheres,* 359–362.

Looking at each of these wavelength regions in detail, it can be seen that the strong methane absorption band which occurs in the 7.2–8.2 μ region should increase the opacity and limit the effective radiating region to high in the atmosphere. GLS have interpreted the high temperature in the 7.2–8.2 μ region to be due to an inversion. Such a temperature structure could be supported by absorption of sunlight in the 3.3 μ

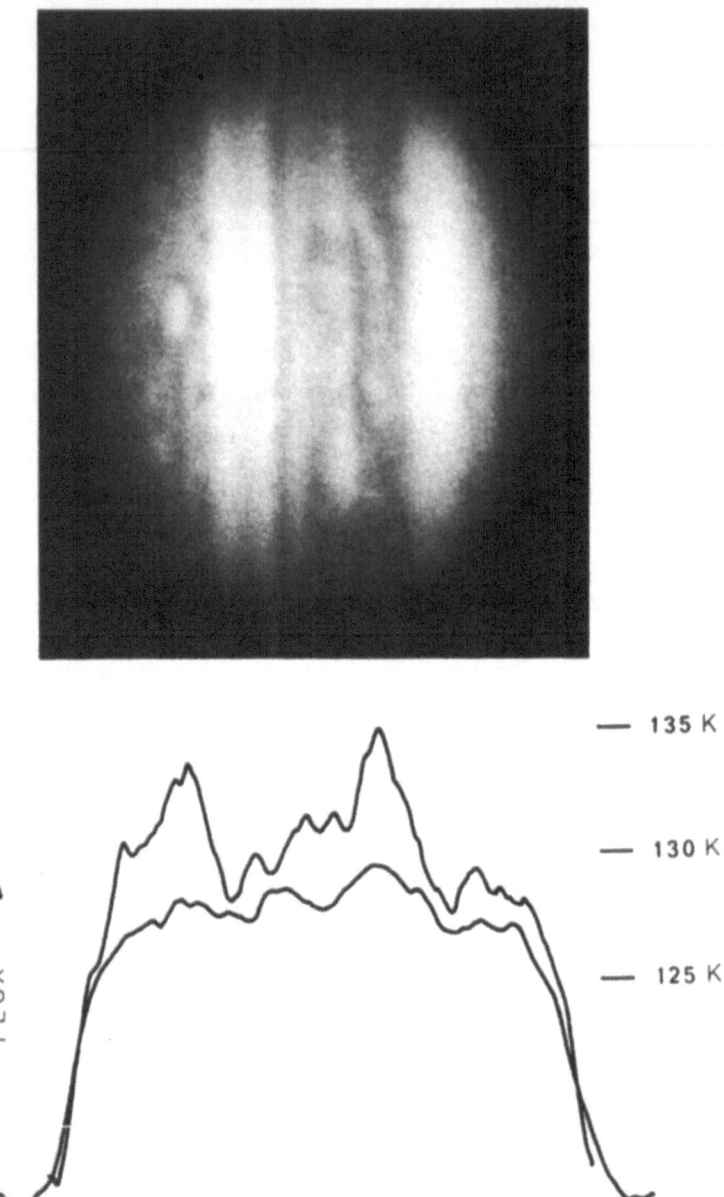

Fig. 1. Polar scans of Jupiter, respectively at 8.2–13.5 μ (lower curve), and 8.2–9.2 μ (upper curve). See text for further description. The photograph shows at the same scale the appearance of Jupiter some four months earlier, with a pattern of cloud belts similar to those on the date of the scans.

methane band. They propose that the temperature decreases with height until it reaches ⩽ 115 K then increases again to ⩾ 145 K.

If such an inversion exists, then the flux observed in region (2) 8.2–9.2 μ and shown in Figure 1, top curve, could come from a region either above or below the 115 K level. If it comes from above the inversion then one must explain the strong correlation with the visible albedo. If however, it comes from below the inversion, it could then come from a region near the cloud tops and the structure seen in Figure 1 would be due to either variations in cloud height, with the dark areas low and the bright areas high, or the structure could be due to localized heating due to albedo variations. A recent paper by Ingersoll and Cuzzi (1969), has suggested that the light bands are clouds in the top of upward moving convection cells and that the dark bands are relatively clear areas in downward moving regions. Thus it is very likely that the white bands would be higher than the dark bands.

In the GLS model, region (3), the flux observed from 9.5–12 μ would come from a region where the ammonia concentration is high and yet near enough to the cloud tops to still have some apparent flux structure. The equatorial scans from 8.2–9.2 μ show about the same limb darkening character as the 8.2–13.5 scans. This observation is compatible with the 8.2–9.2 μ flux coming from about the same region as the 9.5–12 μ flux since most of the energy observed from 8.2–13.5 μ comes from the 10–12 μ part of the spectrum.

An alternate model, in which the 8.2–9.2 μ flux comes from above the proposed inversion will require some special source of opacity which is not effective at longer wavelengths, where there is strong evidence (GLS) that the opacity is due to ammonia, which must be below a 115 K inversion level. Also such a model will require significant 20–50 μ opacity so that the region could be locally heated from below by the flux from the band structure. Although molecular hydrogen and helium (Trafton, 1967) could possibly furnish this opacity, it is difficult to see why the flux contrast would be greater than that observed (lower curve Figure 1) in region (3), which presumably is closer to the clouds in this model and has even higher hydrogen opacity.

Another model would abandon the temperature inversion and simply allow the 7.4–8.2 μ flux to come from even deeper than the 8.2–9.2 μ flux. It is hard to see how this could be the case since it would require the methane absorption in region (1) to be much less than the calculated value; however, this possibility exists since no laboratory measurements under Jovian conditions are available. Measurements of Jupiter in the 7.4–8.2 μ region, with high spatial resolution are possible and will quickly settle this point. If limb brightening is found, then the inversion proposed by GLS is confirmed. If, with this limb brightening, no flux structure correlated with that found in region (2) in this investigation is seen, then it is most likely that the flux in region (2) is coming from below the inversion and below the level of region (3). If on the other hand correlative structure is found, then it is possible that the flux from region (2) is coming from above the temperature minimum.

If no limb brightening is observed in region (1) then it seems likely that the flux in region (1) is coming from below that in regions (2) and (3) and that there is severe difficulty with the methane absorption calculation. In this case the large flux observed by GLS in region (1) may be coming from localized emission from the North Equatorial Belt in a manner similar to that seen at 5 μ (Westphal, 1969).

2. Summary

Large spatial variations in the 8.2–9.2 μ flux from Jupiter which are strongly correlated with the visible albedo have been observed. The most likely source of the flux in this wavelength region is radiation from material near the top of the cloud deck. The variations in flux can be due either to variations in cloud height or to some process closely related to the distribution of visible albedo. Measurements of limb darkening and flux structure in the 7.2–8.2 μ region will allow a choice between several possible models of the Jovian temperature-opacity profile above the cloud deck.

References

Gillett, F. C., Low, F. J., and Stein, W. A.: 1969, *Astrophys. J.* **157**, 925.
Ingersoll, A. P. and Cuzzi, J. N.: 1969, *J. Atmospheric Sci.*
Trafton, L. M.: 1967, *Astrophys. J.* **147**, 765.
Westphal, J. A.: 1969, *Astrophys. J.* **157**, L63.

PHASE BEHAVIOR OF LIGHT GAS MIXTURES
AT HIGH PRESSURES

WILLIAM B. STREETT

*Science Research Laboratory and Department of Chemistry, U.S. Military Academy, West Point,
N.Y., U.S.A.*

Abstract. If solid surfaces exist beneath the visible clouds of the major planets, they may be
expected to exist at depths and pressures at which the component gas mixtures solidify under their
own weight. The elucidation of phase behavior in mixtures of light gases at very high pressures is
therefore essential to the solution of the problem of deep atmosphere structures in these planets.
Available experimental evidence suggests several possible extrapolations of the H_2-He phase diagram to high pressures. These have been used to develop a structural model for a H_2-He atmosphere. In this model, gravitational separation of coexisting phases results in a layered structure,
and it is shown that masses of H_2-rich solid can exist in dynamic and thermodynamic equilibrium
with a fluid layer of equal density but higher He content. This model forms the basis of a new
hypothesis for Jupiter's Red Spot.

1. Introduction

It is generally accepted (DeMarcus, 1958; Öpik, 1962; Peebles, 1964) that Jupiter and
Saturn are composed almost entirely of H_2 and He. If solid surfaces exist beneath the
visible clouds of these planets, they may be expected to exist at depths and pressures
at which the H_2-He mixture solidifies under its own weight. Whether solidification
occurs depends on the equation of path of the atmosphere and the melting behavior
of the mixture at high pressures and temperatures. Neither is known with any certainty,
so the question of whether a solid surface exists remains open.

2. Solidification of Gas Mixtures

Existing models for the interiors of Jupiter and Saturn (DeMarcus, 1958; Opik, 1962)
have been developed on the assumption that the presence of He has little or no effect
on fluid-solid phase transitions in the molecular phase of H_2. In this case, solidification
occurs abruptly at the point at which the equation of path of the atmosphere crosses
the melting curve of hydrogen (Point (a) in Figure 1). Solid and fluid phases coexist
only at this point, which marks the level of the planet's solid surface. The purpose of
this paper is to point out that the presence of He adds a second degree of freedom (in
the context of the phase rule) to the conditions of equilibrium between two phases in a
H_2-He mixture, and that this may produce a more complicated structure in the region
of transition from fluid to solid in a H_2-He body.

    Thermodynamic equilibrium between solid and fluid phases in a H_2-He mixture
is likely to exist over a finite distance along the equation of path (effectively over a
range of depths in the atmosphere) with the compositions of the coexisting phases
varying continuously with pressure and temperature in this region. This behavior is
illustrated in the *P-T* diagram of Figure 2, which shows (schematically) the melting

Sagan et al. (eds.), *Planetary Atmospheres*, 363–370.

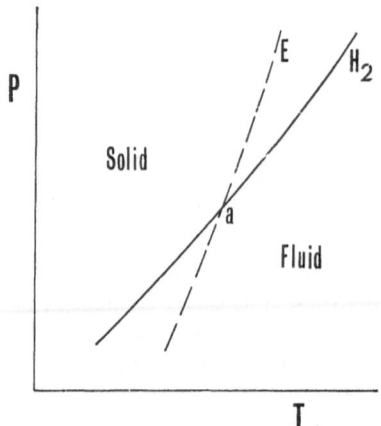

Fig. 1.   Pressure-temperature diagram showing the melting curve of hydrogen (H_2) and the
equation of path (E) for a planetary atmosphere.

curves of H_2 and He, along with several hypothetical equations of path for the
planetary atmosphere. The coexistence region for solid and fluid phases, in a H_2-He
mixture, lies approximately between the melting curves, so that for an equation of path
E_1 (the isothermal case) these phases coexist approximately between points a and b.*

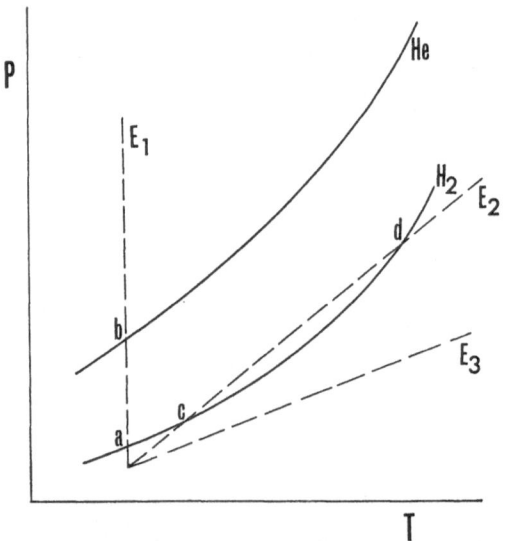

Fig. 2.   Pressure-temperature diagram showing the melting curves of hydrogen and helium (H_2
and He), and several equations of path (E_1, E_2, E_3) representing different temperature gradients
in a planetary atmosphere.

* The limiting boundaries, in P-T space, for two-phase regions in a two-component system are
defined by lines representing conditions of a single degree of freedom. These include three phase
lines and critical lines, in addition to the pure component phase boundaries (melting curves, etc.)
shown in Figure 1. A more complete discussion of phase diagrams in this context has been presented elsewhere (Streett, 1969; Streett and Hill, 1970).

The planet would be entirely solid at pressures above b and fluid at pressures below a. Lines E_2 and E_3 represent equations of path for which there are temperature gradients. If the gradient is large, as in E_3, solidification does not occur, and the planet remains fluid throughout. For the intermediate case, E_2, a region c-d exists in which solid and fluid phases may coexist (in thermodynamic equilibrium), while the planet is entirely fluid above and below.

3. Gravitational Separation of Solid Fluid Phases

Within the region of phase separation, coexisting phases would tend to separate in the gravitational field, producing a layered structure. At first glance, it seems natural to assume that the solid phase, being more dense than the fluid, would sink to the solid surface. A closer examination, however, suggests that this is not necessarily the case for a mixture of H_2 and He. In the case of a pure substance, such as H_2, solidification is accompanied by a slight increase in density as a result of the more efficient packing of the molecules in an ordered lattice structure. Although a similar density increase results from solidification in a mixture, the relative densities of the two phases will be determined largely by their compositions, due to the large difference in the densities of the pure components. The H_2-He system belongs to an unusual class of binary mixtures, in which the more volatile component (He) has the higher molecular weight, and is therefore more dense, at high pressures, than the second component, even though the former may be a gas and the latter a solid or liquid in the pure state. (The estimated density of He at high pressures exceeds that of H_2 by a factor of 3 or more (DeMarcus, 1958).) The result is that a reversal in the sign of the density difference between two coexisting phases occurs with increasing pressure at a fixed temperature. This behavior – known as the barotropic phenomenon – was first observed for gas-liquid mixtures of H_2-He by Kamerlingh-Onnes (1906). He observed that if a gas-liquid mixture of H_2-He is compressed at a temperature of about 20 K, the density of the gas phase (which is mostly He) exceeds that of the liquid phase (which is mostly H_2) at pressures above about 30 atm. As the pressure passes through this value, the liquid phase rises up and floats on top of the gas phase. Similar behavior has been observed in other binary mixtures in which the more volatile component has the higher molecular weight, such as N_2-NH_3 (Krichevskii, 1940) and CO_2-H_2O (Takenouchi and Kennedy, 1964).

4. Phase Behavior of H_2-He Mixtures at High Pressures

Although the behavior described above has not been observed experimentally in gas-solid mixtures of H_2-He, it almost certainly does occur. The equality of density between two phases in a two-component system reduces the number of degrees of freedom to one, and the locus of equal density points appears as a line on the P-T diagram of the system. The probable location of this line can be found if the shape of the phase boundary curves in pressure-composition space are known. Although experimental data on the phase behavior of light gas mixtures at high pressures are meager, the available evidence (Tsiklis, 1946; Streett and Hill, 1970) suggests that the isothermal pressure-composition (P-X) diagram for a mixture of H_2-He will have the form shown in Figure 3.* Area F is a region in which a single homogeneous fluid phase exists, while S_1 and S_2 are regions in which a single solid phase exists. S_1 is a H_2-rich solid and S_2 a He-rich solid. The remaining areas (S_1+F, S_2+F, and S_1+S_2) are regions in which two phases coexist in equilibrium. The behavior illustrated in this diagram is that of a binary mixture in which the components are completely miscible in the fluid phase (F), but solidify to form partially miscible solid phases (S_1 and S_2).

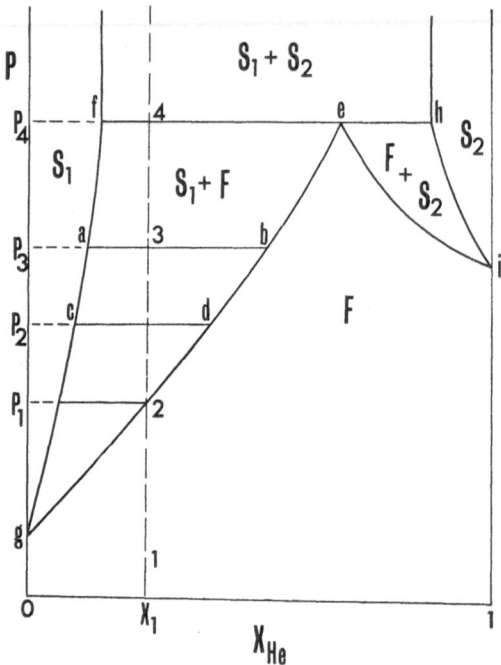

Fig. 3.   Suggested pressure-composition diagram for H_2-He mixtures at constant temperature.

5. Proposed Structure for H_2-He Atmosphere

The sequence of thermodynamic states encountered in descending into the atmosphere of a body of overall composition X_1 would lie along an ascending vertical line 1-2-3-4 in Figure 3. At pressures below P_1 (point 2) a homogeneous fluid phase exists, while at pressures above P_4 (point 4) a mixture of two solid phases exists. P_4 therefore corresponds to the approximate depth of the solid surface. At any pressure between P_1 and P_4 the mixture separates into solid and fluid phases whose compositions are found at points where a horizontal line, at that pressure, intersects the phase boundary lines g-c-a-f and g-d-b-e. At pressure P_3, for example, the mixture of overall composition X_1 separates into a solid phase a and fluid phase b. It is of interest to consider

* There is some evidence (Sneed *et al.*, 1968; Streett and Hill, 1970) that H_2-He mixtures may separate into two distinct fluid phases at high pressures. This would result in a more complicated phase diagram and would lead to a more complex atmospheric structure (Streett, 1969).

the relative densities of the coexisting phases defined by the lines g-c-a-f and g-d-b-e. At pressures just above g (the melting pressure of pure H_2) the fluid phase is only slightly richer in He than the solid phase, and the latter is likely to be more dense by virtue of the volume decrease accompanying solidification. At higher pressures, however, the fluid phase is likely to be more dense by virtue of its higher He content. An example will illustrate this point. At a pressure of 10 kilobars DeMarcus (1958) estimates the molar volumes of cold H_2 and He to be 7.3 and 4.3 cm^3/mole respectively. If these values are taken for the component molar volumes in the solid phase, and values, say, 5% greater for the fluid phase, it can be shown that the fluid will be more dense than the solid if its He content is greater by about fifteen mole percent. Then there must exist a pressure P_2 – the barotropic pressure – at which the coexisting phases c and d have equal densities. At pressures below P_2 the solid is more dense, and above P_2 the fluid is more dense. In this situation the solid phases, which condense

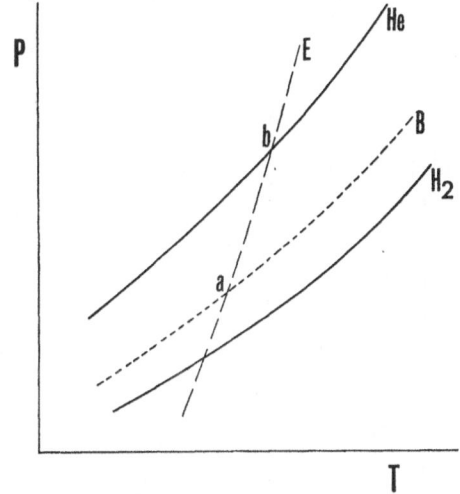

Fig. 4. Pressure-temperature diagram showing the melting curves of hydrogen and helium (H_2 and He), the barotropic line (*B*) at which the solid and fluid phases have equal density, and the equation of path (*E*) of a planetary atmosphere.

between P_1 and P_4, will gravitate toward the level corresponding to P_2, *both from above and below*. Through slow diffusion within these solids, their compositions would eventually reach the value c and they would remain suspended in dynamic and thermodynamic equilibrium with the surrounding fluid of composition d. In other words, the solid phase would behave as a thermodynamic Cartesian diver.

Consideration of the relative densities and compositions of the phases, and the effect of gravity, leads to an equilibrium structure with the following characteristics:

(1) the atmosphere above the level of P_2 has a uniform composition d;

(2) masses of H_2-rich solid, of composition c, float at the level corresponding to P_2, in equilibrium with the fluid d;

(3) at low levels, the fluid composition varies with depth and pressure along the lines d-b-e; and

25—P.A.

(4) at the level corresponding to P_4 a solid surface exists, consisting of a mixture of two solid phases f and h, whose average density exceeds that of the fluid e in contact with it.

This structure has been derived for the special case of an isothermal atmosphere; however, because the phase diagram is likely to have the same form at higher temperatures, the same structure might exist in the presence of a temperature gradient. The locus of points at which the solid and fluid phases have equal densities – the barotropic curve – would appear as a line in the P-T diagram, approximately parallel to, and lying above, the melting curve of H_2, as shown in Figure 4. For an equation of path such as E, floating masses of H_2-rich solid exist at the level of point a, and the solid surface lies in the vicinity of b. Extrapolations of the melting curves of H_2 and He to high pressures suggest that, for temperatures of the order of a few hundred degrees, point a in Figure 4 would lie at a depth of about 800 km and point b at a depth of

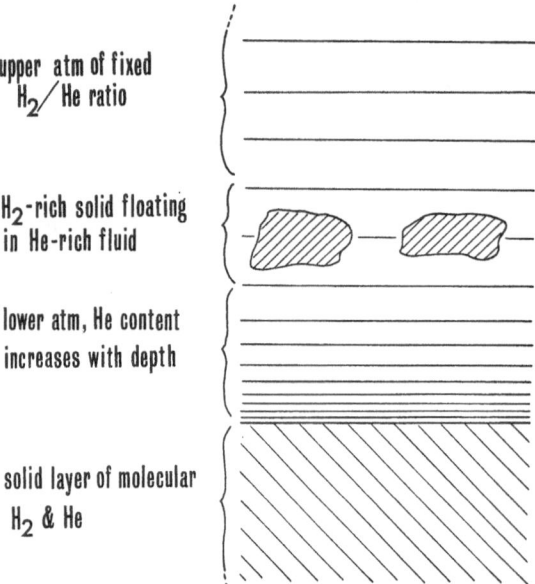

upper atm of fixed H_2/He ratio

H_2-rich solid floating in He-rich fluid

lower atm, He content increases with depth

solid layer of molecular H_2 & He

Fig. 5.   Proposed structure for H_2-He atmosphere, based on the phase diagram of Figure 3.

about 1600 km, below the visible surface of Jupiter. A graphic picture of the structure suggested here is shown in Figure 5. The relevance of this picture to the floating raft concept of the Red Spot is immediately obvious.

6. A New Hypothesis for Jupiter's Red Spot

The structure suggested by Figure 5 forms the basis of a new hypothesis – called the Cartesian diver hypothesis – which seeks to explain the physical nature and observed variations in longitude, size, and intensity of the Red Spot. The following is a brief summary of this hypothesis; full details will be published elsewhere (Streett *et al.*, 1971).

According to the Cartesian diver hypothesis, the Red Spot is a region of contrast in the cloud structure of Jupiter's outer layers, caused by the presence of a mass of H_2-rich solid (the diver) floating within a fluid layer of H_2 and He at some depth below the visible surface. The resulting physical model thus includes some of the qualitative features of both a floating raft and a Taylor column. It can be shown that the floating solid behaves as a thermodynamically stabilized Cartesian diver: that is, as a floating object which seeks an equilibrium level within a fluid layer. In the case of the Red Spot the floating solid is not necessarily continuous, and, indeed, it is unlikely to be do.

Equations of motion of a stable Cartesian diver in a rotating system suggest that the longitudinal motion of the Red Spot consists of several periodic components of different amplitudes and frequencies. These predictions are in agreement with the conclusions of Solberg (1968) based on recent precise measurements of Red Spot longitude. When the effects of rapid rotation of the planet upon the dynamic behavior of the fluid in the vicinity of the Cartesian diver are included, the resulting physical model provides a qualitative explanation not only for the observed variations in size and intensity of the Red Spot, but also for the manner in which these variations correlate with changes in longitude.

7. Conclusions

Only a few of the more obvious consequences of phase separations in H_2-He mixtures at high pressures have been considered here. From a chemical standpoint, the most pressing need is for experimental data on the phase behavior and equation of state of H_2-He mixtures at high pressures. The available data for this system (Sneed *et al.*, 1969) are insufficient even to suggest a choice between several possible extrapolations of the phase diagram to high pressures. It is clear that knowledge of the phase behavior of H_2-He mixtures at high pressures will be valuable, not only for understanding the deep atmosphere structures, but also for interpreting correctly some of the direct observational data for the upper atmosphere. We are preparing to carry out experimental studies of phase behavior in H_2-He mixtures at pressures up to 10 kilobars. The results will be reported in due course.

Acknowledgment

The author is indebted to R. Wildt of Yale University for several useful discussions relating to this work.

References

DeMarcus, W. C.: 1958, *Astron. J.* **63**, 2.
Kamerlingh-Onnes, H. and Keesom, W. H.: 1906, *Comms. Phys. Lab. Univ. Leiden*, No. 96a, 96b, 96c.
Krichevskii, I. R.: 1940, *J. Phys. Chem. (U.S.S.R.)* **12**, 480.
Öpik, E. J.: 1962, *Icarus* **1**, 200.

Peebles, P. J. E.: 1964, *Astrophys. J.* **140**, 328.

Peek, B. M.: 1958, in *The Planet Jupiter*, Faber and Faber, London, Chapter 19.

Reese, E. J. and Solberg, H. G.: 1969, *Latitude and Longitude Measurements of Jovian Features in 1967–68*, U.S. Government Research and Development Report N69-26410.

Sneed, C. M., Sonntag, R. E., and Van Wylen, G. J.: 1968, *J. Chem. Phys.* **49**, 2410.

Solberg, H. G.: 1968, *Icarus* **8**, 82.

Streett, W. B.: 1969, *J. Atmospheric Sci.* **26**, 924.

Streett, W. B. and Hill, J. L. E.: 1970, in *J. Chem. Phys.* **52**, 1402.

Streett, W. B., Ringermacher, H. I., and Veronis, G.: 1971, submitted to *Icarus*.

Takenouchi, S. and Kennedy, G. C.: 1964, *Am. J. Sci.* **262**, 1055.

Tsiklis, D. S.: 1946, *J. Phys. Chem. (U.S.S.R.)* **20**, 181.

Question following Dr. Streett's paper

Question: Can you state more accurately the pressures at which solids might be floating about in Jupiter's atmosphere?

Streett: No. In addition to our ignorance about the melting behavior of H_2 and He at high pressures and temperatures, we have very little information about temperatures of the regions beneath the clouds which make up Jupiter's visible surface. For temperatures of the order of, say, 1000 K, the estimated melting pressures of H_2 and He are 460 kilobars and 800 kilobars, but these are based on extrapolations of experimental data which extend to temperatures of only about 100 K and pressures of about 20 kilobars.

ATMOSPHERIC DEPTHS OF JUPITER, SATURN, AND URANUS

S. F. DERMOTT

The Royal Military College of Science, Shrivenham, Swindon, Wilts, England

Abstract. The presence of numerous near-commensurabilities among pairs of mean motions and the strong correlation between orbital radius and mass in the satellite systems of the three major planets (particularly in the Saturn system) suggest that the orbits of the satellites have evolved considerably under the action of tides. It is shown that the source of dissipation could be boundary layer turbulence at the base of the planetary atmosphere. If this is the source of dissipation then it should be possible to estimate the depths of these atmospheres from the mean rates of energy dissipation.

1. Introduction

It has been shown (Roy and Ovenden, 1954; Goldreich, 1965; Dermott, 1968) that the observed number of near-commensurate pairs of mean motions in the solar system is too great to have arisen from a random distribution of mean motions. Goldreich (1965) has shown that certain near-commensurabilities are stable and considers that the present non-random distribution of mean motions has been brought about by the action of tides. In this paper I indicate how Goldreich's tidal hypothesis can be tested. If it can be substantiated then a knowledge of the depths of the planetary atmospheres of Jupiter, Saturn, and Uranus may result.

The theory can be developed quantitatively and in substantial detail but as some parts of the argument have not yet been completed I consider a brief presentation to be the more appropriate.

2. Simple Tide Theory

In a satellite system which has been acted on considerably by tides there should be a strong correlation between orbital radius and mass. The rate of orbital evolution due to tidal action is mass dependent and thus in a satellite system in which initially there is no correlation between orbital radius and mass, tides will act causing the satellites to collide and grow until they are sorted according to their masses – the biggest satellite being the one furthest from the planet and the smallest the one nearest to it. This process may, of course, take some considerable time. The sorting process will also be checked to some extent by the formation of stable commensurabilities.

Consider the evolution of a single satellite. Standard theory shows that

$$\frac{\mathrm{d}r}{\mathrm{d}t} = f(p) \frac{m}{r^{11/2}} \cdot \frac{1}{Q},\tag{1}$$

where r is the orbital radius and m the mass of the satellite, $f(p)$ is a function of parameters of the planet only and Q^{-1} is the dissipative function (Dermott, 1968). It is probable that Q is amplitude- and frequency-dependent, and the nature of this dependence must be determined before (1) can be integrated. As a first step I consider it reasonable to write

$$Q = f'(p)\,(\text{amplitude})^a\,(\text{frequency})^f,$$

where $f'(p)$ is a function of parameters of the planet only. Goldreich and Soter (1966) have shown that, if the source of dissipation is boundary layer turbulence at the base of the planetary atmosphere, then

$$Q = f'(p)\,(\text{amplitude})^{-1}\,(\text{frequency})^{-2}.$$

The amplitude of the tide is proportional to m/r^3 and the frequency to $1-(r_b/r)^{3/2}$, where r_b is the radius of the tidal barrier (see Dermott (1968)). As $r_b \ll r$, for most satellites, the frequency term can be neglected in the first instance and we can write

$$Q = f'(p)\,r^{-3a}m^a.$$

On substitution of this expression into (1) and integration we obtain

$$r^{13/2-3a}\left\{1 - \left(\frac{r_i}{r}\right)^{13/2-3a}\right\} = f''(p)\,m^{1-a}t,$$

where r_i is the initial orbital radius and t the time of evolution. We have $a \leqslant 0$ ($Q \to \infty$ as amplitude $\to 0$) and therefore $(r_i/r)^{13/2-3a} \ll 1$ and can be neglected if evolution has been appreciable. Thus,

$$\log r = \frac{1-a}{13/2 - 3a}\log m + \frac{\log f''(p)\,t}{13/2 - 3a}. \tag{2}$$

Thus the orbital radius of a satellite which has been acted on appreciably by tides is virtually independent of its initial orbital radius. If we allow that in any one system all the satellites are of one age and that the evolution of a single satellite can be treated independently of all the others, then we can predict that in a tidally evolved satellite system there will be a linear correlation between $\log r$ and $\log m$.

A plot of $\log r$ against visual magnitude for the satellite system of Uranus is shown in Figure 1 (the masses of these satellites are not known). A plot of $\log r$ against $\log m$ for the satellite system of Saturn is shown in Figure 2. The line in the latter diagram is drawn to pass through Titan and has a slope corresponding to $a = -1$. The correlation coefficients of the two plots are respectively 0.005 and 0.001. I take this to be strong supporting evidence that the satellite systems of Saturn and Uranus are tidally evolved systems. The satellite system of Jupiter cannot be treated in the above manner as all of the four major satellites are linked together in a presumably stable manner and thus cannot possibly be treated separately. The very presence of the numerous near-commensurabilities in this system though indicates that it also is a tidally evolved system.

A fuller theory is needed (and has in part been developed) to take account of: (a) the frequency dependence of Q; (b) the formation and evolution of stable commensurabilities; and (c) satellites of varying ages, but the most interesting feature of a tidally

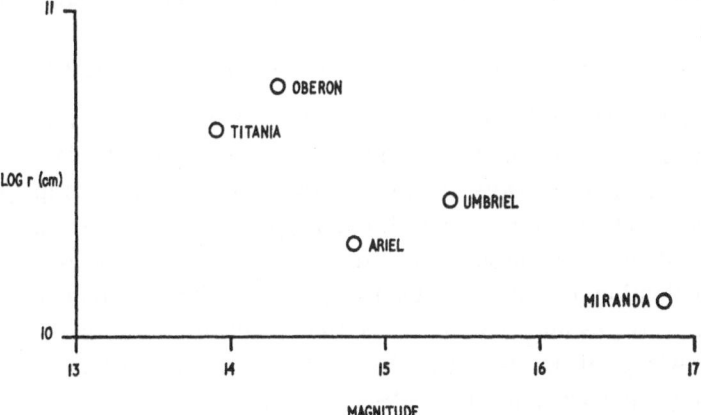

Fig. 1. A plot of log of orbital radius against visual magnitude for the satellite system of Uranus.

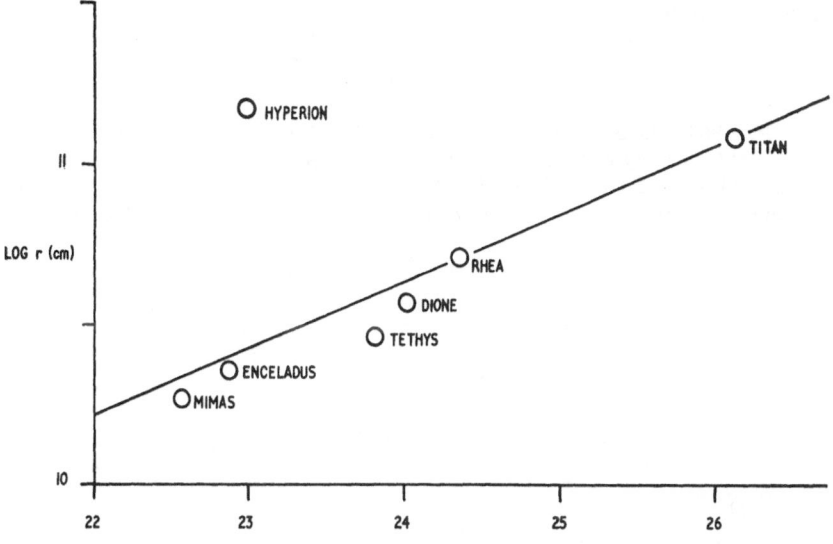

Fig. 2. A plot of log of orbital radius against log of mass for the satellite system of Saturn.

evolved satellite system has been revealed by the simplified discussion given above – the present orbital radius of a tidally evolved satellite does not depend on its initial orbital radius. Because of this it should be possible to correlate the orbital radii and the masses of the satellites in any one system in an exact manner, e.g. as the mass of Titan is known to three figures it should be possible to predict that of Rhea to three figures.

3. Conclusion

The features of the satellite systems of the major planets which suggest they are tidally evolved are:

(i) the preference for near-commensurability among pairs of mean motions (Jupiter and Saturn),

(ii) the linear correlation between log r and log m (Saturn and Uranus).

It would appear that in the Saturn system Q is inversely proportional to amplitude and thus the source of dissipation could be boundary layer turbulence at the base of the planetary atmosphere. If we assume that this is the source of dissipation on Jupiter and Uranus as well as Saturn then it should be possible to estimate parameters associated with the depths and density structures of these atmospheres from the mean rates of energy dissipation (this being related to $f''(p)$ in (2)). It would then have to be shown, of course, that the estimates were in reasonable accord with our present (somewhat limited) knowledge of the structure of these planets and that no other possible source of dissipation could account for the observations.

Perhaps at this stage it is best simply to state that the consequences of tidal evolution can and should be determined in a detailed quantitative manner and that if the predictions of the theory are substantiated then some knowledge of the nature of the major planets should result.

References

Dermott, S. F.: 1968, *Monthly Notices Roy. Astron. Soc.* **141**, 349.
Goldreich, P.: 1965, *Monthly Notices Roy. Astron. Soc.* **130**, 159.
Goldreich, P. and Soter, S.: 1966, *Icarus* **5**, 375.
Roy, A. E. and Ovenden, M. W.: 1954, *Monthly Notices Roy. Astron. Soc.* **114**, 232.

THE SPECTRAL CHARACTERISTICS AND PROBABLE
STRUCTURE OF THE CLOUD LAYER OF SATURN

V. G. TEIFEL, L. A. USOLTZEVA, and G. A. KHARITONOVA

Astrophysical Institute of the Academy of Sciences, Kazakh S.S.R., Alma-Ata, U.S.S.R.

Abstract. Spectrophotometric (photographic and photoelectric) measurements of the intensity of CH_4 absorption bands at 6190 and 7250 Å over different regions of Saturn's disk show an increase in intensity toward the poles and a decrease toward the equatorial limb. In the bright equatorial belt of Saturn the methane absorption is about 25–28% less than in the south temperate belt (latitude about $-20°$).

Absolute photoelectric spectrophotometry of Saturn's disk gives a value for the single-scattering albedo of the aerosol particles $\tilde{\omega}_c \simeq 0.99$ at λ 6200–6500 Å at the center of the disk. Calculations of the curves of growth for the absorption lines formed in the thin gas and in the cloud layer were made and the comparison with observations of Jupiter and Saturn lead to the mean value of the volume scattering coefficient of the aerosol layer $\sigma_a < 5 \times 10^{-6}$ cm^{-1}. In the equatorial region of Saturn σ_a is larger than in the temperate region by a factor 1.3 to 1.8.

The models of Saturn's atmosphere that fit well the observational data preclude the condensation of methane. An aerosol layer of ammonia is more probable in the atmosphere of Saturn. Calculations of the distribution of the ammonia aerosol volume density (Q_a) give $Q_a \simeq 10^{-9}$ to 10^{-7} gm/cm^3 for relative abundances of ammonia $A = 10^{-6}$ to 10^{-4}. Observational estimates of Q_a derived from the values of σ_a give $Q_a < 10^{-9}$ gm/cm^3. Apparently Saturn's atmosphere departs from conditions of ordinary convection.

Some very interesting variations in the spectral reflectivity of the different regions of Saturn are observed, especially in the ultraviolet. These data, as well as a systematic study of the methane absorptions in the near infrared strong bands are needed in future studies of Saturn's atmosphere.

1. Introduction

Spectral observations of Jupiter and Saturn made at the Observatory of the Astrophysical Institute reveal many peculiarities in the distribution of molecular absorption across the disks of these planets, showing the importance of the role of their aerosol layers in forming absorption lines and bands. In this work, we bring together the results of spectral (photographic and photoelectric) observations of Saturn and a few tentative results of theoretical calculations. Although the radiation regime of the atmosphere of Saturn is somewhat more severe than that of Jupiter (the solar constant at Saturn's orbit is 0.02 cal cm^{-2} min^{-1}), a few effects connected with molecular absorption turn out to be expressed more sharply than on Jupiter, giving evidence of greater inhomogeneities in the aerosol layer in the latitudinal direction than occur on the larger planet.

2. The Continuous Spectrum

Photoelectric tracings of the continuous spectrum of the central meridian and east and west ansae of Ring B were obtained with the 70 cm telescope using a spectrometer with dispersion 30.4 Å/mm (resolution ~ 500) in the course of three nights in September, 1968. For absolute calibration of the tracings, spectra of the stars α Aurigae and β Arietis were also made. After the appropriate reductions we obtained the important coefficient of brightness of the center of the disk of Saturn and the rings as presented in Figure 1. The reflection properties of the planet are almost monotonic, decreasing

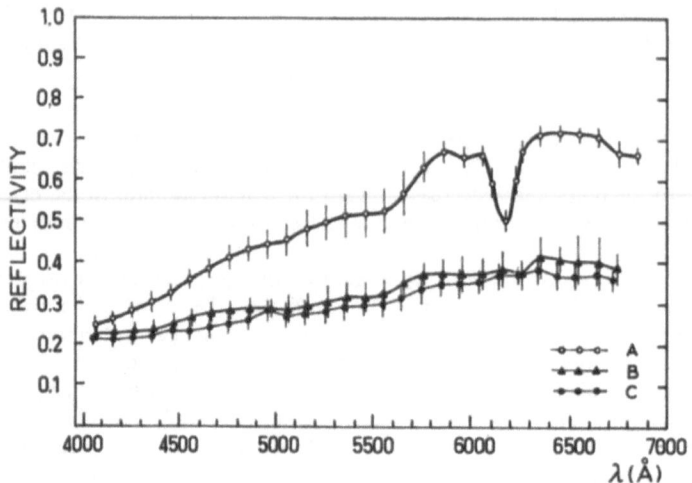

Fig. 1.   Spectral reflectivity of the bright equatorial belt of Saturn near the center of the disk (A), of the west ansa (B) and of the east ansa (C) in 1968.

toward shorter wavelengths, and this must be connected with the existence of true absorption of light by a particulate aerosol layer. Of great interest is the region of the spectrum where methane absorption bands occur. In the region 6200–6600 Å, the brightness coefficient of the center of the disk of Saturn reaches the maximum value $\rho_\lambda = 0.72$. This gives the albedo of single scattering $\omega_c = \sigma_a/(\sigma_a + \kappa_a) = 0.984$ for isotropic scattering and $\omega_c = 0.990$ for the indicatrix of scattering $x(\gamma) = 1 + \cos \gamma$. In the region 4000–4100 Å, $\rho_\lambda = 0.25$ and ω_c is approximately equal to 0.79 or 0.88, depending on the form of the scattering indicatricies. This means that the magnitude of the coefficient of true absorption in the aerosol, k_a, increases in the ultraviolet region of the spectrum by about 14–17 times, if the coefficient of scattering σ_a does not depend on wavelength. In the case of aerosol particles of small size $(r \sim 10^{-1} \mu)$, the volume scattering coefficient must increase in the short wavelength region while the true absorption must increase more strongly.

For different parts of the disk of Saturn the absorption changes non-uniformly with wavelength. Spectrograms of the central meridian of Saturn obtained in August and September, 1969, in the region 3200–6800 Å clearly reveal differences in the colors of individual cloud bands on the planet. The equatorial band of Saturn, the brightest region on the planet at $\lambda < 4500$ Å, was quite dark at $\lambda < 4250$ Å. At that time the north polar region of Saturn, which seems to be the darkest region on spectrograms and photographs taken in red light, was the brightest detail on blue light photographs and on spectrograms in the region $3500 < \lambda < 4300$ (Figure 2). At $\lambda < 3500$ Å the brightness of the disk of Saturn increases, thus confirming the photoelectric measure-

FRAGMENTS OF SATURN'S SPECTRUM

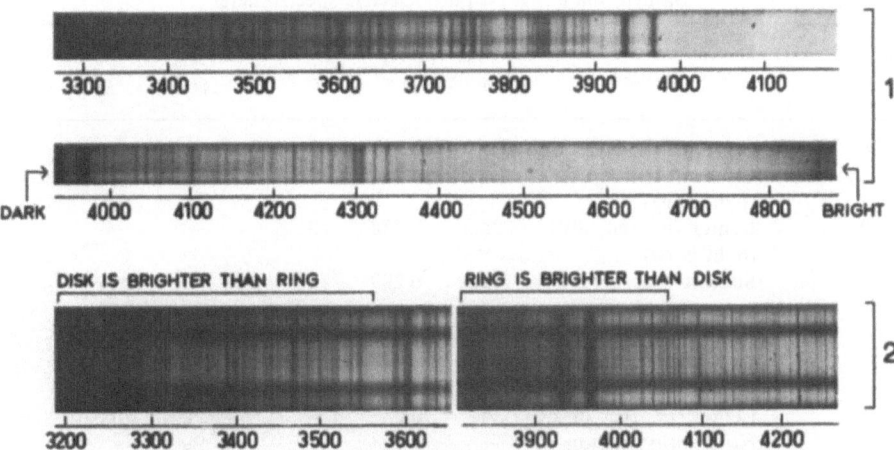

Fig. 2.   Spectrograms of the central meridian (1) and of the equator of intensity (2) of
Saturn in 1969.

ments of Younkin and Münch (1963) and ultraviolet rocket measurements by Bless *et al.* (1968).

3. Absorption Bands of CH_4

Photographic spectral measurements of the relative intensity in the center of the CH_4 6190 Å absorption band, made in 1966 and 1968, and also photoelectric measurements of the contours and equivalent widths of the 6190 Å and 7250 Å bands in 1968, reveal the following peculiarities of the change of intensity of methane absorption across the disk of Saturn. Along the equator of Saturn the absorption gradually decreases toward the edge of the disk. At the same time an increase in absorption was observed toward the poles of the planet, that is, in the light equatorial zone of Saturn the absorption was minimum, but sharply increased in the darker northern temperate regions (Table I and Figure 3).

At latitude $\phi \simeq -20°$ absorption was about 25–28% greater than in the equatorial zone of Saturn. This cannot be explained by a simple secant effect and must be connected with peculiarities in the structure of the aerosol layer in the atmosphere of Saturn. The nature of the variation of absorption with latitude in the relatively weak bands at 6190 and 7250 Å agrees well with the results obtained from photographs of Saturn obtained by Owen and Mason (1969) using a narrow-band interference filter centered in the strong CH_4 band at 8860 Å. This is in disagreement, however, with the results obtained from similar photographs of Jupiter. On Jupiter there is no correlation between the distribution with latitude of the intensity of weak and strong bands of methane.

It is strange that observations in 1966 show a weakening of the CH_4 absorption in all equatorial regions of Saturn. That is, in both components of the equatorial band,

TABLE I

Measurements of the intensities of absorption bands of CH_4 at 6190 and 7250 Å on the disk of Saturn, made with a photoelectric spectrometer

Date 1968	Region on the disk of Saturn	CH_4 6190 Å			CH_4 7250 Å		
		W (Å)	R	n	W (Å)	R	n
8–9.08	Center of the disk	23.5	0.258	2	93.6	0.700	12
19–20.08	Center of the disk (light zone)	26.4	0.254	17	90.5	0.678	35
6–7.09	Center of the disk (light zone)	24.3	0.234	18	—	—	—
	South temperate region	31.0	0.299	16	—	—	—
12–13.09	Equator (light zone)	27.0	0.252	16	87.4	0.670	14
	South temperate region	30.8	0.287	18	100.3	0.726	17
3–4.11	Center of the disk	24.2	0.252	4	78.6	0.630	10
	North polar region	—	—	—	99.3	—	4
5–6.11	Center of the disk	22.6	0.237	5	83.3	0.665	14
6–7.11	Center of the disk (light zone)	—	—	—	70.8	0.602	4
	Latitude −15°	—	—	—	83.4	0.648	5
	Latitude −25°	—	—	—	91.9	0.722	8
	Latitude −35°	—	—	—	106.3	0.730	3
	South polar region	—	—	—	110.3	0.757	6
	North polar region	—	—	—	104.8	0.816	5

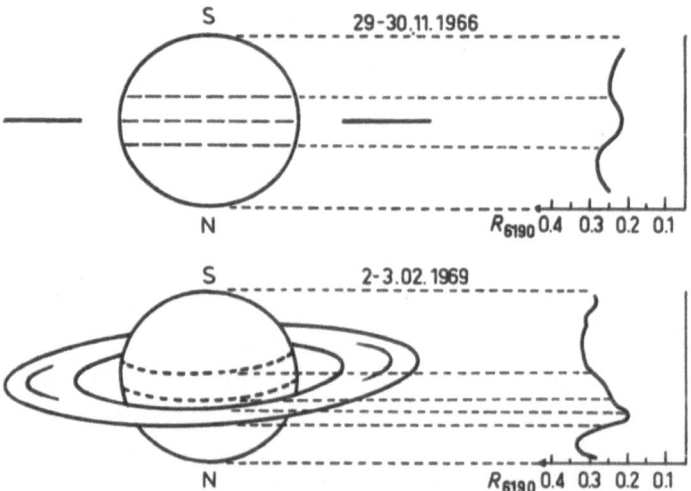

Fig. 3.   Changes of the methane absorption in the band CH_4 6190 Å along the central meridian of Saturn in 1966 and 1969. R_{6190} is the central depth of CH_4 6190 absorption band.

of which, as shown in photographs at λ 3550 Å by Marin (1968), the southern component was light and the northern component was very dark.

We carried out calculations of the curves of growth for absorption lines forming in a pure gas as well as in a gas-aerosol mixture. Calculations for the center of the disk

($\mu = 1.0$) and for limb regions ($\mu = 0.5$) allow us to find the dependence of the relative intensity of lines

$$\beta = \frac{W(0.5)}{W(1.0)}$$

with parameters of the characteristic absorption in the outer, pure gaseous atmosphere

$$u_0 = \frac{N_0 S H_0}{\pi \alpha_0},\tag{1}$$

and in the aerosol layer

$$u_1 = \frac{N_1 S}{\pi \alpha_1 \sigma_a}.\tag{2}$$

Here, N is the number density of absorbing molecules, S is the coefficient of absorption for a single molecule, α is the mean half-width of the line, H_0 is the scale height for the outer atmosphere, and σ_a is the mean volume coefficient of scattering of the aerosol layer.

This dependence is shown in Figure 4 for two cases: $\omega_c = 1.0$ and $\omega_c = 0.975$, for

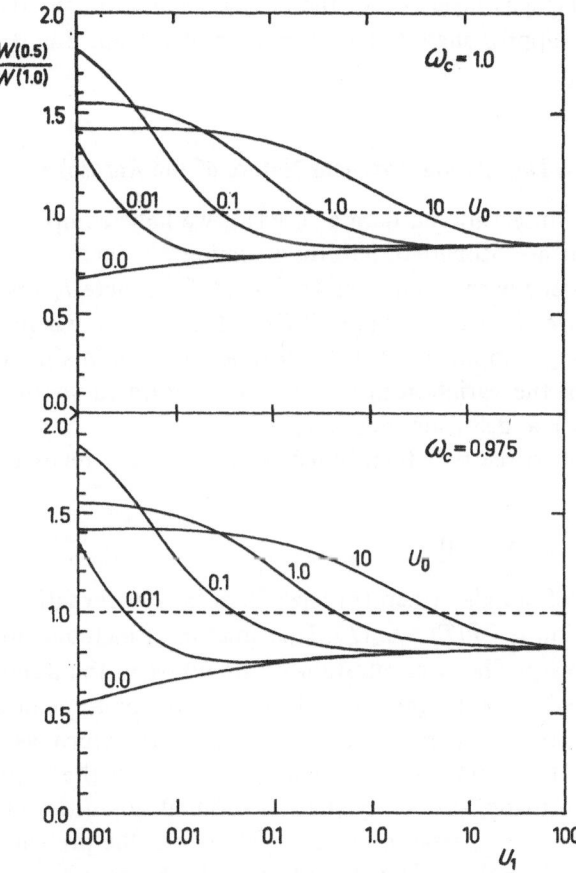

Fig. 4.   Changes of value $\beta = W(0.5)/W(1.0)$ with parameters u_0 and u_1.

the scattering indicatrix $1 + \cos \gamma$, and for several fixed values of the parameter u_0. The figures show that in the case of line and band formation originating only in the aerosol layer $(u_1 \gg u_0)$, the line intensity at the limb is smaller than in the center, such that $0.5 < \beta < 1$. If we know from laboratory measurements and observations the values of N, S, H_0, and α and also β, we can, as is shown in Figure 4, find u_1, from which we can determine σ_a. Analysis of a sufficiently large number of observations of Jupiter leads to the following probable limit of the mean value of σ_a for the aerosol layer:

$$5 \times 10^{-8} < \sigma_a < 5 \times 10^{-6} \text{ cm}^{-1}. \tag{3}$$

The lower limit is obviously too small. But the determination gives a value of σ_a that is somewhat smaller than, for example, terrestrial stratus clouds where $\sigma_a \sim 10^{-4} \text{ cm}^{-1}$. This confirms the conjectural statements in the literature that the cloud layer of Jupiter has its own aerosol haze.

The decrease in intensity of CH_4 bands from the center of the disk toward the equatorial limb of Saturn has been insufficiently studied. It is obvious only that for Saturn $\beta < 1$, but the exact quantity is uncertain, and we may say only that in the aerosol layer of Saturn the magnitude of σ_a cannot be of the nature of that for Jupiter, although the thicker gas layer above the clouds, seen from the data of rocket UV measurements, is approximately two times greater than for Jupiter. Therefore $u_{0 \text{ Saturn}} > u_{0 \text{ Jupiter}}$.

4. Theoretical Density and Physical Nature of the Aerosol Layer of Saturn

Using the existing observational data as a basis, we have computed an approximate 'gray' model of the atmosphere of Saturn, assuming $T_e = 90$ K and $H_2 : He = 5 : 1$, and also a model using one of the models of Trafton (1967), where $T_e = 90$ K and $H_2 : He = 1 : 0$. In Figure 5 are shown the curves $P(T)$ and $A_n \cdot P(T)$, where A_n is the relative concentration of any component of the atmosphere for full mixing. In the same figure are given curves of the variation in pressure of a saturated vapor with temperature for methane, $E_M(T)$, and for ammonia, $E_A(T)$.

It is not difficult to see that there must be a very great relative concentration of methane

$$A(CH_4) > 5 \times 10^{-2} \tag{4}$$

for the formation of clouds or aerosol haze from the condensation of CH_4, in order to fulfil the condition $A \cdot P(T) > E_M(T)$. This amount of methane does not agree with the observed quantity. The concentration of ammonia in the atmosphere of Saturn at the level where $T > 100$ K, that is, below the tropopause, can even be possibly $A(NH_3) = 10^{-5}$ to 10^{-6}. Apparently the aerosol layer on Saturn, as it is on Jupiter, is composed of crystals of NH_3. Approximate calculations of the volume density of the aerosol Q_a may be made, supposing that all residual quantities of ammonia corresponding to the excess of partial pressure of NH_3 over the pressure of the saturated vapor, are transferred to the solid or liquid phase. In this case it can be computed that

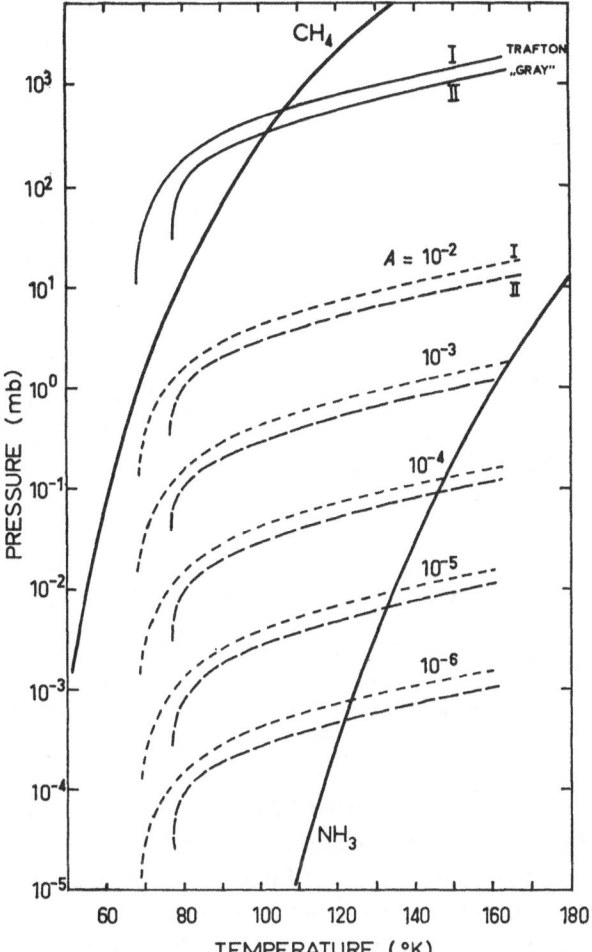

Fig. 5. Atmospheric pressure versus temperature in the atmosphere of Saturn and the pressure
of saturation for CH₄ and NH₃.

the gaseous component is subjected to continuous convective mixing. Then, according
to Obuchov and Golitsyn (1968)

$$Q_a = \frac{\mu_n}{RT_z} [A_n P_z - E_n(T_z)],$$ (5)

where μ_n is the molecular weight of the condensing component. Results of the calculations of Q_a for NH_3 in the atmosphere of Saturn for different values of $A(NH_3)$
are given in Figure 6. Since the function $P(T)$ for the model of Trafton (1967) and the
'gray' model differ in the region of condensation by a constant factor of two, computations were made only for the 'gray' model. Figure 6 shows that after a sharp increase
near the well defined lower boundary of the aerosol layer, the density of the aerosol in
the absence of intensive rising and lowering of flow very slowly diminishes with
height. The upper boundary of the aerosol layer, apparently, should be connected

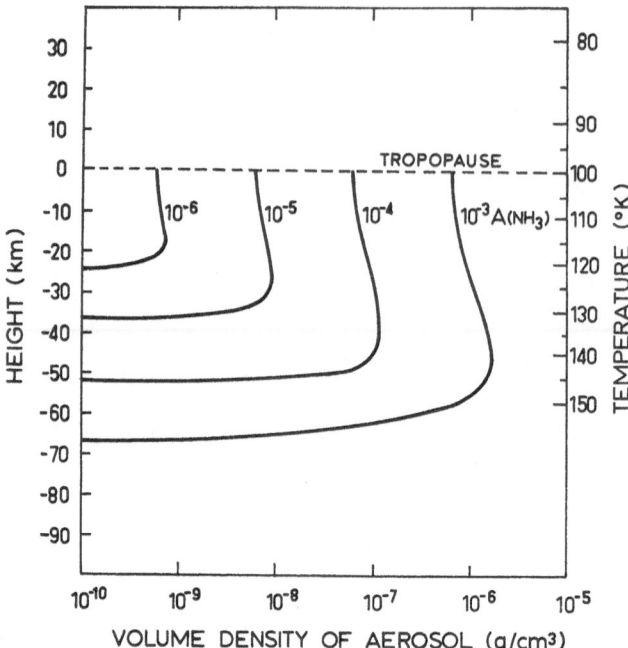

Fig. 6.   Vertical distribution of the volume density of ammonia aerosol layer (Q_a) in the atmosphere of Saturn for different relative abundances of ammonia. The adiabatic lapse rate is
$dT/dz = -0.88$ K/km.

with the level where convective or turbulent mixing becomes insignificant by comparison with the velocity of fall of condensing material.

Continuing, we note that as in the case of Jupiter these calculations for Saturn give a larger value of Q_a than is obtained from the value of the scattering coefficient σ_a. For $\sigma_a = 5 \times 10^{-6}$ cm^{-1} and refractive index $m = 1.33$, the volume density obtained is $Q_a \simeq 2.5 \times 10^{-10}$ to 4.04×10^{-9} gm/cm^3 for particle radii corresponding to $r = 0.05$ to 1.0μ. The reason for the divergences should probably be sought in the differences in the characteristics of the circulation regimes of Jupiter and Saturn, from those which were used as the basis of the calculations introduced above.

On the other hand, the observed difference between the absorption of CH_4 in the light equatorial zone of Saturn and that in the north temperate belt can be explained by a change in the mean coefficient of scattering at the effective level of formation of the absorption bands. Using the curve of growth it is not difficult to find that σ_a is 1.3 to 1.8 times larger in the light equatorial region than in the temperate belt. Such a difference in σ_a is completely possible since a small increase in the upward currents in the equatorial regions can lead to a redistribution of the volume density of the aerosol and to its increase at greater altitude by a factor of approximately 1.5.

5. Conclusions

Spectral variations in absorption bands of methane on the disk of Saturn show that the volume density in the aerosol layer in the atmosphere of the planet must be significantly lower than can exist in a cloud layer analogous to terrestrial stratus clouds. The density of Saturn's aerosol haze can change with latitude, and the denser layer of haze lies near the equator. A series of questions remains yet unanswered. In particular, it is not clear what the increased ultraviolet absorption in the belt near the equator is connected with. The reason for the difference between the values of the volume density determined by spectral observations and those theoretically obtained has not been established. Photographs and spectrograms of Saturn obtained in different years show very clearly the nonstationary character of processes in the atmosphere of this planet. From this fact follows the need for regular observations. Especially important are pictures and spectrograms taken in the extreme regions of the photographically accessible spectrum. It is also necessary to study in detail the distribution of molecular absorptions across the disk of Saturn in the strong infrared bands. These bands, more than others, must be formed in the upper layers of the aerosol layer and in the atmosphere above the clouds. Using calculations analogous to those presented in Figure 5, a sufficiently accurate value of σ_a can be obtained from observations of absorption bands, or even better, individual lines. On the other hand, in investigating effects in the continuum, especially in the ultraviolet, it is probably necessary to take into account the role of scattering and absorption by particles of various sizes. It may be, if only in part, that these effects are caused by the dependence of the scattering factor K_P on the size of particles and on wavelength. At large optical thickness of the aerosol layer this cannot play a determining role. In any case it is necessary to note the absence of a reliable optical model of the cloud layer of Saturn that is able to explain all phenomena observed on the disk of this planet.

Note Added in Proof. Our new measurements of Saturn's spectra give $0.95 < \beta < 1.0$. Using Rozenberg asymptotic formulae (Rozenberg, G. V.: 1962, *Dokl. Akad. Sci. S.S.S.R.* **145**, 775) for the strongly forward scattering function we obtained $u_0 \simeq 0.098$, $\sigma_a \simeq 2.2 \times 10^{-5}$ cm^{-1} for the light equatorial zone, and $u_0 \simeq 0.094$, $\sigma_a \simeq 1.4 \times 10^{-5}$ cm^{-1} for the temperate belt of Saturn.

References

Bless, R. C., Code, A. D., and Taylor, D. J.: 1968, *Astrophys. J.* **154**, 1151.
Marin, M.: 1968, *J. Obs.* **51**, 179.
Obuchov, A. M. and Golitsyn, G. S.: 1968, *Kosmich. Issled.* 6, 759.
Owen, T. C. and Mason, II. P.: 1969, *J. Atmospheric Sci.* **26**, 870.
Trafton, L. M.: 1967, *Astrophys. J.* **147**, 765.
Younkin, R. L. and Münch, G.: 1963, in *La Physique des Planètes*, Liège, p. 125.

ESTIMATE OF THE H_2 ABUNDANCE IN THE ATMOSPHERE
OF URANUS FROM THE PRESSURE INDUCED SPECTRUM

J. D. POLL

Dept. of Physics, University of Toronto, Toronto, Ontario, Canada

Abstract. Theoretical values of matrix elements of the quadrupole moment (Birnbaum and Poll, 1969; Dalgarno and Allison, 1969) and of the polarizability of H_2 (Birnbaum and Poll, to be published) are used to calculate the integrated intensities of a number of lines in the pressure-induced overtone spectra of molecular hydrogen. These calculated intensities are used to obtain an estimate of the H_2 abundance in the atmosphere of Uranus from the equivalent width of a feature in the 4-0 pressure-induced band observed by Giver and Spinrad (1966). An estimate of the equivalent width of the corresponding bands on Jupiter is also given.

1. Introduction

Direct spectroscopic information on the presence and abundance of molecular hydrogen in planetary atmospheres can be obtained from both the quadrupole and the pressure-induced spectra.

The intensities of the quadrupole lines provide, when the line strengths are known, information on the abundance of H_2. The simplest type of interpretation gives rise to the determination of a column density $u_1 = \int \rho(r) \, dr$, where ρ is the number density. The ratios of the intensities of quadrupole lines yield information on the temperature. From the widths of the lines information on the presence of perturbing atoms such as He is obtained. The main difficulty in the interpretation of the observed quadrupole spectra is the problem of saturation.

Similarly, the pressure-induced spectra provide, when the corresponding line strengths are known, information on the H_2 abundance. In this case, a simple analysis yields the quantity $u_2 = \int \rho(r)^2 \, dr$, which, in general, is not related in a simple way to $u_1 = \int \rho(r) \, dr$. The ratios of the intensities of pressure-induced lines again give information on the temperature. However, because the pressure-induced spectra are entirely due to the interaction between molecules, they provide in principle a more sensitive probe for the investigation of perturbers like He than the quadrupole spectra. As it happens, the lines that are most suitable for this purpose appear to be heavily obstructed by methane. Finally, the width of pressure-induced lines also provides direct information on the temperature of the medium. Because the pressure-induced lines are broad and weak, the main observational difficulty is associated with obtaining a good signal to noise ratio.

During the past year, Mr. A. Birnbaum and the present author have calculated matrix elements of the quadrupole moment of H_2 of interest in astrophysical applications (Birnbaum and Poll, 1969). Similar, but more extensive, calculations have been performed by Dalgarno and Allison (1969). These matrix elements are of use in the interpretation of quadrupole spectra. More recently, calculations of matrix elements of

Sagan et al. (eds.), Planetary Atmospheres, 384–391.

the polarizability of H$_2$ have been completed (Birnbaum and Poll, to be published), some of which are given in Table I. The possibility of making reliable calculations of matrix elements for H$_2$ is largely due to the work of Kołos and Wolniewicz (1964, 1965, 1967, 1968). It is the purpose of the present paper to show that from the calculated matrix elements of the quadrupole moment and the polarizability the integrated intensity of a large number of lines in the pressure-induced spectra of H$_2$ can be calculated (see Section 2). These intensities, some of which are given in Table II, are of use in the interpretation of the observed pressure induced bands (Section 3).

For a detailed review of the properties of pressure-induced spectra of interest to the planetary scientist, the reader is referred to a recent article by Welsh (1969). In particular, we will use his notation for double transitions in which, for example, the symbol $S_v(J) + Q_{v'}(J')$ denotes a transition in which one molecule of a pair makes a v-0 $S(J)$ transition, whereas simultaneously, the other molecule makes a v'-0 $Q(J')$ transition.

2. Review of the Theory of the Integrated Intensity of Pressure-Induced Lines

The theory of the integrated intensity of the fundamental band has been given by Van Kranendonk (1957, 1958). In this section we review some aspects of this theory and present expressions for the integrated intensity valid for the overtone bands.

The basic quantity in the theory is the dipole moment associated with a pair of molecules. This dipole moment will be a function of the orientations (θ, ϕ), the internuclear distances (r), and the separation (R) of the two molecules; it can be expanded as a sum of components, each with a definite angular dependence. In a coordinate system with a z-axis along the intermolecular separation we have

$$\mu_\kappa(r_1 r_2; \theta_1\phi_1, \theta_2\phi_2; R) = 4\pi \sum D_\kappa^{\lambda_1\lambda_2}(\mu_1\mu_2; r_1 r_2 R) Y_{\lambda_1\mu_1}(\theta_1\phi_1) Y_{\lambda_2\mu_2}(\theta_2\phi_2)$$

$$(1)$$

where $Y_{\lambda\mu}(\theta\phi)$ denotes a spherical harmonic and the summation is over all $\lambda_1\mu_1\lambda_2\mu_2$; μ_κ denotes a spherical component of the dipole moment. Each of the components $D_\kappa^{\lambda_1\lambda_2}$ will contribute additively to the total intensity. It is clear from (1) that vibrational and rotational transitions can take place either in one or in both molecules simultaneously. Previous experience indicates that the pressure-induced absorption is largely due to two types of components in the expansion (1), viz. the isotropic component ($\lambda_1 = \lambda_2 = 0$) and the quadrupolar components ($\lambda_1 = 2, \lambda_2 = 0$ or $\lambda_1 = 0$, $\lambda_2 = 2$). For the purposes of this paper we will therefore neglect all other contributions to the dipole moment. The isotropic part of the dipole moment contributes only to the intensity of Q-lines, i.e. to transitions of the form $Q_v(J) + Q_{v'}(J')$. Because the matrix elements of this part are not known for the overtone bands we will not consider transitions of this kind. The components of the form D_k^{20} and D_k^{02} correspond to quadrupolar induction, i.e. their long range parts are due to the creation of a dipole moment in one molecule by the quadrupole field of the other and vice versa. For this part of the dipole moment the non-zero components are

$$D_k^{20}(\kappa 0; r_1 r_2 R) = \frac{Q_1(r_1)\alpha_2(r_2)}{R^4} \sqrt{7}\, C(231; \kappa 0\kappa),$$

$$D_k^{02}(0\kappa; r_1 r_2 R) = -\frac{\alpha_1(r_1)Q_2(r_2)}{R^4} \sqrt{7}\, C(231; \kappa 0\kappa),$$

(2)

where $C(231; \kappa 0\kappa)$ is a Clebsch-Gordan coefficient (Rose, 1957) and Q_1 and α_2 denote the quadrupole moment and polarizability (as a function of internuclear distance) of molecules 1 and 2 respectively. This part of the dipole moment contributes to transitions of the form $Q_v(J) + Q_{v'}(J')$ as well as $S_v(J) + Q_{v'}(J')$. The intensity of the $S_v(J) + Q_{v'}(J')$ bands can therefore be considered to be due entirely to quadrupolar induction. The expression (2) for the components D_k^{20} and D_k^{02} is of course not correct when the separation R between the molecules is small. It turns out, however, that the contribution of the long-range part dominates over the contribution of the short-range part which will therefore be ignored. For example, the integrated intensity of the first overtone band calculated using the dipole moment according to (2) agrees to within 5% with the experimental intensity (McKellar and Welsh, private communication).

We conclude therefore that the intensity of bands of the form $S_v(J) + Q_{v'}(J')$ can be calculated in terms of the dipole moment given in (2), i.e. in terms of the matrix elements of Q and α. As it happens, all bands observed so far in planetary atmospheres are of this type.

As mentioned in the Introduction, matrix elements of Q and α are now available. Matrix elements of the quadrupole moment of H_2, and a discussion of their accuracy, are given in Birnbaum and Poll (1969). The relevant matrix elements of the polarizability are presented in Table I. The accuracy of the values in Table I is difficult to assess; we estimate that these matrix elements are reliable to within one unit in the third decimal place. This still allows for a considerable uncertainty in the 4-0 matrix

TABLE I

Matrix elements of the polarizability of the H_2 molecule (in atomic units)

v	J	$\langle vJ \mid \alpha \mid 0J \rangle$
0	0	5.4138
0	1	5.4234
0	2	5.4426
1	0	0.7392
1	1	0.7403
1	2	0.7423
2	0	0.0713
2	1	0.0715
2	2	0.0720
3	0	0.0099
3	1	0.0100
3	2	0.0100
4	0	0.0023
4	1	0.0023
4	2	0.0023

elements and an experimental verification of these numbers would therefore be desirable.

It has been shown by Van Kranendonk (1957) that it is more convenient to calculate a quantity proportional to the integrated transition probability than the integrated absorption coefficient itself. This quantity, which can be expanded in powers of the density, is given by

$$\int \alpha(\nu) \frac{d\nu}{\nu} = a_1 \rho^2 + a_2 \rho^3 + \cdots, \tag{3}$$

where $\alpha(\nu)$ is the absorption coefficient. By a straightforward generalization of the expressions for the fundamental band one finds for a dipole moment as given in (2) the following expression for the coefficient a_1

$$a_1 = \frac{2\pi^2 e^2}{3\hbar c} n_0^2 a_0^5 J^*$$

$$\times \sum \{P_{J_1} P_{J_2} C(J_1 2J_1'; 00)^2 C(J_2 0J_2'; 00)^2 \tag{4}$$

$$\times \langle v_1'J_1' \mid Q_1 \mid v_1 J_1 \rangle^2 \langle v_2'J_2' \mid \alpha_2 \mid v_2 J_2 \rangle^2 + \text{cycl}\}.$$

In this expression n_0 denotes the number density of the gas at 1 atm and 0 °C, e is the electron charge and $a_0 = 0.52917$ Å is the first Bohr radius. The quantity J^* represents the average R dependence of the square of the dipole moment (cf. Van Kranendonk, 1958)

$$J^* = 12\pi \int_0^\infty g(x) x^{-6} \, dx, \tag{5}$$

where $x = R/a_0$ and $g(x)$ is the low density limit of the pair correlation function of the gas. For classical statistical mechanics $g(x) = \exp(-\phi/kT)$ where $\phi(x)$ is the intermolecular potential. At the temperatures appropriate to Uranus and Neptune, quantum effects in $g(x)$ should be taken into account, however. The Boltzmann factors P_J in (4) are defined as

$$P_J = Z^{-1} g_J (2J + 1) \exp\left(-\frac{E_J}{kT}\right), \tag{6}$$

where $g_J = 1$, 3 for J even, odd and Z is the rotational partition function. Note that $\sum_J P_J = 1$. The quantities $C(J\lambda J'; 00)$ are Clebsch-Gordan coefficients and are given by

$$C(J0J'; 00)^2 = \delta_{JJ'},$$

$$C(J2J - 2; 00)^2 = \frac{3J(J - 1)}{2(2J - 1)(2J + 1)},$$

$$C(J2J; 00)^2 = \frac{J(J + 1)}{(2J - 1)(2J + 3)}, \tag{7}$$

$$C(J2J + 2; 00)^2 = \frac{3(J + 1)(J + 2)}{2(2J + 1)(2J + 3)}.$$

The matrix elements in (4) are expressed in atomic units. The abbreviation 'cycl' denotes a term identical to the preceding one except for an interchange of index 1 and

2. The summation sign, $\sum$, in (4) denotes a sum over all initial $(v_1v_2J_1J_2)$ vibrational rotational and all final $(v_1'v_2'J_1'J_2')$ vibrational rotational states of a pair of molecules that contribute to the particular band in which one is interested.

For example, for bands of the type $S_v(J) + Q_{v'}(J')$, which are the ones in which we are mainly interested, the coefficient a_1 becomes

$$a_1 = \frac{4\pi^2 e^2}{3\hbar c}\, n_0^2 a_0^5 J^* P_J P_{J'} C(J2J+2; 00)^2 \langle vJ+2| \ Q \ |0J\rangle^2$$
$$\times \langle v'J'| \ \alpha \ |0J'\rangle^2. \qquad (8)$$

In this expression v, v', J, and J' can take on the values, 0, 1, 2, ... Note that for $v'=0$ the symbol $Q_0(J')$ corresponds to a situation in which one molecule makes no transition (or an orientational one). In such a case all features differing by the value of J' only fall at the same frequency and one can, in addition, sum over all J' to obtain the integrated intensity of the $S_v(J)$ line. As can be seen from Table I, the J' dependence of the matrix elements $\langle 0J' \ | \ \alpha \ | \ 0J'\rangle$ is small. If we neglect this J' dependence the intensity of the $S_v(J)$ line takes on a particularly simple form

$$a_1[S_v(J)] = \frac{4\pi^2 e^2}{3\hbar c}\, n_0^2 a_0^5 J^* P_J C(J2J+2; \ 00)^2$$
$$\times \langle vJ+2 \ | \ Q \ | \ 0J\rangle^2 \langle 00 \ | \ \alpha \ | \ 00\rangle^2. \qquad (9)$$

Finally, we briefly discuss the case in which a certain amount of helium is present. We then have for the integrated probability

$$\int \alpha(\nu)\, \frac{d\nu}{\nu} = a_1^2 \rho_{H_2} + b_1 \rho_{H_2} \rho_{He} + \cdots \qquad (10)$$

The second term in (10), in the approximation that only binary interactions are of importance, describes the enhancement of the intensity due to the presence of helium. Because He atoms are not excited for the bands that we are interested in, the enhancement by He contributes only to the single transitions, i.e. to transitions in which only one of two H_2 molecules changes its energy. The dipole moment in a He-H_2 pair can again be decomposed in an isotropic part, the matrix elements of which are not known for the overtone bands, and an anisotropic part dominated by the quadrupolar induction mechanism also present in pure H_2. The contribution due to quadrupolar induction can again be calculated theoretically; it contributes to the S lines whereas the isotropic induction does not. The quantity b_1 for an S line is, however, about an order of magnitude smaller than the corresponding value of a_1 because of the small polarizability of He. We therefore see that for lines of the form $S_v(J) + Q_{v'}(J')$, He does not contribute at all to the double transitions and only very little to the single transitions, $S_v(J)$, unless the density of He is considerably greater than that of H_2 which does not seem consistent with other information on planetary atmospheres. The transitions considered above are therefore rather insensitive to the He concentration and only very accurate observations will make quantitative conclusions possible.

3. Estimate of the H_2 Abundance in the Atmosphere of Uranus

Pressure-induced spectra have been observed in the atmosphere of Uranus and Neptune by Kuiper (1949) and Spinrad (1963). Kuiper observed a feature at 8270 Å, subsequently shown to be mainly due to the $S_3^{(0)}$ transition (Herzberg, 1952), and Spinrad observed a feature at 6420 Å which is mainly due to the $S_4(0)$ transition. No equivalent widths were given for these features. However, Herzberg (1952) compared Kuiper's results with a laboratory spectrum of the 3-0 band and obtained partial pressures of 2 and 6 atm for H_2 and He respectively. More recently, Giver and Spinrad (1966) investigated the pressure-induced spectrum in the vicinity of 6420 Å again and obtained an equivalent width of 3.0 ± 0.5 Å and a half-width of about 30 Å. In the same paper Giver and Spinrad report a temperature of 124 ± 30 K for Uranus, obtained from the ratio of the equivalent widths of the $S(0)$ and $S(1)$ quadrupole lines neglecting saturation effects. Using improved line strengths for the quadrupole lines (Birnbaum and Poll, 1969) this temperature becomes 98 ± 30 K. For the purpose of the present paper we therefore take the temperature of the atmosphere of Uranus to be 100 K.

TABLE II

The integrated intensity, as defined in (3), for a number of lines in the 3-0 and 4-0 pressure induced overtone bands

Transitions	Wavelength (Å)	a_1 at 100 K (cm^{-1} amagat^{-2})
$S_3(0)$	8275	2.48×10^{-11}
$Q_3(1) + S_0(0)$	8251	0.33×10^{-11}
$Q_3(0) + S_0(0)$	8239	0.20×10^{-11}
$S_3(1)$	8153	2.96×10^{-11}
$S_4(0)$	6437	1.32×10^{-12}
$Q_4(1) + S_0(0)$	6418	0.18×10^{-12}
$Q_4(0) + S_0(0)$	6408	0.11×10^{-12}
$S_4(1)$	6369	1.80×10^{-12}

In Table II we give the integrated intensity, calculated using expressions (8) and (9), of a number of lines of interest in the pressure-induced overtone spectra of Uranus. There are other lines in the spectrum the intensity of which can be calculated from (8) but they appear to be obstructed by methane bands.

The feature observed by Giver and Spinrad is clearly the superposition of the $S_4(0)$, $Q_4(0) + S_0(0)$ and $Q_4(1) + S_0(0)$ bands, each of which has a half-width of about 30 Å. From the ratio of the intensities in Table II, one expects the maximum of the observed spectrum to be somewhat displaced from the $S_4(0)$ frequency in the direction of shorter wavelengths (cf. Welsh, 1969). To interpret the observed equivalent width in terms of an abundance we use a reflecting layer model without scattering, although this is certainly not a good approximation for the atmosphere of Uranus. In this case we have for the equivalent width of a weak absorption feature.

$$W_\lambda = a_1 \bar{\lambda} \eta u_2, \tag{11}$$

where a_1 is the line strength as defined in (3), $\bar{\lambda}$ an average wavelength of the feature defined according to

$$\bar{\lambda} = \int \alpha(\lambda) \frac{d\lambda}{\lambda} \Big/ \int \alpha(\lambda) \frac{d\lambda}{\lambda^2}, \tag{12}$$

and u_2 is defined by

$$u_2 = \int dr\rho(r)^2, \tag{13}$$

where the integral of the square of the number density is taken over a vertical column through the atmosphere. The quantity η denotes the slant path. Using $W_\lambda = 3.0$ Å, $\bar{\lambda} = 6430$ Å, and $a_1 = 1.6 \times 10^{-12}$ cm^{-1} amagat^{-2} we find

$$\eta u_2 = 3000 \pm 500 \text{ km amagat}^2. \tag{14}$$

The quoted uncertainty in ηu_2 is the one associated with W_λ; the uncertainty in a_1 is difficult to estimate but of the same order of magnitude. Within the framework of the model used, ηu_2 is the basic quantity that is determined from observation. (Note that, due to a numerical error, this number is not the same as the one reported at the Symposium.) To compare the result (14) with the column density $u_1 = \int dr\rho(r)$ as obtained from quadrupole spectra we need, in addition, values for η and the scale height H. Taking $\eta = 4$ and $H = 40$ km we find

$$\begin{aligned} \eta u_1 &= 980 \text{ km amagat,} \\ \text{density at 'reflecting surface'} &= 5 \text{ amagat.} \end{aligned} \tag{15}$$

The value for the density is close to the one obtained by Herzberg in 1952. The value for the column density seems rather low. In fact, from a preliminary determination of u_1 from the quadrupole spectrum and using a reflecting layer model, Owen (private communication) finds a value several times larger than the one obtained here. On the other hand Belton, McElroy, and Price (private communication) find, also from the quadrupole spectrum, a value of $\eta u_1 = 1450 \pm 430$ km atm which is rather close to the value quoted in (15). It should be stressed that the results given in (15) depend strongly on the values for η and H that one adopts. In particular, there is evidence that the scale height is greater than 40 km. Giver and Spinrad did not observe a pressure-induced feature at the frequency of the $S_4(1)$ transition, although the calculated intensity of the $S_4(1)$ line is actually larger than that of the $S_4(0)$ line. Giver and Spinrad ascribe this fact to a possible blending with a nearby methane band. The absence of the $S_4(1)$ line might also be interpreted in terms of a much lower temperature (Spinrad, 1963), but this would contradict the results obtained from the quadrupole spectrum.

4. Concluding Remarks

According to the calculations presented in this paper, the equivalent width of the feature at 8270 Å seen by Kuiper, which is the superposition of the $S_3(0)$, $Q_3(0) + S_0(0)$ and $Q_3(1) + S_0(0)$ bands, is (at 100 K) about 25 times larger than the equivalent width of the feature at 6420 Å discussed in the previous section. The feature at 8270 Å is therefore more suitable for an abundance determination than the other one, not only because a more reliable equivalent width can be obtained but also because the theoretical value for the coefficient a_1 is expected to be more accurate for the 3-0 than for the 4-0 band.

From the theory presented in this paper one can estimate the strength of the pressure-induced lines for other planets. For Jupiter, for example, one finds, using $\eta = 2$ and $u_1 = 90$ km amagat (Owen and Mason, 1968), a scale height of 8 km and $T = 150$ K, an equivalent width of the feature at 8270 Å of about 15 Å with a half-width of about 60 Å. It would appear therefore, that pressure-induced features are observable on Jupiter and presumably on Saturn as well. In fact, of course, no such feature has been seen. It is not likely that such a feature would have been overlooked even though it is quite broad. Perhaps the calculation of ηu_2 in terms of a reflecting layer model is not justified for Jupiter. Or finally, the theoretical calculation of the coefficients a_1 may be in error. To clear up this matter, as well as to gain a deeper understanding of the overtone spectra, in particular as regards the contribution of the isotropic part of the dipole moment to the intensity of the Q lines, more laboratory experiments are needed. At the same time, more accurate observations of pressure-induced bands in planetary atmospheres are necessary before reliable quantitative conclusions can be drawn.

Acknowledgments

The author would like to acknowledge helpful discussions with Dr. M. J. S. Belton, Dr. T. C. Owen, Dr. H. L. Welsh, and Mr. A. R. W. McKellar. He is indebted to Dr. M. J. S. Belton, Dr. M. B. McElroy, and Dr. M. J. Price for permission to quote one of their results prior to publication.

References

Birnbaum, A. and Poll, J. D.: 1969, *J. Atmospheric Sci.* **26**, 943.
Dalgarno, A., Allison, A. C., and Browne, J. C.: 1969, *J. Atmospheric Sci.* **26**, 946.
Giver, L. P. and Spinrad, H.: 1966, *Icarus* **5**, 586.
Herzberg, G.: 1952, *Astrophys. J.* **115**, 337.
Kołos, W. and Wolniewicz, L.: 1964, *J. Chem. Phys.* **41**, 3663.
Kołos, W. and Wolniewicz, L.: 1965, *J. Chem. Phys.* **43**, 2429.
Kołos, W. and Wolniewicz, L.: 1967, *J. Chem. Phys.* **46**, 1426.
Kołos, W. and Wolniewicz, L.: 1968, *J. Chem. Phys.* **49**, 404.
Kuiper, G. P.: 1949, *Astrophys. J.* **109**, 540.
Owen, T. C. and Mason, H. P.: 1968, *Astrophys. J.* **154**, 317.
Rose, M. E.: 1957, *Elementary Theory of Angular Momentum*, Wiley, New York.
Spinrad, H.: 1963, *Astrophys. J.* **138**, 1242.
Van Kranendonk, J.: 1957, *Physica* **23**, 825.
Van Kranendonk, J.: 1958, *Physica* **24**, 347.
Welsh, H. L.: 1969, *J. Atmospheric Sci.* **26**, 835.

THE SPECTRUM OF URANUS IN THE REGION 4800–7500 Å

L. S. GALKIN

Crimean Astrophysical Observatory, Academy of Sciences, U.S.S.R., p/o Nauchny, Crimea, U.S.S.R.

and

L. A. BUGAENKO, O. I. BUGAENKO, A. V. MOROZHENKO

Main Astronomical Observatory, Academy of Sciences, Ukraine, S.S.R., Kiev, U.S.S.R.

Spectra of Uranus were obtained with the 122 cm reflector of the Crimean Astro-physical Observatory using the spectrometer of the Main Astronomical Observatory of the Ukrainian Academy of Sciences (Kiev), in March 1969. A photomultiplier (Soviet type 79) was used in a pulse counting mode, and the dispersion of the spectro-meter at the camera focal plane was 15 Å/mm.

The spectrum was scanned in discrete increments of width 11 Å which corresponds to the width of the exit slit. Four scans of the spectrum were made in three nights. Six scans were made of the solar spectrum so that the effects of the Fraunhofer lines could be subtracted. The mean of the four Uranus spectra with the solar spectrum removed is given in Figure 1. The method we used did not permit a precise elimination of the solar spectrum, but the cumulative error does not exceed one-half of a scan increment (5.5 Å).

Apart from well-known absorption bands of methane, the spectrum in Figure 1 shows features at 5210 Å and 5571 Å, as well as a band at 6475 Å that was mentioned

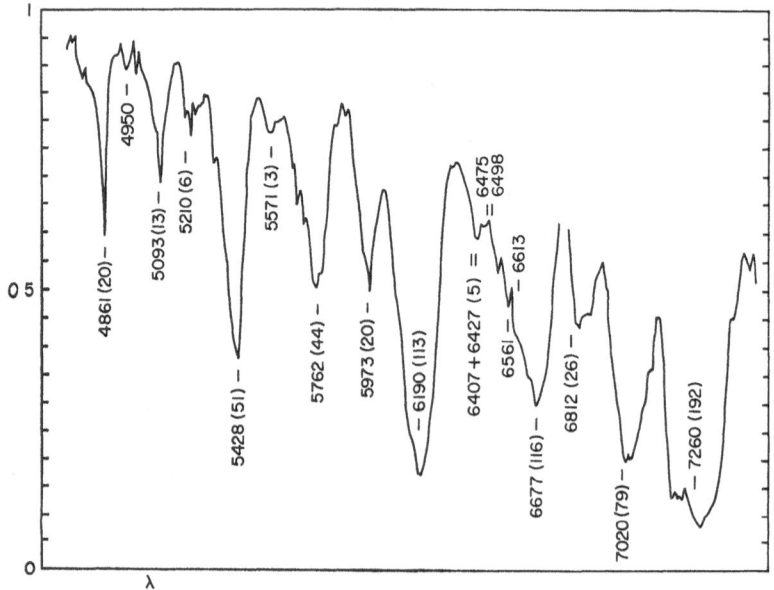

Fig. 1.   Mean of four spectral scans of Uranus, 4800–7400 Å.

Sagan et al. (eds.), Planetary Atmospheres, 392–393.

earlier by V. G. Teifel. Furthermore, near 4950 Å there is a small depression having a residual intensity of 96%, as well as several weak absorption bands lying on the wings of such strong bands as 5428 Å and 6677 Å.

Because of the overlapping of bands, the continuum is practically impossible to define precisely. Therefore the exact equivalent widths of the bands cannot be measured, but for the majority of important bands the lower limit of the equivalent width (in angstroms) is given in Figure 1 in brackets.

Reference

Moroz, V. I.: 1967, in *Physica Planet*, Nauka, Moscow. [English translation: *Physics of the Planets*, NASA TT F-515 (1968)].

UPPER LIMIT OF HYDROGEN AND HELIUM
CONCENTRATIONS ON TITAN

WAYNE E. McGOVERN

Institute of Atmospheric Physics, The University of Arizona, Tucson, Ariz., U.S.A.

Abstract. The satellite Titan is commonly quoted as possessing an atmosphere consisting of at least 2×10^4 cm atm of methane. Plausible additional atmospheric constituents like hydrogen and helium are assumed to have completely escaped from the satellite. However, the employment of recent techniques to the upper atmosphere of Titan permits an improved estimate of the upper limits of the hydrogen to methane and helium to methane mixing ratios existing in the lower atmosphere of Titan, namely $10^{-6 \pm 1}$ and $10^{-3.5 \pm 0.5}$.

1. Introduction

Of the 32 identified satellites in the solar system, only Titan, a satellite of Saturn, is commonly accepted as possessing an atmosphere. Kuiper (1952), in evaluating the spectra of Titan, particularly at 6190 Å, concluded that the atmosphere of Titan consisted of at least 2×10^4 cm atm of CH_4. Proceeding from this information, Kuiper (1952), using an approximation method for computing the exospheric temperature in conjunction with a simplified escape criterion, deduced that all molecules with molecular weights less than 10 must have escaped from Titan. However, since this original study, our knowledge of the physical mechanisms controlling the temperature and density structure of the upper atmosphere of a planetary body has increased significantly. Therefore, a re-examination of the upper atmosphere of Titan, using procedures similar to those suggested by Chamberlain (1962) could result in improved estimates of several physical parameters associated with the atmosphere of Titan, in particular the maximum atmospheric hydrogen concentration.

Spectroscopically, only CH_4 has been positively identified on Titan. Ammonia, due to its low frost point, has probably precipitated out; hence it is not surprising that an analysis of the 6450 Å band of NH_3 by Kuiper (1952) resulted in an ammonia upper limit of only 3×10^2 cm atm.

Plausible additional atmospheric constituents are hydrogen and helium; nevertheless, on account of their low molecular weights, both gases have a strong tendency to dissipate rapidly through gravitational escape. Specifically, the escape rate is a function of both the height and temperature of the exosphere. However, both parameters are in turn dependent upon the relative concentrations of these same gases; this is particularly true for hydrogen. Under such circumstances, as the quantity of these gases decreases, the remaining portions become less susceptible to escape, until a threshold level is reached below which equilibrium concentrations can be expected to be maintained through either outgassing and/or photolysis. The purpose of this article is to determine these threshold levels for H_2 and He through a study of the upper atmosphere of Titan.

2. Upper Atmosphere Composition

The composition of the upper atmosphere depends to a large extent upon the photo-chemistry associated with methane; however, studies of the problem with particular reference to Jupiter have not produced uniform results. Wildt (1937) noted that the absorption spectrum of complex hydrocarbons extends beyond that of CH_4, also that of all the hydrocarbons only CH_4 appears to be stable in the presence of atomic hydrogen. He concluded that, in the presence of sufficient H_2, the formation of C_2H_6 will be almost completely suppressed. In this situation, the photochemical decomposi-tion of CH_4 will be thwarted, and only small amounts of molecules formed as a result of the photochemical processes should be expected. In contrast, Cadle (1962) indicated that when CH_4 undergoes appreciable photolysis its concentration approaches zero and that the vast majority of the carbon from the dissociated methane reappears as ethane (C_2H_6). In a recent numerical study, Strobel (1969a), noting the numerous laboratory studies concerning the photolysis of CH_4 since the article by Cadle, concluded that in the Jovian atmosphere the over-all mixing ratio of ethane to methane was approximately 10^{-4}. Applying this technique to Titan resulted in a similar conclusion, namely, that the ethane-methane mixing ratio is small, approxi-mately 10^{-5} (Strobel, 1969b).

We are therefore presented with two extreme views concerning the photolysis of methane: complete dissociation and, in effect, no dissociation. In the initial series of models in this article, CH_4 will be assumed to be completely dissociated and converted to ethane in the upper atmosphere of Titan. This assumption is based not so much on the probable composition resulting from the photolysis of CH_4, but on the fact that C_2H_6 at 12.2 μ, especially at low temperatures is superior to CH_4 at 7.7 μ as a cooling agent for Titan. This model will therefore produce minimum exospheric temperatures for a given H_2 concentration, and, in essence, an estimate of the upper limit of H_2 on Titan independent of the present uncertainties associated with the photochemistry of methane. Consequently, above the mesopause, cooling was limited to this single vibrational energy exchange, since pure rotational exchanges will be of minor import-ance for molecules with a center of symmetry like CH_4, C_2H_2, C_2H_6, C_3H_8, and H_2 (Goody, 1964). Therefore, on Titan in this initial series, equilibrium conditions will consist of a CH_4-dominated lower atmosphere associated with a small amount of free hydrogen. At the mesopause, a mixture of mainly C_2H_6, CH_4, and molecular and/or atomic hydrogen is assumed to prevail, with cooling restricted to C_2H_6. Above the mesopause, atomic hydrogen is produced from molecular hydrogen by dissociative recombination following the ionization of molecular hydrogen. Through 3-body recombination, molecular hydrogen is reformed at lower altitudes. Based upon these processes, Zabriskie (1960) has concluded that, in the vicinity of the Jovian lower thermosphere, hydrogen is predominantly in the molecular form ($H/H_2 \sim 10^{-3}$). The photochemical analysis by Strobel indicates that, in the region of the Jovian

27—P.A.

mesopause, molecular hydrogen likewise predominates ($H/H_2 \lesssim 10^{-6}$). Applying these methods to Titan resulted in the conclusion that in the lower thermosphere, for hydrogen to methane mixing ratios in excess of 10^{-3}, the H to H_2 ratio is less than one, but, through diffusion, atomic hydrogen will become the dominant constituent in the upper thermosphere and exosphere of Titan. For mixing ratios less than 10^{-3} atomic hydrogen dominates over molecular hydrogen throughout the entire upper atmosphere. The specific mesopause H/H_2 mixing ratios adopted are found in Tables I and II. Above the mesopause, under normal conditions, the constituents are assumed to be in diffusive equilibrium, while below the mesopause mixing dominates.

3. Models and Results

In reconstructing the temperature and density structure of the upper atmosphere, the heat conduction equation, the hydrostatic equation, the cooling term, and the assumptions associated with these equations as well as the procedure for applying these equations are similar to those outlined by McGovern (1969), with the following exceptions: (1) selection of a higher heating efficiency (0.30); (2) higher cooling rate by C_2H_6 at low temperatures by reducing the cooling constant x_i from 0.95 to 0.75; so that the relative variation in the cooling as a function of temperature is similar to that for CO_2; (3) maximum solar flux limited to the wavelength interval between 31 and 964 Å; (4) selected a mesopause height and temperature of 200 km and 80 K, respectively; (5) computed average day side instead of sub-solar point temperature; and (6) physical parameters pertinent to Titan, for example, solar flux and gravity, were incorporated.

The results of this numerical program for various hydrogen to ethane mixing ratios at the mesopause are tabulated in Table I. The percentage of hydrogen in the atomic state at the mesopause is listed in column 2, the remaining portion is H_2.

For the particular range of mixing ratios selected, the mesopause density (10^{11} to 10^{12} molecules cm^{-3}) and the temperature of the exosphere (column 3) were found to be relatively stable. In contrast, the base of the exosphere (column 4) varies considerably, which is the primary reason for the sizable variations in the escape flux L in column 5. The escape flux (L) was computed using the following equation given by Spitzer (1949):

$$L = \pi R_e^2 n_e \bar{v} \, e^{-R_e/H_e} \, (1 + R_e/H_e),$$

where $\bar{v}$ is the average thermal velocity, while R_e/H_e, and n_e are, respectively, the geocentric or planetocentric radius, scale height and number density of the escaping constituent at the base of the exosphere. In column 6 are the production rates of hydrogen through outgassing and/or photolysis which are necessary in order to balance the loss of hydrogen through escape.

The maximum photon flux reaching Titan capable of dissociating CH_4 is approximately 1.4×10^{10} photons cm^{-2} sec^{-1} (Hinteregger et al., 1965); therefore, the maximum possible escape flux through photolysis alone is approximately 10^{10} cm^{-2} sec^{-1}.

TABLE I

Exospheric parameters associated with various mesopause hydrogen-to-ethane mixing ratios

Mesopause hydrogen-ethane mixing ratio	Percentage of mesopause hydrogen in atomic state	Exospheric temperature (K)	Base of exosphere (km) ($n_e \equiv 10^6$ cm^{-3})	Hydrogen escape flux (L) (atoms cm^{-2} sec^{-1})	Equilibrium rate (Q) (atoms cm^{-2} sec^{-1})	$(R_e/H_e)_H$	Blow-off level $\times 10^3$ (km) H	H$_2$
10^{-6}	100	120	500	1.4×10^9	2.0×10^9	3.1	4	10
10^{-5}	100	120	1400	1.5×10^{10}	3.9×10^{10}	2.3	4	10
10^{-4}	100	125	9400	$> 2.3 \times 10^{10}$	$> 6.0 \times 10^{11}$	< 1.5	4	10
10^{-3}	10	150	14000	$> 2.5 \times 10^{10}$	$> 1.2 \times 10^{12}$	< 1.5	4	9

Exospheric escape fluxes higher than 10^{10} cm^{-2} sec^{-1} seem improbable, independent of the mixing ratio, because the time necessary for upward diffusion of hydrogen is such as to limit the escape rate. An estimate of the molecular diffusion time constant (τ) is given by $\tau = H^2/D$, where the scale height is represented by H and D is the molecular diffusion coefficient which can be approximated by $10^{17} H^{1/2}/n$. This results in a τ of 10^5 to 10^6 sec, and based upon a total exospheric concentration of 10^{15} atoms cm^{-2}, indicates that, under normal conditions, the maximum escape rate for atomic hydrogen is 10^{10} atoms cm^{-2} sec^{-1}.

At mixing ratios higher than 10^{-5}, the upper atmosphere would appear to enter a chaotic stage in which τ may no longer be a limiting factor on the escape rate. Öpik (1963) has indicated that, when the thermal energy of translation exceeds the gravitational energy, the top portion of the atmosphere will 'blow-off', and that the escape of gases will be both extremely high and indiscriminate as to species. An equivalent manner of expressing this criterion is by noting that, when the ratio of the exospheric planetocentric distance to the exospheric scale height falls below 1.5, the kinetic energy will exceed the gravitational potential energy and 'blow-off' occurs. Column 7 indicates the value of this ratio for atomic hydrogen at the exospheric base. In columns 8 and 9 for the particular exospheric temperature found in column 3, the level at which the planetocentric distance to scale height ratio equals 1.5 is listed for atomic and molecular hydrogen. These results indicate that, when the mixing ratio approaches 10^{-4}, then the blow-off level falls below the theoretical base of the exosphere and the entire upper atmosphere would be in a chaotic state. Under such circumstances, the escape rate for hydrogen would be very high, requiring outgassing rates from Titan significantly higher than anticipated ($> 10^{11}$ molecules cm^{-2} sec^{-1}), in order to maintain equilibrium. Thus, based upon the assumption of the dominance of C_2H_6 in the upper atmosphere of Titan, it is concluded that the maximum hydrogen to methane mixing ratio in this atmosphere is in the vicinity of 10^{-5}.

The alternative model, namely, assuming that the dissociation of methane is very small, means that the net cooling in the upper atmosphere will decrease. As noted previously, CH_4 at 7.7 μ and 150 K is several orders of magnitude ($\sim$3 or 4) less effective as a radiator than C_2H_6 at 12.2 μ and 150 K; however, the cooling rates associated with CH_4 and C_2H_6 are both strongly dependent upon temperature. For example, CH_4 cooling per molecule at 250 K is comparable to the cooling rate of C_2H_6 at 150 K. This results from selecting a cooling constant (x_i) for CH_4 of 1.37. Under these circumstances, it is not surprising that the exospheric temperature for a pure CH_4 atmosphere approaches 250 K in contrast to 115 K for a pure C_2H_6 atmosphere; in fact, as seen from Table II, the permissible atmospheric hydrogen contents are so low that the exospheric temperature for these models and a pure CH_4 atmosphere are essentially the same. Another aspect of these results is that, when the mixing ratio (w) becomes equal to or less than 10^{-7} the base of the exosphere is no longer dominated by atomic hydrogen, in which case the escape flux of atomic hydrogen becomes dependent upon the hydrogen concentration at the mesopause and can be estimated by $L_H \simeq 10^{16} w$ for a CH_4-dominated atmosphere. A similar situation existed for a C_2H_6-dominated upper atmosphere for $w \leqslant 10^{-6}$, then $L_H \simeq 10^{15} w$. In both cases,

TABLE II

Exospheric parameters associated with various mesopause hydrogen-to-methane mixing ratios

Mesopause hydrogen-methane mixing ratio	Percentage of mesopause hydrogen in atomic state	Exospheric temperature (K)	Base of exosphere (km) ($n_e \equiv 10^6$ cm^{-3})	Hydrogen escape flux (L) (atoms cm^{-2} sec^{-1})	Equilibrium rate (Q) (atoms cm^{-2} sec^{-1})
10^{-8}	100	245	1900	3.3×10^8	1.0×10^9
10^{-7}	100	245	1900	3.3×10^9	1.1×10^{10}
10^{-6}	100	250	2500	$> 3.0 \times 10^{10}$	$> 1.3 \times 10^{11}$

when w falls below 10^{-10} then the escape of hydrogen occurs through CH_4 molecule and the above simple relationships are no longer valid.

The sensitivity of these methods decreases rapidly with increasing molecular weight; for the application of Öpik's criterion to a pure helium-ethane upper atmosphere results in a maximum He/C_2H_6 mixing ratio between 5×10^{-2} and 10^{-2}. However, even at 10^{-2} the outgassing rate needed to maintain equilibrium with the escape is over an order of magnitude greater than the outgassing rate for the earth. Comparable outgassing rates are obtained when the mixing ratio approaches 10^{-3} for a C_2H_6-dominated atmosphere and 10^{-4} for an upper atmosphere consisting mainly of CH_4; therefore, a reasonable upper limit for the helium-methane mixing ratio on Titan is $10^{-3.5 \pm 0.5}$.

4. Conclusions

Based upon the molecular diffusion time constant (τ), it was concluded that the maximum escape rate for atomic hydrogen under non-chaotic conditions was 10^{10} atoms $cm^{-2} sec^{-1}$. Contingent upon this escape flux, the upper limit on the hydrogen-methane mixing ratio on Titan was found to be 10^{-7} if the net dissociation of methane is small. In comparison, if the upper atmosphere is dominated by C_2H_6, the mixing ratio upper limit is 10^{-5}, while a similar analysis yields a helium-methane upper limit of $10^{-3.5 \pm 0.5}$.

Acknowledgment

I am grateful to Dr. Darrell F. Strobel, not only for the use of his methane photochemistry computer program, but also for several helpful discussions.

References

Cadle, R. D.: 1962, 'The Photochemistry of the Upper Atmosphere of Jupiter', *J. Atmospheric Sci.* **19**, 281–285.

Chamberlain, J. W.: 1962, 'Upper Atmosphere of the Planets', *Astrophys. J.* **136**, 582–593.

Goody, R. M.: 1964, *Atmospheric Radiation*, Vol. 1, Clarendon Press, Oxford, 436 pp.

Hinteregger, H. E., Hall, L. A., and Schmidtke, G.: 1965, 'Solar XUV Radiation and Neutral Particle Distribution in July, 1963 Thermosphere', *Space Research* **5**, North-Holland Publ. Co., Amsterdam, pp. 1175–1190.

Kuiper, G. P.: 1952, 'Planetary Atmospheres and Their Origin', in *The Atmospheres of the Earth and Planets*, University of Chicago Press, pp. 306–405.

McGovern, W. E.: 1969, 'The Primitive Earth: Thermal Models of the Upper Atmosphere for a Methane-Dominated Environment', *J. Atmospheric Sci.* **26**, 623–635.

Öpik, E. J.: 1963, 'Selective Escape of Gases', *Geophys. J.* **7**, 490–509.

Spitzer, L.: 1949, 'The Terrestrial Atmosphere Above 300 km', in *The Atmospheres of the Earth and Planets*, University of Chicago Press, pp. 211–247.

Strobel, D. F.: 1969a, 'The Photochemistry of Methane in the Jovian Atmosphere', *J. Atmospheric Sci.* **26**, 906–911.

Strobel, D. F.: 1969b, private communication.

Wildt, R.: 1937, 'Photochemistry of Planetary Atmospheres', *Astrophys. J.* **86**, 321–336.

Zabriskie, F. R.: 1960, 'Studies on the Atmosphere of Jupiter', Ph.D. dissertation, Dept. of Astronomy, Princeton University, 48 pp.

PART IV

SCIENTIFIC DEDICATION OF THE 107-INCH
REFLECTOR

REMARKS MADE AT THE SCIENTIFIC DEDICATION OF THE 107-INCH REFLECTOR, OCTOBER 30, 1969

Dr. Harlan Smith: Not many large telescopes in the world have been dedicated twice. Yet this may not be such a bad thing to do. The convention has long existed of christening a baby at birth and baptizing it later; in this sense the 107-inch telescope was christened almost a year ago, and today might be considered its baptism. Indeed, the parallel is rather apt, because christening is a naming – a formal recognition of coming into existence – whereas a baptism in the traditional sense of the word is a dedication to a certain path of life.

Our first dedication was a christening, a coming-into-existence affair. Those who have been closely involved in the construction of a large telescope can appreciate the number of problems that arise, the moments when there is even doubt whether the job will be finished. So the first completion in a real sense was the birth, the proper christening time of the instrument. At the November, 1968, dedication the telescope was essentially complete, but several problems delayed the start of observing until March 7, 1969, after which time it was appropriate to consider a scientific baptism – its dedication to research.

A more practical rationale for our second dedication was the large number of people who were in some way connected with this telescope. There was scarcely room for so many here; to have a substantial scientific delegation in addition would have overcrowded the facilities and changed the character of the first program. On the other hand, this second dedication reflects the facts that we have present a quorum of the world's planetary astronomers, and are baptizing the world's first very large telescope conceived and dedicated in considerable measure for planetary research.

It is also appropriate that we should be here dedicating this telescope in part for planetary research because McDonald Observatory was, during the lean years of planetary work, one of the few strongholds of continued effort along such lines. Beginning about the time Professor Kuiper came to Chicago and continuously to the present, the McDonald Observatory has devoted an appreciable fraction of its 82-inch time to solar system problems, a statement which could probably not be made of any other large telescope in the world. But the large group of astronomers present at this meeting gives heartening evidence for the rebirth of interest in planetary subjects, this time with a strongly astrophysical flavor which usually requires large telescopes to gather enough light for the necessary types of observation. Such studies at McDonald using the 82-inch Struve reflector (and the 36-inch reflector to a lesser degree) helped greatly to put on a firm foundation the basic astrophysics of the planets and satellites of the solar system: colors, spectra, atmospheres, and surface properties of the moon, planets, satellites, asteroids, and comets – even zodiacal light and gegenschein.

About five years ago we came to feel that the increasing amount of planetary work needing to be done was more than could be handled by the 82- and 36-inch reflectors, along with increasing needs also in stellar, interstellar, and extragalactic astronomy. The proper solution to the shortage of planetary observing time was of course to

Sagan et al. (eds.), *Planetary Atmospheres*, 403–408.

augment the world's supply of large telescopes. The National Aeronautics and Space Administration concurred; two men who especially contributed during this early stage at NASA were Dr. Urner Liddell and Dr. Ronald Schorn.

Thus, late in 1964, it was agreed between the University of Texas and NASA that an additional large telescope should be erected here at the McDonald Observatory, which would normally devote at least a quarter of its time to solar system – mainly planetary – research. A year and a half were needed to work out arrangements and detailed design; three and a half years ago fabrication began on the telescope and the dome. That work took only two and a half years, and the telescope began to be installed in mid-1968. Beginning March 7, 1969, the drives were effectively turned on, and spectra of Mars showing water vapor were obtained on the first several nights. Since then the 107-inch has been in almost continuous use, principally at the coude focus where the planetary instruments are concentrated.

The first of these coude instruments is the giant horizontal coude spectrograph which fills most of the floor directly beneath us. In the coude slit room is the Connes-type interferometic spectrometer of Dr. Reinhard Beer (JLP), currently working from 3 to 5 μ with extremely high resolution. Medium-dispersion but therefore faster interferometric spectrometers are also used in the slit room (Dr. Rudi Hanel of Goddard and Dr. Andrew Potter of Manned Spacecraft Center).

This past summer we added a new coude focus area at the south end of the observing floor, to provide space for the principal equipment for laser ranging on the lunar target left by the Apollo 11 astronauts in July. The laser beam sent out almost daily from the 107-inch reflector has a divergence typically of only several seconds of arc; the retroreflector on the moon returns about 10^{-8} to 10^{-9} of this light toward the earth, diverging in a wider cone of which again a comparably small fraction is intercepted by the telescope. The few remaining photons are recorded in the laser room by a photocell and their time of travel measured to about a nanosecond. This system has been in operation since August, initially under the supervision of Dr. Brian Warner.

In addition to such solar-system coude programs, the telescope obviously has many other potential uses. In particular at the four Cassegrain focal positions (three of them reached by rotating the first coude flat mirror inside the tube) we propose to do stellar and extragalactic work. The $f/18$ Cassegrain has been brought into initial operation over the last few weeks; the $f/9$ will follow shortly. To refer to only one example of our Cassegrain plans, those of you who on the way to this session visited the University of Texas Radio Astronomy Observatory under the direction of Dr. James Douglas, saw an installation which among other activities will soon begin to produce thousands of radio source positions accurate to about a second of arc. A major use of the 107-inch telescope during dark of the moon will be the optical study of these radio sources, including Seyfert galaxies, quasars, and similar objects. Over the next few decades this telescope is thus not likely to be idle.

But I want to emphasize again that more than half of the support necessary to build the 107-inch reflector was provided by NASA (principally the Lunar and Planetary Office); the University of Texas provided more than half of the remainder, and the National Science Foundation the rest. This support over the last five years has been deeply appreciated; and has made possible the telescope and associated new McDonald facilities. Now, with its successful coming into operation, there exists the first large telescope planned with one of its major roles in support of planetary research. As long as there are such programs which can usefully be done with a large ground-based telescope, the 107-inch will be deeply involved.

It is especially appropriate at this time to introduce the man who over the last three years has had the most to do with supporting the telescope administratively and financially, Dr. William Brunk of the National Aeronautics and Space Administration, who through his direction of the planetary research support programs of NASA has been our chief reliance in this area.

Dr. Brunk: After that introduction, I must admit that there were several times when I wondered whether I would actually stand under this telescope. It's very nice to see it here, and better still to know that we're already beginning to get some important and interesting information from it. A further bonus is the fact that we have been able on such short notice to come in with an entirely new program such as the lunar laser and have it operating so soon in a relatively smooth manner.

I do want to emphasize that we at NASA have realized more and more with the passage of time that to really gain a thorough understanding of our solar system, the moon and planets, it is necessary to rely not only on spacecraft which can give unique answers but also on observations made from the earth and from the vicinity of the earth (from the earth's atmosphere, or rockets or satellites immediately above the atmosphere). For this reason we have put a great deal of effort into trying to foster the maximum ground-based program in planetary astronomy, and one of the fruits of the program has been this telescope. It happens to be the largest we have been able to undertake; it is not the only one, as the astronomers here are well aware. Another major NASA-supported instrument is approaching completion near the top of Mauna Kea on the island of Hawaii – an 88-inch telescope which looks quite different from this one, and stands at an altitude of nearly 14 000 ft. Also we were able a few years ago to build for Dr. Kuiper a very fine instrument for doing primarily photography and photometry of the planets. In addition to this we have been fortunate enough to support construction of a few minor telescopes, as mentioned in the discussion on planetary patrols this morning, comprising about six stations scattered around the world. And we look forward to being able to help astronomers who are interested in a few other areas that have not been worked on enough in the past. One of these areas is the far infrared, and we accordingly hope to be able to support some major instrumentation for infrared astronomy. We are also very interested in making maximum use of radio and radar astronomy, which have certainly shown great benefit for planetary programs.

Concerning these major instruments in which the major portion of the NASA investment has been primarily for planetary research (even though of course they are also useful for stellar and other work), we are anxious to have these instruments available to guest observers when possible. And that has already been true here. The lunar laser experiment is really a guest-investigator experiment. Reinhard Beer with his large interferometer is from the Jet Propulsion Laboratory. Likewise Ron Schorn from JPL comes to McDonald to observe planets, mainly with the 82-inch reflector to be sure, and people from NASA-Goddard have also been using these instruments. Indeed, as a matter of principle we hope in all cases where we are able to help an institution with an instrument, that it becomes accessible to people with good, scientifically sound programs in planetary astronomy, regardless of the institution from which they come.

Dr. Harlan Smith: Thank you, Dr. Brunk; we agree with these sentiments.

This telescope is new in several ways. In particular it was designed to be integrated with a computer from the beginning, both for control and data-taking functions. To operate the ever-increasingly sophisticated equipment, both of the telescope and its auxiliary instruments, needs a larger and more sophisticated staff than was formerly the case. The staff burden is further increased because we now schedule both large telescopes nearly 24 hours a day, so much work being done in the infrared where the difference between day and night is almost immaterial. Finally, as all of you know, the Observatory works every day of the year, which further adds to the burden. It would be quite impossible to run a facility like this with our relatively small staff on such a schedule, if it were not for the dedication of a number of people, of whom I should mention particularly the Superintendant, Curtis Laughlin, and his three capable assistants, David Dittmar, George Grubb, and Dr. Maurice Marin; also Dr. Edwin Barker, Resident Supervisor of the planetary programs here. I want to take this rare opportunity to thank them for their extraordinarily devoted work, and to make a passing comment on it. A federal wage-hour law forbids people engaged in interstate commerce – and apparently our photons come from out-of-state – to be worked more than 40 hours a week. Fortunately for the sake of the observatory, this rule does not apply to supervisors!

I commented earlier that McDonald has had a long history of planetary work. One man, above all others together, made that possible – in fact, did most of the work himself either directly or through his immediately co-working graduate students. I refer of course to Professor Gerard Kuiper, now of the University of Arizona, Director of the Lunar and Planetary Laboratory. It is totally appropriate for this dedication that Professor Kuiper should be here to address us in a baptismal sense about the role of ground-based astronomy.

Dr. Kuiper: Mrs. Kuiper and I are happy to be back – almost home. We spent a lot of time here from about 1938 until about 1960, and even after moving to the University of Arizona have had the privilege of using the 82-inch from time to time.

A few comments might be especially appropriate on this occasion.

With the pressures of the large national space programs, a redoubled effort of course is needed in ground-based planetary astronomy. During the past dozen years or so the space programs, from the point of view of a scientist interested in planets, have matured in a very interesting way, in that a balanced work load can now be assigned to the six successively more complex modes of planetary research: ground-based, aircraft-based, balloon-based, rocket-based, satellite-based, and deep space probes. Each of these six technologies can make its own special contribution. Each simpler one can and should be used to the maximum effectiveness in its workload, because of the comparative economy, smaller required manpower, and quicker results frequently obtainable with the less complex techniques.

So, to a scientist who has watched the development of planetary studies for the past twenty years or more, I think we can truly say that the present age is a very fortunate one in which all six approaches are open to the scientist where necessary, and the individual interest can best be served, and the maximum contribution made.

The 107-inch here outclasses technically its 82-inch neighbor on the mountain. But the 82-inch Struve reflector has its own history, and perhaps I may be permitted to comment about some of the planetary programs that have been carried out with it (not to mention the tremendous output of stellar spectroscopy and programs of galactic structure carried out by Struve and dozens of collaborators since the inauguration of the 82-inch just thirty years ago).

Let me then reflect briefly on some of the contributions of this older and smaller sister here on the mountain. In the first place, an atmosphere of methane was discovered with it on the satellite Titan in 1944, the only satellite for which a gaseous atmosphere has been shown to exist. Carbon dioxide was identified on Mars with the 82-inch in 1947; the polar caps of Mars were found to show the reflection spectrum of very fine, very cold snow with it around 1950. The fifth satellite of Uranus and the second satellite of Neptune were discovered with it in 1947–48. The reflection spectrum of the dark areas of Mars was found to be not that of vegetation containing water, nor in fact were the dark areas of Mars found to be green. This was done in the late 40's and early 50's. The 8270 band of the pressure-induced dipole spectrum of hydrogen was discovered on Uranus and Neptune in the early 50's. The fifth satellite of Jupiter was photographed for the first time with the 82-inch; it had been seen but not photographed; from this we could get accurate astrometric positions. An extensive program of lunar studies was begun with the 82-inch in 1953 and carried through 1954, which led to a major 1954 publication in the Proceedings of the National Academy of Sciences, in which lunar timescales were developed. It put the age of the maria at 4.5 billion years, and indicated that the maria may not have been entirely caused by impacts but rather by magmas released as a result of major impacts; these impacts may not have come from meteorites or asteroids or comets, but from a satellite ring around the earth. Finally, a program of lunar photography was begun in the middle 50's that led to a series of lunar atlases; systems of coordinates were combined with these photographs. This led to the production of the orthographic atlas of the moon, which has been the basis of modern front-side lunar cartography. As a personal interjection, when I was recently in Houston with the Apollo 11 astronauts, I was naturally pleased that the only atlas in their office was the orthographic atlas of the moon. (It was an unannounced visit!)

I am sure that the opportunities of this mighty 107-inch will be larger than they were for the 82-inch. Already the instrument has begun to show its power in several major programs to which Dr. Smith has made reference.

Now since these proceedings are in a sense part of the 40th IAU Symposium, I might make just one brief remark about some thoughts that occurred to me during this Symposium, that have a definite bearing on the purposes of a telescope such as this.

If the 40th IAU Symposium may have left the impression with some of you that even incorrect observations, such as the microwave phase-curves of Venus, can lead to correct answers by good theoreticians, then I think I should point out a counterpart conclusion that greatly outweighs that impression. Specifically, much of planetary astronomy in the last decade has been the clearing away of incorrect ghosts left by earlier publications. Fortunately this audience is largely composed of planetary scientists, so this is really criticism in the family, but nevertheless let me mention a few of the skeletons in our planetary closet that we had to remove: the rotation of Mercury; the surface temperature and the rotation of Venus; the pressure and the rotational temperature of the Venus atmosphere; the canals and the surface vegetation on Mars; the spectroscopic rotation of Jupiter; the degree of penetration of the infrared radiation into the Jupiter atmosphere; dust and ice on the moon; the acceleration of Phobos around Mars. Other examples could be cited of hasty conclusions based on inadequate data. I think this strengthens the central convictions which must have led to the erecting of this mighty telescope; namely that, in the first place, planetary astronomy is an empirical science of great complexity rather like geophysics in its many branches. Thus with the large national investments now being made in the space programs, it behooves planetary astronomers to make sure that the data they produce are relevant to the great scientific problems and that they are reliable. The 61-inch telescope of the Lunar and Planetary Laboratory was the first telescope sponsored by NASA; the McDonald 107-inch and the Hawaii 88-inch followed. The completion of this mighty 107-inch on this high and dry mountain is a great event for planetary science.

I congratulate NASA and the University of Texas that their wise decisions have led to construction of this instrument, and look forward to great results from the various programs to which it is dedicated.